AF358819

Editor
Waldo Cerpa
Pontificia Universidad
Católica de Chile
Santiago
Chile

Editorial Office
MDPI AG
Grosspeteranlage 5
4052 Basel, Switzerland

This is a reprint of articles from the Special Issue published online in the open access journal *Antioxidants* (ISSN 2076-3921) (available at: https://www.mdpi.com/journal/antioxidants/special_issues/Oxidative_Stress_in_Brain_Function).

For citation purposes, cite each article independently as indicated on the article page online and as indicated below:

Lastname, A.A.; Lastname, B.B. Article Title. *Journal Name* **Year**, *Volume Number*, Page Range.

ISBN 978-3-7258-1149-6 (Hbk)
ISBN 978-3-7258-1150-2 (PDF)
doi.org/10.3390/books978-3-7258-1150-2

Cover image courtesy of Rodrigo Mira

Contents

About the Editor

Waldo Cerpa

I am Waldo Cerpa and I am an associate professor in the faculty of biological science at the Pontificia Universidad Católica de Chile. The main interest of my laboratory lies in studying brain function from the perspective of the pathophysiological processes generated by external damage. In particular, traumatic brain injury and alcohol intake are the current models that my team and I use. Using these models, we study the contribution of different cellular processes associated with glutamate receptors, their signaling, and their relationship with cellular homeostasis. Additionally, we are looking for pharmacological (or other) tools that can modulate the cellular processes described above.

Preface

Studying the functioning of brain cells continues to be the main approach for solving some of the main diseases that affect an increasingly elderly population. Our brain has a series of cellular tools to defend itself against both intrinsic (genetic) and external (environmental) damage, which endanger brain function and, therefore, human life. Antioxidant defenses and the ability to regulate them are one of the most comprehensive approaches to preventing brain damage that is associated with a series of pathological conditions. We wanted to bring together different views and approaches that aim to prevent/revert oxidative damage in our brains from the perspective of the cellular mechanisms that regulate essential processes in different brain cells. We hope that the variety of cellular processes and targets addressed in this reprint will be of interest to a broad audience fascinated by the regulation of brain function and how our brain defends itself against oxidative damage, which is typical of a wide range of brain pathologies.

Waldo Cerpa
Editor

 *antioxidants*

Article

Mild Traumatic Brain Injury Induces Mitochondrial Calcium Overload and Triggers the Upregulation of NCLX in the Hippocampus

Rodrigo G. Mira [1,2], Rodrigo A. Quintanilla [3] and Waldo Cerpa [1,2,*]

[1] Laboratorio de Función y Patología Neuronal, Departamento de Biología Celular y Molecular, Facultad de Ciencias Biológicas, Pontifica Universidad Católica de Chile, Santiago 8331150, Chile
[2] Centro de Excelencia en Biomedicina de Magallanes (CEBIMA), Universidad de Magallanes, Punta Arenas 6213515, Chile
[3] Laboratory of Neurodegenerative Diseases, Universidad Autónoma de Chile, Santiago 8910060, Chile
* Correspondence: wcerpa@bio.puc.cl

Abstract: Traumatic brain injury (TBI) is brain damage due to external forces. Mild TBI (mTBI) is the most common form of TBI, and repeated mTBI is a risk factor for developing neurodegenerative diseases. Several mechanisms of neuronal damage have been described in the cortex and hippocampus, including mitochondrial dysfunction. However, up until now, there have been no studies evaluating mitochondrial calcium dynamics. Here, we evaluated mitochondrial calcium dynamics in an mTBI model in mice using isolated hippocampal mitochondria for biochemical studies. We observed that 24 h after mTBI, there is a decrease in mitochondrial membrane potential and an increase in basal matrix calcium levels. These findings are accompanied by increased mitochondrial calcium efflux and no changes in mitochondrial calcium uptake. We also observed an increase in NCLX protein levels and calcium retention capacity. Our results suggest that under mTBI, the hippocampal cells respond by incrementing NCLX levels to restore mitochondrial function.

Keywords: traumatic brain injury; mitochondrial calcium; NCLX; hippocampus

Citation: Mira, R.G.; Quintanilla, R.A.; Cerpa, W. Mild Traumatic Brain Injury Induces Mitochondrial Calcium Overload and Triggers the Upregulation of NCLX in the Hippocampus. *Antioxidants* **2023**, *12*, 403. https://doi.org/10.3390/antiox12020403

Academic Editor: Stanley Omaye

Received: 26 December 2022
Revised: 3 February 2023
Accepted: 6 February 2023
Published: 7 February 2023

1. Introduction

Traumatic brain injury (TBI) is brain damage due to external forces produced by direct hits, acceleration, and deacceleration, among others [1]. The main causes of TBI are self-perpetrated harm, vehicle accidents, falls, and contact sports [2]. TBI is one of the leading causes of injury-related deaths and disability, with the male population being more affected by TBI than females [3]. Depending on the severity, TBI could be classified as mild, moderate, or severe. The first one is the most common form of TBI, comprising more than 80% of total cases [2]. mTBI is characterized by the absence of skull fracture (Glasgow Coma Score 13–15), the loss of consciousness that could be present briefly or absent, and headache, among other symptoms [4]. Importantly, in a low percentage of patients, symptoms could persist for up to 1 year, and they could develop post-concussive syndrome, i.e., behavioral changes and psychological symptoms [5]. mTBI and repeated mTBI have acquired more attention in recent years, given the importance of the long-term consequences and the fact that repeated mTBI has been recognized as a risk factor for developing neurodegenerative diseases, including Alzheimer's disease (AD) and chronic traumatic encephalopathy (CTE) [1,6].

The physical damage to the head affects the cortex and spreads to subcortical regions such as the hippocampus through gradients of pressure that cause vascular and axonal damage [7,8]. The hippocampus is a critical brain region involved in complex processing such as episodic and semantic memories, avoidance learning, and anxiety [9,10]. Thus, the hippocampal damage after TBI is related to psychological symptoms and memory problems in patients, and degeneration of the hippocampus is crucial in AD and other types of dementia.

TBI is characterized by developing two phases of brain damage, primary and secondary damage [11]. Primary damage involves the mechanisms triggered by the hit, such as hemorrhages and vasculature and axonal tract damage by the tensile and stress forces, among others [7]. Secondary damage develops in the next hours and days after the impact [11]. Neuronal damage mechanisms include the release of glutamate and excitotoxicity [12,13], neuroinflammation [14,15], oxidative stress [16,17], and mitochondrial dysfunction [18,19]. Mitochondrial bioenergetics is impaired soon after TBI in cortical and hippocampal mitochondria, with different severities, including mild [18,19]. Different chemical compounds have shown protective effects on mitochondrial bioenergetics and cell viability under TBI models, including the antioxidant MitoQ [20] and inhibitors of the mitochondrial permeability transition pore (mPTP) cyclosporine A [21] and NIM811 [22], indicating the role of calcium influx to the mitochondria in neuronal dysfunction. An important route of calcium entry into hippocampal neurons is the glutamate receptor N-methyl-D-aspartate receptor (NMDAR). Our lab has previously shown that NMDAR signaling and intracellular distribution are altered by mild TBI [23], suggesting that calcium dyshomeostasis is a good candidate for the promotion of mitochondrial dysfunction.

Calcium influx from the cytoplasm or membrane contact sites with the ER is mediated by the voltage-dependent anion channel (VDAC) in the outer mitochondrial membrane (OMM). In contrast, in the inner mitochondrial membrane (IMM), the main route for calcium entry into mitochondria is the mitochondrial calcium uniporter (MCU) complex [23]. It is composed of the channel protein MCU and its paralogue MCUb, as well as the essential MCU regulator (EMRE) and the regulatory proteins MICU1, MICU2, and MICU3 in the brain. The channel is a low-affinity calcium channel that drives the calcium influx depending on the mitochondrial membrane potential as a driving force and the calcium binding to the EF-hand domains in the regulatory proteins MICU1–MICU2 [24]. On the other hand, the calcium efflux is mainly driven by the mitochondrial sodium/calcium exchanger known as NCLX. Although the existence of a mitochondrial proton/calcium exchanger has been established, the molecular identity has been controversial, although Letm1 protein has been proposed for this role [24]. In the brain, the calcium efflux from the mitochondria is mainly mediated by NCLX [25,26], with a minor contribution from other exchangers. In some cases, the transient opening of the mPTP has also been proposed to regulate calcium efflux [27].

Until now, there has been no available information about either mitochondrial calcium uniporter (MCU) complex function or mitochondrial calcium dynamics after TBI in hippocampal mitochondria. Thus, we wanted to evaluate the changes in mitochondrial calcium dynamics after mild TBI, the most common TBI case, in the hippocampus. To do this, we used a mouse model of mild repeated TBI and evaluated mitochondrial calcium dynamics in hippocampal isolated mitochondria. We observed that intramitochondrial calcium levels are increased, and the mitochondrial membrane potential is decreased, two hallmarks of mitochondrial dysfunction. Interestingly, we observed that hippocampal mitochondria respond to this damage by increasing NCLX protein levels, suggesting regulatory mechanisms to alleviate calcium overload.

2. Materials and Methods

2.1. Animals

Male C57BL/6J mice, 8 weeks old, obtained from the animal care unit of Pontificia Universidad Católica de Chile (CIBEM) were housed in groups of between 3 and 5 animals per cage and maintained at 23 °C on a 12 h:12 h light–dark cycle with food and water *ad libitum*. The animals were treated and handled according to the National Institutes of Health guidelines for the care and use of laboratory animals (NIH Publications No. 8023, revised 1978, Baltimore, MD, USA).

2.2. Antibodies and Reagents

The primary antibodies used were mouse anti-cyclophilin F (sc-376061, Sta. Cruz Biotech.), rabbit anti-MCU (HPA016480, Sigma, Atlas Antibodies, Bromma, SE), rabbit anti-MCUb (HPA048776, Sigma, Atlas Antibodies, Bromma, SE), rabbit anti-MICU2 (ab101465, abcam, Cambridge, UK), rabbit anti-MICU1 (HPA037480, Sigma, Atlas Antibodies, Bromma, SE), rabbit anti-MICU3 (PA5107178, Invitrogen, Carlsbad, CA, USA), rabbit anti-SLC24A6 (ab83551, abcam, Cambridge, UK), anti-TOM20 (sc-17764, Sta. Cruz Biotechnology, Inc., Dallas, TX, USA.), rabbit anti-COX IV (4844S, Cell Signaling, Danvers, MA, USA), rabbit anti-GAPDH (sc-25778, Santa Cruz Biotechnology, Inc., Dallas, TX, USA), mouse anti-VDAC1 (B-6) (sc-390996, Santa Cruz Biotechnology, Inc., Dallas, TX, USA), mouse anti-PSD-95 (7E3, sc-32290, Sta. Cruz Biotechnology, Inc., Dallas, TX, USA), rabbit anti-lamin B1 (ab16048, abcam, Cambridge, UK), mouse anti-PDI (A-1, sc-376370, Sta. Cruz Biotechnology, Inc., Dallas, TX, USA), mouse anti-cytochrome C (556432, BD Pharmingen, San Diego, CA, USA), mouse anti-Letm1 (sc-271235, Sta. Cruz Biotechnology, Inc., Dallas, TX, USA), and mouse anti-OSCP (sc-365162, Sta. Cruz Biotechnology, Inc., Dallas, TX, USA). All secondary antibodies used were obtained from Jackson Immunoresearch. Chemicals used: Tetramethyl rhodamine ethylester perchlorate (TMRE, T669 Invitrogen, Carlsbad, CA, USA), calcium indicator Calcium Green-5N Hexapotassium salt (C3737, Invitrogen, Carlsbad, CA, USA), Ru360 (557440 Merck Millipore, Burlington, MA, USA), ruthenium red (1439, Tocris, UK) cyclosporine A (CsA, 1101, Tocris, UK), CGP-37157 (1114, Tocris, UK).

2.3. Mild Traumatic Brain Injury Induction

To induce mTBI, we adapted Maryland's weight drop model used for rats [28] to fit mouse anatomy as previously described [29–31]. Animals were randomly assigned to receive either sham or mTBI. Mice were subjected to 5 sessions of 3 blasts each with a 2-day interval in a frontal weight impact device. Sham animals were subjected to all procedures except injury induction.

2.4. Mitochondrial Isolation

Mitochondrial isolation was performed as previously described [32] (Supplementary Figure S1). Briefly, both hippocampi from a mouse brain were dissected and lysed in MSH-BSA buffer (mannitol 225 mM, sucrose 75 mM, HEPES 5 mM, EGTA 1 mM, and BSA 0.2 mg/mL supplemented with protease inhibitor cocktail). The lysates were centrifuged at 500 g for 5 min. The supernatant was then centrifuged at $14{,}000 \times g$ for 10 min. The pellet was resuspended in 200 µL of Percoll 12% dissolved in MSH (without BSA) and then gently transferred to 1 mL Percoll 24% dissolved in MSH. The gradient was centrifuged at $18{,}000 \times g$ for 15 min. Then, the pellet was washed twice, and the final pellet was resuspended in MSH. The protein concentration was determined by a BCA Protein Assay Kit (Pierce) [29,32,33].

2.5. Immunoblot

As previously described, immunoblots were performed with isolated mitochondria or whole hippocampal lysates. The hippocampi of treated or sham animals were dissected on ice and immediately processed. Briefly, the hippocampal tissue was homogenized in RIPA buffer (25 mM Tris-Cl, pH 7.6, 150 mM NaCl, 1% NP-40, 1% sodium deoxycholate, and 0.1% SDS) supplemented with a protease inhibitor mixture and phosphatase inhibitors (25 mM NaF, 100 mM Na_3VO_4, and 30 µM $Na_4P_2O_7$) using a homogenizer. The protein samples were centrifuged at 13,500 rpm for 15 min at 4 °C. The protein concentrations were determined using the BCA Protein Assay Kit (Pierce). The samples were resolved by SDS-PAGE, followed by immunoblotting on PVDF membranes. The membranes were incubated with the primary antibodies and corresponding peroxidase-conjugated antibodies (Jackson Immunoresearch, Inc.) and developed using an ECL kit (Westar Sun, Cyanagen, Bologna, Italy; Westar Supernova, Cyanagen, Bologna, Italy).

2.6. Indirect Mitochondrial Membrane Potential Measurement

Mitochondrial membrane potential was determined by TMRE exclusion as previously described [34]. Briefly, 20 µg of mitochondrial protein in 100 µL of experimental buffer (KCl 125 mM, HEPES 20 mM, $MgCl_2$ 2 mM, KH_2PO_4 2.5 mM, BSA 0.1%, glutamate 5 mM, and malate 5 mM) was incubated for 10 min at 37 °C to allow the energization of mitochondria. Then, 100 µL 2 µM TMRE was added (final volume 200 µL and final TMRE concentration 1 µM) and incubated for 10 min at 37 °C. Finally, the suspension was centrifuged at $14,000 \times g$ for 5 min. For supernatant measurements, 100 µL of the supernatant was measured. The remaining supernatant was discarded, the pellet resuspended in a new 100 µL of buffer, and charged into a plaque for measurement. TMRE was measured at 514/570 nm excitation/emission in a fluorometer.

2.7. Intramitochondrial Calcium Levels

Intramitochondrial calcium levels were determined as previously described [35,36]. In brief, 100 µg of hippocampal mitochondria was isolated in the presence of 10 µM ruthenium red and without EGTA. Washes were performed in the absence of EGTA. The mitochondrial pellet was diluted in 0.6 N HCl, homogenized, and sonicated. Then, samples were heated at 95 °C for 30 min and then centrifuged at $10,000 \times g$ for 5 min. The supernatants were recovered, and the calcium content was determined spectrophotometrically using the O-Cresolphtalein Complexone Calcium Assay Kit (Cayman Chemical, Ann Arbor, MI, USA). The absorbance was measured at 570 nm.

2.8. Calcium Uptake Assays

A mitochondrial calcium uptake assay was performed as previously described [36]. Briefly, 20 µg of isolated mitochondria was resuspended in experimental buffer (KCl 125 mM, HEPES 20 mM, $MgCl_2$ 2 mM, KH_2PO_4 2.5 mM, BSA 0.1%, glutamate 5 mM, and malate 5 mM). Cell-impermeable calcium indicator Green-5N 5 µM was added to the buffer. Fluorescence was measured at 506 nm excitation and 532 nm emission on a plate reader. First, the fluorescence was measured for 2 min every 20 s. Then, $CaCl_2$ 25 µM (final concentration) was added and measured for 8 min every 20 s. Data were adjusted to basal and maximum values to calculate the slope (MCU complex activity). For the assessment of calcium efflux, we registered calcium uptake for 6 min, and then the suspension was incubated with both Ru360 5 µM (final concentration) and NaCl 12 mM (final concentration). CGP-37157 20 µM (final concentration) was added as a control when indicated.

For calcium retention capacity, we used a first calcium challenge with $CaCl_2$ 25 µM (final concentration) and then $CaCl_2$ 10 µM (final concentration). Fluorescence was measured for 2 min at the baseline and after each calcium challenge every 20 s. We performed the experiments using one sham and one mTBI animal each time. To avoid differences between readings we analyzed results with a paired *t*-test.

2.9. Swelling Assay

A swelling assay was performed as previously described [35] with minor modifications. Mitochondrially induced swelling was measured spectrophotometrically as a decrease in absorbance at 540 nm. Fifty micrograms of isolated hippocampal mitochondria was resuspended in a total volume of 100 µL of swelling buffer (KCl 120 mM, Tris-HCl 10 mM, MOPS 5 mM, Na_2HPO_4 5 mM, glutamate 10 mM, malate 2 mM, and EGTA 0.1 mM). The swelling was induced by the addition of 100 µM $CaCl_2$ while monitoring absorbance. When indicated, ruthenium red (10 µM) or CsA (1 µM) was added.

2.10. Electron Microscopy

The sample was fixed in glutaraldehyde 2.5% and prepared in sodium cacodylate buffer 0.1 M pH 7.0 for 16 h. Then, the samples were washed 3 times for 20 min each and then were postfixed with osmium tetroxide 1% in water for 90 min. Then, samples were washed 3 times for 20 min each and stained in blocks with uranyl acetate 1% in water for 60 min. Samples were dehydrated in acetone battery: 50, 70, 95, 100, and 100% for 20 min each, and left overnight in a mixture of epoxide:acetone 1:1, and then pure epoxide for 4 h. The samples were embedded in pure resin and polymerized in a stove at 60 °C for 48 h. Thin slices (70–80 nm) were obtained with a Leica Ultracut R ultramicrotome, placed on a copper grid, and stained with uranyl acetate 4% in methanol for 1 min and lead citrate according to Reynolds for 4 min. Images were taken using a Philips Tecnai 12 electron microscope at 80 kV in the Unidad de Microscopía Avanzada belonging to the Biological Sciences Faculty at Pontificia Universidad Católica de Chile.

2.11. Statistical Analysis

Data and statistical analysis were performed using Prism8 software (GraphPad 8 Software). All data are expressed as mean $\pm$ SEM and relativized to sham animals when indicated. Student's *t*-test was used when two groups were compared. Two-way ANOVA with Bonferroni post hoc analysis was used in the swelling assay when two experimental conditions and time were evaluated simultaneously. The number of independent experiments "*n*" is indicated in every figure. A $p < 0.05$ value was considered significant.

3. Results

3.1. Mitochondrial Isolation

First, we performed mitochondrial isolation from hippocampal tissue to evaluate the mitochondrial calcium dynamic. Mitochondria in the brain could be experimentally separated into "synaptic mitochondria" and "non-synaptic mitochondria". Synaptic mitochondria correspond to presynaptic mitochondria contained in synaptic boutons and experimentally obtained from synaptosomes. On the other hand, non-synaptic mitochondria correspond to those mitochondria found in the soma, dendrites, and axons in neurons, but also mitochondria found in glial cells [37] (Figure 1A). Considering that, in neurons, the calcium influx through glutamate receptors is of particular interest in the context of mTBI, we decided to perform the experiments in the "non-synaptic" mitochondrial fraction, containing dendritic and glial mitochondria, directly exposed to glutamate and glutamate spillover. We assessed the purity and structure of our fraction using immunoblot and electron microscopy (Figure 1B,C). We evaluated the total lysate from hippocampal tissue, pellet P1 containing nuclei and cell debris, supernatant S2, containing cytosol and microsomes, and the pellet P2 was divided between mitochondrial fraction (Mito) and the upper fraction after a Percoll gradient containing synaptosomes and myelin (SM). We evaluated different protein markers: PSD-95 for synaptosomes, lamin B1 for nuclei, GAPDH for cytosol, MCU and cytochrome C (Cyt C) for mitochondria, and PDI for ER. We observed that the mitochondrial fraction is enriched in mitochondrial proteins MCU and Cyt C with minor contamination of synaptosomes (PSD-95) (Figure 1B). We did not observe Cyt C in S2, suggesting mitochondrial integrity. As expected, we observed mitochondrial proteins in synaptosomes containing presynaptic mitochondria. We performed transmission electron microscopy of the obtained mitochondrial fractions to evaluate the mitochondrial structure. We observed that most structures correspond with mitochondria, containing both outer and inner membranes with cristae (Figure 1C). We also observed synaptosomes in a minor proportion, as expected by immunoblots results.

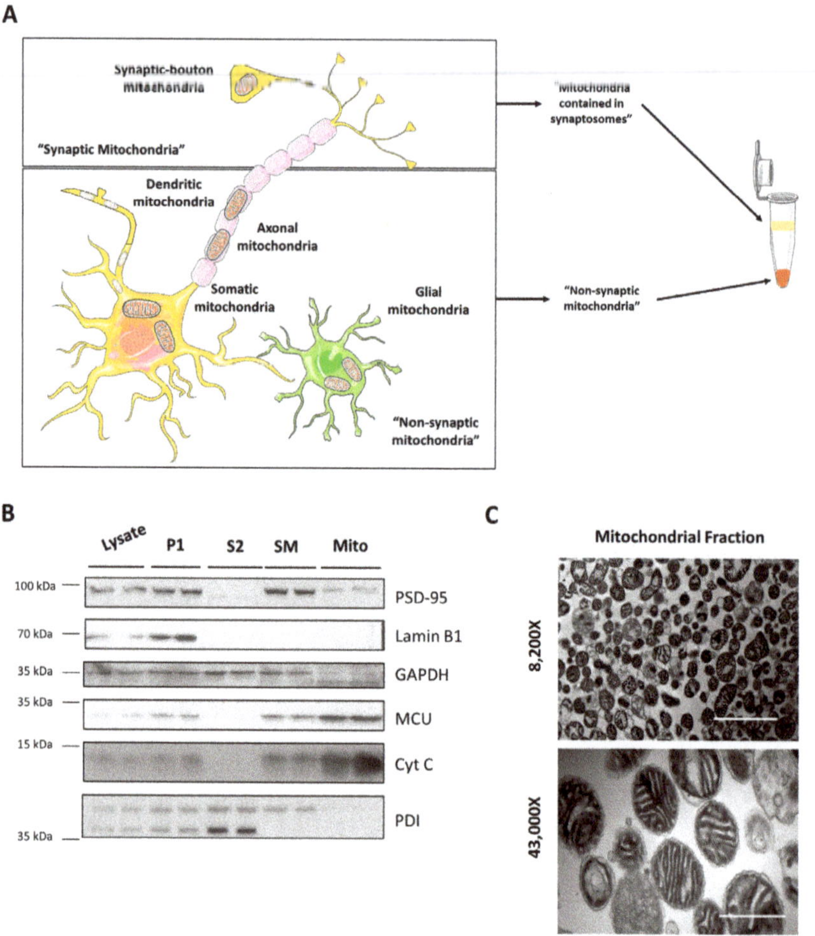

Figure 1. Mitochondrial isolation. (**A**) Scheme of mitochondrial distribution in the brain cells. (**B**) Immunoblot assessment of isolation protocol. It shows total lysate, first precipitant (P1), second supernatant (S2), synaptosomes + myelin after Percoll gradient (SM), and mitochondrial pellet (Mito). It shows synaptosomal marker PSD95, nuclear marker lamin B1, cytosolic marker GAPDH, two mitochondrial markers MCU and Cyt C, and endoplasmic reticulum marker PDI. (**C**) Electron microscopy of mitochondrial fraction seen at 8200× and 43,000×. Upper image scale bar 3.0 μm. Lower image scale bar 0.6 μm.

3.2. Mitochondrial Membrane Potential Decreases Soon after mTBI

Then, we indirectly evaluated the mitochondrial membrane potential of the isolated mitochondria derived from our sham or mTBI-submitted mice. To do this, we used the exclusion of the positively charged dye tetramethylrhodamine (TMRE). As we have previously published, there was cognitive impairment one week after the mTBI protocol [30]. So, we evaluated mitochondrial membrane potential at this point, and we did not observe differences between experimental groups (Figure 2A) (TMRE in the supernatant, unpaired two-tailed t-test $t = 0.04425$, $p = 0.9659$; TMRE in mitochondrial pellet, unpaired two-tailed t-test $t = 0.4725$, $p = 0.6509$; sham $n = 4$, mTBI $n = 5$). Since other studies have suggested that mitochondrial bioenergetic impairment occurs early [18,19] in mTBI pathophysiology, we decided to evaluate 24 h after the mTBI protocol. We observed a decrease in TMRE signal in the mitochondrial fraction with a consistent increase in the supernatant, suggesting decreased mitochondrial membrane potential (Figure 2B) (TMRE in the supernatant, unpaired two-tailed *t*-test $t = 3.083$, $p = 0.0177$; TMRE in mitochondrial pellet, unpaired two-tailed *t*-test $t = 3.580$, $p = 0.0373$; sham $n = 5$, mTBI $n = 4$).

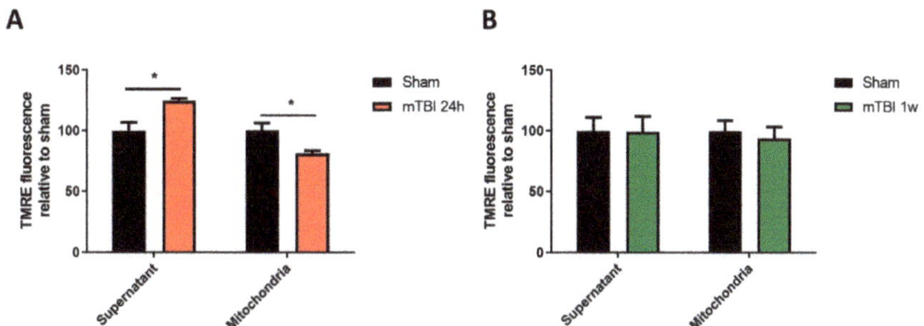

Figure 2. mTBI induces early and partial mitochondrial depolarization. (**A**) TMRE exclusion assay in mTBI-derived mitochondria 24 h after protocol indicating relative levels of TMRE in supernatant and mitochondria. (**B**) TMRE exclusion assay in mTBI-derived mitochondria one week after protocol. Student's *t*-test for supernatant and mitochondria separately. * $p < 0.05$.

3.3. mTBI Increases Intramitochondrial Calcium Content and Calcium Efflux in Hippocampal Mitochondria

Two hallmarks of mitochondrial damage are decreased membrane potential and mitochondrial calcium overload. Thus, we evaluated intramitochondrial calcium (IMC) levels in our mitochondrial fraction. To do this, we isolated our mitochondrial fraction in the presence of ruthenium red, a known non-selective inhibitor of the MCU channel, to avoid calcium influx during fraction preparation. Once the mitochondrial fraction was obtained, we obtained lysates containing matrix components, and we spectrophotometrically measured calcium levels on the mitochondrial matrix. We observed a marked increase of around 50% in IMC levels (Figure 3A) (unpaired two-tailed *t*-test $t = 3.232$, $p = 0.0231$, sham $n = 4$, mTBI $n = 3$). We then wanted to directly explore the activity of the MCU complex using the non-permeable calcium indicator Calcium Green-5N. We stimulated calcium uptake using 25 µM $CaCl_2$, and we fluorometrically monitored extramitochondrial calcium.

Traces showed calcium uptake in mitochondrial fraction derived from both groups, sham and mTBI (Figure 3B). Error bars were removed for better visualization (graphs with error bars are found in Supplementary Figure S2A) We used ruthenium red as an MCU blocker to corroborate that a decrease in fluorescence is mediated by the MCU activity (Figure 3B). To quantify MCU activity, we measured the slope of the register after calcium stimuli. We observed a slight decrease in MCU activity in the mTBI-derived mitochondrial fraction, despite no statistical differences (Figure 3C) (paired two-tailed *t*-test, $t = 1.210$, $p = 0.1565$, sham and mTBI $n = 4$). Then, to analyze the mitochondrial calcium efflux in our mitochondrial fraction, we used the same paradigm with the calcium indicator Calcium Green-5N and stimulated calcium uptake with 25 µM $CaCl_2$, but after calcium uptake, we stopped calcium influx with the MCU inhibitor Ru360 and stimulated calcium efflux by adding sodium to the milieu. We observed an increase in the fluorescence product of the calcium release (Figure 3D and Supplementary Figure S2B). To quantify calcium efflux, we assessed the slope of fluorescence increase. We observed an increase in calcium efflux in mitochondrial fractions derived from mTBI-submitted animals (Figure 3E) (unpaired two-tailed *t*-test, $t = 3.086$, $p = 0.0367$, sham and mTBI $n = 3$). Since the mitochondrial sodium/calcium exchanger NCLX is the main route of calcium efflux, we assessed the source of the calcium with the NCLX inhibitor CGP-37157 (Figure 3D,E). These results indicate that 24 h after our mTBI protocol, hippocampal mitochondria increased NCLX activity while MCU activity remained unchanged.

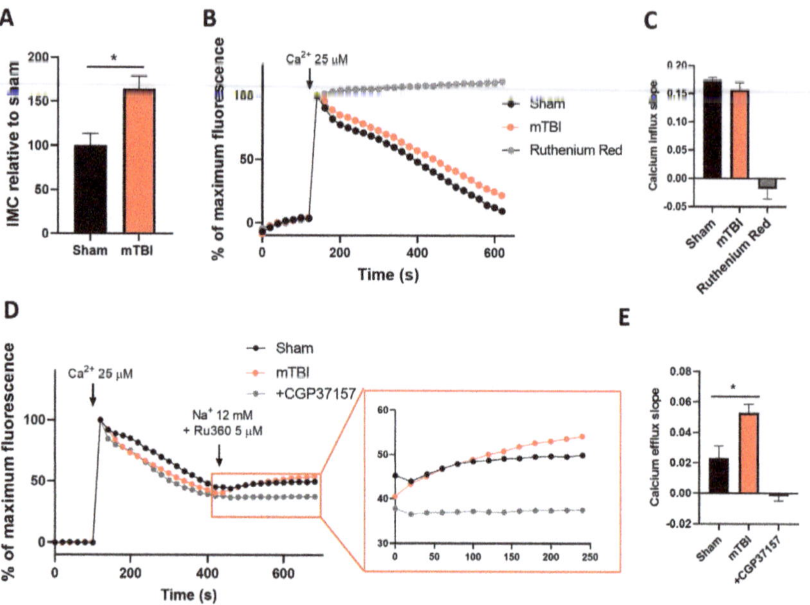

Figure 3. mTBI induces calcium overload and increased calcium efflux. (**A**) IMC relative to sham levels was measured spectrophotometrically. $n = 4$ for sham and 3 for mTBI. (**B**) Mitochondrial calcium uptake in isolated mitochondria measuring extramitochondrial calcium with impermeable sensor Calcium Green-5N. (**C**) Slope quantification of B. $n = 4$. (**D**) Mitochondrial calcium uptake and efflux measured fluorometrically as performed in B. The red square indicates the magnification of efflux curves. (**E**) Slope quantification of D. $n = 3$. Student's *t*-test for IMC and slopes. * $p < 0.05$.

3.4. mTBI Increases the Protein Levels of the Mitochondrial Sodium/Calcium Exchanger NCLX in Hippocampal Mitochondria

To determine if changes in calcium influx/efflux have a molecular correlation, we evaluated the respective protein levels in the mitochondrial fraction (Figure 4A,B). We evaluated the protein components of the MCU complex: MCU, MCUb, MICU1, MICU2, and MICU3. We observed that none of the core protein components of the MCU complex changed their expression after mTBI, in accordance with the absence of changes in MCU activity (Figure 4A,B) (unpaired two-tailed *t*-test. For MCU $t = 0.2344$, $p = 0.8262$; for MCUb $t = 0.02810$, $p = 0.9789$; for MICU1 $t = 0.6869$, $p = 0.5299$; for MICU2 $t = 1.249$, $p = 0.2797$ sham and mTBI $n = 3$). However, the brain-enriched MICU protein, MICU3, is the only protein that shows an increase after mTBI (unpaired two-tailed *t*-test, $t = 3.744$, $p = 0.0028$ sham and mTBI $n = 7$) (Figure 4A,B and Supplementary Figure S3). We also evaluated the mitochondrial sodium/calcium exchanger NCLX and observed an increased protein expression in mitochondrial fractions (unpaired two-tailed *t*-test, $t = 2.808$, $p = 0.0484$, sham and mTBI $n = 3$), consistent with increased activity of the exchanger. We also evaluated Letm1, a proton/calcium antiporter of the inner mitochondrial membrane. We observed a slight increase in Letm1 protein levels which are not statistically significant (Figure 4A,B) (unpaired two-tailed *t*-test, $t = 2.017$, $p = 0.1139$, sham and mTBI $n = 3$). To confirm that the above-mentioned results are not the product of changes in mitochondrial mass, we assessed mitochondrial housekeeping proteins in whole-hippocampal lysates. We evaluated the cytochrome c oxidase subunit IV (COX IV), the voltage-dependent anion channel (VDAC), and the import-machinery protein Tom20. We did not observe differences in any of the mitochondrial housekeeping proteins (Figure 4C,D) (unpaired two-tailed *t*-test. For COX IV, $t = 0.3471$, $p = 0.7357$ sham and mTBI $n = 6$; for Tom20 $t = 0.4280$, $p = 0.6907$; for VDAC

$t = 0.1582$, $p = 0.8820$, sham and mTBI $n = 3$), indicating that NCLX protein levels are not increased because mitochondrial mass changed.

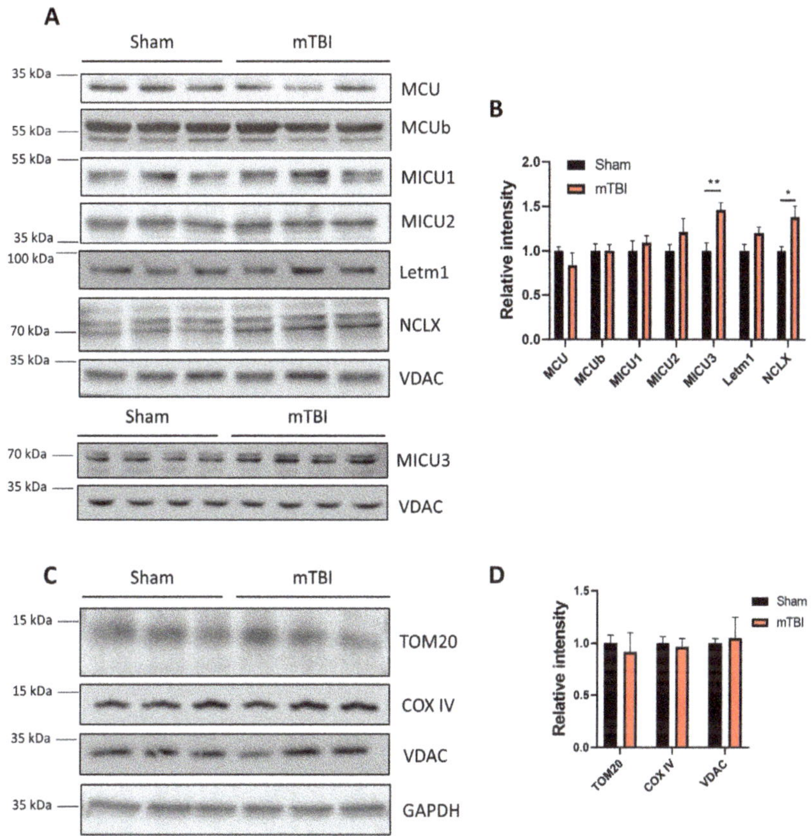

Figure 4. NCLX levels increased in mTBI mitochondria. (**A**) Immunoblot of mitochondrial fractions. It shows different MCU complex proteins: MCU, MCUb, MICU1, MICU2, and MICU3. It also shows sodium/calcium exchanger NCLX and proton/calcium exchanger Letm1. VDAC was used as the loading control. (**B**) Quantification of A. $n = 3$. For MICU3, $n = 7$. (**C**) Immunoblot of total hippocampal lysates. It shows two mitochondrial mass markers: TOM20, COX IV, and VDAC. GAPDH was used as the loading control. (**D**) Quantification of C. $n = 3$. For COX IV $n = 6$. Student's *t*-test for every protein measured. * $p < 0.05$; ** $p < 0.01$.

3.5. mTBI Increases the Calcium Retention Capacity in Hippocampal Mitochondria

We next decided to evaluate the sensitivity of our mitochondrial fractions to trigger mPTP, or the calcium retention capacity (CRC), given that calcium overload is one of the signals that triggers mPTP, and it is associated with cell death signaling pathways. Using the same paradigm measuring extramitochondrial calcium with the calcium indicator Calcium Green-5N, we performed several stimulations every 2 min. We started with stimulation of 25 µM $CaCl_2$, and then, subsequent stimulations were 10 µM $CaCl_2$ (Figure 5A and Supplementary Figure S4A). To quantify the CRC, we plotted the inverse of the area under the curve, and surprisingly we observed a mild, but significant, increase in CRC (Figure 5B) (paired two-tailed *t*-test, $t = 10.22$, $p = 0.0020$, sham and mTBI $n = 4$). We also evaluated the mPTP opening using a stronger calcium stimulation in a swelling assay. We monitored absorbance at 540 nm before and after stimulation with 100 µM $CaCl_2$ and observed that both sham and mTBI-derived mitochondrial fractions decreased absorbance, indicating mitochondrial swelling (Figure 5C and Supplementary Figure S4B). There was no difference between the registers (two-way ANOVA, time: F = 7.730, $p < 0.0001$; treatment: F = 1.215, $p = 0.3322$; interaction: F = 1.188, $p = 0.2696$; subject: F = 19.52, $p < 0.0001$; Bonferroni post hoc analysis, $p > 0.05$ for every point; $n = 3$ for sham and mTBI). To identify any change in molecular players of the mPTP opening, we analyzed protein levels of cyclophilin D (CypD), the gatekeeper of mPTP opening, and oligomycin-sensitive conferring protein (OSCP), an ATPase subunit that regulates mPTP opening. We did not observe changes in either protein, CypD and OSCP, but a tendency to increase in CypD is reported (Figure 5D) (unpaired two-tailed *t*-test. For CypD $t = 1.883$, $p = 0.1329$; for OSCP $t = 0.4528$, $p = 0.6666$, sham and mTBI $n = 3$ for CypD and $n = 4$ for OSCP). All these data suggest mild effects on mPTP opening, with a mild delay in opening in mTBI-derived isolated mitochondria.

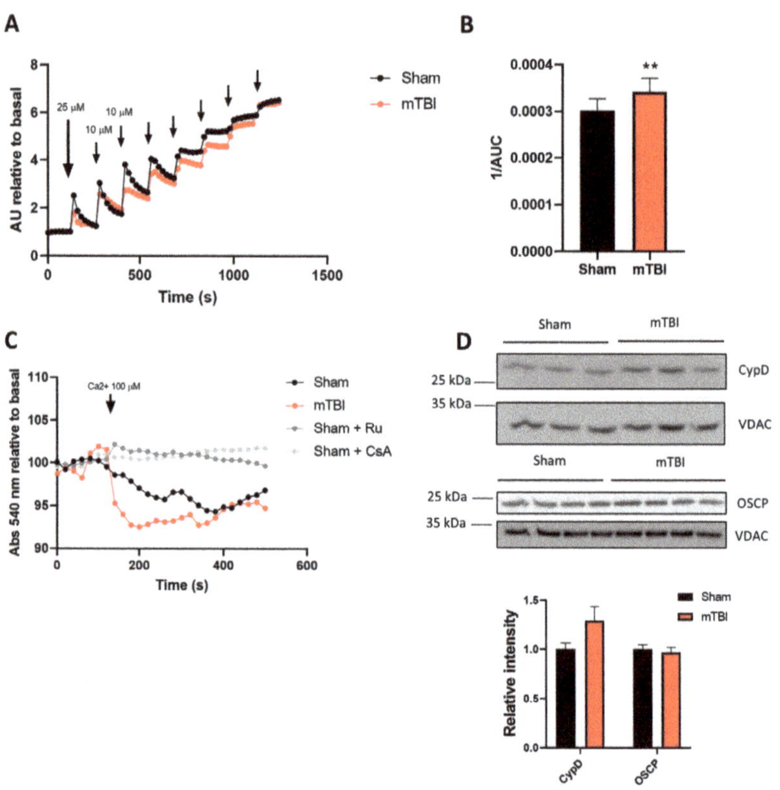

Figure 5. mTBI increases calcium retention capacity. (**A**) Calcium retention capacity (CRC). Isolated mitochondria were submitted to several calcium challenges starting with 25 µM, and then 10 µM stimuli. (**B**) The CRC was quantified as the inverse of the area under the curve of the entire register. $n = 4$. (**C**) Swelling assay. Isolated mitochondria were stimulated with 100 µM calcium and followed spectrophotometrically at 540 nm. When indicated, ruthenium red 10 µM or cyclosporine A 1 µM was used. $n = 3$. (**D**) Immunoblot analysis for two proteins involved in mPTP opening: CypD ($n = 3$) and OSCP ($n = 4$). Student's *t*-test for 1/AUC and immunoblots; two-way ANOVA, and Bonferroni post hoc analysis for swelling assay. ** $p < 0.01$.

3.6. mTBI Increases NCLX Protein Levels in Mitochondria Contained in Synaptosomes

Finally, we decided to evaluate if the mitochondria contained in synaptosomes also showed the same response to mTBI as our mitochondrial fraction. Therefore, we evaluated NCLX and MCU protein levels in these preparations, and we observed an increase in NCLX protein levels while MCU levels remain unaltered (Figure 6A,B) (unpaired two-tailed t-test. For NCLX $t = 2.544$, $p = 0.0438$; for MCU $t = 1.448$, $p = 0.1977$, sham and mTBI $n = 4$), suggesting that NCLX upregulation could be a global response in hippocampal cells.

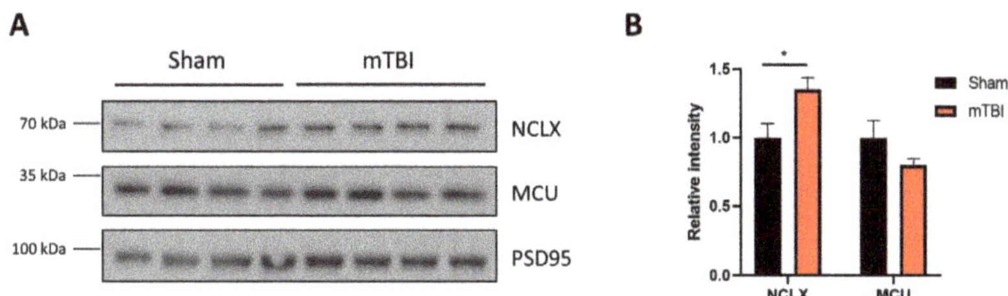

Figure 6. mTBI increases NCLX protein levels in synaptic mitochondria. (**A**) Immunoblot of mitochondrial proteins contained in synaptosomes. It shows NCLX and MCU. PSD-95 was used as a loading control. (**B**) Quantification of NCLX and MCU. $n = 4$. * $p < 0.05$.

All these data suggest that after mTBI, the mild and early alterations in mitochondrial function characterized by decreased mitochondrial membrane potential and increased calcium content may trigger a cellular response to increase NCLX protein levels to decrease calcium content.

4. Discussion

Using an mTBI mouse model, we found signals of early mitochondrial dysfunction such as decreased mitochondrial membrane potential and increased basal intramitochondrial calcium levels. We also found an increase in MICU3 and NCLX expression in the mouse hippocampus and increased NCLX activity. Given that mitochondrial membrane potential is restored one week after our mTBI protocol, we believe that NCLX upregulation is a compensatory mechanism of hippocampal cells to restore mitochondrial function.

It has been described that NCLX is a key player in mitochondrial calcium homeostasis. In fact, NCLX transports calcium slower than the MCU complex, becoming the rate-limiting step in calcium transients in the mitochondria [38]. The importance of NCLX in cell physiology is evident in the heart, where cardiomyocytes display many mitochondria. The conditional deletion of NCLX in cardiomyocytes produces premature cell death by heart failure [39]. On the other hand, the MCU conditional KO in cardiomyocytes did not show basal phenotype in the CD1 mouse strain [35], possibly explained by other calcium channels in the IMM.

Moreover, in AD patients and mouse models, the protein levels of NCLX are downregulated while MCU protein levels remained unchanged [40], suggesting that in the brain, the loss of calcium efflux from the mitochondria is critical for proper organ function. In our results, acute pathology such as mTBI increases NCLX protein levels, contrary to chronic pathology such as AD [40]. It is expected that under acute and mild pathology, the compensatory mechanisms reestablished normal cell function such as mitochondrial bioenergetics, which was reestablished 96 h after mTBI [18], while in the AD model, the mitochondrial energy production is persistently impaired [41,42].

NCLX KO mice have been generated, and interestingly the brain slices from these mice showed impaired synaptic transmission [43]. Considering these data, we believe that upregulation of NCLX could be a reasonable response of the hippocampal cell to acute mitochondrial damage by mTBI over other possible mechanisms, such as the regulation of MCU activity. The regulatory mechanism of the NCLX gene remains unknown. The transcription factors and coregulators that govern NCLX gene (*Slc8b1*) expression remain undescribed. However, there are several signaling pathways altered after mTBI that could contribute to a shift in the activation state of several transcription factors; for example, our lab has previously described alterations in NMDAR signaling 1 week after mTBI [30].

The increase in NCLX protein levels as a compensatory mechanism to acute damage seems to agree with other compensatory mechanisms. The severity of TBI produces different outcomes in mitochondrial dynamics proteins. A study revealed that under mTBI the fusion protein machinery Opa1, Mfn1, and Mfn2 are increased, while the fission protein machinery Drp1 and Fis1 are decreased. On the other hand, severe TBI produces the net opposite effect, an increase in fission protein machinery and a decrease in fusion protein machinery [44]. The prevalence of mitochondrial fusion after mTBI suggests a compensatory mechanism to avoid apoptosis and increase ETC activity [44]. We believe that NCLX upregulation is in the same line, avoiding calcium overload and apoptosis. Moreover, in our mTBI system, we reproduced the changes in mitochondrial dynamics proteins using whole hippocampal lysates with increased Mfn2 and Opa1 protein levels and decreased Drp1 protein levels (Supplementary Figure S5), arguing in favor of this hypothesis. Mitochondrial fusion helps to alleviate mitochondrial stress and avoid apoptotic cell death [45]. Mfn2, in fact, has been suggested to play a crucial role in the mitochondrial network balance in neurons after oxygen/glucose deprivation, coordinating mitophagy and mitochondrial biogenesis [46]. Indeed, the mitochondrial biogenesis master regulator peroxisome proliferator-activated receptor γ coactivator 1α (PGC-1 α) increases protein expression of Mfn2 [47], which in turn, regulates Parkin clustering to mitochondria by PINK1-mediated phosphorylation [48]. We did not observe changes in mitochondrial mass (Figure 4C,D) after our mTBI protocol, which could be explained by the coordination of both mitochondrial biogenesis and mitophagy to renew the mitochondrial network, however, if both processes are cooccurring remains to be determined.

Indeed, the mechanisms of mitochondrial response to mTBI are complex and involve different aspects of mitochondrial physiology. For example, it has been described that the NCLX function is dependent on mitochondrial membrane potential [49] and MCU function. As we observed a decreased mitochondrial potential, the driving force for calcium influx is decreased, but also the driving force for calcium efflux. Therefore, NCLX upregulation seems logical as a compensatory mechanism to extrude the excess calcium. There is also posttranslational regulation in both the MCU complex and NCLX [24]. Since we did not observe changes in calcium influx after mTBI, MCU posttranslational regulation seems unlikely, although we could not discard that option. In the counterpart, the phosphorylation of Ser-258 on NCLX by PKA increases the exchanger's activity [50]. In this study, we could not immunoprecipitate NCLX to evaluate its phosphorylation state, and currently, there are no commercial antibodies for NCLX-pS258. However, we believe that increased phosphorylation of the exchanger after mTBI is unlikely, given that it has been reported that cAMP signaling and PKA activity after TBI are decreased [51], although we have not evaluated this signaling pathway in our mTBI model.

We observed that increased NCLX protein levels are also present in mitochondria contained in synaptosomes (Figure 6). The mitochondrial fraction used in this study contains mitochondria derived from neurons (dendrites, axons, and soma), but also glial cells. Mitochondria contained in synaptosomes are neuronally derived exclusively. Thus, we could confirm that NCLX upregulation is occurring in neurons. However, the astroglial contribution might be important and remains to be explored. In fact, disrupting the astroglial expression of NCLX in vitro make neurons more vulnerable to excitotoxic stimuli [52], suggesting that NCLX function in astrocytes impacts neuronal viability. Furthermore, NCLX in astrocytes regulates gliotransmission and proliferation [53]. Hence, the contribution of increased NCLX in astrocytes in the context of mTBI is of particular interest for future studies.

We also observed an increase in MICU3 protein levels. MICU3 has been described as an enriched MICU protein in the brain compared to other tissues, and it has been described as a potentiator of mitochondrial calcium uptake [54]. Surprisingly, we did not observe changes in mitochondrial calcium uptake. We suggest two possibilities. First, posttranslational modifications in MCU protein could decrease calcium uptake that is contra-rested with the increased expression of MICU3, reestablishing MCU function. Second, the enrichment of

MICU3 in the brain has not been extensively studied yet, so, we could not rule out other functions for the MICU3 protein.

Interestingly, we observed an increase in CRC, although we observed a slight increase in CypD and no differences in the swelling assay. The relevance of mPTP in the pathophysiology of TBI has been studied using mPTP inhibitors such as cyclosporine A [21] or NIM811 [22], which improve memory performance and mitochondrial bioenergetics, but also using CypD knockout mice [55]. The KO of CypD when submitted to mTBI showed partial amelioration of synaptic impairment produced by mTBI in the somatosensory cortex. This study suggests that mPTP opening contributes partially to synaptic dysfunction [55], although there are no data about mitochondrial performance. In our model, consisting of repetitive mild traumas, the effect of mPTP in the pathophysiology is not clearly observed. We can speculate that if mPTP is crucial in cognitive impairment and cell death, it might play a role very early in cellular events. The main reason we did not observe changes is the temporal resolution of our study. As we are observing a cellular response to increase NCLX protein levels, other mechanisms could be driving an mPTP inhibition not directed by CypD. In this way, recent evidence points to Drp1 as a new contributor to mPTP opening and overopening in hypoxia in vitro models [56]. Notably, we observe a decrease in Drp1 protein levels that may regulate mPTP to a closed state in our model and in our window of time. Recently described mechanisms include circular RNAs in mPTP opening regulation [48,49], a poorly described regulatory pathway, and absent in mTBI research. On the contrary, with chronic pathology such as AD where CRC is decreased, in our acute mTBI model CRC is slightly increased. The overexpression of NCLX in the AD model also helped to increase CRC [40], although we could not explain the phenomenon observed by us regarding NCLX expression given subtle differences in experimental procedures.

5. Conclusions and Perspectives

Taken together, we showed that under mild TBI, mitochondrial membrane potential decreases, basal intramitochondrial calcium increases, and NCLX is upregulated as a compensatory mechanism. We describe for the first time that NCLX protein could be upregulated under acute pathology in the hippocampus and emerge as a new therapeutic target for neuropathology as a key regulatory element in mitochondrial calcium homeostasis.

Supplementary Materials: The following supporting information can be downloaded at: https://www.mdpi.com/article/10.3390/antiox12020403/s1, Supplementary Figure S1: Methodological figure, Supplementary Figure S2: Extended Data Figure 3. Supplementary Figure S3: Extended Data Figure 4. Supplementary Figure S4: Extended Data Figure 5. and Supplementary Figure S5: mTBI induces upregulation of fusion proteins and decrease Drp1 protein levels.

Author Contributions: R.G.M. and W.C. designed the experiments and R.G.M. performed the experimental procedures, the data analysis, and wrote the manuscript. W.C. and R.A.Q. revised the manuscript. All authors have read and agreed to the published version of the manuscript.

Funding: This work has been supported by the following grants: Fondo Nacional de Desarrollo Científico y Tecnológico (FONDECYT) 1190620, and Center for Excellence in Science and Technology (AFB 170005, PFB 12/2007, ACE210009) to WC, a Ph.D. fellowship was granted by Agencia Nacional de Investigación y Desarrollo (ANID) to RGM. None of the funding sources had any involvement in study design, data collection, preparing the manuscript, or in the decision to submit the manuscript for publication.

Institutional Review Board Statement: The animal study protocol was approved by the Ethics Committee of the Pontificia Universidad Católica de Chile (protocol code 180814017 and approved 21 February 2021).

Data Availability Statement: The original contributions presented in this study are included in the article/Supplementary Material, further inquiries can be directed to the corresponding author.

Acknowledgments: We thank to the laboratory of cell differentiation and pathology of the Pontificia Universidad Católica de Chile for making his laboratory equipment available.

Conflicts of Interest: The authors declare no conflict of interest.

References

1. Blennow, K.; Brody, D.L.; Kochanek, P.M.; Levin, H.; Moser, A.; Bigler, G.M.; Yaffe, K.; Zetterberg, H. Traumatic brain injuries. *Nat. Rev. Dis. Primers* **2016**, *2*, 16084. [CrossRef]
2. Capizzi, A.; Woo, J.; Verduzco-Gutierrez, M. Traumatic Brain Injury: An Overview of Epidemiology, Pathophysiology, and Medical Management. *Med. Clin. N. Am.* **2020**, *104*, 213–238. [CrossRef]
3. Mollayeva, T.; Mollayeva, S.; Colantonio, A. Traumatic brain injury: Sex, gender and intersecting vulnerabilities. *Nat. Rev. Neurol.* **2018**, *14*, 711–722. [CrossRef]
4. Mccrory, P.; Meeuwisse, W.; Aubry, M.; Cantu, B.; Dvorak, J.; Echemendia, R.; Engebretsen, L.; Johnston, K.; Kutcher, J.; Raftery, M.; et al. Consensus statement on Concus-sion in—The 4th International Conference on Concussion in Sport held in Zurich, November 2012. *Phys. Ther. Sport* **2013**, *14*, e1–e13. [CrossRef]
5. Silverberg, N.D.; Duhaime, A.C.; Iaccarino, M.A. Mild Traumatic Brain Injury in 2019–2020. *JAMA* **2020**, *323*, 177–178. [CrossRef]
6. Gupta, R.; Sen, N. Traumatic brain injury: A risk factor for neurodegenerative diseases. *Rev. Neurosci.* **2016**, *27*, 93–100. [CrossRef]
7. Rule, G.T.; Bocchieri, R.T.; Burns, J.M.; Young, L.A. Biophysical Mechanisms of Traumatic Brain Injuries. *Semin. Neurol.* **2015**, *35*, 005–011. [CrossRef]
8. Sharp, D.J.; Scott, G.; Leech, R. Network dysfunction after traumatic brain injury. *Nat. Rev. Neurol.* **2014**, *10*, 156–166. [CrossRef]
9. Cominski, T.P.; Jiao, X.; Catuzzi, J.E.; Stewart, A.L.; Pang, K.C.H. The Role of the Hippocampus in Avoidance Learning and Anxiety Vulnerability. *Front. Behav. Neurosci.* **2014**, *8*, 273. [CrossRef]
10. Carlesimo, G.A. The temporal lobes and memory. *Handb. Clin. Neurol.* **2022**, *187*, 319–337.
11. Walker, K.R.; Tesco, G. Molecular mechanisms of cognitive dysfunction following traumatic brain injury. *Front. Aging Neurosci.* **2013**, *5*, 29. [CrossRef] [PubMed]
12. Arundine, M.; Tymianski, M. Molecular mechanisms of glutamate-dependent neurodegeneration in ischemia and traumatic brain injury. *Cell Mol. Life Sci.* **2004**, *61*, 657–668. [CrossRef] [PubMed]
13. Bullock, R.; Zauner, A.; Woodward, J.J.; Myseros, J.; Choi, S.C.; Ward, J.D.; Marmarou, A.; Young, H.F. Factors affecting excitatory amino acid release following severe human head injury. *J. Neurosurg.* **1998**, *89*, 507–518. [CrossRef] [PubMed]
14. Acosta, S.A.; Tajiri, N.; Shinozuka, K.; Ishikawa, H.; Grimmig, B.; Diamond, D.M.; Sanberg, P.R.; Bickford, P.C.; Kaneko, Y.; Borlongan, C.V. Long-term upregulation of inflammation and suppression of cell proliferation in the brain of adult rats exposed to traumatic brain injury using the controlled cortical impact model. *PLoS ONE* **2013**, *8*, e53376. [CrossRef]
15. Mira, R.G.; Lira, M.; Cerpa, W. Traumatic Brain Injury: Mechanisms of Glial Response. *Front. Physiol.* **2021**, *12*, 740939. [CrossRef]
16. Ohta, M.; Higashi, Y.; Yawata, T.; Kitahara, M.; Nobumoto, A.; Ishida, E.; Tsuda, M.; Fujimoto, Y.; Shimizu, K. Attenuation of axonal injury and oxidative stress by edaravone protects against cognitive impairments after traumatic brain injury. *Brain Res.* **2013**, *1490*, 184–192. [CrossRef] [PubMed]
17. Xu, M.; Li, L.; Liu, H.; Lu, W.; Ling, X.; Gong, M. Rutaecarpine Attenuates Oxidative Stress-Induced Traumatic Brain Injury and Reduces Secondary Injury via the PGK1/KEAP1/NRF2 Signaling Pathway. *Front. Pharmacol.* **2022**, *13*, 807125. [CrossRef]
18. Hubbard, W.B.; Joseph, B.; Spry, M.; Vekaria, H.J.; Saatman, K.E.; Sullivan, P.G. Acute Mito-chondrial Impairment Underlies Prolonged Cellular Dysfunction after Repeated Mild Traumatic Brain Injuries. *J. Neurotrauma* **2019**, *36*, 1252–1263.
19. Gilmer, L.K.; Roberts, K.N.; Joy, K.; Sullivan, P.G.; Scheff, S.W. Early mitochondrial dysfunction after cortical contusion injury. *J. Neurotrauma* **2009**, *26*, 1271–1280. [CrossRef]
20. Zhou, J.; Wang, H.; Shen, R.; Fang, J.; Yang, Y.; Dai, W.; Zhu, Y.; Zhou, M. Mitochondrial-targeted antioxidant MitoQ provides neuroprotection and reduces neuronal apoptosis in experimental traumatic brain injury pos-sibly via the Nrf2-ARE pathway. *Am. J. Transl. Res.* **2018**, *10*, 1887–1899.
21. Kulbe, J.R.; Hill, R.L.; Singh, I.N.; Wang, J.A.; Hall, E.D. Synaptic Mitochondria Sustain More Damage than Non-Synaptic Mitochondria after Traumatic Brain Injury and Are Protected by Cyclosporine A. *J. Neurotrauma* **2017**, *34*, 1291–1301. [CrossRef]
22. Readnower, R.D.; Pandya, J.D.; McEwen, M.L.; Pauly, J.R.; Springer, J.E.; Sullivan, P.G. Post-Injury Administration of the Mitochondrial Permeability Transition Pore Inhibitor, NIM811, Is Neuroprotective and Improves Cognition after Traumatic Brain Injury in Rats. *J. Neurotrauma* **2011**, *28*, 1845–1853. [CrossRef]
23. Mira, R.G.; Cerpa, W. Building a Bridge Between NMDAR-Mediated Excitotoxicity and Mitochondrial Dys-function in Chronic and Acute Diseases. *Cell Mol. Neurobiol.* **2021**, *41*, 1413–1430. [PubMed]
24. Giorgi, C.; Marchi, S.; Pinton, P. The machineries, regulation and cellular functions of mitochondrial calcium. *Nat. Rev. Mol. Cell Biol.* **2018**, *19*, 713–730. [CrossRef] [PubMed]
25. Palty, R.; Silverman, W.F.; Hershfinkel, M.; Caporale, T.; Sensi, S.L.; Parnis, J.; Nolte, C.; Fishman, D.; Shoshan-Barmatz, V.; Herrmann, S.; et al. NCLX is an essential com-ponent of mitochondrial Na+/Ca2+ exchange. *Proc. Natl. Acad. Sci. USA* **2010**, *107*, 436–441.
26. Nita, L.I.; Hershfinkel, M.; Sekler, I. Life after the birth of the mitochondrial Na+/Ca2+ exchanger, NCLX. *Sci. China Life Sci.* **2015**, *58*, 59–65.
27. Yalamanchili, K.; Afzal, N.; Boyman, L.; Mannella, C.A.; Lederer, W.J.; Jafri, M.S. Under-standing the Dynamics of the Transient and Permanent Opening Events of the Mitochondrial Permeability Transition Pore with a Novel Stochastic Model. *Membranes* **2022**, *12*, 494.

28. Kilbourne, M.; Kuehn, R.; Tosun, C.; Caridi, J.; Keledjian, K.; Bochicchio, G.; Scalea, T.; Gerzanich, V.; Simard, J.M. Novel Model of Frontal Impact Closed Head Injury in the Rat. *J. Neurotrauma* **2009**, *26*, 2233–2243. [CrossRef]
29. Mira, R.G.; Lira, M.; Quintanilla, R.A.; Cerpa, W. Alcohol consumption during adolescence alters the hippocampal response to traumatic brain injury. *Biochem. Biophys. Res. Commun.* **2020**, *528*, 514–519. [CrossRef]
30. Carvajal, F.J.; Cerpa, W. Regulation of Phosphorylated State of NMDA Receptor by STEP$_{61}$ Phosphatase after Mild-Traumatic Brain Injury: Role of Oxidative Stress. *Antioxidants* **2021**, *10*, 1575. [CrossRef]
31. Lira, M.; Zamorano, P.; Cerpa, W. Exo70 intracellular redistribution after repeated mild traumatic brain injury. *Biol. Res.* **2021**, *54*, 5. [CrossRef]
32. Chinopoulos, C.; Zhang, S.F.; Thomas, B.; Ten, V.; Starkov, A.A. Isolation and Functional Assessment of Mitochondria from Small Amounts of Mouse Brain Tissue. *Neurodegener. Methods Protoc.* **2011**, *793*, 311–324. [CrossRef]
33. Smith, P.K.; Krohn, R.I.; Hermanson, G.T.; Mallia, A.K.; Gartner, F.H.; Provenzano, M.D.; Fujimoto, E.K.; Goeke, N.M.; Olson, B.J.; Klenk, D.C. Measurement of protein using bicinchoninic acid. *Anal. Biochem.* **1985**, *150*, 76–85. [CrossRef]
34. Lampl, T.; Crum, J.A.; Davis, T.A.; Milligan, C.; Del Gaizo Moore, V. Isolation and functional analysis of mitochondria from cultured cells and mouse tissue. *J. Vis. Exp.* **2015**, *23*, e52076.
35. Pan, X.; Liu, J.; Nguyen, T.; Liu, C.; Sun, J.; Teng, Y.; Fergusson, M.M.; Rovira, I.I.; Allen, M.; Springer, D.A.; et al. The physiological role of mito-chondrial calcium revealed by mice lacking the mitochondrial calcium uniporter. *Nat. Cell Biol.* **2013**, *15*, 1464–1472. [CrossRef]
36. Liu, J.C.; Liu, J.; Holmström, K.M.; Menazza, S.; Parks, R.J.; Fergusson, M.M.; Yu, Z.-X.; Springer, D.A.; Halsey, C.; Liu, C.; et al. MICU1 Serves as a Molecular Gatekeeper to Prevent In Vivo Mitochondrial Calcium Overload. *Cell Rep.* **2016**, *16*, 1561–1573. [CrossRef]
37. Lai, J.C.K.; Walsh, J.M.; Dennis, S.C.; Clark, J.B. Synaptic and non-synaptic mitochondria from rat brain: Isolation and characterization. *J. Neurochem.* **1977**, *28*, 625–631. [CrossRef]
38. Rudolf, R.; Mongillo, M.; Magalhaes, P.J.; Pozzan, T. In vivo monitoring of Ca(2+) uptake into mi-tochondria of mouse skeletal muscle during contraction. *J. Cell Biol.* **2004**, *166*, 527–536. [CrossRef]
39. Luongo, T.S.; Lambert, J.P.; Gross, P.; Nwokedi, M.; Lombardi, A.A.; Shanmughapriya, S.; Carpen-Ter, A.C.; Kolmetzky, D.; Gao, E.; Van Berlo, J.H.; et al. The mitochondrial Na(+)/Ca(2+) exchanger is essential for Ca(2+) homeostasis and viability. *Nature* **2017**, *545*, 93–97. [CrossRef]
40. Jadiya, P.; Kolmetzky, D.W.; Tomar, D.; Di Meco, A.; Lombardi, A.A.; Lambert, J.P.; Luongo, T.S.; Ludtmann, M.H.; Praticò, D.; Elrod, J.W. Impaired mitochondrial calcium efflux contributes to disease progression in models of Alzheimer's disease. *Nat. Commun.* **2019**, *10*, 3885. [CrossRef]
41. Hauptmann, S.; Scherping, I.; Dröse, S.; Brandt, U.; Schulz, K.; Jendrach, M.; Leuner, K.; Eckert, A.; Müller, W. Mitochondrial dysfunction: An early event in Alzheimer pathology accumulates with age in AD transgenic mice. *Neurobiol. Aging* **2009**, *30*, 1574–1586. [CrossRef]
42. Chen, L.; Xu, S.; Wu, T.; Shao, Y.; Luo, L.; Zhou, L.; Ou, S.; Tang, H.; Huang, W.; Guo, K.; et al. Studies on APP metabolism related to age-associated mitochondrial dysfunction in APP/PS1 transgenic mice. *Aging* **2019**, *11*, 10242–10251. [CrossRef]
43. Stavsky, A.; Stoler, O.; Kostic, M.; Katoshevsky, T.; Assali, E.A.; Savic, I.; Amitai, Y.; Prokisch, H.; Leiz, S.; Daumer-HAAS, C.; et al. Aberrant activity of mitochondrial NCLX is linked to impaired synaptic transmission and is associated with mental retardation. *Commun. Biol.* **2021**, *4*, 666. [CrossRef]
44. Di Pietro, V.; Lazzarino, G.; Amorini, A.M.; Signoretti, S.; Hill, L.J.; Porto, E.; Tavazzi, B.; Lazzarino, G.; Belli, A. Fusion or Fission: The Destiny of Mitochondria in Traumatic Brain Injury of Different Severities. *Sci. Rep.* **2017**, *7*, 9189. [CrossRef]
45. Youle, R.J.; Van Der Bliek, A.M. Mitochondrial fission, fusion, and stress. *Science* **2012**, *337*, 1062–1065. [CrossRef]
46. Wojtyniak, P.; Boratynska-Jasinska, A.; Serwach, K.; Gruszczynska-Biegala, J.; Zablocka, B.; Jaworski, J.; Kawalec, M. Mitofusin 2 Integrates Mitochondrial Network Remodelling, Mitophagy and Re-newal of Respiratory Chain Proteins in Neurons after Oxygen and Glucose Deprivation. *Mol. Neurobiol.* **2022**, *59*, 6502–8618. [CrossRef]
47. Soriano, F.X.; Liesa, M.; Bach, D.; Chan, D.C.; Palacin, M.; Zorzano, A. Evidence for a mitochondrial regulatory pathway defined by peroxisome proliferator-activated receptor-gamma coactivator-1 alpha, estrogen-related receptor-alpha, and mitofusin 2. *Diabetes* **2006**, *55*, 1783–1791. [CrossRef]
48. Chen, Y.; Dorn, G.W., 2nd. PINK1-phosphorylated mitofusin 2 is a Parkin receptor for culling damaged mi-tochondria. *Science* **2013**, *340*, 471–475. [CrossRef]
49. Kostic, M.; Katoshevski, T.; Sekler, I. Allosteric Regulation of NCLX by Mitochondrial Membrane Potential Links the Metabolic State and Ca^{2+} Signaling in Mitochondria. *Cell Rep.* **2018**, *25*, 3465–3475.e4. [CrossRef]
50. Kostic, M.; Ludtmann, M.H.; Bading, H.; Hershfinkel, M.; Steer, E.; Chu, C.T.; Abramov, A.Y.; Sekler, I. PKA Phosphorylation of NCLX Reverses Mitochondrial Calcium Overload and Depolarization, Promoting Survival of PINK1-Deficient Dopaminergic Neurons. *Cell Rep.* **2015**, *13*, 376–386. [CrossRef]
51. Atkins, C.M.; Oliva, A.A.; Alonso, O.F.; Pearse, D.D.; Bramlett, H.M.; Dietrich, W.D. Modulation of the cAMP signaling pathway after traumatic brain injury. *Exp. Neurol.* **2007**, *208*, 145–158. [CrossRef]
52. Hagenston, A.M.; Yan, J.; Bas-Orth, C.; Tan, Y.; Sekler, I.; Bading, H. Disrupted expression of mi-tochondrial NCLX sensitizes neuroglial networks to excitotoxic stimuli and renders synaptic activity toxic. *J. Biol. Chem.* **2022**, *298*, 101508. [CrossRef]

53. Parnis, J.; Montana, V.; Delgado-Martinez, I.; Matyash, V.; Parpura, V.; Kettenmann, H.; Sekler, I.; Nolte, C. Mitochondrial Exchanger NCLX Plays a Major Role in the Intracellular Ca^{2+} Signaling, Gliotransmission, and Proliferation of Astrocytes. *J. Neurosci.* **2013**, *33*, 7206–7219. [CrossRef]
54. Patron, M.; Granatiero, V.; Espino, J.; Rizzuto, R.; De Stefani, D. MICU3 is a tissue-specific enhancer of mitochondrial calcium uptake. *Cell Death Differ.* **2018**, *26*, 179–195. [CrossRef]
55. Sun, J.; Jacobs, K.M. Knockout of Cyclophilin-D Provides Partial Amelioration of Intrinsic and Synaptic Prop-erties Altered by Mild Traumatic Brain Injury. *Front. Syst. Neurosci.* **2016**, *10*, 63. [CrossRef]
56. Duan, C.; Kuang, L.; Hong, C.; Xiang, X.; Liu, J.; Li, Q.; Peng, X.; Zhou, Y.; Wang, H.; Liu, L.; et al. Mitochondrial Drp1 recognizes and induces excessive mPTP opening after hypoxia through BAX-PiC and LRRK2-HK2. *Cell Death Dis.* **2021**, *12*, 1050. [CrossRef]

Article

Human Microglia Synthesize Neurosteroids to Cope with Rotenone-Induced Oxidative Stress

Chiara Lucchi [1], Alessandro Codeluppi [2], Monica Filaferro [1], Giovanni Vitale [2], Cecilia Rustichelli [2], Rossella Avallone [2], Jessica Mandrioli [1,3] and Giuseppe Biagini [1,*]

1   Department of Biomedical, Metabolic and Neural Sciences, University of Modena and Reggio Emilia, 41125 Modena, Italy; lucchi.chiara86@gmail.com (C.L.); monica.filaferro@unimore.it (M.F.); jessica.mandrioli@unimore.it (J.M.)
2   Department of Life Sciences, University of Modena and Reggio Emilia, 41125 Modena, Italy; alle.codeluppi@gmail.com (A.C.); giovanni.vitale@unimore.it (G.V.); cecilia.rustichelli@unimore.it (C.R.); rossella.avallone@unimore.it (R.A.)
3   Department of Neurosciences, Ospedale Civile di Baggiovara, Azienda Ospedaliero-Universitaria di Modena, 41126 Modena, Italy
*   Correspondence: gbiagini@unimore.it

Abstract: We obtained evidence that mouse BV2 microglia synthesize neurosteroids dynamically to modify neurosteroid levels in response to oxidative damage caused by rotenone. Here, we evaluated whether neurosteroids could be produced and altered in response to rotenone by the human microglial clone 3 (HMC3) cell line. To this aim, HMC3 cultures were exposed to rotenone (100 nM) and neurosteroids were measured in the culture medium by liquid chromatography with tandem mass spectrometry. Microglia reactivity was evaluated by measuring interleukin 6 (IL-6) levels, whereas cell viability was monitored by the 3-(4,5-dimethylthiazol-2-yl)-2,5-diphenyltetrazolium bromide assay. After 24 h (h), rotenone increased IL-6 and reactive oxygen species levels by approximately +37% over the baseline, without affecting cell viability; however, microglia viability was significantly reduced at 48 h ($p < 0.01$). These changes were accompanied by the downregulation of several neurosteroids, including pregnenolone, pregnenolone sulfate, 5α-dihydroprogesterone, and pregnanolone, except for allopregnanolone, which instead was remarkably increased ($p < 0.05$). Interestingly, treatment with exogenous allopregnanolone (1 nM) efficiently prevented the reduction in HMC3 cell viability. In conclusion, this is the first evidence that human microglia can produce allopregnanolone and that this neurosteroid is increasingly released in response to oxidative stress, to tentatively support the microglia's survival.

Keywords: allopregnanolone; microglia; neuroinflammation; neurosteroids; reactive oxygen species; rotenone

Citation: Lucchi, C.; Codeluppi, A.; Filaferro, M.; Vitale, G.; Rustichelli, C.; Avallone, R.; Mandrioli, J.; Biagini, G. Human Microglia Synthesize Neurosteroids to Cope with Rotenone-Induced Oxidative Stress. *Antioxidants* **2023**, *12*, 963. https://doi.org/10.3390/antiox12040963

Academic Editor: Waldo Cerpa

Received: 10 March 2023
Revised: 5 April 2023
Accepted: 18 April 2023
Published: 19 April 2023

1. Introduction

Neurosteroids are a family of molecules produced in the brain by metabolizing pregnenolone or converting peripherally synthesized steroids [1]. Allopregnanolone is considered the major representative steroid of the family and displays neuromodulatory properties mediated by membrane receptors, such as the γ-aminobutyric acid type A receptor (GABA$_A$) [2]. Apart from generating chloride inhibitory currents in neurons, GABA$_A$ has been proposed to mediate some anti-inflammatory properties of allopregnanolone [3]. However, the GABA$_A$ mechanism appeared to be just one of those possibly involved in modulating the anti-inflammatory properties of neurosteroids [4]. Additionally, allopregnanolone has been shown to possess neuroprotective properties [5], and a reduction in allopregnanolone availability has been implied in various neurological disorders, including multiple sclerosis [6], Parkinson's disease [7], Alzheimer's disease [8], and epilepsy [9].

Although the presence of neurosteroids in the human brain has been well established in early studies using radioimmunoassays [10–12], the sources of human neurosteroids are still poorly characterized, and the current knowledge is mainly based on information from in vitro studies with human cell lines [13,14] and postmortem analyses of supposedly healthy donors or patients with different neurological disorders [15–17]. Based on these experiments, it has been suggested that neurosteroids are produced in the human brain thanks to oligodendrocytes that synthesize and release pregnenolone to fuel the production of other cognate molecules by neurons and astrocytes [18,19]. In particular, evidence supporting the synthesis of allopregnanolone in the human brain has been obtained by experiments using surgical specimens from patients with epilepsy [20]. However, it is undetermined which human brain cell type may produce allopregnanolone, although in rats, neurons have been indicated as the probable source of this neurosteroid [21].

Mouse microglia were found to express some steroidogenic enzymes capable of metabolizing dehydroepiandrosterone and androstenediol [22]. Subsequently, evidence was obtained to suggest that human microglia can produce pregnenolone, but it remained undetermined if other neurosteroids could be metabolized by these cells [14,23]. This is a relevant issue because microglia are one of the possible targets involved in the beneficial properties of allopregnanolone, which could be released by microglia to also exert autocrine effects. Recently, allopregnanolone properties have been characterized in cultured BV2 microglial cells and also in primary microglia cultures, disclosing an array of modulatory effects on phagocytosis and morphology, which markedly changed in the experimental conditions able to reproduce the interruption of the blood–brain barrier functions, as it occurs in several neurological disorders characterized by neuroinflammation [24]. Indeed, microglia are a major player not only in protecting the brain from aggression by pathogens [25], but also in mediating neuroinflammation [26], especially when the inflammatory process becomes chronic [27].

In the presence of a lesion, microglia become reactive, which means that microglial cells change their morphology to become ameboid with shorter and thicker pseudopodia and increase their expression of an array of proteins [28], of which the most popular to characterize their reactivity is the ionized calcium-binding adapter molecule 1 (Iba1) [29]. Indeed, the most important products synthesized and released to promote neuroinflammation by reactive microglia include interleukins (ILs), such as IL-1β and IL-6, tumor necrosis factor-α, interferon-γ, chemokine (C-C motif) ligand 2, chemokine (C-X3-C motif) ligand 1, and the C-X-C motif chemokine ligand 10; also, glutamate and nitric oxide are released by microglia during inflammation to promote tissue damage [30].

Rotenone is an inhibitor of the mitochondrial complex I, which produces oxidative damage [31]. This property has been exploited to reproduce features of Parkinson's disease both in vitro and in vivo [32]. We recently described the cytotoxic effects of rotenone at 100 nM on cultured mouse BV2 microglial cells, which were related to the induction of reactive oxygen species (ROS) able to reduce microglia survival by 50% [33]. In such conditions, we described a profound derangement of neurosteroid metabolism, which was characterized by a remarkable reduction in pregnenolone sulfate and allopregnanolone levels and, at variance, increased levels of 5α-dihydroprogesterone (5α-DHP) and pregnanolone. These findings represented the first evidence of the capability to synthesize neurosteroids by murine microglia, but it is unknown whether human microglia could operate similarly, as questioned by other investigators who found that human microglia cell lines were able to produce pregnenolone [23].

Thus, in the present experiments, we aimed at evaluating if the human microglia could also be able to synthesize neurosteroids and if this property may be modified by challenging microglial cells with rotenone at the same dose used in our previous work to induce ROS production and cell death [33]. To disclose the activating effects of rotenone on the human microglial clone 3 (HMC3) cell line, we evaluated the changes in IL-6 levels and ROS in the culture medium. We also assessed HMC3 viability after 24 h (h) and 48 h of exposure to rotenone and the modifying effects of allopregnanolone added to cell cultures. These experiments revealed that human microglia can produce neurosteroids and that this property can be modified by rotenone-induced oxidative stress.

2. Materials and Methods

2.1. Chemicals and Reagents

Eagle's minimum essential medium (EMEM) was purchased from ATCC (Manassas, VA, USA). Penicillin, streptomycin, fetal bovine serum (HyClone Laboratories, Logan, UT, USA), 3-(4,5-dimethylthiazol-2-yl)-2,5-diphenyltetrazolium bromide (MTT), rotenone, and the human IL-6 enzyme-linked immunosorbent assay (ELISA) kit were purchased from Merck Life Science (RAB0306, Milan, Italy); 2,7 dichlorodihydrofluorescein diacetate (H2DCFDA) was from Invitrogen (Thermo Fisher Scientific, Waltham, MA, USA); Hank's Balanced Salt Solution (HBSS) was obtained from Gibco (Thermo Fisher Scientific, Waltham, MA, USA). All other chemical reagents were HPLC grade. The certified standard for the investigated neurosteroids and the isotope-labeled internal standards (allopregnanolone-2,2,3,4,4-d5 and pregnenolone-20,21-13C2-16,16-d2 sulfate sodium salt) were supplied as pure substances or solutions (100 µg/mL) from Merck Life Science. AmplifexTM-Keto Reagent Kit and Discovery DSC-18 SPE cartridges (100 mg; 1 mL) were also purchased from Merck Life Science. Acetonitrile (ACN), methanol, formic acid (FA), and ammonium formate (AmmF) were of liquid chromatography–mass spectrometry (LC-MS) purity grade (Merck Life Science); ultra-pure water was obtained by a Milli-Q Plus185 system (Millipore, Milford, MA, USA).

2.2. Cell Culture

HMC3 microglia (ATCC, Manassas, VA, USA)were grown in EMEM supplemented with 100 U/mL penicillin, 10 µg/mL streptomycin, and 10% fetal bovine serum (FBS) and kept in a humidified incubator at 37 °C with 95% O_2/5% CO_2. For all experiments, cells were grown to 80–90% confluency and then subjected to no more than 20 cell passages.

2.3. IL-6 Quantification

To evaluate the production and release of IL-6, we exposed the HMC3 cells to rotenone (100 nM) for 24 h and 48 h. For quantification, we used the IL-6 ELISA kit. Briefly, 100 µL of each standard and samples (culture media) were added to wells of a 96-well plate. The plate was covered and incubated for 2.5 h at room temperature. After discarding the standard and samples, the plate was cleaned four times with a washing solution. After that, the detection antibody (100 µL) was added to each well, and the plate was incubated at room temperature for 1 h with gentle shaking. Following four washes with washing solution, 100 µL of horseradish peroxidase–streptavidin was added to each well. The plate was subsequently incubated for 45 minutes (min) at room temperature and washed four times with the washing solution. After that, 100 µL of the ELISA colorimetric 3,3,5,5–tetramethylbenzidine reagent was added to each well for 30 min at room temperature and covered. A total of 50 µL of stop solution was added, and the plate was read at 450 nm immediately on the microplate reader (Multiskan FC Microplate, Thermo Fisher Scientific, Waltham, MA, USA). All the values obtained with the ELISA assay were normalized with the total protein content of each sample (cell lysate according to the Bradford method).

2.4. Determination of ROS

The ROS generation was determined using the fluorogenic probes H2DCFDA according to the manufacturer's instructions. Briefly, HMC3 cells were seeded in a 96-well plate (at a density of 10,000 cells/well) and maintained in complete culture media for 24 h. Afterward, cells were washed once with HBSS and incubated for 45 min at 37 °C with 10 µM H2DCFDA dye. After incubation, the dye was removed and cells were treated for 24 h with rotenone 100 nM, control medium, or H_2O_2 used as a positive control. Cell staining was performed in HBSS. The emitted fluorescence intensity was measured using a Fluoroskan FL Microplate Fluorometer (Thermo Fisher Scientific, Waltham, MA, USA) with wavelengths of 485 nm (excitation) and 520 nm (emission).

2.5. Sample Processing for Neurosteroid Quantification

A total of 100,000 HMC3 cells/well were seeded in 24 well plates in serum-free Roswell Park Memorial Institute (RPMI) 1640 medium without phenol red for the neurosteroid quantification. Then, cells were treated with rotenone (100 nM) for at least 24 h. As previously published [33], the medium was aspirated and centrifuged at the end of treatment to remove cells in suspension, and the supernatant was used to process the neurosteroid analysis. Aliquots (700 μL) of cell medium were spiked with 50 μL of the internal standard solution, vortexed (90 s, s), and purified using the C-18 SPE procedure. The obtained eluates were dried using an Eppendorf Concentrator Plus AG5305 (Eppendorf AG, Hamburg, Germany) and derivatized with Amplifex Keto Reagent (50 μL) for 1 h at room temperature and in the dark. Finally, the obtained samples were resuspended with 50 μL methanol/water (70/30), centrifuged, and analyzed using LC-MS/MS.

2.6. Working Solutions and Calibrators

To obtain the working solutions at 10 concentration levels, a stock solution containing all of the examined neurosteroids was serially diluted with methanol. Moreover, a stock solution of the isotope-labeled internal standards (ISs) was prepared in methanol at a concentration of 1000 fg/mL for both ISs. All solutions were kept at −20 °C until use. To obtain calibration samples ($n = 10$) in the range of 5.0 ± 1250 fg/700 μL for pregnenolone and 5α-DHP, 1.0 ± 250 fg/700 μL for pregnanolone, and 0.2 ± 50 fg/700 μL for pregnenolone sulfate, progesterone, and allopregnanolone; aliquots (700 μL) of blank cell medium were spiked with 50 μL of the ISs solution and 50 μL of the working solutions. The calibrators were prepared as previously reported [33] and evaluated in triplicate on three separate days.

2.7. LC-MS/MS Analysis

LC analyses were performed as described previously [33]. Briefly, we used a Kinetex XB-C18 column (100 mm length × 2.1 mm inner diameter; 2.6 μm particle size) equipped with a UHPLC C18 SecurityGuard cartridge (2.1 mm) (Phenomenex, Torrance, CA, USA). Mass spectrometric detection was performed using an Agilent QQQ-MS/MS (6410B) triple quadrupole operating in electrospray positive ionization mode. In the case of progesterone and 5α-DHP, the presence of two keto groups led to cis/trans derivatives, which eluted as two separable LC peaks; therefore, to quantitate progesterone and 5α-DHP in medium samples, the summation of both peaks was used.

2.8. Viability of HMC3 Microglial Cells Treated with Rotenone

To evaluate the viability of HMC3 cells exposed to rotenone (100 nM), the 3-(4,5-dimethylthiazol-2-yl)-2,5-diphenyltetrazolium bromide (MTT) test was performed. The rotenone 100 nM concentration was selected because it was shown to be below IC_{50} in our previous experiment, with no progression in cell death over an observation period of 72 h [33]. A total of 10,000 HMC3 cells were seeded in 96 well plates. After 24 h, cells were exposed to rotenone or the combination of rotenone (100 nM) and allopregnanolone (1 nM) and incubated for 24 h and 48 h. Subsequently, the MTT solution was added to each well (10 μL) and incubated for 2 h. Finally, a 10% sodium dodecyl sulfate solution was added to dissolve formazan, and absorbance was measured at 570 nm, using 620 nm as a wavelength reference.

2.9. Statistical Analysis

Data were analyzed using SigmaPlot 11 (Systat Software, San Jose, CA, USA). The evaluation of IL-6 was carried out using the two-way analysis of variance (ANOVA) and Holm–Šídák as post hoc tests. Data analysis of ROS assay was compared using one-way ANOVA and Bonferroni post hoc test. The neurosteroid levels were compared using the Mann–Whitney test. The data on viability at 24 h were analyzed using the Student's *t*-test. The data on viability at 48 h were compared using one-way ANOVA and Holm–Šídák as post hoc tests. A *p*-value < 0.05 was considered to be statistically significant. Results are illustrated by mean values and standard error of the mean (SEM) or median values and interquartile range.

3. Results

3.1. Oxidative Stress Activates HMC3 Microglia to Release IL-6

As previously shown by others [34] and illustrated in Figure 1, we confirmed that HMC3 cells can produce and release IL-6 in resting conditions. As expected, after 24 h of stimulation, rotenone activated the HMC3 microglia inasmuch as they produced a significant increase in the release of IL-6 (+37%) compared to the basal condition ($p < 0.001$, two-way ANOVA and Holm–Šídák post hoc test). On the other hand, prolonged exposure to rotenone did not lead to an additional release of IL-6, whose levels remained significantly higher when compared to the control group ($p < 0.05$, two-way ANOVA and Holm–Šídák post hoc test) (Figure 1a).

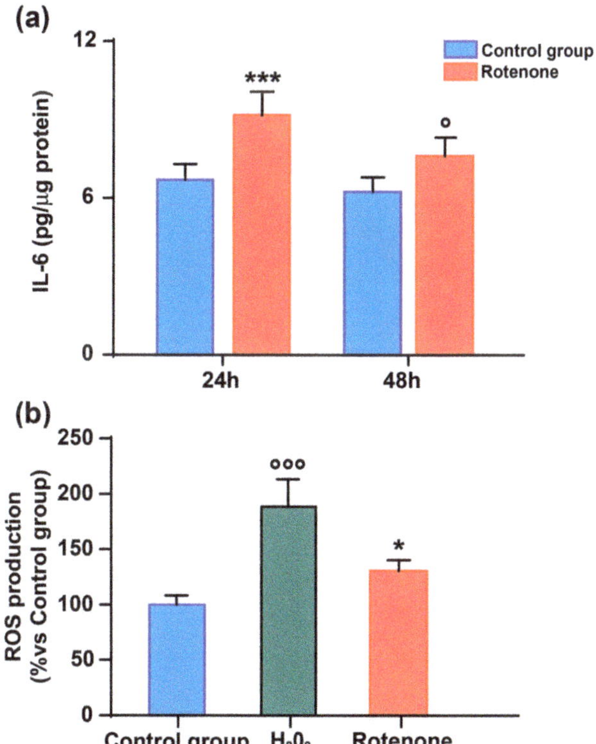

Figure 1. (**a**) Effect of oxidative stress on IL-6 production by HMC3 cells. After 24 h, the treatment with rotenone (100 nM) significantly increased the release of IL6 when compared to unstimulated cells. Furthermore, prolonged oxidative stress did not lead to additional IL-6 release but its levels remain significantly higher when compared to the control group. Each column represented the average ± SEM ($n = 5$) normalized by the total protein content measured per well. Data were analyzed by two-way ANOVA followed by Holm–Šídák post hoc test. ° $p < 0.05$ vs. control group 48 h; *** $p < 0.001$ vs. control group 24 h. In (**b**), bars illustrate the ROS production in HMC3 cells following rotenone treatment (100 nM). Each bar represents the mean ± SEM ($n = 5$) of the percentage of the fluorescence, normalized with the respective value of the control. Data were analyzed by one-way ANOVA followed by Bonferroni post hoc test. * $p < 0.05$ vs. control group; °°° $p < 0.001$ H_2O_2 vs. rotenone and control groups.

As already demonstrated for BV2 cells, rotenone led to a significant increase in ROS production in HMC3 after 24 h ($p < 0.05$ rotenone vs. controls; one-way ANOVA followed by Bonferroni post hoc test) using, this time, a more ubiquitous fluorogenic probe. However, this increase was significantly lower when compared to the changes in ROS levels caused by the exposure to strong oxidants, such as H_2O_2 (500 μM) ($p < 0.001$, H_2O_2 vs. control or rotenone levels, one-way ANOVA followed by Bonferroni correction) (Figure 1b).

3.2. Effect of Rotenone on Neurosteroid Levels in Culture Medium of HMC3 Cells

To evaluate if human microglia could produce neurosteroids, we measured pregnenolone, pregnenolone sulfate, progesterone, 5α-DHP, pregnanolone, and allopregnanolone by LC-MS/MS in the culture medium of HMC3 cells, both at rest and after 24 h of exposure to rotenone (100 nM).

First, we established that all the evaluated neurosteroids were detectable, although in low amounts, in basal conditions (Figures 2–4). In particular, pregnenolone sulfate was the most synthesized neurosteroid (about 0.03 pg/100,000 cells), followed by pregnenolone (at 0.004 pg/100,000 cells). Lower amounts were also found for pregnanolone (approximately 0.008 pg/100,000 cells), 5α-DHP (0.002), progesterone (0.0001 pg/100,000 cells), and, especially, allopregnanolone, which was barely detectable (0.000002 pg/100,000 cells).

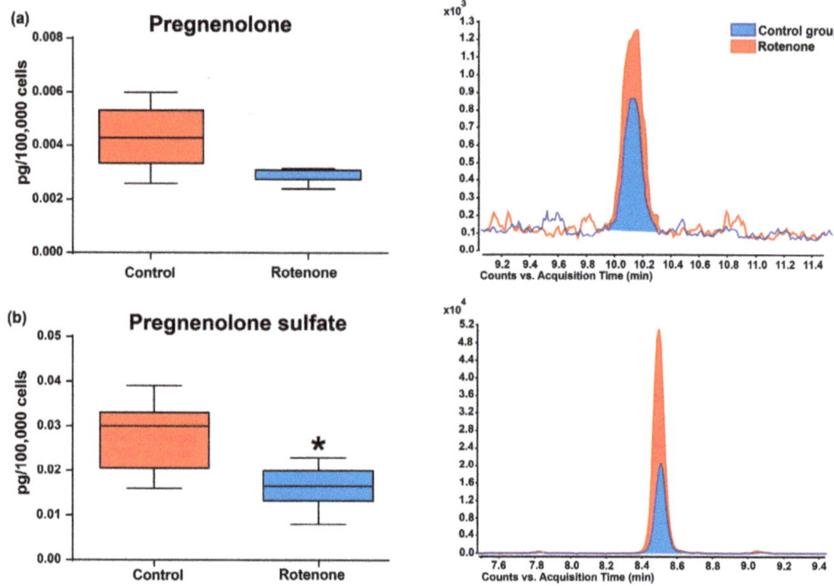

Figure 2. Pregnenolone (**a**) and pregnenolone sulfate (**b**) concentrations in HMC3 cells' medium determined by LC-MS/MS in microglia in resting condition (Control) or after activation by exposure to rotenone for 24 h. In (**a**), pregnenolone levels, illustrated in the box plot, after stimulation with rotenone were lower but not significantly different when compared to the control group. Peak areas corresponding to the respective median values are illustrated on the right. In (**b**), pregnenolone sulfate levels, illustrated in the box plot, were significantly lower (* $p < 0.05$, Mann–Whitney rank sum test) after stimulation with rotenone. Peak areas corresponding to the respective median values are illustrated on the right. The neurosteroid concentrations are expressed in pg/100,000 cells.

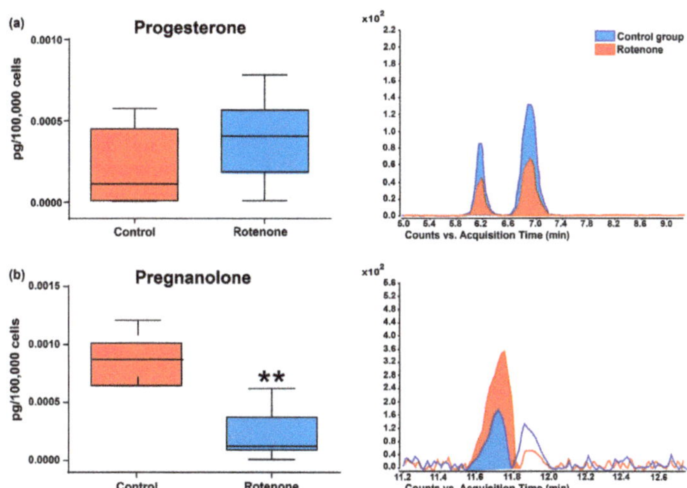

Figure 3. Progesterone (**a**) and pregnanolone (**b**) concentrations in HMC3 cells' medium determined by LC-MS/MS in microglia in resting condition (Control) or after activation by exposure to rotenone for 24 h. In (**a**), progesterone levels, illustrated in the box plot, after stimulation with rotenone increased but not significantly when compared to the control group. Peak areas corresponding to the respective median values are illustrated on the right. In (**b**), pregnanolone levels, illustrated in the box plot, were significantly lower (** $p < 0.01$, Mann–Whitney rank sum test) after stimulation with rotenone. Peak areas corresponding to the respective median values are illustrated on the right. The neurosteroid concentrations are expressed in pg/100,000 cells.

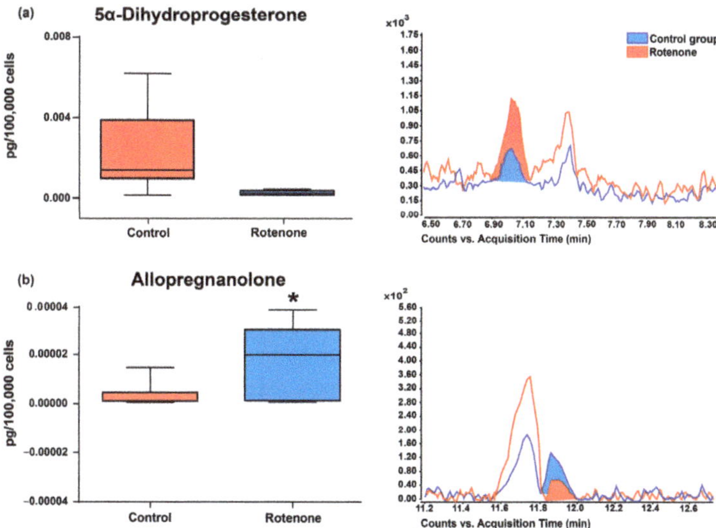

Figure 4. 5α-Dihydroprogesterone (5α-DHP) (**a**) and allopregnanolone (**b**) concentrations in HMC3 cells' medium determined by LC-MS/MS in microglia in resting condition (Control) or after activation by exposure to rotenone for 24 h. In (**a**), 5α-DHP levels, illustrated in the box plot, after stimulation with rotenone were lower but not significantly different when compared to the control group. Peak areas corresponding to the respective median values are illustrated on the right. In (**b**), allopregnanolone levels, illustrated in the box plot, were significantly reduced (* $p < 0.05$, Mann–Whitney rank sum test) after stimulation with rotenone. Peak areas corresponding to the respective median values are illustrated on the right. The neurosteroid concentrations are expressed in pg/100,000 cells.

This scenario was completely modified by the treatment of HMC3 cells with rotenone. Rotenone induced a remarkable reduction in pregnenolone sulfate levels with respect to the basal condition ($p < 0.01$, Mann–Whitney rank sum test; Figure 2b). In addition, pregnenolone levels were slightly decreased in comparison to the control group ($p = 0.052$; Figure 2a).

Progesterone levels increased but not significantly ($p = 0.310$; Figure 3a). Pregnanolone levels were highly reduced by rotenone treatment (-85%), especially when rotenone-exposed cells were compared to the unstimulated microglia ($p < 0.01$; Figure 3b).

Levels of 5α-DHP were only slightly reduced in the medium of rotenone-treated HMC3 ($p = 0.052$; Figure 4a). Notably, the main progesterone metabolite, allopregnanolone, presented a remarkable increase in response to rotenone treatment ($p < 0.05$ vs. unstimulated cells; Figure 4b).

3.3. Exogenous Allopregnanolone Preserved HMC3 Cell Viability from Rotenone-Induced Damage

We evaluated whether rotenone could influence the survival of HMC3 microglia using the MTT test, as we previously found a cytotoxic effect on murine microglia [33]. At variance with the previous results, exposure to rotenone (100 nM) did not affect HMC3 cell viability (control condition: $90 \pm 1.5\%$ vs. rotenone treatment: $88 \pm 2.8\%$) after 24 h (Figure 5), but significantly decreased the survival of human microglia after 48 h ($p < 0.01$, control condition: $97 \pm 4.5\%$ vs. rotenone treatment: $74 \pm 4.7\%$), even if at a lower extent in respect to BV2 cells [33].

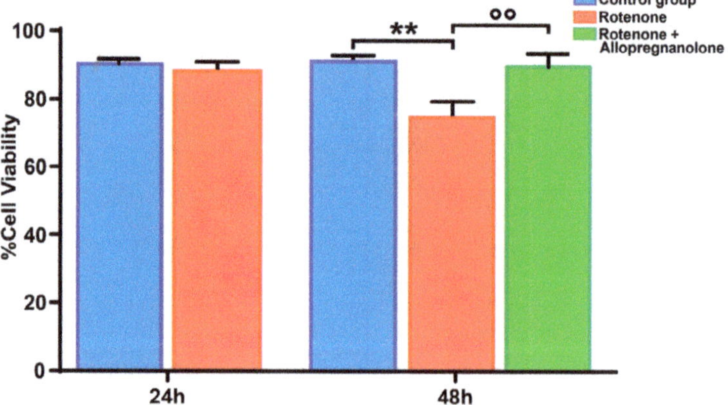

Figure 5. Viability of HMC3 cells exposed to rotenone (100 nM) at different time intervals (24 h and 48 h) and rescued by allopregnanolone (1 nM). The data represent the average $\pm$ SEM of two independent experiments. ** $p < 0.01$ rotenone vs. control group; °° $p < 0.01$ rotenone + allopregnanolone vs. rotenone, according to one-way analysis of variance (ANOVA) followed by Holm–Šídák post hoc test for multiple comparisons.

Since we interpreted the difference observed at 24 h in the response to rotenone of HMC3 vs. BV2 microglia as possibly related to the altered allopregnanolone metabolism, which led to reduced allopregnanolone levels in BV2 [33], we hypothesized that it could be possible to limit the HMC3 cell death found at the 48 h time interval by further increasing allopregnanolone in the culture medium. To this aim, we decided to combine rotenone treatment with allopregnanolone, exogenously added to HMC3 cultures at 1 nM; thus, we observed that HMC3 cell viability was maintained at values similar to those of the control condition ($p < 0.01$, Holm–Šídák post hoc test; rotenone treatment: $74 \pm 4.7\%$ vs. rotenone + allopregnanolone treatment: $93.8 \pm 5.4\%$).

4. Discussion

We previously found that murine microglia were able to produce and release a variety of neurosteroids other than pregnenolone in BV2 cell cultures [33] and that this property could be markedly modified by oxidative stress able to affect cell survival. Indeed, our previous work elicited some criticism, which we already mentioned in the introduction. Precisely, Germelli and collaborators [23] highlighted that the ability of BV2 cells to produce neurosteroids was a novel finding requiring confirmation in the human microglia so as to possibly design new therapeutic approaches to neurological disorders, especially when characterized by neuroinflammation. Thus, in the present work, we explored the possibility that: (i) human microglia could be able to synthesize and release neurosteroids; (ii) synthesis of these steroidal molecules could be modified in response to oxidants such as rotenone. We also evaluated the possibility that the major neurosteroid, allopregnanolone, could produce a protective effect on human microglia exposed to the toxic activity of rotenone. The main findings of our study show that also the human microglia is able to produce and release neurosteroids, as previously found for the mouse microglia [33], and that this ability could be altered by neurotoxicants such as rotenone. Furthermore, as an additional novel finding, we showed that a high dose of exogenous allopregnanolone could result in rescue effects for the human microglia endangered by metabolic challenges, as in the case of exposure to rotenone, thus suggesting a possible therapeutic use of this neurosteroid to protect the microglia.

Although HMC3 microglia were able to produce neurosteroids in basal conditions, the metabolic profile of the evaluated molecules was different from that found in mouse BV2 cells [33]. Indeed, in our previous study, we found that neurosteroids were present at much higher levels in the culture medium of BV2 cells. For instance, pregnenolone was at a concentration 500-fold higher than that found in the HMC3 culture medium. Consequently, pregnenolone sulfate was also approximately 67-fold higher in the BV2 culture medium. Similar or even more remarkable differences could be mentioned for the other neurosteroids, including progesterone, 5α-DHP, and allopregnanolone, with the only exception of pregnanolone, which was produced at very low levels also by resting BV2 cells.

This scenario was notably changed by rotenone, which induced in HMC3 culture medium changes that were partly different from those observed for BV2 cells. Despite the fact that a 24 h exposure to rotenone did not significantly modify the levels of pregnenolone, pregnenolone sulfate, progesterone, 5α-DHP, pregnanolone, and allopregnanolone in the mouse BV2 cell cultures, human HMC3 microglia responded to rotenone by reducing the levels of pregnenolone sulfate and pregnanolone and by increasing the levels of allopregnanolone. Thus, human microglia appeared to react more rapidly to oxidative damage caused by rotenone than mouse microglia, which instead modified the release of neurosteroids in the culture medium only 48 h after the rotenone treatment [33]. Interestingly, the late changes in neurosteroid release of BV2 cells were also qualitatively different from those evidenced in HMC3 cells because rotenone produced an increase in 5α-DHP and pregnanolone levels and reduced pregnenolone levels in the culture medium of BV2 cells with no changes in allopregnanolone levels. Thus, the ability to modify allopregnanolone metabolism in response to the oxidant rotenone appears to be specific for the human microglia, at least in the case of HMC3 cells. Alternatively, it could be proposed that the differences in allopregnanolone levels in HMC3 and BV2 cell cultures were dependent on the different extent of rotenone toxicity, which was more pronounced for BV2 microglia, but this interpretation is at odds with the changes observed for the other neurosteroids, 5α-DHP and pregnanolone, which increased in the BV2 culture medium.

Allopregnanolone is a protective molecule [5], but the changes we observed in response to rotenone were ineffective in promoting the survival of HMC3 cells for more than 24 h. However, by adding allopregnanolone exogenously, we were able to block the detrimental effects of rotenone on the survival of HMC3 cells, as observed at 48 h. The extremely increased concentration of allopregnanolone required to obtain this beneficial result may explain why the changes in allopregnanolone synthesis and release by HMC3 microglia were not sufficient to afford protection at the 48 h time interval. Anyway, the effective allopregnanolone concentration was close to the range of levels measured in the cerebrospinal fluid of healthy humans (females: 0.09 nM; males: 0.2 nM) [35], suggesting that the contribution of microglia to neurosteroid synthesis may be limited because other sources can provide adequate levels of allopregnanolone to protect microglial cells. Moreover, allopregnanolone concentration could be increased in the brain tissue by drugs that modify the steroid metabolism in the adrenal gland, such as trilostane [36], to pave the way for the growing field of microglia pharmacology [37].

We did not investigate any possible mechanisms activated by allopregnanolone to afford protection from the oxidative effects of rotenone. Indeed, this is a limitation of our study, and further experiments will be required to address this issue. Interestingly, an in vivo study using mice treated with pilocarpine to induce a status epilepticus evidenced a reduction in ROS cerebral tissue levels and oxidative damage by treatment with allopregnanolone. This effect was associated with an increased expression of superoxide dismutase 2 in the hippocampus of allopregnanolone-treated mice [38]. It is also interesting to note that the oxidative effect of silver nanoparticles disrupting the antioxidant defense in the hippocampus of Wistar rats was associated with a reduction in progesterone, 17α-hydroxyprogesterone, and testosterone hippocampal levels, whereas allopregnanolone, dehydroepiandrosterone, dehydroepiandrosterone sulfate, androstenedione, dihydrotestosterone, and 17β-estradiol hippocampal levels were increased or unmodified, depending on the silver formulation used [39]. It is also worth noting that both progesterone and allopregnanolone exerted beneficial effects in the Wobbler mouse model of amyotrophic lateral sclerosis by preserving mitochondrial respiratory complex I activity, reducing the mitochondrial expression and activity of nitric oxide synthase, and inducing the Mn-dependent superoxide dismutase [40].

5. Conclusions

Our work provides the first evidence that human microglia produce and release neurosteroids and that this metabolic activity could be modulated in response to a damaging event. We also found that the anti-inflammatory neurosteroid allopregnanolone could provide microglia protection at physiological levels [41], which might be further increased by pharmacological approaches. These results constitute a pathophysiological background to encourage the investigators to consider neurosteroids and, especially, allopregnanolone as a possible useful therapeutic tool to cope with brain diseases in which oxidative stress and neuroinflammation play a pathological role, as in the case of multiple sclerosis, amyotrophic lateral sclerosis, Parkinson's disease, Alzheimer's disease, and epilepsy.

Author Contributions: Concept and design of the study: C.L., A.C., M.F., G.V., C.R., R.A. and G.B. Experiments, data acquisition, and analysis: C.L., A.C., C.R. and R.A. Drafting the manuscript and figures: C.L. and G.B. Writing—review and editing: C.L., G.V., C.R., R.A., J.M. and G.B. Supervision: G.V., J.M. and G.B. Funding: J.M. and G.B. All authors have read and agreed to the published version of the manuscript.

Funding: This research was funded by the University of Modena and Reggio Emilia, (Fondo di Ateneo per la Ricerca, FAR 2021 "Enhancing Neuroprotection against nEuroinflammation: the role of neURosteroids in Amyotrophic Lateral Sclerosis (NEURALS)" to J.M. and C.L.), and by the Minister of Health (Ministero della Salute, RF-2021-12373036 "Neurosteroids against Neuroinflammation: a way to fight Neurodegeneration in Amyotrophic Lateral Sclerosis? (NeurALStop) to J.M. and G.B.").

Institutional Review Board Statement: Not applicable.

Informed Consent Statement: Not applicable.

Data Availability Statement: The data presented in this study are available on request from the corresponding author.

Conflicts of Interest: The authors declare no conflict of interest.

References

1. Giatti, S.; Garcia-Segura, L.M.; Barreto, G.E.; Melcangi, R.C. Neuroactive Steroids, Neurosteroidogenesis and Sex. *Prog. Neurobiol.* **2019**, *176*, 1–17. [CrossRef]
2. Gunn, B.G.; Cunningham, L.; Mitchell, S.G.; Swinny, J.D.; Lambert, J.J.; Belelli, D. GABAA Receptor-Acting Neurosteroids: A Role in the Development and Regulation of the Stress Response. *Front. Neuroendocrinol.* **2015**, *36*, 28–48. [CrossRef] [PubMed]
3. Yilmaz, C.; Karali, K.; Fodelianaki, G.; Gravanis, A.; Chavakis, T.; Charalampopoulos, I.; Alexaki, V.I. Neurosteroids as Regulators of Neuroinflammation. *Front. Neuroendocrinol.* **2019**, *55*, 100788. [CrossRef]
4. Balan, I.; Beattie, M.C.; O'Buckley, T.K.; Aurelian, L.; Morrow, A.L. Endogenous Neurosteroid (3α, 5α)3-Hydroxypregnan-20-One Inhibits Toll-like-4 Receptor Activation and Pro-Inflammatory Signaling in Macrophages and Brain. *Sci. Rep.* **2019**, *9*, 1220. [CrossRef]
5. Diviccaro, S.; Cioffi, L.; Falvo, E.; Giatti, S.; Melcangi, R.C. Allopregnanolone: An Overview on Its Synthesis and Effects. *J. Neuroendocrinol.* **2022**, *34*, e12996. [CrossRef] [PubMed]
6. Noorbakhsh, F.; Baker, G.B.; Power, C. Allopregnanolone and Neuroinflammation: A Focus on Multiple Sclerosis. *Front. Cell. Neurosci.* **2014**, *8*, 134. [CrossRef]
7. di Michele, F.; Luchetti, S.; Bernardi, G.; Romeo, E.; Longone, P. Neurosteroid and Neurotransmitter Alterations in Parkinson's Disease. *Front. Neuroendocrinol.* **2013**, *34*, 132–142. [CrossRef]
8. Irwin, R.W.; Solinsky, C.M.; Brinton, R.D. Frontiers in Therapeutic Development of Allopregnanolone for Alzheimer's Disease and Other Neurological Disorders. *Front. Cell. Neurosci.* **2014**, *8*, 203. [CrossRef]
9. Lévesque, M.; Biagini, G.; Avoli, M. Neurosteroids and Focal Epileptic Disorders. *Int. J. Mol. Sci.* **2020**, *21*, 9391. [CrossRef]
10. Lanthier, A.; Patwardhan, V.V. Sex Steroids and 5-En-3 Beta-Hydroxysteroids in Specific Regions of the Human Brain and Cranial Nerves. *J. Steroid Biochem.* **1986**, *25*, 445–449. [CrossRef] [PubMed]
11. Lacroix, C.; Fiet, J.; Benais, J.P.; Gueux, B.; Bonete, R.; Villette, J.M.; Gourmel, B.; Dreux, C. Simultaneous Radioimmunoassay of Progesterone, Androst-4-Enedione, Pregnenolone, Dehydroepiandrosterone and 17-Hydroxyprogesterone in Specific Regions of Human Brain. *J. Steroid Biochem.* **1987**, *28*, 317–325. [CrossRef] [PubMed]
12. Bixo, M.; Andersson, A.; Winblad, B.; Purdy, R.H.; Bäckström, T. Progesterone, 5alpha-Pregnane-3,20-Dione and 3alpha-Hydroxy-5alpha-Pregnane-20-One in Specific Regions of the Human Female Brain in Different Endocrine States. *Brain Res.* **1997**, *764*, 173–178. [CrossRef] [PubMed]
13. Brown, R.C.; Cascio, C.; Papadopoulos, V. Pathways of Neurosteroid Biosynthesis in Cell Lines from Human Brain: Regulation of Dehydroepiandrosterone Formation by Oxidative Stress and Beta-Amyloid Peptide. *J. Neurochem.* **2000**, *74*, 847–859. [CrossRef] [PubMed]
14. Lin, Y.C.; Cheung, G.; Porter, E.; Papadopoulos, V. The Neurosteroid Pregnenolone Is Synthesized by a Mitochondrial P450 Enzyme Other than CYP11A1 in Human Glial Cells. *J. Biol. Chem.* **2022**, *298*, 102110. [CrossRef]
15. Stoffel-Wagner, B.; Beyenburg, S.; Watzka, M.; Blümcke, I.; Bauer, J.; Schramm, J.; Bidlingmaier, F.; Elger, C.E. Expression of 5alpha-Reductase and 3alpha-Hydroxisteroid Oxidoreductase in the Hippocampus of Patients with Chronic Temporal Lobe Epilepsy. *Epilepsia* **2000**, *41*, 140–147. [CrossRef]
16. Yu, L.; Romero, D.G.; Gomez-Sanchez, C.E.; Gomez-Sanchez, E.P. Steroidogenic Enzyme Gene Expression in the Human Brain. *Mol. Cell. Endocrinol.* **2002**, *190*, 9–17. [CrossRef]
17. Giatti, S.; Diviccaro, S.; Serafini, M.M.; Caruso, D.; Garcia-Segura, L.M.; Viviani, B.; Melcangi, R.C. Sex Differences in Steroid Levels and Steroidogenesis in the Nervous System: Physiopathological Role. *Front. Neuroendocrinol.* **2020**, *56*, 100804. [CrossRef]
18. Schumacher, M.; Guennoun, R.; Robert, F.; Carelli, C.; Gago, N.; Ghoumari, A.; Gonzalez Deniselle, M.C.; Gonzalez, S.L.; Ibanez, C.; Labombarda, F.; et al. Local Synthesis and Dual Actions of Progesterone in the Nervous System: Neuroprotection and Myelination. *Growth Horm. IGF Res. Off. J. Growth Horm. Res. Soc. Int. IGF Res. Soc.* **2004**, *14*, 18–33. [CrossRef]
19. Lin, Y.C.; Papadopoulos, V. Neurosteroidogenic Enzymes: CYP11A1 in the Central Nervous System. *Front. Neuroendocrinol.* **2021**, *62*, 100925. [CrossRef]
20. Steckelbroeck, S.; Watzka, M.; Reichelt, R.; Hans, V.H.; Stoffel-Wagner, B.; Heidrich, D.D.; Schramm, J.; Bidlingmaier, F.; Klingmüller, D. Characterization of the 5alpha-Reductase-3alpha-Hydroxysteroid Dehydrogenase Complex in the Human Brain. *J. Clin. Endocrinol. Metab.* **2001**, *86*, 1324–1331. [CrossRef]
21. Agís-Balboa, R.C.; Pinna, G.; Zhubi, A.; Maloku, E.; Veldic, M.; Costa, E.; Guidotti, A. Characterization of Brain Neurons That Express Enzymes Mediating Neurosteroid Biosynthesis. *Proc. Natl. Acad. Sci. USA* **2006**, *103*, 14602–14607. [CrossRef] [PubMed]
22. Gottfried-Blackmore, A.; Sierra, A.; Jellinck, P.H.; McEwen, B.S.; Bulloch, K. Brain Microglia Express Steroid-Converting Enzymes in the Mouse. *J. Steroid Biochem. Mol. Biol.* **2008**, *109*, 96–107. [CrossRef] [PubMed]
23. Germelli, L.; Da Pozzo, E.; Giacomelli, C.; Tremolanti, C.; Marchetti, L.; Wetzel, C.H.; Barresi, E.; Taliani, S.; Da Settimo, F.; Martini, C.; et al. De Novo Neurosteroidogenesis in Human Microglia: Involvement of the 18 kDa Translocator Protein. *Int. J. Mol. Sci.* **2021**, *22*, 3115. [CrossRef] [PubMed]
24. Jolivel, V.; Brun, S.; Binamé, F.; Benyounes, J.; Taleb, O.; Bagnard, D.; De Sèze, J.; Patte-Mensah, C.; Mensah-Nyagan, A.-G. Microglial Cell Morphology and Phagocytic Activity Are Critically Regulated by the Neurosteroid Allopregnanolone: A Possible Role in Neuroprotection. *Cells* **2021**, *10*, 698. [CrossRef]

25. Vidal-Itriago, A.; Radford, R.A.W.; Aramideh, J.A.; Maurel, C.; Scherer, N.M.; Don, E.K.; Lee, A.; Chung, R.S.; Graeber, M.B.; Morsch, M. Microglia Morphophysiological Diversity and Its Implications for the CNS. *Front. Immunol.* **2022**, *13*, 997786. [CrossRef]
26. Salter, M.W.; Stevens, B. Microglia Emerge as Central Players in Brain Disease. *Nat. Med.* **2017**, *23*, 1018–1027. [CrossRef]
27. Vainchtein, I.D.; Alsema, A.M.; Dubbelaar, M.L.; Grit, C.; Vinet, J.; van Weering, H.R.J.; Al-Izki, S.; Biagini, G.; Brouwer, N.; Amor, S.; et al. Characterizing Microglial Gene Expression in a Model of Secondary Progressive Multiple Sclerosis. *Glia* **2023**, *71*, 588–601. [CrossRef]
28. Bennett, M.L.; Viaene, A.N. What Are Activated and Reactive Glia and What Is Their Role in Neurodegeneration? *Neurobiol. Dis.* **2021**, *148*, 105172. [CrossRef]
29. Drago, F.; Sautière, P.-E.; Le Marrec-Croq, F.; Accorsi, A.; Van Camp, C.; Salzet, M.; Lefebvre, C.; Vizioli, J. Microglia of Medicinal Leech (*Hirudo Medicinalis*) Express a Specific Activation Marker Homologous to Vertebrate Ionized Calcium-Binding Adapter Molecule 1 (Iba1/alias Aif-1). *Dev. Neurobiol.* **2014**, *74*, 987–1001. [CrossRef]
30. Jurga, A.M.; Paleczna, M.; Kuter, K.Z. Overview of General and Discriminating Markers of Differential Microglia Phenotypes. *Front. Cell. Neurosci.* **2020**, *14*, 198. [CrossRef] [PubMed]
31. Sanders, L.H.; Timothy Greenamyre, J. Oxidative Damage to Macromolecules in Human Parkinson Disease and the Rotenone Model. *Free Radic. Biol. Med.* **2013**, *62*, 111–120. [CrossRef]
32. Ibarra-Gutiérrez, M.T.; Serrano-García, N.; Orozco-Ibarra, M. Rotenone-Induced Model of Parkinson's Disease: Beyond Mitochondrial Complex I Inhibition. *Mol. Neurobiol.* **2023**, *60*, 1929–1948. [CrossRef]
33. Avallone, R.; Lucchi, C.; Puja, G.; Codeluppi, A.; Filaferro, M.; Vitale, G.; Rustichelli, C.; Biagini, G. BV-2 Microglial Cells Respond to Rotenone Toxic Insult by Modifying Pregnenolone, 5α-Dihydroprogesterone and Pregnanolone Levels. *Cells* **2020**, *9*, 2091. [CrossRef]
34. Dello Russo, C.; Cappoli, N.; Coletta, I.; Mezzogori, D.; Paciello, F.; Pozzoli, G.; Navarra, P.; Battaglia, A. The Human Microglial HMC3 Cell Line: Where Do We Stand? A Systematic Literature Review. *J. Neuroinflammation* **2018**, *15*, 259. [CrossRef] [PubMed]
35. Meletti, S.; Lucchi, C.; Monti, G.; Giovannini, G.; Bedin, R.; Trenti, T.; Rustichelli, C.; Biagini, G. Decreased Allopregnanolone Levels in Cerebrospinal Fluid Obtained during Status Epilepticus. *Epilepsia* **2017**, *58*, e16–e20. [CrossRef] [PubMed]
36. Costa, A.M.; Gol, M.; Lucchi, C.; Biagini, G. Antiepileptogenic Effects of Trilostane in the Kainic Acid Model of Temporal Lobe Epilepsy. *Epilepsia* **2023**. [CrossRef]
37. Šimončičová, E.; Gonçalves de Andrade, E.; Vecchiarelli, H.A.; Awogbindin, I.O.; Delage, C.I.; Tremblay, M.-È. Present and Future of Microglial Pharmacology. *Trends Pharmacol. Sci.* **2022**, *43*, 669–685. [CrossRef]
38. Cho, I.; Kim, W.-J.; Kim, H.-W.; Heo, K.; Lee, B.I.; Cho, Y.-J. Increased Superoxide Dismutase 2 by Allopregnanolone Ameliorates ROS-Mediated Neuronal Death in Mice with Pilocarpine-Induced Status Epilepticus. *Neurochem. Res.* **2018**, *43*, 1464–1475. [CrossRef]
39. Dziendzikowska, K.; Wilczak, J.; Grodzicki, W.; Gromadzka-Ostrowska, J.; Węsierska, M.; Kruszewski, M. Coating-Dependent Neurotoxicity of Silver Nanoparticles-An In Vivo Study on Hippocampal Oxidative Stress and Neurosteroids. *Int. J. Mol. Sci.* **2022**, *23*, 1365. [CrossRef] [PubMed]
40. De Nicola, A.F.; Meyer, M.; Garay, L.; Kruse, M.S.; Schumacher, M.; Guennoun, R.; Gonzalez Deniselle, M.C. Progesterone and Allopregnanolone Neuroprotective Effects in the Wobbler Mouse Model of Amyotrophic Lateral Sclerosis. *Cell. Mol. Neurobiol.* **2022**, *42*, 23–40. [CrossRef]
41. Mtchedlishvili, Z.; Bertram, E.H.; Kapur, J. Diminished allopregnanolone enhancement of GABAA receptor currents in a rat model of chronic temporal lobe epilepsy. *J. Physiol.* **2001**, *537*, 453–465. [CrossRef] [PubMed]

Article

Distinct Roles of CK2- and AKT-Mediated NF-κB Phosphorylations in Clasmatodendrosis (Autophagic Astroglial Death) within the Hippocampus of Chronic Epilepsy Rats

Ji-Eun Kim, Duk-Shin Lee, Tae-Hyun Kim, Hana Park and Tae-Cheon Kang *

Department of Anatomy and Neurobiology and Institute of Epilepsy Research, College of Medicine, Hallym University, Chuncheon 24252, Republic of Korea; hyun1028@hallym.ac.kr (T.-H.K.)
* Correspondence: tckang@hallym.ac.kr; Tel.: +82-33-248-2524; Fax: +82-33-248-2525

Abstract: The downregulation of glutathione peroxidase-1 (GPx1) plays a role in clasmatodendrosis (an autophagic astroglial death) in the hippocampus of chronic epilepsy rats. Furthermore, N-acetylcysteine (NAC, a GSH precursor) restores GPx1 expression in clasmatodendritic astrocytes and alleviates this autophagic astroglial death, independent of nuclear factor erythroid-2-related factor 2 (Nrf2) activity. However, the regulatory signal pathways of these phenomena have not been fully explored. In the present study, NAC attenuated clasmatodendrosis by alleviating GPx1 downregulation, casein kinase 2 (CK2)-mediated nuclear factor-κB (NF-κB) serine (S) 529 and AKT-mediated NF-κB S536 phosphorylations. 2-[4,5,6,7-Tetrabromo-2-(dimethylamino)-1H-benzo[d]imidazole-1-yl]acetic acid (TMCB; a selective CK2 inhibitor) relieved clasmatodendritic degeneration and GPx1 downregulation concomitant with the decreased NF-κB S529 and AKT S473 phosphorylations. In contrast, AKT inhibition by 3-chloroacetyl-indole (3CAI) ameliorated clasmatodendrosis and NF-κB S536 phosphorylation, while it did not affect GPx1 downregulation and CK2 tyrosine (Y) 255 and NF-κB S529 phosphorylations. Therefore, these findings suggest that seizure-induced oxidative stress may diminish GPx1 expression by increasing CK2-mediated NF-κB S529 phosphorylation, which would subsequently enhance AKT-mediated NF-κB S536 phosphorylation leading to autophagic astroglial degeneration.

Keywords: 3CAI; astrocyte; autophagy; GPx1; NAC; oxidative stress; seizure; TMCB

Citation: Kim, J.-E.; Lee, D.-S.; Kim, T.-H.; Park, H.; Kang, T.-C. Distinct Roles of CK2- and AKT-Mediated NF-κB Phosphorylations in Clasmatodendrosis (Autophagic Astroglial Death) within the Hippocampus of Chronic Epilepsy Rats. *Antioxidants* **2023**, 12, 1020. https://doi.org/10.3390/antiox12051020

Academic Editor: Stanley Omaye

Received: 27 March 2023
Revised: 19 April 2023
Accepted: 27 April 2023
Published: 28 April 2023

1. Introduction

Clasmatodendrosis is an autophagic and non-apoptotic type II programmed death in astrocytes. Clasmatodendritic degeneration is characterized by lysosome-derived vacuolization in hypertrophic cell body and fragmentation/vanishing of processes [1–3]. Clasmatodendrosis was first reported by Alzheimer and termed by Cajal [4,5]. Clasmatodendrosis is detected in aging and various pathophysiological conditions in ischemia [6], acidosis [7], dementia [6,8,9], head trauma [10], infection [11] and demyelination disease [12]. In epilepsy rats, clasmatodendritic degeneration is restrictedly detected in astrocytes within the stratum radiatum of the CA1 region (CA1 astrocytes) [13–15]. Although the roles of these astroglial degenerations in pathogenesis of various neurological diseases remain an open issue, clasmatodendrosis influences the duration of spontaneous seizures in chronic epilepsy rats [14,16].

The underlying mechanisms of clasmatodendrosis are relevant to oxidative stress, the impaired ATP production induced by acidosis and/or energy-consuming events, aberrant chaperone accumulation and neuroinflammation [7,13,17], although the regulatory signal pathways are largely unknown. Interestingly, nuclear factor-κB (NF-κB) is one of the upstream molecules evoking autophagic cell death of astrocytes [18,19]. Indeed, tumor necrosis factor-α (TNF-α) neutralization alleviates clasmatodendrosis by inhibiting NF-κB serine (S) 529 phosphorylation [2]. Furthermore, AKT and its downstream effector glycogen synthase kinase-3β (GSK-3β) also play a pivotal role in this process [2,13,14,20,21]. However, the possibility of integration between AKT- and NF-κB-mediated signaling pathways during clasmatodendritic degeneration has not been reported.

Glutathione peroxidase-1 (GPx1) is the first identified selenoprotein that scavenges reactive oxygen species (ROS) through the reduction of H_2O_2 by using glutathione (GSH, an endogenous antioxidant) as a cofactor [22,23]. GPx1 regulates the induction of autophagy in response to ROS [24,25]. GPx1 downregulation is relevant to clasmatodendrosis in the hippocampus of chronic epilepsy rats, which is regulated by GSH biosynthesis. Briefly, GPx1 is significantly decreased in clasmatodendritic CA1 astrocytes, while it is increased in reactive CA1 astrocytes. N-acetylcysteine (NAC, a GSH precursor) restores GPx1 expression in clasmatodendritic astrocytes and alleviates clasmatodendrosis. In contrast, L-buthionine sulfoximine (BSO, an inducer of GSH depletion) aggravates clasmatodendrosis accompanied by GPx1 downregulation, independent of nuclear factor erythroid-2-related factor 2 (Nrf2) activity [15]. Since GPx1 plays an important role in astroglial viability against ROS-mediated toxicity [26,27], we have suggested that GSH-mediated GPx1 regulation may be related to clasmatodendrosis in CA1 astrocytes, although the regulatory signal pathways of these phenomena have not been fully explored.

As aforementioned, NAC induces GPx1 upregulation [15], which can inhibit NF-κB S536 phosphorylation [28,29]. In addition, GPx1 expression is transiently reduced in CA1 astrocytes at 3 day after status epilepticus (SE) when NF-κB transactivation is increased [15,30]. NAC also inhibits TNF-α-induced AKT activation and AKT-mediated NF-κB S536 phosphorylation [31]. Furthermore, GPx1 silencing drives the ROS-mediated AKT activation [32,33]. Considering these previous studies, it is likely that (1) NF-κB and GPx1 may reciprocally regulate each other and/or that (2) GPx1 may integrate between NF-κB- and AKT-mediated signaling pathways during clasmatodendritic degeneration, which have not been reported yet. Thus, we conducted the present study to elucidate these hypotheses.

Here, we demonstrate that NAC attenuated clasmatodendrosis by alleviating GPx1 downregulation, casein kinase 2 (CK2)-mediated NF-κB S529 and AKT-mediated NF-κB S536 phosphorylations. 2-[4,5,6,7-tetrabromo-2-(dimethylamino)-1H-benzo[d]imidazole-1-yl]acetic acid (TMCB; a selective CK2 inhibitor) relieved clasmatodendritic degeneration and GPx1 downregulation concomitant with the decreased NF-κB S529, S536 and AKT S473 phosphorylations. However, AKT inhibition by 3-chloroacetyl-indole (3CAI) did not affect GPx1 downregulation and CK2 tyrosine (Y) 255 and NF-κB S529 phosphorylations, although it mitigated clasmatodendrosis and NF-κB S536 phosphorylation. These unreported data suggest that seizure-induced oxidative stress may diminish GPx1 expression by increasing CK2-mediated NF-κB S529 phosphorylation, which would subsequently enhance NF-κB S536 phosphorylation by AKT hyperactivation, leading to autophagic astroglial degeneration.

2. Materials and Methods

2.1. Experimental Animals and Chemicals

Male Sprague Dawley (SD) rats (200–250 g) were cared under controlled environmental conditions (23–25 °C, 12 h light/dark cycle) and freely accessed to water and conventional rat diets. All experimental protocols described below were approved by the Institutional Animal Care and Use Committee of Hallym University (Hallym 2021-3, approval date: 17 May 2021). All reagents were obtained from Sigma-Aldrich (St. Louis, MO, USA), except as noted.

2.2. Chronic Epilepsy Rat Model

One day before pilocarpine treatment, rats were given LiCl (127 mg/kg, i.p.). The following day, animals were injected with atropine methylbromide (5 mg/kg i.p.) 20 min before pilocarpine (30 mg/kg, i.p.) treatment. To cease status epilepticus (SE), diazepam (Valium; Hoffmann-la Roche, Neuilly-sur-Seine, France; 10 mg/kg, i.p.) was administered

2 h after SE on-set and repeated, as needed. Control animals were given saline instead of pilocarpine. After SE induction, rats were monitored 8 h/day to identify chronic epilepsy activity [14,34].

2.3. NAC Treatment

Chronic epilepsy rats were given N-acetylcysteine (NAC, 70 mg/kg/day, i.p.) over a 7-day period [17]. Five hours after the last injection, the animals were used for experiments.

2.4. Infusion of TMCB and 3CAI

Animals were implanted with an infusion needle (Brain infusion kit 1, Alzet, Cupertino, CA, USA) into the right lateral ventricle (coordinates: 1 mm posterior; 1.5 mm lateral; 3.5 mm depth) under isoflurane anesthesia (3% induction, 1.5–2% for surgery, and 1.5% maintenance in a 65:35 mixture of $N_2O:O_2$), and connected with an Alzet 1007D osmotic pump (Alzet, Cupertino, CA, USA) containing (1) the vehicle, (2) TMCB (0.5 μM) or (3) 3CAI (25 μM). Seven days after surgery, the animals were used for experiments [14,30].

2.5. Western Blot

Under urethane anesthesia (1.5 g/kg, i.p.), rats were decapitated, and the hippocampus was rapidly dissected out and homogenized in lysis buffer containing protease inhibitor cocktail (Roche Applied Sciences, Branford, CT, USA) and phosphatase inhibitor cocktail (PhosSTOP®, Roche Applied Science, Branford, CT, USA). The protein concentration was measured using a Micro BCA Protein Assay Kit (Pierce Chemical, Dallas, TX, USA). Thereafter, Western blotting was performed by the standard protocol (n = 7 rats in each group). After electrophoresis, proteins were transferred to polyvinylidene fluoride membranes that were subsequently incubated with a blocking solution followed by immunoblotting with the primary antibody (Table 1). For chemiluminescent detection and analysis, an ImageQuant LAS4000 system (GE Healthcare Korea, Seoul, South Korea) was used. The β-actin value was used for the normalization of each protein value. The phosphoprotein/total protein ratio was represented as the phosphorylation ratio [14,30,34].

Table 1. Primary antibodies used in the present study.

Antigen	Host	Manufacturer (Catalog Number)	Dilution Used
AKT	Rabbit	Cell signaling (#9272)	1:1000 (WB)
CK2	Mouse	Millipore (#05-1431)	1:1000 (WB)
GFAP	Mouse	Millipore (#MAB3402)	1:2000 (IH)
GPx1	Sheep	Biosensis (#S-072-100)	1:2000 (IH) 1:10,000 (WB)
NF-κB	Rabbit	Abcam (#ab16502)	1:500 (IH) 1:2000 (WB)
NF-κB S529	Rabbit	Abcam (#ab47395)	1:100 (IH) 1:1000 (WB)
NF-κB S536	Rabbit	Abcam (#ab28856)	1:100 (IH) 1:1000 (WB)
p-AKT S473	Rabbit	Cell signaling (#4060)	1:250 (IH) 1:1000 (WB)
p-CK2 Y255	Rabbit	Invitrogen (#PA5-38831)	1:1000 (WB)
β-actin	Mouse	Sigma (#A5316)	1:5000 (WB)

IH: Immunohistochemistry; WB: Western blot.

2.6. Tissue Preparation and Immunohistochemistry

Animals were anesthetized with urethane anesthesia (1.5 g/kg, i.p.) and perfused with 4% paraformaldehyde in 0.1 M phosphate buffer (PB, pH 7.4) through the left ventricle followed by post-fixation in the same fixative overnight. After immersion with 30% sucrose overnight, brains were sectioned at 30 μm. Sections were blocked with 3% bovine serum albumin in PBS for 30 min, and later incubated overnight with mixtures of primary antibodies (Table 1) in PBS containing 0.3% Triton X-100. After washing, tissues were reacted with Brilliant Violet-421, Cy2- or Cy3-fluorescent dye conjugated secondary antibodies (Jackson Immuno Research Laboratories, West Grove, PA, USA). The fluorescent intensity was quantified in the randomly selected 2–3 reactive astrocytes or clasmatodendritic astrocytes in the stratum radiatum of the CA1 region ($n = 7$ rats in each group) with AxioVision Rel. 4.8 (Carl Zeiss Korea, Seoul, Republic of Korea) and ImageJ software. For quantification of clasmatodendritc astrocytes, cell counts were conducted in areas of interest (1×10^4 μm^2) of 10 sections per each animal [14,34].

2.7. Data Analysis

The Mann–Whitney test was applied to analyze statistical significance of data obtained from two groups. The Kruskal–Wallis test with Dunn–Bonferroni post hoc test was used for the comparison of data obtained four groups. The Spearman test was applied to identify the relationship between two variables. A p-value less than 0.05 was considered significant.

3. Results

3.1. NAC Restores GPx1 Expression and Inhibits CK2-Mediated NF-κB S529 Phosphorylation in Clasmatodendritic CA1 Astrocytes

NF-κB signaling pathway activates autophagy. In particular, NF-κB S529 phosphorylation is involved in clasmatodendritic astrocytes [2,35]. Since NF-κB at S529 site is phosphorylated by CK2 [36], we investigated the effects of NAC on GPx1 expression and CK2-mediated NF-κB S529 phosphorylation in clasmatodendritic astrocytes.

Compatible with a previous study [16], NAC ameliorated clasmatodendritic degeneration of CA1 astrocytes (Figure 1A,B). Clasmatodendritic CA1 astrocytes showed GPx1 downregulation, although reactive CA1 astrocytes exhibited GPx1 upregulation (Figure 1A,C). Compared to the vehicle, NAC increased GPx1 expression in clasmatodendritic (vacuolized) astrocytes, but not in reactive astrocytes (Figure 1A,C). In contrast to GPx1, NF-κB S529 phosphorylation was enhanced in clasmatodendritic astrocytes, as compared reactive astrocytes (Figure 1A,D). NAC abolished NF-κB S529 phosphorylation in clasmatodendritic CA1 astrocytes, but not in reactive astrocytes (Figure 1A,D). Thus, NF-κB S529 phosphorylation showed an inverse correlation with GPx1 expression (Figure 1E).

Compatible with immunofluorescent studies, Western blot data also revealed that NAC augmented GPx1 expression, but reduced NF-κB S529 phosphorylation level, as compared to the vehicle (Figures 2A–C and S1). Since NF-κB S529 phosphorylation is regulated by CK2, which increases NF-κB-mediated nuclear transcriptional activity [36], we further evaluated the effect of NAC on CK2 phosphorylation (activity). Western blot data revealed that CK2 Y255 phosphorylation was reduced in the hippocampus of chronic epilepsy rats, as compared to control animals (Figures 2A,D and S1). Compared to the vehicle, NAC further diminished CK2 Y255 phosphorylation without altering its total protein level (Figures 2A,D and S1). Considering that inhibition of the Src/CK2 signaling pathway is one of the insufficient adaptive responses to seizures [14,37,38], our findings indicated that CK2-mediated NF-κB S529 phosphorylation may lead to clasmatodendrosis, accompanied by GPx1 downregulation.

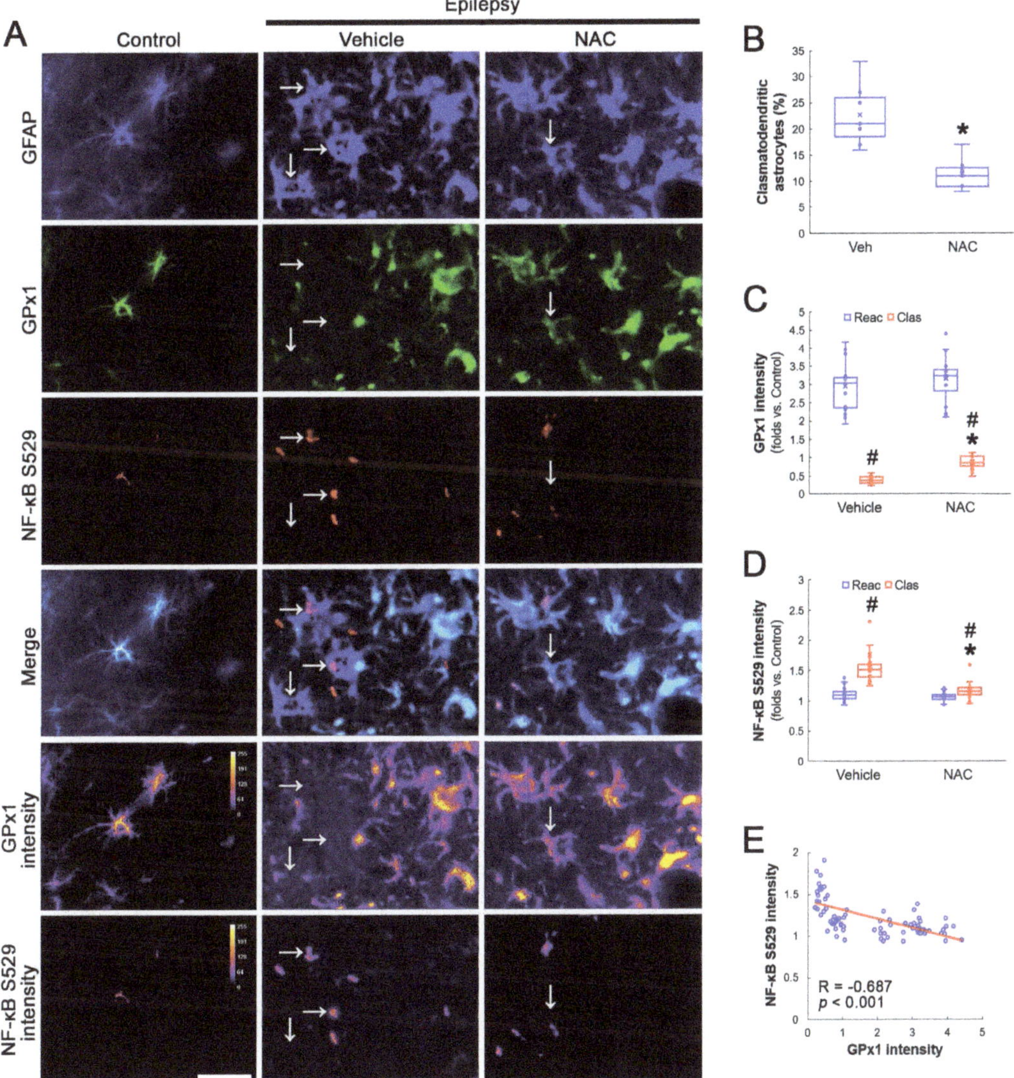

Figure 1. Effects of NAC on GPx1 expression and NF-κB S529 phosphorylation in CA1 astrocytes. Compared to control rats, GPx1 expression is increased in reactive CA1 astrocytes (Reac), while it is reduced in clasmatodendritic (vacuolized) CA1 astrocytes (Clas, arrows). However, the NF-κB S529 signal is enhanced only in clasmatodendritic CA1 astrocytes. CK2 Y255 phosphorylation is diminished in the whole hippocampus of chronic epilepsy rats. Compared to the vehicle (Veh), NAC ameliorates GPx1 downregulation and NF-κB S529 phosphorylation in clasmatodendritic astrocytes, accompanied by the reduced CK2 Y255 phosphorylation. (**A**) Representative photos of GPx1 expression, NF-κB S529 signal and their intensities. Bar = 25 μm. (**B**) Quantification of clasmatodendritic degeneration in CA1 astrocytes (* $p < 0.05$ vs. vehicle, $n = 7$ rats, respectively; Mann–Whitney test). (**C,D**) Quantification of GPx1 and NF-κB S529 intensities in CA1 astrocytes (*,# $p < 0.05$ vs. vehicle and reactive astrocytes, respectively, $n = 20$ cells in 7 rats, respectively; Kruskal–Wallis test with Dunn–Bonferroni post hoc test). (**E**) Linear regression analysis between GPx1 and NF-κB S529 intensities in reactive and clasmatodendritic CA1 astrocytes of chronic epilepsy rats ($n = 80$ cells in 14 rats; Spearman test).

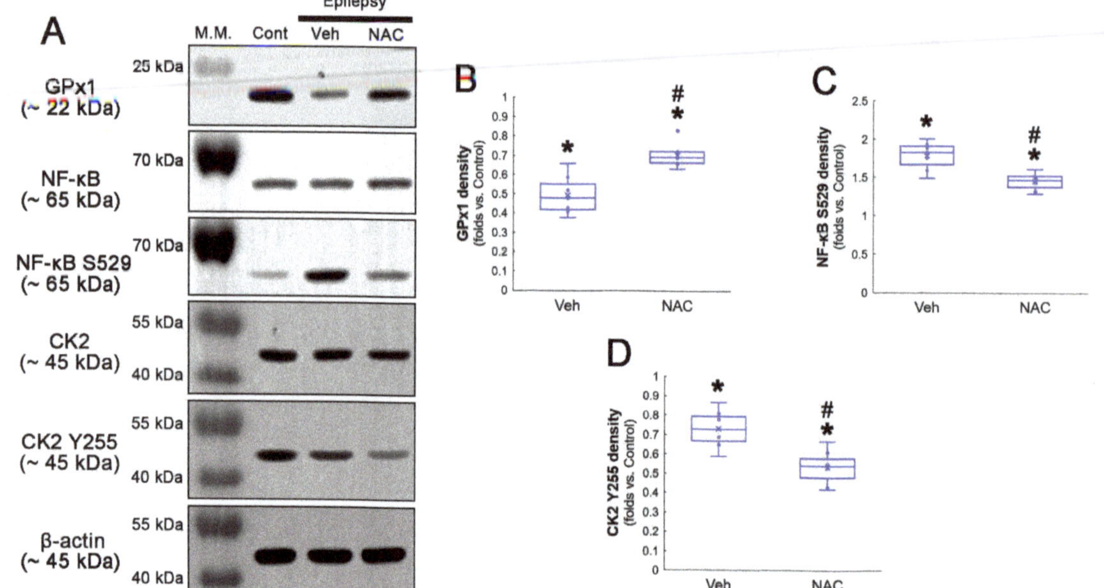

Figure 2. Western blot data representing the effects of NAC on GPx1 expression, NF-κB S529, and CK2 Y255 phosphorylations. Consistent with the immunofluorescent study (Figure 1), NAC increases GPx1 expression, but diminishes NF-κB S529 phosphorylation level, as compared to the vehicle (Veh). In addition, CK2 Y255 phosphorylation is decreased in the whole hippocampus of chronic epilepsy rats, which is further reduced by NAC. (**A**) Representative Western blot of GPx1, NF-κB, NF-κB S529, CK2 and CK2 Y255 levels. (**B–D**) Quantification of GPx1, NF-κB S529 and CK2 Y255 phosphorylation levels based on Western blot data (*,# $p < 0.05$ vs. control rats and vehicle-treated epilepsy rats, $n = 7$ rats, respectively; Kruskal–Wallis test with Dunn–Bonferroni post hoc test).

3.2. NAC Diminished AKT-Mediated NF-κB S536 Phosphorylation in Clasmatodendritic CA1 Astrocytes

AKT S473 hyperphosphorylation causes bax-interacting factor 1 (Bif-1)-mediated astroglial autophagy [13,14]. AKT activation also stimulates NF-κB S536 phosphorylation [39]. Interestingly, deletion or inhibition of GPx1 increases NF-κB S536 phosphorylation [28,29] and NFκB S536 phosphorylation is critical for autophagy in response to oxidative stress [40,41]. Furthermore, NAC inhibits TNF-α-induced AKT S473 and NF-κB S536 phosphorylation [31]. Thus, we explored whether NF-κB S536 phosphorylation is involved in AKT-mediated clasmatodendrosis and NAC abolishes this pathway. Compared to intact astrocytes, both reactive astrocytes and clasmatodendritic CA1 astrocytes showed AKT S473 hyperphosphorylation (Figure 3A,B). However, AKT S473 intensity in clasmatodendritic astrocytes was higher than that in reactive astrocytes (Figure 3A,B). NAC attenuated AKT S473 hyperphosphorylation in clasmatodendritic astrocytes, but not in reactive astrocytes (Figure 3A,B). AKT S473 phosphorylation showed an inverse proportion with GPx1 expression (Figure 3C). Similar to the case of AKT S473 phosphorylation, NF-κB S536 phosphorylation was enhanced in clasmatodendritic astrocytes, compared to reactive astrocytes (Figure 4A,B). NAC abolished NF-κB S536 phosphorylation in clasmatodendritic CA1 astrocytes, but not in reactive astrocytes (Figure 4A,B). Linear regression analysis showed an inverse proportional relationship between GPx1 and NF-κB S536 phosphorylation (Figure 4C).

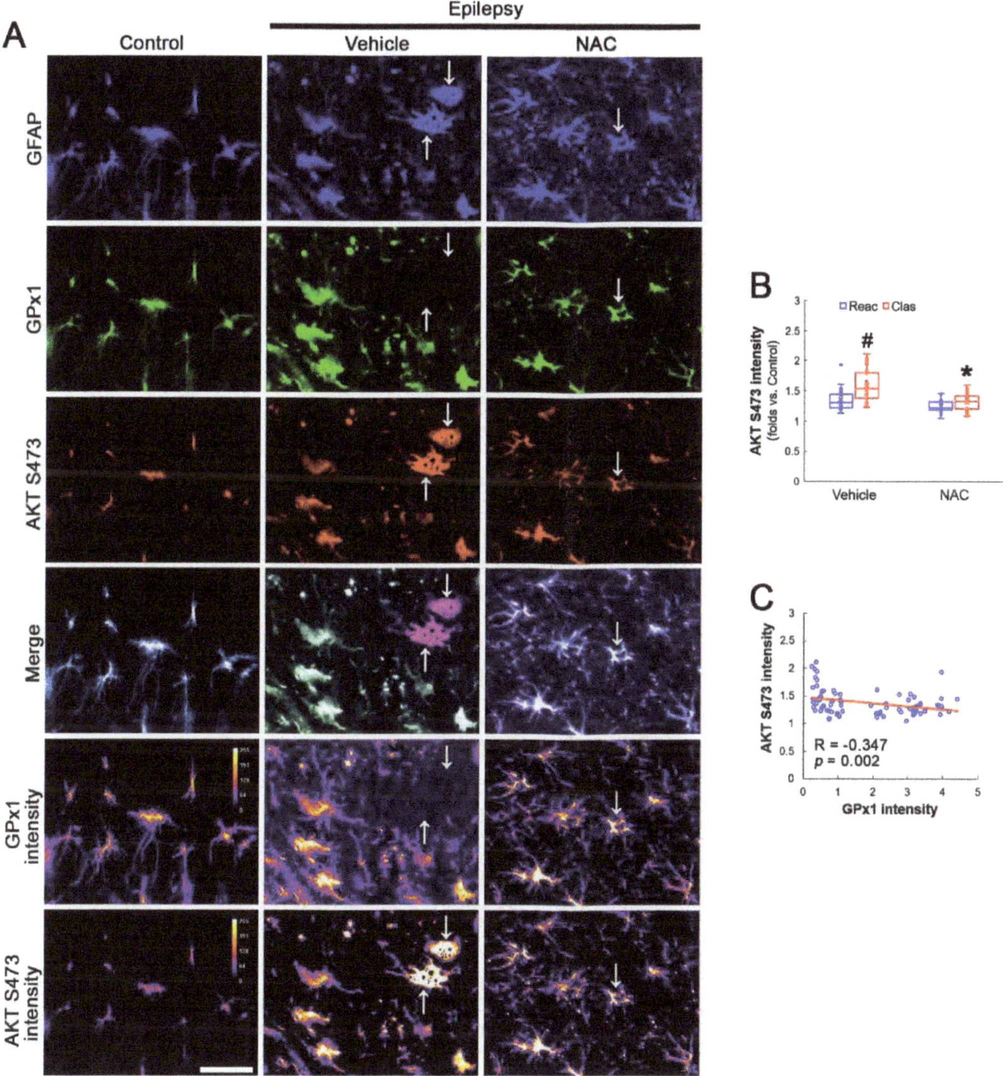

Figure 3. Effects of NAC on GPx1 expression and AKT S473 phosphorylation in CA1 astrocytes.
Compared to control rats, AKT S473 phosphorylation is enhanced in clasmatodendritic (vacuolized)
CA1 astrocytes (Clas, arrows) more than reactive CA1 astrocytes (Reac), which is attenuated by NAC
treatment. (**A**) Representative photos of GPx1 expression, AKT S473 signal and their intensities.
Bar = 25 µm. (**B**) Quantification of AKT S473 intensity in CA1 astrocytes (*,# $p < 0.05$ vs. vehicle and
reactive astrocytes, respectively, $n = 20$ cells in 7 rats, respectively; Kruskal–Wallis test with Dunn–
Bonferroni post hoc test). (**C**) Linear regression analysis between GPx1 and AKT S473 intensities
in reactive and clasmatodendritic CA1 astrocytes of chronic epilepsy rats ($n = 80$ cells in 14 rats;
Spearman test).

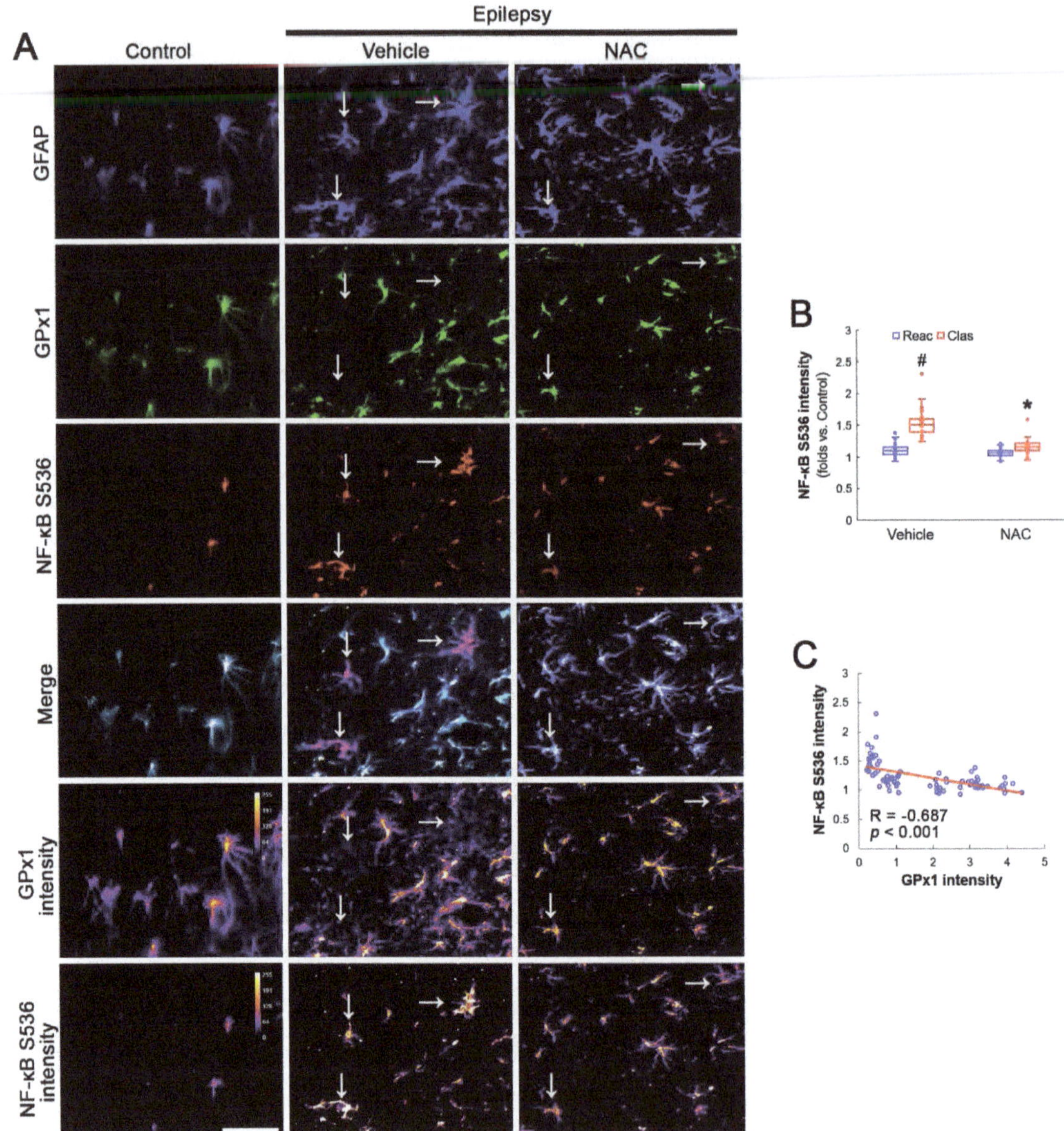

Figure 4. Effects of NAC on GPx1 expression and NF-κB S536 phosphorylation in CA1 astrocytes. Compared to control rats, NF-κB S536 signal is increased in clasmatodendritic (vacuolized) CA1 astrocytes (Clas, arrows), but not reactive CA1 astrocytes (Reac), which is attenuated by NAC treatment. (**A**) Representative photos of GPx1 expression, NF-κB S536 signal and their intensities. Bar = 25 μm. (**B**) Quantification of NF-κB S536 intensity in CA1 astrocytes (*,# $p < 0.05$ vs. vehicle and reactive astrocytes, respectively, n = 20 cells in 7 rats, respectively; Kruskal–Wallis test with Dunn–Bonferroni post hoc test). (**C**) Linear regression analysis between GPx1 and NF-κB S536 intensities in reactive and clasmatodendritic CA1 astrocytes of chronic epilepsy rats (n = 80 cells in 14 rats; Spearman test).

Compatible with immunofluorescent studies, Western blot data also revealed that NAC augmented AKT S473 and NF-κB S536 phosphorylation levels, as compared to the vehicle (Figures 5A–C and S2). These findings indicate that AKT-mediated NF-κB S536 phosphorylation may participate in clasmatodendritic degeneration, and that NAC may ameliorate clasmatodendrosis by inhibiting this pathway as well as CK2-mediated NF-κB S529 phosphorylation.

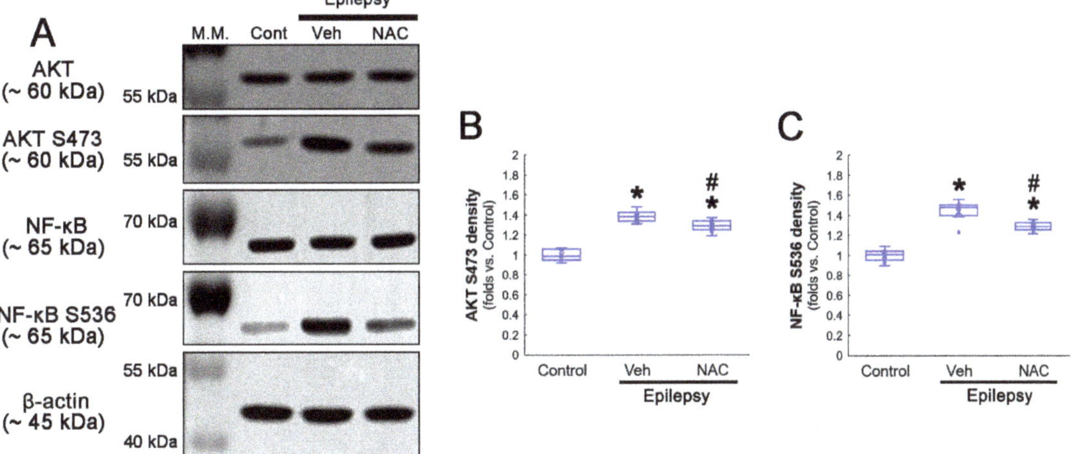

Figure 5. Western blot data representing the effects of NAC on AKT S473 and NF-κB S536 phosphorylations. Consistent with immunofluorescent study (Figures 3 and 4), NAC diminishes AKT S473 and NF-κB S536 phosphorylation levels, as compared to the vehicle (Veh). (**A**) Representative Western blot of AKT, AKT S473, NF-κB and NF-κB S536 levels. (**B,C**) Quantification of AKT S473 and NF-κB S536 phosphorylation levels based on Western blot data (*,# $p < 0.05$ vs. control animals and vehicle-treated epilepsy rats, respectively, $n = 7$ rats, respectively; Kruskal–Wallis test with Dunn–Bonferroni post hoc test).

3.3. CK2 Inhibition Restores GPx1 Upregulation and Attenuates NF-κB and AKT Phosphorylations in Clasmatodendritic CA1 Astrocytes

Next, we applied TMCB (a selective CK2 inhibitor) to identify whether the CK2 signaling pathway would induce GPx1 downregulation during clasmatodendritic degeneration. Similar to the case of NAC, TMCB attenuated clasmatodendritic degeneration of CA1 astrocytes (Figure 6A,B). Compared to the vehicle, TMCB increased GPx1 expression in clasmatodendritic astrocytes, but not in reactive astrocytes (Figure 6A,C). TMCB abolished NF-κB S529 phosphorylation in clasmatodendritic CA1 astrocytes, but not in reactive astrocytes (Figure 6A,D). Furthermore, TMCB abrogated AKT S473 hyperphosphorylation in clasmatodendritic astrocytes, but not in reactive astrocytes (Figure 7A,B). TMCB also diminished NF-κB S536 phosphorylation in clasmatodendritic CA1 astrocytes, but not in reactive astrocytes (Figure 7C,D).

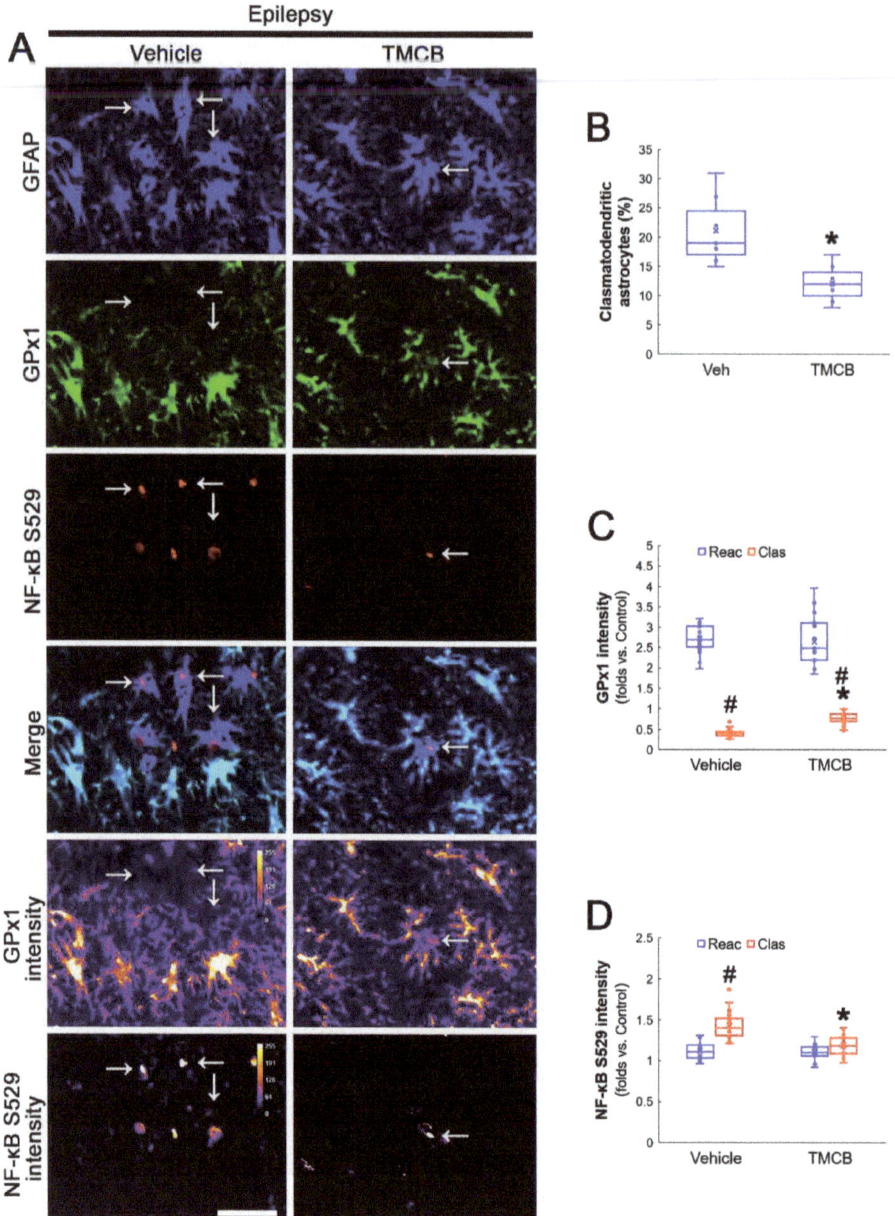

Figure 6. Effects of TMCB on GPx1 expression and NF-κB S529 phosphorylation in CA1 astrocytes. Compared to the vehicle, TMCB attenuates clasmatodendritic degeneration concomitant with the enhanced GPx1 expression and the decreased NF-κB S529 phosphorylation in clasmatodendritic (vacuolized) CA1 astrocytes (Clas, arrows), but not reactive CA1 astrocytes (Reac). (**A**) Representative photos of GPx1 expression, NF-κB S529 signal and their intensities. Bar = 25 μm. (**B**) Quantification of clasmatodendritic degeneration in CA1 astrocytes (* $p < 0.05$ vs. vehicle, $n = 7$ rats, respectively; Mann–Whitney test). (**C,D**) Quantification of GPx1 and NF-κB S529 intensities in CA1 astrocytes (*,# $p < 0.05$ vs. vehicle and reactive astrocytes, respectively, $n = 20$ cells in 7 rats, respectively; Kruskal–Wallis test with Dunn–Bonferroni post hoc test).

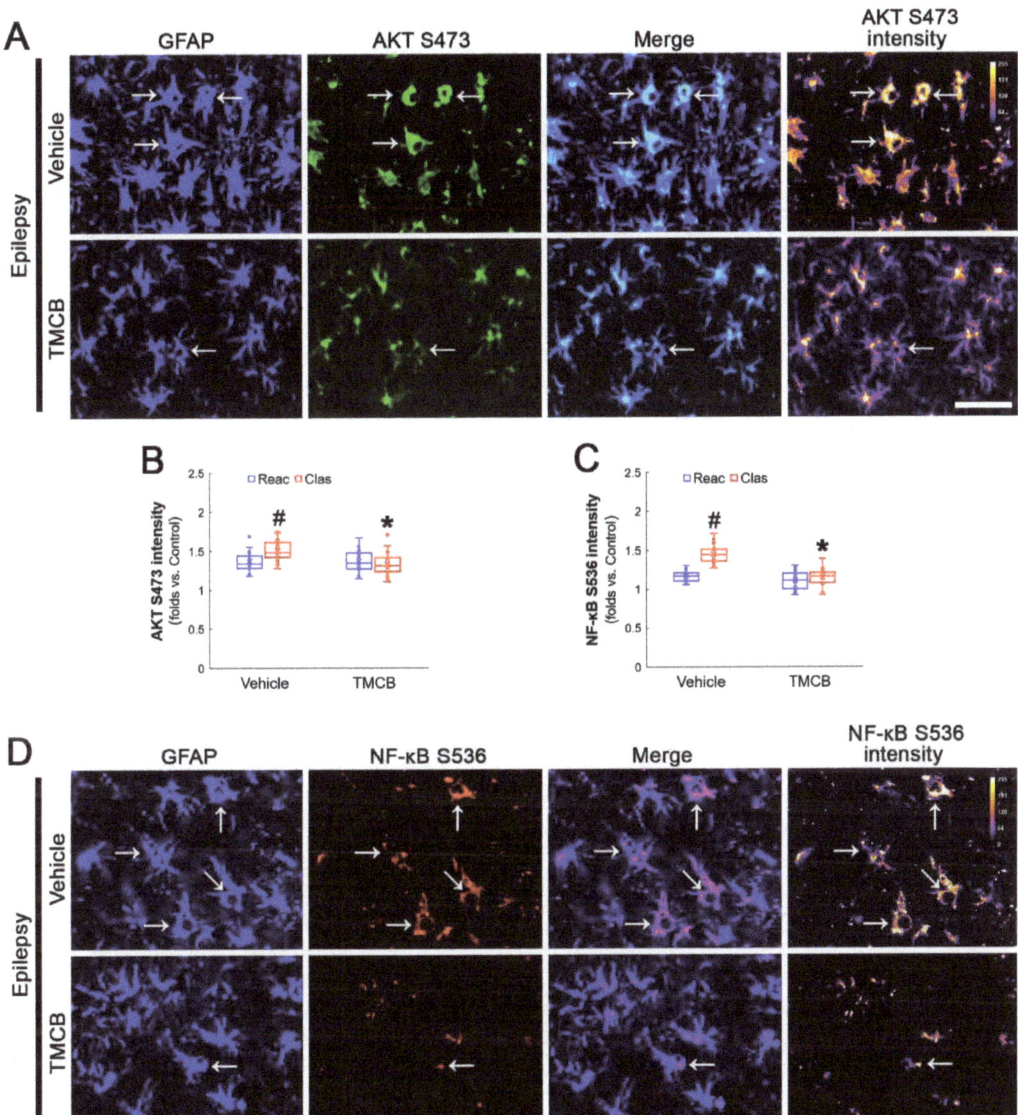

Figure 7. Effects of TMCB on AKT S473 and NF-κB S536 phosphorylations in CA1 astrocytes. Compared to the vehicle, TMCB ameliorates NF-κB S536, but not AKT S473, phosphorylation in clasmatodendritic (vacuolized) CA1 astrocytes (Clas, arrows), but not reactive CA1 astrocytes (Reac). (**A**) Representative photos of AKT S473 phosphorylation and its intensities. Bar = 25 μm. (**B,C**) Quantification of AKT S473 and NF-κB S536 intensity in CA1 astrocytes (*,# $p < 0.05$ vs. vehicle and reactive astrocytes, respectively, $n = 20$ cells in 7 rats, respectively; Kruskal–Wallis test with Dunn–Bonferroni post hoc test). (**D**) Representative photos of NF-κB S536 phosphorylation and its intensities. Bar = 25 μm.

Western blot data also demonstrated that TMCB increased GPx1 expression, but decreased AKT S473, NF-κB S529 and NF-κB S536 phosphorylation levels, as compared to the vehicle (Figures 8A–E and S3). These findings indicate that CK2-mediated NF-κB S529 phosphorylation may diminish GPx1 expression during clasmatodendrosis, and that AKT-mediated NF-κB S536 phosphorylation may be a consequence of GPx1 downregulation induced by this pathway.

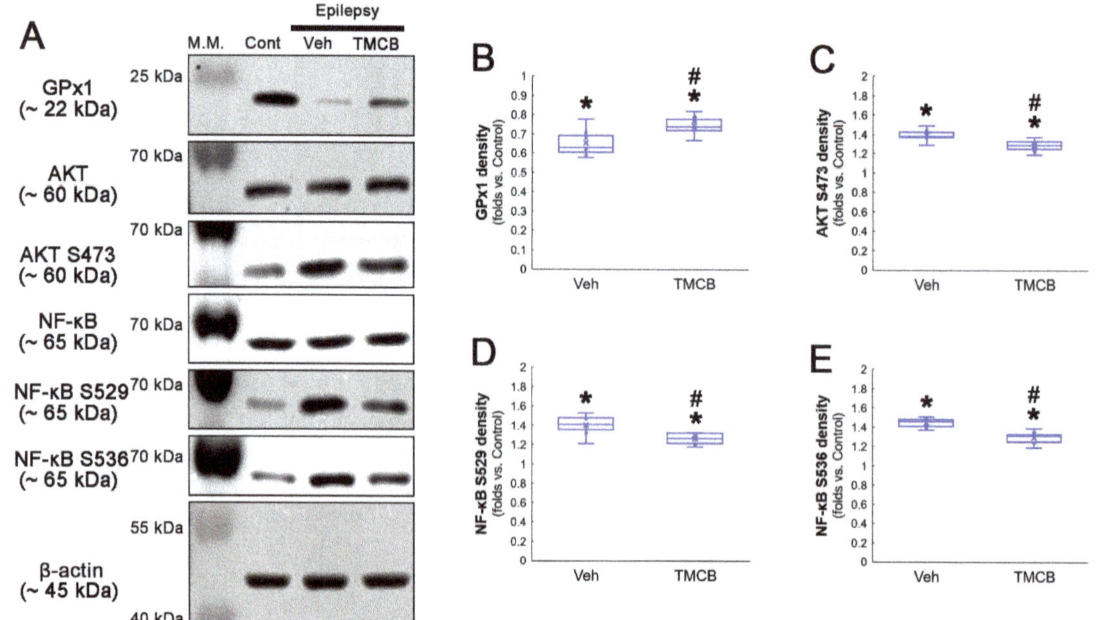

Figure 8. **Western blot data representing the effects of TMCB on GPx1 expression, AKT S473, NF-κB S529 and NF-κB S536 phosphorylations**. Consistent with immunofluorescent study (Figures 6 and 7), TMCB increases GPx1 expression, but reduces AKT S473, NF-κB S529 and NF-κB S536 phosphorylation levels, as compared to the vehicle (Veh). (**A**) Representative Western blot of GPx1, AKT, AKT S473, NF-κB, NF-κB S529 and NF-κB S536 levels. (**B–E**) Quantification of GPx1 expression, AKT S473, NF-κB S529 and NF-κB S536 phosphorylation levels based on Western blot data (*,# $p < 0.05$ vs. control rats and vehicle-treated epilepsy rats, respectively, $n = 7$ rats, respectively; Kruskal–Wallis test with Dunn–Bonferroni post hoc test).

3.4. AKT Inhibition Attenuates Clasmatodendrosis and NF-κB S536 Phosphorylation without Affecting GPx1 Level and CK2-Mediated NF-κB S529 Phosphorylation in Clasmatodendritic CA1 Astrocytes

To confirm the role of AKT-mediated NF-κB S536 phosphorylation in clasmatodendritic degeneration, we applied 3CAI to chronic epilepsy rats. 3CAI ameliorated clasmatodendritic degeneration of CA1 astrocytes (Figure 9A,B). 3CAI also decreased NF-κB S536 phosphorylation in clasmatodendritic astrocytes, but not in reactive astrocytes (Figure 9A,C). However, 3CAI could not affect reduced GPx1 expression in clasmatodendritic astrocytes (Figure 9A,D). In addition, 3CAI did not influence increased NF-κB S529 phosphorylation in CA1 astrocytes (Figure 10A,B).

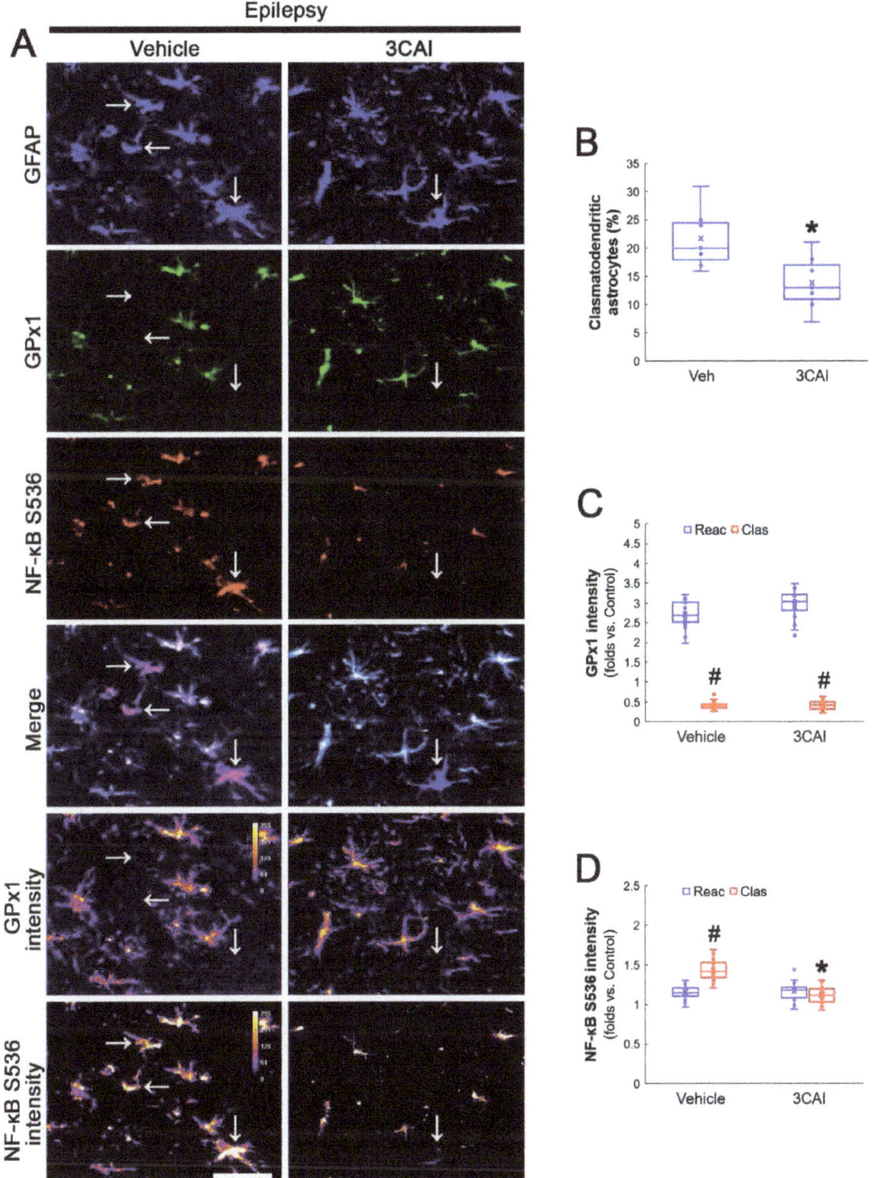

Figure 9. Effects of 3CAI on GPx1 expression and NF-κB S536 phosphorylation in CA1 astrocytes. Compared to the vehicle, 3CAI attenuates clasmatodendritic degeneration concomitant and the increased NF-κB S536 phosphorylation in clasmatodendritic (vacuolized) CA1 astrocytes (Clas, arrows), but not reactive CA1 astrocytes (Reac), while it does not affect GPx1 expression level. (**A**) Representative photos of GPx1 expression and NF-κB S536 signal and their intensities. Bar = 25 μm. (**B**) Quantification of clasmatodendritic degeneration in CA1 astrocytes (* $p < 0.05$ vs. vehicle, $n = 7$ rats, respectively; Mann–Whitney test). (**C,D**) Quantification of GPx1 and NF-κB S536 intensities in CA1 astrocytes (*,# $p < 0.05$ vs. vehicle and reactive astrocytes, respectively, $n = 20$ cells in 7 rats, respectively; Kruskal–Wallis test with Dunn–Bonferroni post hoc test).

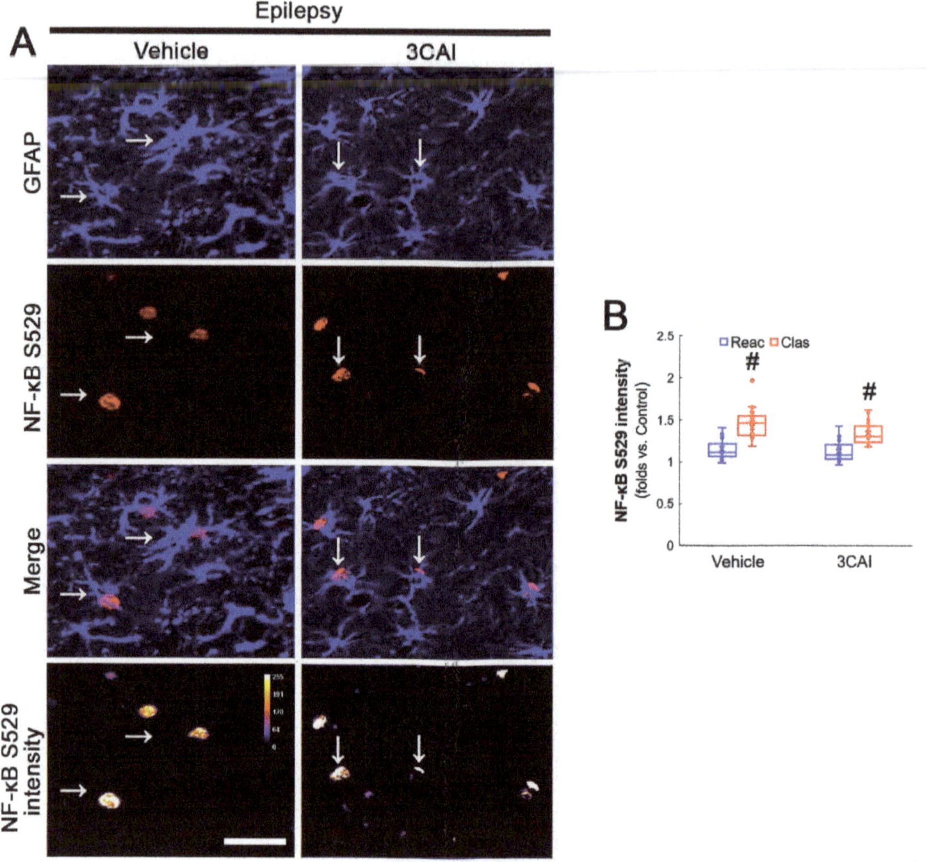

Figure 10. Effects of 3CAI on NF-κB S529 and CK2 Y255 phosphorylations in CA1 astrocytes.
Compared to the vehicle, 3CAI does not influence NF-κB S529 in clasmatodendritic (vacuolized)
CA1 astrocytes (Clas, arrows) and reactive CA1 astrocytes (Reac). CK2 Y255 phosphorylation in
the whole hippocampus is also unaffected by 3CAI treatment. (**A**) Representative photos of the
NF-κB S529 signal and its intensities. Bar = 25 μm. (**B**) Quantification of NF-κB S529 intensity in CA1
astrocytes ($^\#$ $p < 0.05$ vs. reactive astrocytes, n = 20 cells in 7 rats, respectively; Kruskal–Wallis test
with Dunn–Bonferroni post hoc test).

Western blot data revealed that 3CA1 reduced the NF-κB S536 phosphorylation
level without affecting GPx1 expression, NF-κB S529 and CK2 Y255 phosphorylation
(Figures 11A–E and S4). Regarding the GPx1-mediated inhibition of NF-κB S536 phosphorylation [25,26], these findings indicate that the CK2-NF-κB S529-GPx1 signaling pathway
may be an upstream regulator of AKT-mediated NF-κB S536 phosphorylation during
clasmatodendritic degeneration.

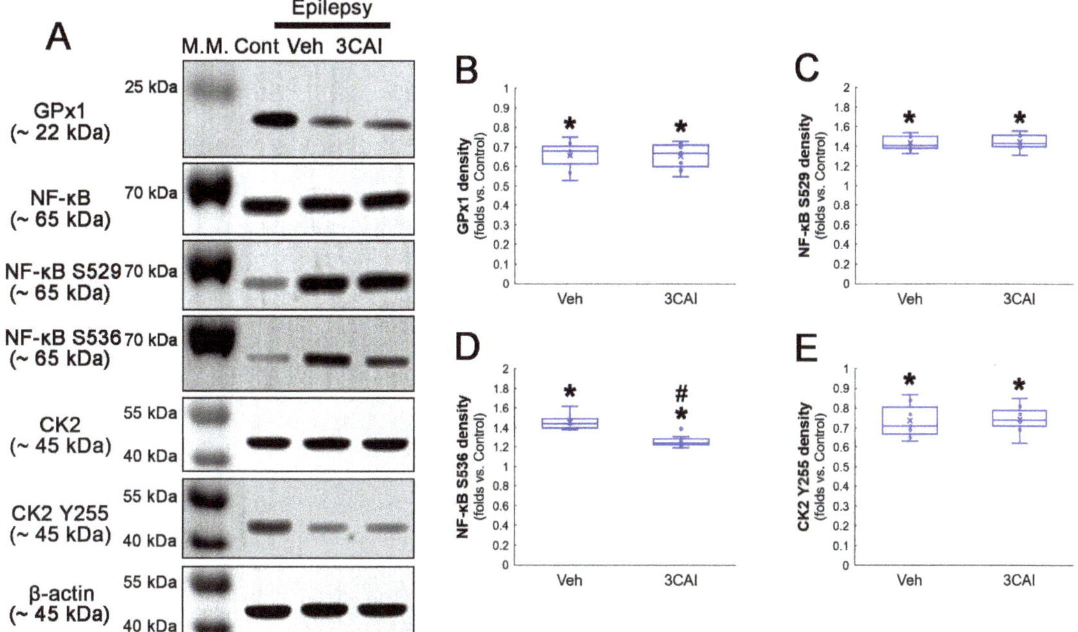

Figure 11. Western blot data representing the effects of 3CAI on GPx1 expression, NF-κB S529, NF-κB S536 and CK2 Y255 phosphorylations. Consistent with the immunofluorescent study (Figures 9 and 10), 3CAI reduces only the NF-κB S536 phosphorylation level without affecting GPx1 expression, NF-κB S529 and CK2 Y255 phosphorylations, as compared to the vehicle (Veh). (**A**) Representative Western blot of GPx1, NF-κB, NF-κB S529, NF-κB S536, CK2 and CK2 Y255 levels. (**B–E**) Quantification of GPx1 expression, AKT S473, NF-κB S529 and NF-κB S536 phosphorylation levels based on Western blot data ($*,^{\#}$ $p < 0.05$ vs. control rats and vehicle-treated epilepsy rat, respectively, $n = 7$ rats, respectively; Kruskal–Wallis test with Dunn–Bonferroni post hoc test).

4. Discussion

Astroglial activation generates H_2O_2 that evokes an imbalance of redox homeostasis in the brain [42]. Therefore, the defense system removing H_2O_2 is essential for astroglial viability. GPx1 plays an important role in GSH-mediated H_2O_2 elimination [22,23]. Indeed, GPx expression is increased in glial cells around surviving neurons [43] and GPx1 inhibits the ROS-mediated AKT activation [32,33]. In the present study, GPx1 was upregulated in reactive CA1 astrocytes, suggesting that increased GPx1 expression in reactive astrocytes may be an adaptive response against oxidative stress. However, GPx1 expression was significantly diminished in clasmatodendritic CA1 astrocytes concomitant with increased NF-κB S529 phosphorylation, which was recovered by NAC. NF-κB signaling pathway activates autophagy after heat shock [35]. Indeed, NF-κB S529, but not S276 and S311, phosphorylation is involved in clasmatodendritic astrocytes [2]. Since NAC acts as a direct ROS scavenger *per se* as well as a GSH precursor leading to increased GPx1/2 expression [44–46], our findings suggest that the antioxidant properties of NAC may improve GPx1 downregulation in clasmatodendritic astrocytes by inhibiting NF-κB S529 phosphorylation.

S529 phosphorylation increases NF-κB-mediated nuclear transcriptional activity, which is regulated by CK2 [36]. CK2 is a highly conserved and constitutively active serine/threonine kinase that promotes cell viability, proliferation and differentiation [47,48]. CK2 activity is enhanced by phosphorylation of Y255 and T360/S362 sites, which are modulated by the Src family and extracellular signal-regulated kinase 1/2 (ERK1/2), respectively [49,50]. In the epileptic hippocampus, CK2 Y255, but not T360/S362, phosphorylation is decreased as an insufficient and maladaptive response to inactivation/downregulation of phosphatase and tensin homolog deleted on chromosome 10 (PTEN). Furthermore, inhibition of Src-mediated CK2 Y255 phosphorylation further ameliorates PTEN downregulation/phosphorylation and clasmatodendrosis [14,38]. Furthermore, Src inhibition enhances GPx1 levels [51] and Src kinase upregulation inhibits GPx1 activity [52]. Most of all, NF-κB activation increases proinflammatory cytokines, including TNF-α, which abrogates the compensatory GPx1 induction following oxidative stress [27,53]. Indeed, NAC suppresses ROS-mediated NF-κB and subsequent mRNA expression of chemokines in human astrocytes [54] and fully blocks ROS-induced CK2 upregulation that induces NF-κB activation [55,56]. Compatible with these reports, the present data show that CK2 inhibition by NAC and TMBC effectively enhanced GPx1 expression in vacuolized CA1 astrocytes concomitant with reduced NF-κB S529 phosphorylation. Therefore, our findings indicate that CK2-mediated NF-κB S529 phosphorylation may be an upstream pathway of GPx1 downregulation.

The present data show AKT S473 hyperphosphorylation in clasmatodendritic CA1 astrocytes exhibiting low GPx1 intensity. Oxidative stress triggers AKT activation [57], which inhibits ROS-induced GPx1 upregulation [58]. Therefore, the present data are simply interpreted as that AKT may be one of the upstream molecules to suppress GPx1 expression in clasmatodendritic astrocytes. In the present study, however, AKT inhibition by 3CAI did not improve GPx1 downregulation in clasmatodendritic astrocytes, although it attenuated clasmatodendrosis. Therefore, our findings indicate that AKT S473 hyperphosphorylation may not be relevant to reduced GPx1 expression during clasmatodendritic degeneration.

On the other hand, CK2 also activates AKT by phosphorylation at S129 site [59–61]. In addition, the present study reveals that both CK2 inhibition by TMCB and AKT inhibition by 3CAI attenuated clasmatodendritic degeneration. Considering these, it is plausible that CK2-mediated AKT S129 phosphorylation would also elicit clasmatodendrosis by NF-κB S536 phosphorylation. However, CK2-mediated AKT S129 phosphorylation is necessary for the cell viability in HEK-293T cells [59]. Indeed, CX-4945 (a CK2 inhibitor) exerts strong anti-proliferative activity by blocking AKT S129 phosphorylation in cancer cells [60,61]. Therefore, it is likely that CK2-mediated AKT S129 phosphorylation may not be involved in clasmatodendritic degeneration or astroglial viability in the epileptic hippocampus.

The decreased GPx1 expression also elicits the activation of the redox-sensitive NF-κB canonical pathway and increases autophagic flux [62]. Indeed, GPx1 deletion increases NF-κB S536 phosphorylation [28], which is critical for autophagy in response to oxidative stress [40,41]. Consistent with a previous study demonstrating NAC-induced AKT and NF-κB inhibition [31], the present study demonstrates that NF-κB S536 phosphorylation was also enhanced in clasmatodendritic CA1 astrocytes showing AKT S473 hyperphosphorylation, which were attenuated by NAC and TMCB. However, AKT inhibition by 3CAI did not affect the reduced GPx1 level and the enhanced NF-κB S529 phosphorylation in clasmatodendritic astrocytes, although it attenuated clasmatodendrosis and NF-κB S536 phosphorylation. Considering AKT-mediated NF-κB S536 phosphorylation [39,63], our findings indicate that AKT-mediated NF-κB S536 phosphorylation may be also involved in clasmatodendritic degeneration, accompanied by the AKT/GSK-3β/Bif-1 signaling pathway. Since GPx1 inhibits NF-κB S536 phosphorylation [29] and CK2 inhibition diminishes AKT S473 phosphorylation [64,65], the present data also suggest that the enhanced AKT-mediated S536 phosphorylation may be a consequence from CK2-NF-κB S529-mediated GPx1 downregulation. Therefore, it is likely that that antioxidative capacity of NAC may be attributed to GPx1 upregulation by inhibiting CK2-NF-κB S529-mediated signaling pathway in clasmatodendritic astrocytes, independent of the AKT-NF-κB S536-mediated signaling pathway.

Astrocytes contribute to the slow afterhyperpolarizing potential (sAHP), which is a major intrinsic mechanism of neuronal inhibition and its termination [66]. 4,5,6,7-Tetrabromotriazole (TBB, a CK2 inhibitor) augments sAHP [67]. Since the inhibition of clasmatodendrosis shortens seizure duration in chronic epilepsy rats [14], the present data provide evidence that clasmatodendrosis may be an epiphenomenon maintaining prolonged seizure duration in the epileptic hippocampus.

5. Conclusions

The present study demonstrates for the first time that CK2-mediated NF-κB S529 phosphorylation evoked GPx1 downregulation in clasmatodendritic astrocytes, which subsequently led to AKT-mediated NF-κB S536 phosphorylation facilitating this autophagic astroglial degeneration (Figure 12). Therefore, our findings suggest that GPx1 may integrate between CK2- and AKT-mediated signaling pathways during clasmatodendrosis induced by oxidative stress.

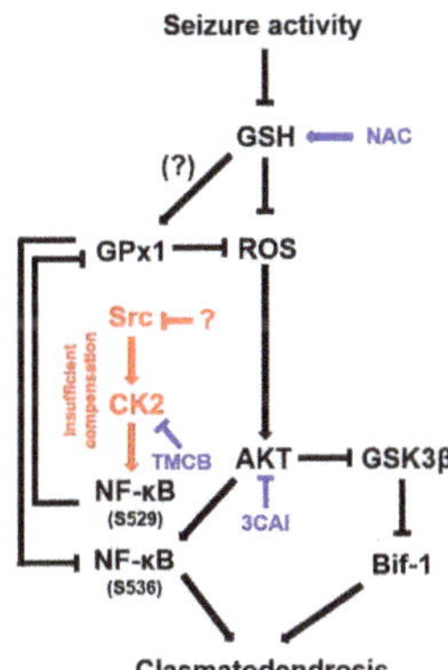

Figure 12. Schematic depiction representing the distinct role of NF-κB phosphorylation in clasmatodendritic CA1 astrocytes based on the present data and previous reports. Seizure activity decreases the GSH level and subsequently increases the ROS level. Aberrant CK2-mediated NF-κB S529 phosphorylation participates in GPx1 downregulation, which abolishes the GPx1-mediated inhibition of NF-κB S536 phosphorylation induced by AKT hyperactivation. In turn, the enhanced NF-κB S536 phosphorylation is involved in clasmatodendritic degeneration concomitant with AKT-mediated Bif-1 activation.

Supplementary Materials: The following supporting information can be downloaded at: https://www.mdpi.com/article/10.3390/antiox12051020/s1. Figure S1: Full-length images of Western blots in Figure 2A. Figure S2: Full-length images of Western blots in Figure 5A. Figure S3: Full-length images of Western blots in Figure 8A. Figure S4: Full-length images of Western blots in Figure 11A.

Author Contributions: T.-C.K. designed the experiments. J.-E.K., D.-S.L., T.-H.K., H.P. and T.-C.K. performed the experiments described in the manuscript. J.-E.K. and T.-C.K. analyzed the data and wrote the manuscript. All authors have read and agreed to the published version of the manuscript.

Funding: This study was supported by a grant of the National Research Foundation of Korea (NRF), grant (No. 2021R1A2B5B01001482).

Institutional Review Board Statement: The animal study protocol was approved by the Institutional Animal Care and Use Committee of Hallym University (No. Hallym 2021-3, approval date: 17 May 2021).

Informed Consent Statement: Not applicable.

Data Availability Statement: Data sharing is not applicable to this article.

Conflicts of Interest: The authors declare no conflict of interest.

References

1. Thorburn, A. Apoptosis and autophagy: Regulatory connections between two supposedly different processes. *Apoptosis* **2008**, *13*, 1–9. [CrossRef]
2. Ryu, H.J.; Kim, J.E.; Yeo, S.I.; Kang, T.-C. p65/RelA-Ser529 NF-κB subunit phosphorylation induces autophagic astroglial death (Clasmatodendrosis) following status epilepticus. *Cell. Mol. Neurobiol.* **2011**, *31*, 1071–1078. [CrossRef]
3. Ryu, H.J.; Kim, J.E.; Yeo, S.I.; Kim, D.W.; Kwon, O.S.; Choi, S.Y.; Kang, T.C. F-actin depolymerization accelerates clasmatodendrosis via activation of lysosome-derived autophagic astroglial death. *Brain Res. Bull.* **2011**, *85*, 368–373. [CrossRef]
4. Penfield, W. Neuroglia and microglia—The interstitial tissue of the central nervous system. In *Special Cytology, the Form and Function of the Cell in Health and Disease*; Cowdry, E.V., Ed.; Hoeber: New York, NY, USA, 1928; pp. 1033–1068.
5. Duchen, L.W. General pathology of neurons and neuroglia. In *Greenfield's Neuropathology*; Adams, J.H., Duchen, L.W., Eds.; Oxford University Press: New York, NY, USA, 1992; pp. 1–68.
6. Tomimoto, H.; Akiguchi, I.; Wakita, H.; Suenaga, T.; Nakamura, S.; Kimura, J. Regressive changes of astroglia in white matter lesions in cerebrovascular disease and Alzheimer's disease patients. *Acta Neuropathol.* **1997**, *94*, 146–152. [CrossRef]
7. Hulse, R.E.; Winterfield, J.; Kunkler, P.E.; Kraig, R.P. Astrocytic clasmatodendrosis in hippocampal organ culture. *Glia* **2001**, *33*, 169–179. [CrossRef]
8. Sahlas, D.J.; Bilbao, J.M.; Swartz, R.H.; Black, S.E. Clasmatodendrosis correlating with periventricular hyperintensity in mixed dementia. *Ann. Neurol.* **2002**, *52*, 378–381. [CrossRef]
9. Mercatelli, R.; Lana, D.; Bucciantini, M.; Giovannini, M.G.; Cerbai, F.; Quercioli, F.; Zecchi-Orlandini, S.; Delfino, G.; Wenk, G.L.; Nosi, D. Clasmatodendrosis and β-amyloidosis in aging hippocampus. *FASEB J.* **2016**, *30*, 1480–1491. [CrossRef]
10. Sakai, K.; Fukuda, T.; Iwadate, K. Beading of the astrocytic processes (clasmatodendrosis) following head trauma is associated with protein degradation pathways. *Brain Inj.* **2013**, *27*, 1692–1697. [CrossRef]
11. Tachibana, M.; Mohri, I.; Hirata, I.; Kuwada, A.; Kimura-Ohba, S.; Kagitani-Shimono, K.; Fushimi, H.; Inoue, T.; Shiomi, M.; Kakuta, Y.; et al. Clasmatodendrosis is associated with dendritic spines and does not represent autophagic astrocyte death in influenza-associated encephalopathy. *Brain Dev.* **2019**, *41*, 85–95. [CrossRef]
12. Bouchat, J.; Gilloteaux, J.; Suain, V.; Van Vlaender, D.; Brion, J.P.; Nicaise, C. Ultrastructural Analysis of Thalamus Damages in a Mouse Model of Osmotic-Induced Demyelination. *Neurotox. Res.* **2019**, *36*, 144–162. [CrossRef]
13. Kim, J.E.; Ko, A.R.; Hyun, H.W.; Min, S.J.; Kang, T.C. P2RX7-MAPK1/2-SP1 axis inhibits MTOR independent HSPB1-mediated astroglial autophagy. *Cell Death Dis.* **2018**, *9*, 546. [CrossRef] [PubMed]
14. Kim, J.E.; Kang, T.C. CDDO-Me Attenuates Astroglial Autophagy via Nrf2-, ERK1/2-SP1- and Src-CK2-PTEN-PI3K/AKT-Mediated Signaling Pathways in the Hippocampus of Chronic Epilepsy Rats. *Antioxidants* **2021**, *10*, 655. [CrossRef] [PubMed]
15. Kim, J.E.; Lee, D.S.; Kim, T.H.; Kang, T.C. Glutathione Regulates GPx1 Expression during CA1 Neuronal Death and Clasmatodendrosis in the Rat Hippocampus following Status Epilepticus. *Antioxidants* **2022**, *11*, 756. [CrossRef] [PubMed]
16. Kim, J.E.; Lee, D.S.; Kang, T.C. Sp1-Mediated Prdx6 Upregulation Leads to Clasmatodendrosis by Increasing Its aiPLA2 Activity in the CA1 Astrocytes in Chronic Epilepsy Rats. *Antioxidants* **2022**, *11*, 1883. [CrossRef]
17. Friede, R.L.; van Houten, W.H. Relations between postmortem alterations and glycolytic metabolism in the brain. *Exp. Neurol.* **1961**, *4*, 197–204. [CrossRef]
18. Wu, X.; Dong, H.; Ye, X.; Zhong, L.; Cao, T.; Xu, Q.; Wang, J.; Zhang, Y.; Xu, J.; Wang, W.; et al. HIV-1 Tat increases BAG3 via NF-κB signaling to induce autophagy during HIV-associated neurocognitive disorder. *Cell Cycle* **2018**, *17*, 1614–1623. [CrossRef]
19. Hwang, J.; Lee, H.J.; Lee, W.H.; Suk, K. NF-κB as a common signaling pathway in ganglioside-induced autophagic cell death and activation of astrocytes. *J. Neuroimmunol.* **2010**, *226*, 66–72.
20. Huang, Y.; Liao, Y.; Zhang, H.; Li, S. Lead exposure induces cell autophagy via blocking the Akt/mTOR signaling in rat astrocytes. *J. Toxicol. Sci.* **2020**, *45*, 559–567. [CrossRef]
21. Yang, J.; Takahashi, Y.; Cheng, E.; Liu, J.; Terranova, P.F.; Zhao, B.; Thrasher, J.B.; Wang, H.G.; Li, B. GSK-3beta promotes cell survival by modulating Bif-1-dependent autophagy and cell death. *J. Cell. Sci.* **2010**, *123*, 861–870. [CrossRef]
22. Marinho, H.S.; Antunes, F.; Pinto, R.E. Role of glutathione peroxidase and phospholipid hydroperoxide glutathione peroxidase in the reduction of lysophospholipid hydroperoxides. *Free Radic. Biol. Med.* **1997**, *22*, 871–883. [CrossRef]
23. de Haan, J.B.; Bladier, C.; Griffiths, P.; Kelner, M.; O'Shea, R.D.; Cheung, N.S.; Bronson, R.T.; Silvestro, M.J.; Wild, S.; Zheng, S.S.; et al. Mice with a homozygous null mutation for the most abundant glutathione peroxidase, Gpx1, show increased susceptibility to the oxidative stress-inducing agents paraquat and hydrogen peroxide. *J. Biol. Chem.* **1998**, *273*, 22528–22536. [CrossRef] [PubMed]
24. Meng, Q.; Xu, J.; Liang, C.; Liu, J.; Hua, J.; Zhang, Y.; Ni, Q.; Shi, S.; Yu, X. GPx1 is involved in the induction of protective autophagy in pancreatic cancer cells in response to glucose deprivation. *Cell Death Dis.* **2018**, *9*, 1187. [CrossRef] [PubMed]
25. Ni, B.; Shen, H.; Wang, W.; Lu, H.; Jiang, L. TGF-β1 reduces the oxidative stress-induced autophagy and apoptosis in rat annulus fibrosus cells through the ERK signaling pathway. *J. Orthop. Surg. Res.* **2019**, *14*, 241. [CrossRef] [PubMed]
26. Liddell, J.R.; Hoepken, H.H.; Crack, P.J.; Robinson, S.R.; Dringen, R. Glutathione peroxidase 1 and glutathione are required to protect mouse astrocytes from iron-mediated hydrogen peroxide toxicity. *J. Neurosci. Res.* **2006**, *84*, 578–586. [CrossRef]
27. Shin, E.J.; Hwang, Y.G.; Pham, D.T.; Lee, J.W.; Lee, Y.J.; Pyo, D.; Jeong, J.H.; Lei, X.G.; Kim, H.C. Glutathione peroxidase-1 overexpressing transgenic mice are protected from neurotoxicity induced by microcystin-leucine-arginine. *Environ. Toxicol.* **2018**, *33*, 1019–1028. [CrossRef]

28. Crack, P.J.; Taylor, J.M.; Ali, U.; Mansell, A.; Hertzog, P.J. Potential contribution of NF-kappaB in neuronal cell death in the glutathione peroxidase-1 knockout mouse in response to ischemia-reperfusion injury. *Stroke* **2006**, *37*, 1533–1538. [CrossRef]
29. Wang, X.; Han, Y.; Chen, F.; Wang, M.; Xiao, Y.; Wang, H.; Xu, L.; Liu, W. Glutathione Peroxidase 1 Protects Against Peroxynitrite-Induced Spiral Ganglion Neuron Damage Through Attenuating NF-κB Pathway Activation. *Front. Cell. Neurosci.* **2022**, *16*, 841731. [CrossRef]
30. Kim, J.E.; Park, H.; Kang, T.C. CDDO-Me Distinctly Regulates Regional Specific Astroglial Responses to Status Epilepticus via ERK1/2-Nrf2, PTEN-PI3K-AKT and NFκB Signaling Pathways. *Antioxidants* **2020**, *9*, 1026. [CrossRef]
31. Usui, T.; Yamawaki, H.; Kamibayashi, M.; Okada, M.; Hara, Y. Mechanisms underlying the anti-inflammatory effects of the Ca2+/calmodulin antagonist CV-159 in cultured vascular smooth muscle cells. *J. Pharmacol. Sci.* **2010**, *113*, 214–223. [CrossRef]
32. Handy, D.E.; Lubos, E.; Yang, Y.; Galbraith, J.D.; Kelly, N.; Zhang, Y.Y.; Leopold, J.A.; Loscalzo, J. Glutathione peroxidase-1 regulates mitochondrial function to modulate redox-dependent cellular responses. *J. Biol. Chem.* **2009**, *284*, 11913–11921. [CrossRef]
33. Meng, Q.; Shi, S.; Liang, C.; Liang, D.; Hua, J.; Zhang, B.; Xu, J.; Yu, X. Abrogation of glutathione peroxidase-1 drives EMT and chemoresistance in pancreatic cancer by activating ROS-mediated Akt/GSK3β/Snail signaling. *Oncogene* **2018**, *37*, 5843–5857. [CrossRef] [PubMed]
34. Kim, J.E.; Park, H.; Kang, T.C. Peroxiredoxin 6 Regulates Glutathione Peroxidase 1-Medited Glutamine Synthase Preservation in the Hippocampus of Chronic Epilepsy Rats. *Antioxidants* **2023**, *12*, 156. [CrossRef] [PubMed]
35. Nivon, M.; Richet, E.; Codogno, P.; Arrigo, A.P.; Kretz-Remy, C. Autophagy activation by NFkappaB is essential for cell survival after heat shock. *Autophagy* **2009**, *5*, 766–783. [CrossRef] [PubMed]
36. Wang, D.; Westerheide, S.D.; Hanson, J.L.; Baldwin, A.S., Jr. Tumor necrosis factor alpha-induced phosphorylation of RelA/p65 on Ser529 is controlled by casein kinase II. *J. Biol. Chem.* **2000**, *275*, 32592–32597. [CrossRef]
37. Zhu, J.M.; Li, K.X.; Cao, S.X.; Chen, X.J.; Shen, C.J.; Zhang, Y.; Geng, H.Y.; Chen, B.Q.; Lian, H.; Zhang, J.M.; et al. Increased NRG1-ErbB4 signaling in human symptomatic epilepsy. *Sci. Rep.* **2017**, *7*, 141. [CrossRef]
38. Kim, J.E.; Lee, D.S.; Park, H.; Kang, T.C. Src/CK2/PTEN-Mediated GluN2B and CREB Dephosphorylations Regulate the Responsiveness to AMPA Receptor Antagonists in Chronic Epilepsy Rats. *Int. J. Mol. Sci.* **2020**, *21*, 9633. [CrossRef]
39. Madrid, L.V.; Mayo, M.W.; Reuther, J.Y.; Baldwin, A.S., Jr. Akt stimulates the transactivation potential of the RelA/p65 Subunit of NF-kappa B through utilization of the Ikappa B kinase and activation of the mitogen-activated protein kinase p38. *J. Biol. Chem.* **2001**, *276*, 18934–18940. [CrossRef]
40. Song, C.; Mitter, S.K.; Qi, X.; Beli, E.; Rao, H.V.; Ding, J.; Ip, C.S.; Gu, H.; Akin, D.; Dunn, W.A., Jr.; et al. Oxidative stress-mediated NFκB phosphorylation upregulates p62/SQSTM1 and promotes retinal pigmented epithelial cell survival through increased autophagy. *PLoS ONE* **2017**, *12*, e0171940. [CrossRef]
41. Chen, Z.; Nie, S.D.; Qu, M.L.; Zhou, D.; Wu, L.Y.; Shi, X.J.; Ma, L.R.; Li, X.; Zhou, S.L.; Wang, S.; et al. The autophagic degradation of Cav-1 contributes to PA-induced apoptosis and inflammation of astrocytes. *Cell Death Dis.* **2018**, *9*, 771. [CrossRef]
42. Waldbaum, S.; Liang, L.P.; Patel, M. Persistent impairment of mitochondrial and tissue redox status during lithium-pilocarpine-induced epileptogenesis. *J. Neurochem.* **2010**, *115*, 1172–1182. [CrossRef]
43. Damier, P.; Hirsch, E.C.; Zhang, P.; Agid, Y.; Javoy-Agid, F. Glutathione peroxidase, glial cells and Parkinson's disease. *Neuroscience* **1993**, *52*, 1–6. [CrossRef] [PubMed]
44. Zafarullah, M.; Li, W.Q.; Sylvester, J.; Ahmad, M. Molecular mechanisms of N-acetylcysteine actions. *Cell. Mol. Life Sci.* **2003**, *60*, 6–20. [CrossRef] [PubMed]
45. Krifka, S.; Hiller, K.A.; Spagnuolo, G.; Jewett, A.; Schmalz, G.; Schweikl, H. The influence of glutathione on redox regulation by antioxidant proteins and apoptosis in macrophages exposed to 2-hydroxyethyl methacrylate (HEMA). *Biomaterials* **2012**, *33*, 5177–5186. [CrossRef] [PubMed]
46. Gallorini, M.; Petzel, C.; Bolay, C.; Hiller, K.A.; Cataldi, A.; Buchalla, W.; Krifka, S.; Schweikl, H. Activation of the Nrf2-regulated antioxidant cell response inhibits HEMA-induced oxidative stress and supports cell viability. *Biomaterials* **2015**, *56*, 114–128. [CrossRef] [PubMed]
47. Guerra, B.; Issinger, O.G. Protein kinase CK2 in human diseases. *Curr. Med. Chem.* **2008**, *15*, 1870–1886. [CrossRef]
48. St-Denis, N.A.; Litchfield, D.W. Protein kinase CK2 in health and disease: From birth to death: The role of protein kinase CK2 in the regulation of cell proliferation and survival. *Cell. Mol. Life Sci.* **2009**, *66*, 1817–1829. [CrossRef]
49. Donella-Deana, A.; Cesaro, L.; Sarno, S.; Ruzzene, M.; Brunati, A.M.; Marin, O.; Vilk, G.; Doherty-Kirby, A.; Lajoie, G.; Litchfield, D.W.; et al. Tyrosine phosphorylation of protein kinase CK2 by Src-related tyrosine kinases correlates with increased catalytic activity. *Biochem. J.* **2003**, *372*, 841–849. [CrossRef]
50. Ji, H.; Wang, J.; Nika, H.; Hawke, D.; Keezer, S.; Ge, Q.; Fang, B.; Fang, X.; Fang, D.; Litchfield, D.W.; et al. EGF-induced ERK activation promotes CK2-mediated disassociation of alpha-Catenin from beta-Catenin and transactivation of beta-Catenin. *Mol. Cell.* **2009**, *36*, 547–559. [CrossRef]
51. Dabo, A.J.; Ezegbunam, W.; Wyman, A.E.; Moon, J.; Railwah, C.; Lora, A.; Majka, S.M.; Geraghty, P.; Foronjy, R.F. Targeting c-Src Reverses Accelerated GPX-1 mRNA Decay in Chronic Obstructive Pulmonary Disease Airway Epithelial Cells. *Am. J. Respir. Cell. Mol. Biol.* **2020**, *62*, 598–607. [CrossRef]

52. Tran, T.V.; Shin, E.J.; Nguyen, L.T.T.; Lee, Y.; Kim, D.J.; Jeong, J.H.; Jang, C.G.; Nah, S.Y.; Toriumi, K.; Nabeshima, T.; et al. Protein Kinase Cδ Gene Depletion Protects Against Methamphetamine-Induced Impairments in Recognition Memory and ERK$_{1/2}$ Signaling via Upregulation of Glutathione Peroxidase-1 Gene. *Mol. Neurobiol.* **2018**, *55*, 4136–4159. [CrossRef]

53. Liu, W.C.; Wu, C.W.; Fu, M.H.; Tain, Y.L.; Liang, C.K.; Hung, C.Y.; Chen, I.C.; Lee, Y.C.; Wu, C.Y.; Wu, K.L.H. Maternal high fructose-induced hippocampal neuroinflammation in the adult female offspring via PPARγ-NF-κB signaling. *J. Nutr. Biochem.* **2020**, *81*, 108378. [CrossRef]

54. Park, J.; Choi, K.; Jeong, E.; Kwon, D.; Benveniste, E.N.; Choi, C. Reactive oxygen species mediate chloroquine-induced expression of chemokines by human astroglial cells. *Glia* **2004**, *47*, 9–20. [CrossRef] [PubMed]

55. Kim, K.J.; Cho, K.D.; Jang, K.Y.; Kim, H.A.; Kim, H.K.; Lee, H.K.; Im, S.Y. Platelet-activating factor enhances tumour metastasis via the reactive oxygen species-dependent protein kinase casein kinase 2-mediated nuclear factor-κB activation. *Immunology* **2014**, *143*, 21–32. [CrossRef] [PubMed]

56. Wang, R.; Wang, Y.; Qu, L.; Chen, B.; Jiang, H.; Song, N.; Xie, J. Iron-induced oxidative stress contributes to α-synuclein phosphorylation and up-regulation via polo-like kinase 2 and casein kinase 2. *Neurochem. Int.* **2019**, *125*, 127–135. [CrossRef]

57. Chen, Y.W.; Huang, C.F.; Tsai, K.S.; Yang, R.S.; Yen, C.C.; Yang, C.Y.; Lin-Shiau, S.Y.; Liu, S.H. The role of phosphoinositide 3-kinase/Akt signaling in low-dose mercury-induced mouse pancreatic beta-cell dysfunction in vitro and in vivo. *Diabetes* **2006**, *55*, 1614–1624. [CrossRef]

58. Duarte, A.I.; Santos, P.; Oliveira, C.R.; Santos, M.S.; Rego, A.C. Insulin neuroprotection against oxidative stress is mediated by Akt and GSK-3beta signaling pathways and changes in protein expression. *Biochim. Biophys. Acta.* **2008**, *1783*, 994–1002. [CrossRef] [PubMed]

59. Ponce, D.P.; Maturana, J.L.; Cabello, P.; Yefi, R.; Niechi, I.; Silva, E.; Armisen, R.; Galindo, M.; Antonelli, M.; Tapia, J.C. Phosphorylation of AKT/PKB by CK2 is necessary for the AKT-dependent up-regulation of β-catenin transcriptional activity. *J. Cell Physiol.* **2011**, *226*, 1953–1959. [CrossRef]

60. Chon, H.J.; Bae, K.J.; Lee, Y.; Kim, J. The casein kinase 2 inhibitor, CX-4945, as an anti-cancer drug in treatment of human hematological malignancies. *Front. Pharmacol.* **2015**, *6*, 70. [CrossRef]

61. Siddiqui-Jain, A.; Drygin, D.; Streiner, N.; Chua, P.; Pierre, F.; O'Brien, S.E.; Bliesath, J.; Omori, M.; Huser, N.; Ho, C.; et al. CX-4945, an orally bioavailable selective inhibitor of protein kinase CK2, inhibits prosurvival and angiogenic signaling and exhibits antitumor efficacy. *Cancer Res.* **2010**, *70*, 10288–10298. [CrossRef]

62. Lam, C.S.; Tipoe, G.L.; So, K.F.; Fung, M.L. Neuroprotective mechanism of Lycium barbarum polysaccharides against hippocampal-dependent spatial memory deficits in a rat model of obstructive sleep apnea. *PLoS ONE* **2015**, *10*, e0117990. [CrossRef]

63. Liu, J.; Yoshida, Y.; Yamashita, U. DNA-binding activity of NF-kappaB and phosphorylation of p65 are induced by N-acetylcysteine through phosphatidylinositol (PI) 3-kinase. *Mol. Immunol.* **2008**, *45*, 3984–3989. [CrossRef] [PubMed]

64. Torres, J.; Pulido, R. The tumor suppressor PTEN is phosphorylated by the protein kinase CK2 at its C terminus. Implications for PTEN stability to proteasome-mediated degradation. *J. Biol. Chem.* **2001**, *276*, 993–998. [CrossRef] [PubMed]

65. Olsen, B.B.; Svenstrup, T.H.; Guerra, B. Downregulation of protein kinase CK2 induces autophagic cell death through modulation of the mTOR and MAPK signaling pathways in human glioblastoma cells. *Int. J. Oncol.* **2012**, *41*, 1967–1976. [CrossRef] [PubMed]

66. Hertz, L.; Xu, J.; Song, D.; Yan, E.; Gu, L.; Peng, L. Astrocytic and neuronal accumulation of elevated extracellular K(+) with a 2/3 K(+)/Na(+) flux ratio-consequences for energy metabolism, osmolarity and higher brain function. *Front. Comput. Neurosci.* **2013**, *7*, 114. [CrossRef] [PubMed]

67. Brehme, H.; Kirschstein, T.; Schulz, R.; Köhling, R. In vivo treatment with the casein kinase 2 inhibitor 4,5,6,7- tetrabromotriazole augments the slow afterhyperpolarizing potential and prevents acute epileptiform activity. *Epilepsia* **2014**, *55*, 175–183. [CrossRef] [PubMed]

Article

Effect of N-Acetylcysteine on Sleep: Impacts of Sex and Time of Day

Priyanka N. Bushana [1], Michelle A. Schmidt [1], Kevin M. Chang [1], Trisha Vuong [1], Barbara A. Sorg [2] and Jonathan P. Wisor [1],*

[1] Elson S. Floyd College of Medicine, Washington State University, Spokane, WA 99202, USA; priyanka.bushana@wsu.edu (P.N.B.); maschmidt@wsu.edu (M.A.S.); kevin.m.chang@wsu.edu (K.M.C.); trisha.vuong@wsu.edu (T.V.)

[2] R.S. Dow Neurobiology Laboratories, Legacy Research Institute, Portland, OR 97232, USA; bsorg@downeurobiology.org

* Correspondence: jonathan.wisor@wsu.edu

Abstract: Non-rapid eye movement sleep (NREMS) is accompanied by a decrease in cerebral metabolism, which reduces the consumption of glucose as a fuel source and decreases the overall accumulation of oxidative stress in neural and peripheral tissues. Enabling this metabolic shift towards a reductive redox environment may be a central function of sleep. Therefore, biochemical manipulations that potentiate cellular antioxidant pathways may facilitate this function of sleep. N-acetylcysteine increases cellular antioxidant capacity by serving as a precursor to glutathione. In mice, we observed that intraperitoneal administration of N-acetylcysteine at a time of day when sleep drive is naturally high accelerated the onset of sleep and reduced NREMS delta power. Additionally, N-acetylcysteine administration suppressed slow and beta electroencephalographic (EEG) activities during quiet wake, further demonstrating the fatigue-inducing properties of antioxidants and the impact of redox balance on cortical circuit properties related to sleep drive. These results implicate redox reactions in the homeostatic dynamics of cortical network events across sleep/wake cycles, illustrating the value of timing antioxidant administration relative to sleep/wake cycles. A systematic review of the relevant literature, summarized herein, indicates that this "chronotherapeutic hypothesis" is unaddressed within the clinical literature on antioxidant therapy for brain disorders such as schizophrenia. We, therefore, advocate for studies that systematically address the relationship between the time of day at which an antioxidant therapy is administered relative to sleep/wake cycles and the therapeutic benefit of that antioxidant treatment in brain disorders.

Keywords: electroencephalography; sleep; antioxidants; schizophrenia; N-acetylcysteine

Citation: Bushana, P.N.; Schmidt, M.A.; Chang, K.M.; Vuong, T.; Sorg, B.A.; Wisor, J.P. Effect of N-Acetylcysteine on Sleep: Impacts of Sex and Time of Day. *Antioxidants* **2023**, *12*, 1124. https://doi.org/ 10.3390/antiox12051124

Academic Editor: Waldo Cerpa

Received: 26 April 2023
Revised: 5 May 2023
Accepted: 11 May 2023
Published: 19 May 2023

1. Introduction

Sleep propensity and its electroencephalographic manifestations increase as a function of time spent awake and decline as a function of time spent asleep. The accumulation of sleep needs during wakefulness and its discharge in sleep is thus said to be a homeostatic process [1]. The biochemical nature of this homeostatic process is not known. The homeostat may relate to the fundamental shift in cerebral metabolism away from the consumption of glucose to other energy sources. Glucose utilization requires a series of oxidation/reduction (redox) reactions, and it is thus conceivable that the homeostat is embodied, at least in part, by cellular redox status in the brain. The brain accumulates oxidative stress as a consequence of wakefulness, and this oxidative stress is reversed by sleep [2–5]. If brain redox status does indeed contribute to sleep homeostasis, then perturbation of brain redox status should alter the sleep homeostatic process. The goal of the current study was to measure the effects of systemic manipulations of redox substrates on EEG readout of sleep homeostasis.

As an electron carrier, nicotinamide adenine dinucleotide in its unphosphorylated (NAD) or phosphorylated (NAD phosphate; NADP) forms, is necessary for glycolytic metabolism and numerous other cellular biochemical pathways that rely on the transfer of electrons via redox reactions. NAD-dependent reactions occur at a high rate during wakefulness, as evidenced by the accumulation of NADH in brain tissues during sleep deprivation [6]. Parallel increases in markers of oxidative stress during sleep deprivation demonstrate that the capacity to undergo redox reactions is biochemically constrained during time spent awake by the availability of NAD+ as an oxidizing substrate [7,8].

Several strategies are available to increase the availability of NAD and related molecules as oxidizing substrates. The current study employed systemic administration of N-acetylcysteine (NAC), which directly increases the capacity for glutathione synthesis (Figure 1) to achieve this goal. NAC crosses the blood–brain barrier and, therefore, can undergo conversion to glutathione within cells of the brain parenchyma. We hypothesized that intraperitoneal (ip) injection of NAC via increased antioxidant capacity (i.e., glutathione) would decrease oxidative stress and subsequently decrease sleep needs. If, indeed, oxidative stress contributes to sleep homeostasis, this manipulation would be expected to manifest as an alteration in the homeostatic regulation of sleep. We tested this hypothesis by measuring sleep state timing and EEG parameters in animals subjected to NAC administration.

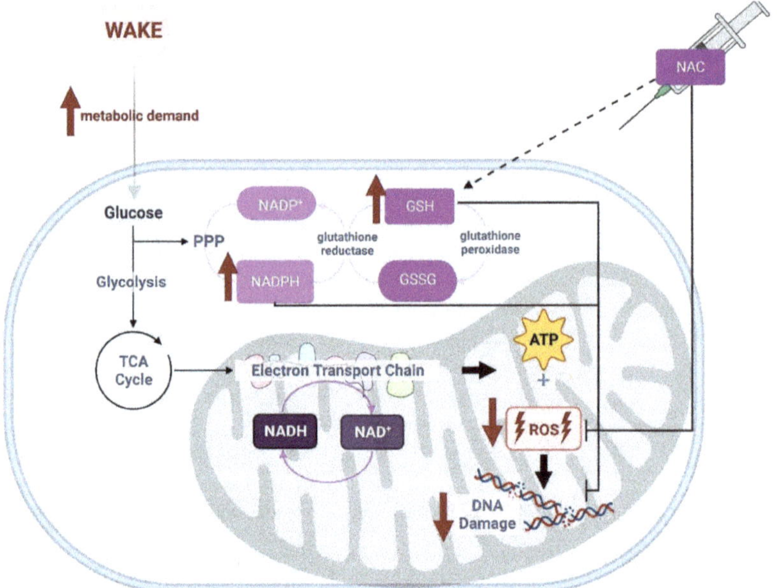

Figure 1. NAD(P)+/H are cofactors for hundreds of redox reactions involved in cellular metabolism. Exogenous application of NAC increases the pool of glutathione available in the cell for NADP(H)-dependent redox reactions. In turn, increased glutathione should produce additional opportunities for production of NADPH via the hexose monophosphate shunt [9]. Wakefulness challenges this system by promoting the accumulation of oxidative stress (i.e., reactive oxygen species [ROS] and DNA damage) during cellular metabolism. We propose that the increased availability of NADPH and glutathione as a result of NAC supplementation directly neutralizes the oxidative burden of wakefulness. Image created with BioRender.com.

2. Materials and Methods

2.1. Ethical Approval

This study was approved by the Institutional Animal Care and Use Committee of Washington State University and conducted in accordance with National Institutes of Health's Guidelines for the Care and Use of Laboratory Animals. All efforts were made to minimize the number of animals used in the experiments and to reduce the amount of pain and suffering.

2.2. Animals and Surgery

Twelve adult C57BL/6J mice ($n = 12$, 6 females, 6 males; all aged 10–12 weeks) were anesthetized using isoflurane (5% induction; 1–3% to maintain 0.5–1 Hz respiration rate) and placed in a stereotaxic frame. A 1-cm midline incision was made in the skin over the dorsal surface of the skull, and the skull was exposed to allow two holes, roughly 0.5 mm in diameter, to be drilled over predetermined coordinates targeting the medial prefrontal cortex (mPFC; A/P + 1.94; M/L ± 0.5; D/V − 1.3). At this location, stainless-steel polyimide-insulated depth electrodes (Plastics One part #E363/1/SPC diameter: 0.25 mm) were implanted bilaterally for local field potential (LFP) measurements. Prior to surgery, depth electrodes were cut to 1.5 mm. Mice were additionally implanted with EEG and electromyographic (EMG) electrodes as described previously [10] and diagrammed in Figure 2B. Briefly, two stainless-steel EEG screw electrodes were implanted over the parietal cortices, and two EMG electrodes were implanted in the nuchal muscles. All electrodes were soldered to a six-pin head mount connector and secured to the skull with dental cement. Buprenorphine SR (1.0 mg/kg) was administered once as a post-operative analgesic. After surgery, all mice were singly housed in a vivarium which remained between 70° and 75°F at a relative humidity of 50%, on an LD 12:12 cycle for 12–13 days during recovery from surgical procedures.

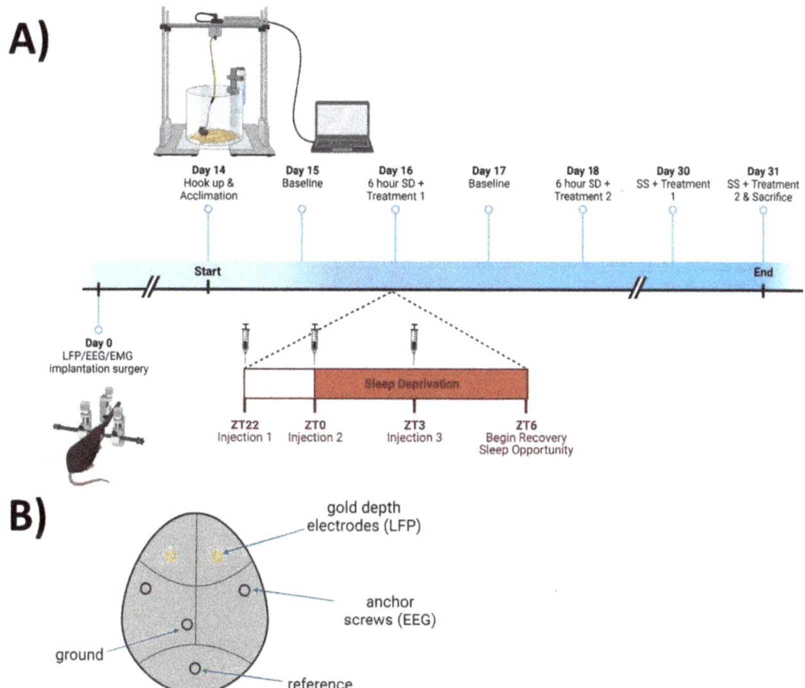

Figure 2. Experimental protocol for N-acetylcysteine (NAC) manipulation, as described in detail in the Section 2.3. (**A**) Schematic representation of experimental setting and timeline for experimentation. (**B**) Schematic representation of electrode placement. Image created with BioRender.com.

2.3. Experimental Design

Animals were subjected to saline (vehicle) and NAC treatments in a within-subjects crossover design, as described in Figure 2A. Two weeks after surgery, mice were connected to recording cables via a head mount and individually re-housed in cylindrical acrylic plastic cages 25 cm in diameter and 20 cm tall. Mice were allowed to habituate to the environment and head mount tethers overnight. Mice then underwent 24 h of undisturbed baseline LFP/EEG and EMG recording, starting 2 h before light onset (ZT22). After the baseline, three treatments of either NAC (600 mg/kg) or saline (12 mL/kg) intraperitoneal (i.p.) injections were applied (randomly assigned as Treatment 1 or Treatment 2) at ZT22, ZT0, and ZT3 (Figure 2A). Mice were concurrently sleep deprived (SD) for 6 continuous hours, from ZT0 to ZT6. Recovery sleep was recorded for the following 16 h before initiating a second session with the other treatment (at ZT 22, 0, and 3) and SD from ZT0-6. Two weeks later, mice were subjected to both treatments on the same time schedule as during the SD experiments but were allowed to sleep undisturbed between the injections (SS; spontaneous sleep). Each animal received both treatments concurrent with the two SD protocols in a repeated-measures, counterbalanced design (day 16 and 18; n = 12) and concurrent with the two SS protocols in a repeated-measures, counterbalanced design (day 30 and 31; n = 10) then was euthanized after the second spontaneous sleep recording (schematized in Figure 2A).

NAC is known to reach peak plasma levels quickly [11], with a half-life reported to range between 11 min and 6 h, depending on the animal and route of administration [12,13]. When administered intraperitoneally, NAC is directly taken up by the hepatic portal system and undergoes extensive first-pass metabolism through the liver before reaching main circulation and entering the brain [14]. The high dose used here (600 mg/kg) and the multiple injections delivered before and throughout sleep deprivation ensured that NAC was present systemically throughout the SD period. NAC was prepared in phosphate-buffered saline the day prior, balanced for pH (using NaOH), stored under nitrogen, and opened immediately prior to each injection to minimize oxidation.

2.4. Data Collection and Processing

Data from LFP, EEG, and EMG potentials were collected, extracted, and processed as described previously [10]. Briefly, LFP, EEG, and EMG signals from the head mount were fed through a PCB-based preamplifier (Part #8406-SL, Pinnacle Technology, Inc., Lawrence, KS, USA) to a commutator (Part #8408, Pinnacle Technology, Inc.), which was read into a PC-based acquisition system (Pinnacle Technology, Inc.; Part #8401). Signals were further amplified 50-fold and sampled at 400 Hz. LFP and EMG potentials were extracted using Sirenia software, version 2.2 from Pinnacle Technology, Inc. Sleep recording files were extracted in European Data format (.edf) and contained data from two frontal LFPs and one nuchal EMG.

Sleep recording files were scored through the online computational tool SPINDLE (sleep phase identification with neural networks for domain-invariant learning) [15]. SPINDLE allows .edf files to be uploaded to a web-based platform and processes signals automatically with a 4-s epoch resolution. The algorithm classifies each epoch as one of three vigilance states: wakefulness, non-rapid eye movement sleep (NREM sleep; NREMS; or slow wave sleep; SWS, which is assumed to be synonymous with NREMS in this manuscript), or rapid-eye-movement sleep (REM sleep; REMS) and additionally designates each epoch as likely or unlikely to contain artifacts. Briefly, the SPINDLE algorithm processes raw signals by windowed Fourier transforms, amongst other pre-processing methods, then feeds these data through a convolutional neural network to detect sleep states. In general, wakefulness was defined by high EMG activity for more than 50% of epoch duration. NREMS was defined by reduced EMG activity and increased LFP power below 4 Hz. REMS was defined by intermediate muscle tone, low LFP power over 4 Hz, and high LFP power between 6–9 Hz. A hidden Markov model is integrated into the SPINDLE scoring process to help to define the dynamics of vigilance states and suppress physiologically implausible sleep transitions. Further, an additional convolutional neural network was applied to mark unclear stages and technical artifacts so that they could be excluded from analysis. SPINDLE vigilance state classifications are exported as .csv files.

The SPINDLE scoring algorithm was validated against manually scored datasets: three independent laboratories achieved average agreement rates between manual and SPINDLE state classification of 93–99% [15]. We additionally validated the algorithm in our own laboratory, where total agreement between manual and SPINDLE state classification for three 24-h mouse polysomnographic recordings was 95%.

EEG spectral data (.edfs) and vigilance state classifications (from .csv files) were processed together with MATLAB as previously described [2]. EEG spectral data were separated into the following bands: delta (1–4 Hz), theta (5–8 Hz), alpha (9–12 Hz), beta (15–35 Hz), and gamma (35–120 Hz). The gamma range was further subdivided into low gamma (35–60 Hz), gamma (60–90 Hz), and high gamma (90–120 Hz). Wakefulness was subdivided into quiet wakefulness (QW) or active wakefulness (AW) using EMG peak-to-peak amplitude of all wake epochs across the entire recording. QW was defined as the 33rd percentile or less and AW as the 66th percentile or higher of all wake EMG peak-to-peak amplitude values [16].

2.5. Statistical Analysis

Statistical analysis was performed with STATISTICA software (version 12.0, StatSoft, Tulsa, Oklahoma). Differences between means of sleep timings or EEG spectra were estimated by repeated-measures analysis of variance (RM ANOVA), with significance levels set to $\alpha \leq 0.05$. Partial eta squared ($\eta^2{}_p$) is reported as a measure of effect size for each significant effect in the results section. Independent variables assessed include treatment (NAC or saline injectate), sex, and time of day (12 intervals for non-cumulative sleep timing measures; 6 intervals for EEG spectral power measures). Data analysis was subjected to sigma-restricted parameterization and effective hypothesis decomposition methods by the software. Significant results were further tested by Fisher's LSD post hoc test. While appropriate when the statistical interaction assessed is supported at $\alpha \leq 0.05$, it does not correct for multiple comparisons.

3. Results

3.1. NAC Increases Time Spent in NREM Sleep at the Cost of Wakefulness

We first assessed within-subjects differences in sleep architecture between EEG recordings made after NAC or saline injections during spontaneous sleep (ZT22–ZT10, days 30 & 31 in Figure 2A; data shown in Figure 3). Due to the short half-life of NAC [12,13], only the first 12 h after the first injection are displayed. RM ANOVAs indicated significant interactions of treatment x time during wakefulness ($F_{11,88} = 2.51$, $p = 0.009$, $\eta^2{}_p = 0.24$, Figure 3A) and NREMS ($F_{11,88} = 2.7$, $p = 0.005$, $\eta^2{}_p = 0.25$, Figure 3B). Post hoc analysis shows that NREM sleep is increased at the expense of wakefulness within the first 7 h of the injection protocol (Figure 3A,B). REMS did not show significant differences for treatment x time interaction (Figure 3C). State classifications assessed during and after sleep deprivation did not produce any significant differences between treatments (ZT0–ZT12, days 16 and 18; data not shown).

Increased time spent in NREM sleep after NAC may be explained by the reduced latency to NREM sleep onset after each of the three injections, which produced a treatment x time interaction ($F_{2,16} = 9.61$, $p = 0.002$, $\eta^2{}_p = 0.55$, Figure 4). NAC treatment reduced latency to sleep onset after the first injection by 66% relative to saline (24 min for NAC vs. 71 min for saline; $p < 0.001$ post hoc: Fisher's LSD). Latency to sleep decreased progressively after injections 2 and 3 regardless of treatment and was not affected by treatment at either of these time points. No order effect (i.e., which treatment was received on which day) was observed in latency to sleep onset.

We further probed differences in the duration of wake and NREM sleep by assessing state consolidation during the SS recordings. In the first 1-h interval beginning immediately after the first injection, wake bout duration and REMS bout duration were reduced by NAC treatment relative to saline (wake $F_{5,40} = 5.01$, $p = 0.001$, $\eta^2{}_p = 0.39$; REMS $F_{5,40} = 3.07$, $p = 0.019$, $\eta^2{}_p = 0.28$; data not shown). NREMS bout duration was not affected by treatment (not significant. [n.s.], data not shown). Neither the number of bouts of each state nor the number of brief awakenings was different across treatments (n.s., data not shown).

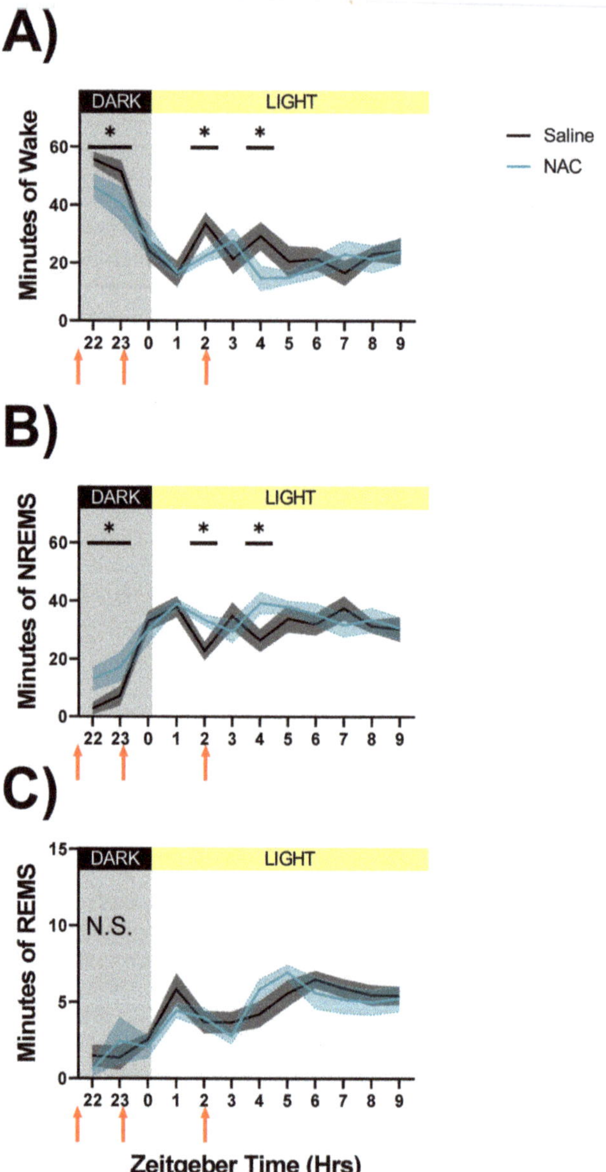

Figure 3. Sleep state classifications derived from SPINDLE-scoring during NAC (blue line) and saline (black line) injections on spontaneous sleep (days 30 and 31, ZT22–ZT10) recording days. Injection times are indicated by red arrows along the X-axis. Light and dark phases are denoted by the color of the background in each panel and the bar along the top of each panel. Data show the mean total number of minutes spent in each sleep state—wake (**A**), NREMS (**B**), and REMS (**C**)—during 1-h bins. Significant differences were found in wake and NREM sleep durations between NAC and saline injections during SS recordings (**A,B**), and indicated by asterisks where $p < 0.05$, Fisher's LSD. Grey and light blue shaded areas signify SEM. N.S: treatment effect on REMS not significant.

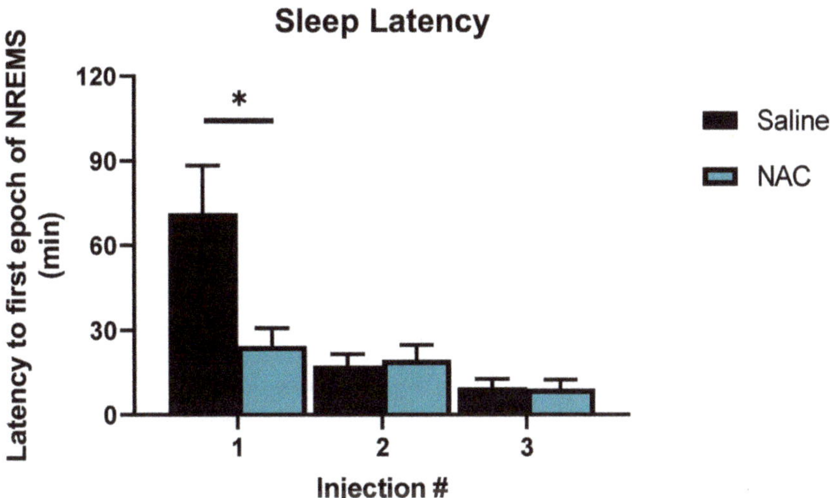

Figure 4. Sleep latency after NAC (blue bars) and saline (black bars) injections on SS days (days 30 and 31). Bars represent minutes elapsed after injections at ZT 22, ZT0, and ZT3, respectively, until the first epoch of NREM sleep was detected. Significant differences were found between the elapsed time after the first NAC and saline injections (at ZT 22), as indicated by the asterisks. Asterisks denote $p < 0.05$, as assessed by Fisher's LSD post hoc test. These differences were not modulated by sex. Error bars signify SEM.

3.2. NAC Accelerates Changes in the Dissipation of Sleep Pressure

To gain a better understanding of the increases in NREM sleep duration displayed in Figure 3B, we conducted a power analysis of low-frequency oscillations (i.e., SWA, theta, alpha, and low beta) during SS injections. When the spectral activity was separated into sleep states, significant power differences were observed in 1–20 Hz bands between NAC and saline recordings during NREM sleep (days 30 and 31 as displayed in Figure 2; RM ANOVA treatment x 2-h time interval x frequency, $F_{95,760} = 1.52$, $p = 0.002$, $\eta^2_p = 0.16$; statistics describe all data displayed in Figure 5).

To account for time-of-day effects in response to treatments, panels in Figure 5 are separated to display low-frequency data in six two-hour time intervals from ZT 22 to ZT 10. Injections occurred 2 h prior to the start of the light phase (at ZT22; Figure 5A), at the start of the light phase (ZT 0; Figure 5B), and 3 h into the light phase (ZT3; Figure 5C). The first panel (Figure 5A) displays LFP power in low-frequency bands during the last two hours of the dark phase. This is the post-siesta period, when wakefulness generally increases, and low-frequency activity is expected to build along with sleep pressure and drowsiness. NAC recordings demonstrate increased delta power when compared to saline recordings, while alpha and low beta power are decreased (Figure 5A). Overall, this indicates that NAC accelerates the discharge of sleep drive (delta oscillations) during the dark phase.

During the light phase (Figure 5B–F), power in low-frequency bands generally decreases as increased NREM sleep duration alleviates sleep pressure and drowsiness. The first 6 h of the light phase demonstrate this effect, regardless of treatment, in Figure 5B–D. However, NAC recordings demonstrate further decreases in power across the 1–20 Hz range when compared with saline recordings. This difference suggests that sleep pressure is discharged at an accelerated rate during the early light phase when NAC is administered. Later in the light phase, a transient elevation of delta power in NAC recordings relative to saline recordings can be noted (Figure 5E), demonstrating a rebound effect after the initial acceleration of SWA discharge.

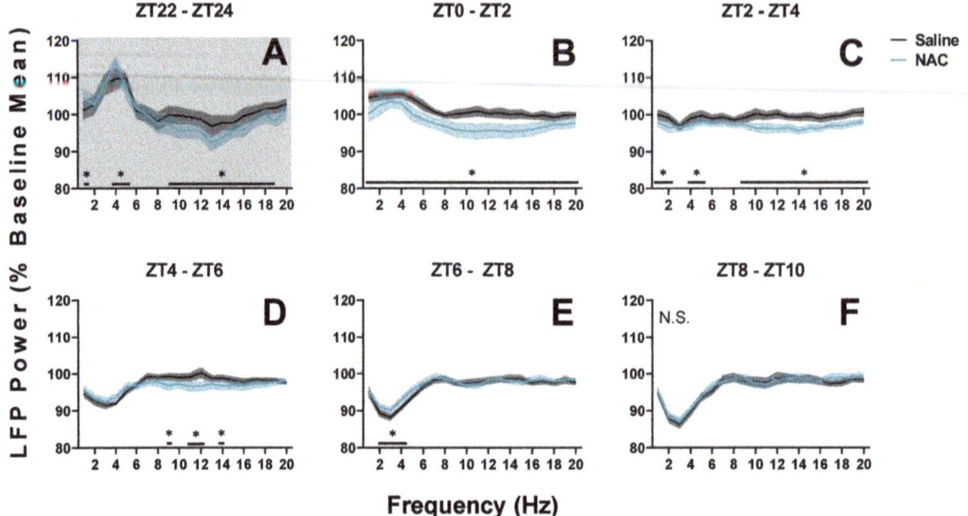

Figure 5. Changes in LFP spectral power during NREM sleep in mice during spontaneous sleep recordings (SS; days 30 and 31) from ZT22 to ZT10. (**A–F**) Panels display power in 1–20 Hz bands in sequential 2-h intervals from days when mice received either NAC (blue) or saline (black/grey) injections. Three injections were administered in this time frame at 0, 120, and 300 min; within each of the first three panels (**A–C**). Treatment x 2-h time interval (panel) x frequency differences between NAC and saline recordings were indicated by RM ANOVA, and individual frequency band differences were derived via post hoc assessment by Fisher's LSD. Significance is indicated by asterisks between these groups where $p < 0.05$. These differences were not modulated by sex. Light and dark phases are denoted by the color of the background in each panel. Grey and blue shaded areas signify SEM. N.S.: No significant effect of treatment on EEG power spectra from ZT8 to ZT10.

When sleep pressure is increased after sleep deprivation, from ZT 6 to ZT 9 (post-SD; days 16 and 18 as displayed in Figure 2), the patterns described above are repeated (RM ANOVA treatment x 30-min time interval x frequency, $F_{95,950} = 1.32$, $p = 0.028$, $\eta^2{}_p = 0.12$; statistics describe all data displayed in Figure 6). As expected, delta power is high after sleep deprivation, regardless of treatment, indicating increased sleep discharge. As we observed during SS (Figure 5), delta power initially increases and is discharged at an accelerated rate in NAC recordings when compared with saline recordings; at the same time, power in other low-frequency sub-bands (i.e., theta, alpha, and low-beta) is suppressed (Figure 6A–C). Eventually, delta power rebounds, decreasing during NAC recordings when compared with saline recordings (Figure 6F).

Overall, these data appear to suggest that NAC accelerates the discharge of delta power during NREMS when sleep need is elevated (i.e., during the dark phase or directly after sleep deprivation). However, in periods when sleep need has already been dissipated, animals that have been subjected to NAC injections display suppressed sleep pressure. Additionally, the initial attenuation of alpha power in NAC recordings suggests decreased cortical EEG synchronization, as alpha power is expected to increase during typical post-SD recovery sleep [17].

3.3. NAC Induces Sex-Specific Effects on the Accumulation of Sleep Need and Drowsiness during Enforced Wakefulness

We next determined whether NAC impacts the dynamics of sleep needs and waking EEG activity throughout the day. To do this, we assessed cumulative LFP energies for gamma, beta, and delta oscillations during active wake (AW), quiet wake (QW), and NREM sleep during NAC and saline injections. No differences in overall LFP energies were apparent between NAC- and saline-treated animals during AW or NREMS during SD (n.s., data not shown). Cumulative gamma power during AW was also unaffected during SD (n.s., data not shown). However, NAC was found to modulate delta and beta LFP energies within QW in a sex-dependent manner during SD. In females, NAC accelerated the accumulation of delta and beta energy across QW during SD, whereas NAC suppressed the accumulation of delta and beta energy across QW during SD in males (days 16 and 18 as displayed in Figure 2; beta: $F_{5,40} = 4.73$, $p = 0.002$, $\eta^2_p = 0.37$; delta: $F_{5,45} = 6.66$, $p < 0.001$, $\eta^2_p = 0.42$; RM ANOVA treatment x sex x hour, assessed for post hoc differences, indicated in Figure 7). These differences indicate that NAC decreases the build-up of sleep pressure across SD in males, whereas NAC accelerates the accumulation of sleep pressure during SD in female mice. Data were also significant for treatment x sex interaction.

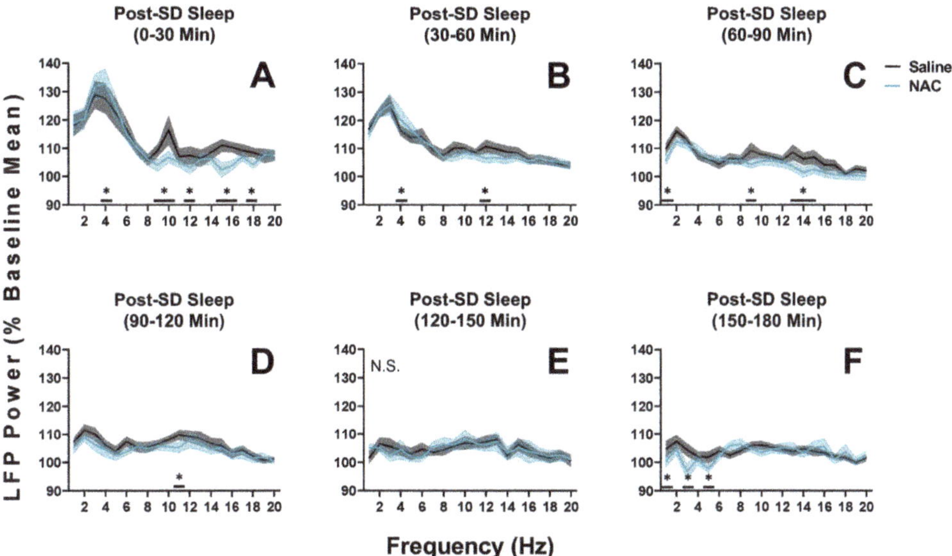

Figure 6. Changes in LFP spectral power during NREM sleep in mice during the hours after sleep deprivation recordings (SD; days 16 and 18; ZT6 to ZT9). (**A–F**) Panels display average power in 1–20 Hz bands in sequential 30-min intervals from days when mice received either NAC (blue) or saline (black/grey) injections. 30-min bins were utilized to ensure that the accelerated dynamics of SWS after SD were captured. Times displayed at the top of each panel refer to the amount of time elapsed after the sleep deprivation protocol was ended, allowing the recovery sleep opportunity to begin. No injections were administered in the timeframe of these recordings. The last injection was administered 3 h before the recordings displayed in panel A (final injection at ZT3). Treatment x 30-min time interval (panel) x frequency differences between NAC and saline recordings were indicated by RM ANOVA, and individual frequency band differences were derived via post hoc assessment by Fisher's LSD. Significance is indicated by asterisks between these groups where $p < 0.05$. These differences were not modulated by sex. Shaded areas signify SEM.

3.4. NAC Attenuates Sleep Need and Drowsiness in a Sex-Independent Manner during Quiet Wakefulness in Spontaneous Sleep

Finally, we assessed cumulative LFP energies during SS recordings (days 30 and 31). Differences were not detected between NAC- and saline-treatment recordings in cumulative delta or beta energies during NREMS in the 6-h interval after the second injection during SS recordings (n.s., not shown). Cumulative gamma energy in the active wake was also unaffected during SS recordings (n.s., not shown). However, NAC-treatments were found to attenuate cumulative beta (RM ANOVA Time x treatment interaction $F_{5,40} = 2.91$, $p = 0.024$, $\eta^2_p = 0.27$; Figure 8A,B) and delta (RM ANOVA Time x treatment interaction $F_{5,40} = 3.36$,

$p = 0.014, \eta^2{}_p = 0.32$; Figure 8C,D) energy during QW in spontaneous sleep/wake recordings in a time-dependent but sex-independent manner. There were no effects of NAC on cumulative beta energy in female mice, but the trend was in the same direction as in male mice, which demonstrated suppression of beta energy when NAC was on board (Figure 8A,B).

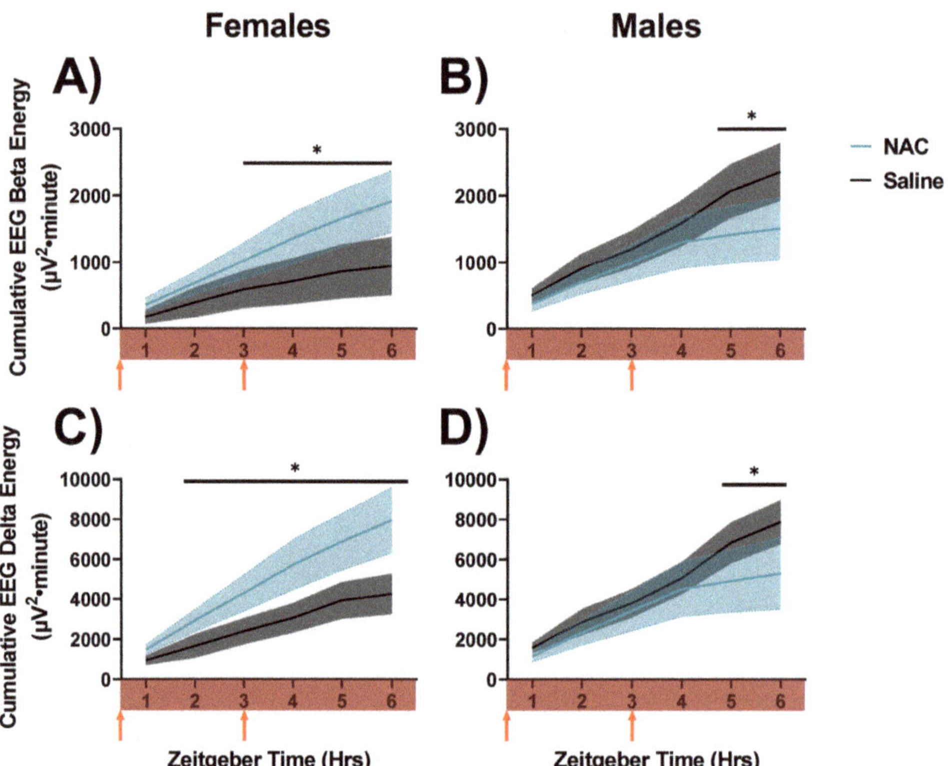

Figure 7. Changes in cumulated delta and beta activities during QW in SD recordings in which saline (black/grey) and NAC (blue) were administered to mice. These recordings were taken during the sleep deprivation period, as indicated by the red bar across the x-axes. Data are displayed in 1-h bins that occur from ZT0-6 on days 16 and 18 of the experimental protocol. These differences are displayed as cumulative LFP energies. Cumulative LFP energy is displayed on the left panels (**A,C**) for female mice and on the right (**B,D**) for male mice. Cumulative beta energy is displayed in the top panels (**A,B**), and cumulative delta energy is displayed on the bottom panels (**C,D**). Treatment x sex x hour differences were indicated by RM ANOVA, and differences between NAC and saline in specific intervals were derived via post hoc assessment by Fisher's LSD. Significance is indicated by asterisks between these groups where $p < 0.05$. Injection times are indicated by red arrows. Grey and blue shaded areas signify SEM.

Overall, delta and beta energies in NAC-treated recordings were lower than in saline-treated recordings, suggesting that NAC suppresses the accumulation of sleep need and drowsiness in QW during SS in the light phase [16].

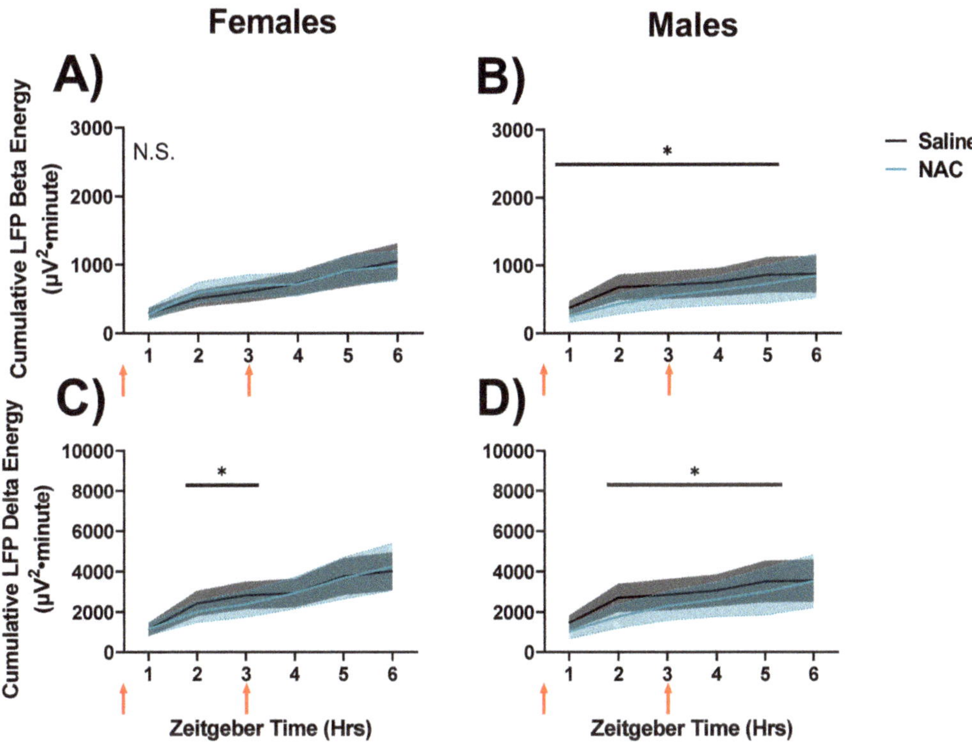

Figure 8. Changes in cumulated delta and beta activities during QW between SS recordings in which saline (black/grey) and NAC was administered (blue). Data are displayed in 1-h bins which occur from ZT0-6 on days 30 and 31 of the experimental protocol. Cumulative LFP energy is displayed on the left panels (**A,C**) for female mice and on the right (**B,D**) for male mice. Cumulative beta energy is displayed in the top panels (**A,B**), and cumulative delta energy is displayed on the bottom panels (**C,D**). Treatment, x time differences, were indicated by RM ANOVA, and differences between NAC and saline in specific intervals were derived via post hoc assessment by Fisher's LSD. Significance is indicated by asterisks between these groups where $p < 0.05$. Injection times are indicated by red arrows. Grey and blue shaded areas signify SEM.

4. Discussion

Here, we describe the effects of systemic redox manipulation via the glutathione precursor NAC on sleep timing and the EEG features associated with sleep homeostasis. We hypothesized that increasing the antioxidant capacity of the brain would facilitate sleep-dependent decreases in oxidative stress, decreasing the time required to dissipate sleep needs during NREMS. The effects of NAC on sleep timing and EEG parameters related to sleep homeostasis confirmed this effect, as animals exposed to NAC fell asleep faster (i.e., reduced latency to SWS) and dissipated delta power more rapidly. Overall, NAC increased the time that mice spent in NREMS at the cost of wakefulness during baseline sleep.

These effects could be a consequence of the perturbation of molecular processes that underlie sleep homeostasis, as glutathione is a regulator of sleep/wake cycles. The literature on this subject has previously described glutathione as a sleep-promoting substance and has shown that intracerebroventricular injections of oxidized glutathione (GSSG) increase time spent asleep [18,19]. It has been hypothesized that the somnogenic effects of GSSG are due to its ability to attenuate glutamatergic neurotransmission in the brain and stimulate nitric oxide synthase or general oxidative stress signaling mechanisms [19,20]. Ultimately, NAC produced the same effect, as it induced a shift in redox status that limits the ability of animals to stay awake. Additionally, NAC was observed to accelerate changes in delta power associated with sleep homeostasis. That is, when sleep need (delta power) is already increasing (post-siesta wakefulness or during SD), NAC further augments it; when sleep needs are decreasing (during sleep, lights on), NAC facilitates accelerated dissipation of delta power. Along with a general decline in EEG synchrony, this suggests that NAC changes the distribution of sleep depth by increasing initial sleep intensity without changing the timing of the accumulation of sleep needed during wakefulness.

4.1. N-Acetylcysteine Transiently Perturbs the Sleep Homeostat

In animals undergoing uninterrupted, spontaneous sleep/wake cycles during the light phase, delta power during NREMS declines from near peak levels at the light onset to a minimum value by the end of the light phase. Here, we have shown that NAC accelerates the rate of decline of NREMS delta power across the light phase, relative to vehicle injection.

The underlying biochemical processes which facilitate this effect of NAC are likely a direct result of increased glutathione (GSH) availability. Generally, systemic NAC is rapidly converted to glutathione via a multi-step enzymatic pathway that includes its direct conversion to L-cysteine, a rate-limiting substrate in glutathione synthesis [21]. Increased glutathione should produce additional opportunities for the production of NADPH [9], as reduced glutathione (GSH) serves as a redox substrate by undergoing oxidation to GSSG, coupled with reduction of NADP+ to NADPH, via glutathione reductase. The generation of NADPH then secondarily impacts NAD+/NADH via pathways as schematized in Figure 1. Through interconversion of NADP/NAD, the availability of GSH thus allows the cell to stabilize the NAD+/NADH ratio and NAD levels in the face of increased NAD+ consumption by enzymatic processes during wakefulness. Since fluctuations in cellular antioxidant capacity are largely due to the balance of the redox couple GSH/GSSG, this metabolic pathway is believed to be involved in mediating oxidative stress in the cell and responsible for providing the antioxidant benefits of NAC [22,23]. These data are compatible with our hypothesis that increased NAC and glutathione generally reduce sleep needs by decreasing cellular oxidative stress.

The biochemical pathway linking NAC to NAD(P):H homeostasis is especially important during prolonged wakefulness: sleep deprivation elevates brain expression of glutathione peroxidase [24] and the concentration of oxidized glutathione (GSSG) [25,26] while lowering the concentration of reduced glutathione (GSH) [27]. Accumulation of GSSG over time spent awake should decelerate glutathione peroxidase activity through product-dependent inhibition and thereby increase NADP+ concentration in the brain (as indeed occurs in sleep deprivation; [6]). Additionally, As NADPH is the primary source of reducing equivalents for glutathione, NADP+:NADPH dysregulation associated with protracted wake could endanger the efficacy of the cell's most robust antioxidant system [28]. Ultimately, a shift in redox state limits the ability to stay awake: intracerebroventricular infusion of oxidized glutathione into the brain increases time spent asleep [19,25]. By serving as a biochemical precursor for GSH, NAC increases the pool of GSH available as a NADP+/NADPH buffering substrate.

The effects of NAC on sleep are not necessarily an exclusive consequence of its participation in glutathione-related redox and antioxidant reactions. The therapeutic potential of NAC has also been proposed to involve the modulation of glutamatergic neurotransmission in the brain [29]. The NAC metabolite L-cysteine is oxidized to L-cystine once it is taken up by tissue. Interactions with the cystine/glutamate antiporter exchange extracellular cystine for intracellular L-glutamate, especially across glial membranes [30]. This increase in extrasynaptic glutamate activates mGluR2/3 receptors, decreasing synaptic glutamate release [29]. Such decreases in excitatory activity may decrease the signaling and metabolic demands of neurons in the presence of NAC [31]. This signaling change, in addition to the conversion of NAC to glutathione, may contribute to the sleep-related changes observed here. NAC also confers other neuroprotective effects, including the stabilization of proteins and DNA by cross-linking cysteine disulfide molecules. Given that DNA damage induces sleep and that sleep has been implicated in DNA repair [32], this mechanism cannot be dismissed in considering the possible effects of NAC related to sleep. In order to use NAC to its full efficacy, further studies should explore the underlying mechanisms that specifically contribute to its sleep-related effects. Future studies should verify the extent to which acute NAC elevates cerebral glutathione in order to understand if the effects observed here are glutathione specific. Other relevant repair/oxidative stress protection mechanisms of NAC may also be relevant and worth exploring [33], including mechanisms by which NAC scavenges free radicals [34], induces neurogenesis [35], reduces mitochondrial apoptosis [36], reduces glutamatergic neurotransmission [29], and chelates metals as a form of oxidative stress protection [37].

4.2. Sex Differences in the Response to N-Acetylcysteine

We revealed sex differences in the impact of NAC on SD when assessing the spectral dynamics of cumulative delta and beta effects during QW (Figure 7). In females, NAC accelerated the accumulation of drowsiness and sleep needs across QW during SD, whereas in males, NAC suppressed the accumulation of drowsiness and sleep needs across QW during SD. The sex differences observed in accumulated EEG beta power during QW (i.e., drowsiness; [16]) could be the result of sex differences in antioxidant mechanisms, as males are more robustly impacted by surges of antioxidants than females. Studies assessing the capacity of human brain mitochondrial respiration have shown that mitochondria derived from males create two-fold more reactive oxygen species (ROS) than those of females [38]. An influx of glutathione via NAC would consequently be expected to have a greater impact in terms of alleviating oxidative stress in males. On the other hand, the same increase in glutathione may create reductive stress in females. If female mitochondria accumulate an excess of antioxidant capacity, further elevating GSH might free up excess NADPH to serve as a substrate for NADPH oxidase 2 (NOX2) in the generation of superoxide ions. Further, imbalances in NADP+/H and, subsequently, the dysregulated activity of NOX2 during sleep fragmentation create excessive superoxide production and subsequent inflammation. In support of this theory, NOX2-deficient mice are protected from cognitive decline associated with sleep fragmentation [39].

Final accumulated QW beta power at hour 6 during SD in the saline control condition was more than 2-fold higher in males (2358 $mV^2 \cdot min$; Figure 7B) than in females (944 $mV^2 \cdot min$; Figure 7A) according to post hoc comparisons of these two values ($p = 0.027$, Fisher's LSD). This sex difference is effectively nullified by NAC, in the sense that the final accumulated beta power at hour 6 during SD in the NAC condition was (non-significantly) higher in females (1917 $mV^2 \cdot min$; Figure 7A) than in males (1508 $mV^2 \cdot min$; Figure 7B).

This pattern was replicated by QW delta power: Final accumulated QW delta power at hour 6 during SD in the saline control condition was higher in males (7873 $mV^2 \cdot min$; Figure 7D) than in females (4268 $mV^2 \cdot min$; Figure 7C; $p = 0.049$, Fisher's LSD). This sex difference is effectively nullified by NAC, as the final accumulated delta power at hour 6 during SD in the NAC condition was (non-significantly) higher in females (7972 $mV^2 \cdot min$; Figure 7C) than in males (5305 $mV^2 \cdot min$; Figure 7D). The nullification of these sex differences by NAC implicates underlying sex differences in wake-dependent redox dynamics. Interestingly, during QW during SS in the light phase, similar sex differences were not observed, as NAC slowed the accumulation of delta oscillations in both males and females (Figure 8). This suggests that these sex differences may be masked under conditions where oxidative stress is not exacerbated (such as during SD).

4.3. Limitations of C57BL/6J as an Experimental Model

C57BL/6J mice are known to have mutations in the nicotinamide nucleotide transhydrogenase (NNT) gene, which results in mitochondrial redox abnormalities. Although this mutation spontaneously arose nearly four decades ago, it was only discovered in 2005, and C57BL/6J mice are still commonly used in laboratory experiments [40]. NNT is an enzyme that is localized to the inner mitochondrial membrane. Its main role is to reduce NADP+ to

NADPH at the expense of NADH oxidation and H+ re-entry to the mitochondrial matrix. In doing so, it provides a major pathway for NADP+/H and NAD+/H interconversion within the cell. In the absence of NNT, redox-related imbalances abound with relatively minimal impact on overall health. Redox challenges of NNT deficient mice that are pertinent to our manipulations include higher rates of hydrogen peroxide (i.e., ROS) release and poorer ability to metabolize peroxide; spontaneous NADPH oxidation; and increased ratio of oxidized glutathione to reduced glutathione (GSSG:GSH). Overall, this results in increased oxidative stress and decreased glutathione-based antioxidant capacity in these mice [40].

NAC has demonstrated neuroprotective effects in C57BL/6J mice in the literature [41–43]. The pathways which utilize nicotinamide-based substrates in the body are rich with redundancies—about 200 enzymes in total can process nicotinamide-based substrates; these redundancies may explain why NNT deficiency does not produce major adverse health effects in mice [40]. A comprehensive phenotypic comparison of C57BL6/J mice against C57BL6/N mice, in which NNT is repaired, additionally demonstrates that NNT-deficient C57BL6/J mice perform better in certain neurobehavioral assays of cognitive function [44]. As the NNT mutation decreases the effectiveness of NADPH-related antioxidant pathways, the metabolic impairments produced by the NNT mutation in C57BL6/J mice might be able to be considered a constitutive model of oxidative stress. Demonstrating the efficacy of NAC in this model allows for a specific understanding of the utility of such treatments in the context of increased oxidative stress in many neurological diseases. Future studies should assess whether the effects of NAC on EEG and sleep differ between C57BL/6J and other strains with intact nicotinamide pathways (such as C57BL/6N mice).

4.4. Translational Relevance of the Findings

The effects of NAC on sleep are likely to have translational relevance. NAC has been applied in human subject trials in neurologic and psychiatric conditions in which oxidative stress is believed to contribute to pathogenesis [45–47]. A prominent line of reasoning about the pathogenesis of schizophrenia, for instance, is that the schizophrenic brain is constitutively vulnerable to oxidative insults and that typically benign oxidative challenges can cause changes to the schizophrenic brain that manifest as symptoms [45]. This line of reasoning led to the hypothesis that pharmaceutical and/or nutraceutical manipulations that increase the availability of antioxidant substrates in the brain could be of therapeutic benefit. Despite the clear rationale for the use of antioxidants such as NAC as adjunct therapies in schizophrenia, clinical trials involving antioxidant administration to schizophrenic patients have led to equivocal results, with only a subset of trials finding statistically and clinically significant benefits [48–50]. Reasons for discrepancies in the outcome of clinical trials are not known, but results of our studies suggest that both time of day (i.e., the timing of antioxidant administration relative to sleep timing) and sex ought to be considered in the application of NAC.

Time of day significantly impacts oxidation-reduction reactions in the brain [51–53], which may influence the efficacy of antioxidant therapeutics in the treatment of neurologic and psychiatric conditions. Sleep is associated with dramatic shifts in the oxidative status of biochemical substrates in the brain and supports mechanisms that reduce accumulated oxidative stress that builds during waking activities [54–56]. Disruption of sleep and its associated neuronal network activities is a hallmark of schizophrenia [57,58]. For example, sleep acutely attenuates abnormalities in sensory gating in schizophrenia [59] and similar sensory gating deficits that emerge in non-schizophrenics voluntarily undergoing habitual sleep restriction [60]. Our experiment demonstrates the somnogenic qualities of NAC, which is replicated in experimental infusions of GSSG [18,19].

Because the oxidative insult of sleep disruption exacerbates the symptoms of schizophrenia, manipulations that increase the availability of antioxidant substrates may benefit from being timed relative to the sleep/wake cycle. Specifically, we hypothesize that for antioxidants taken for the purpose of targeting and protecting the brain (as is the case of NAC for schizophrenia), maximal therapeutic efficacy will occur when the antioxidant is administered at a time and in a manner that delivers the antioxidant to the brain when sleep need is highest, at sleep onset. Conversely, therapeutics administered without regard for sleep-dependent antioxidant processes in the brain may not deliver positive results. We attempted a meta-analysis of studies to test this hypothesis. Pubmed searches were conducted for the following term sets, with filtering for clinical trials: "schizophrenia nicotinamide," "schizophrenia niacin", "schizophrenia acetylcysteine". Collectively, the three searches yielded access to 28 published studies, the PMIDs of which are included as supplementary materials with this manuscript (Supplementary Table S1). Trials involved the administration of five agents: N-acetylcysteine ($n = 17$); niacin or niacinamide ($n = 4$); nicotinic acid ($n = 2$); nicotinamide adenine dinucleotide ($n = 3$); nicotinamide ($n = 2$). Of the 28 trials accessed, only three specified the time of day at which the agent was administered (Figure 9A). Two trials involved one morning dosing and one evening dosing, and one trial involved a morning dose exclusively. No trials engaged in a systematic comparison of exclusive morning vs. exclusive evening dosing; no trials employed exclusively evening dosing. Thus, 89% of clinical studies examining the efficacy of N-acetylcysteine and related antioxidants in schizophrenia do not report the time of day of N-acetylcysteine administration, and no studies appear to have systematically varied time of day as part of the design. Of those 28 studies, only 20 studies (71%) assessed study outcomes in both sexes, and only 7 (25%) reported statistical analysis related to sex differences (Supplementary Table S1), with many studies claiming an inability to properly assess sex differences, given small sample sizes.

A more general Pubmed review of clinical trials examining the efficacy of nicotinamide-based antioxidants without regard to the target disease state indicated that 76% do not report the time of day of administration (Supplementary Table S2). Pubmed searches were conducted for the following term sets, with filtering for clinical trials: "nicotinamide supplement", "nicotinamide supplementation", "dietary nicotinamide", "n-acetylcysteine supplement", "dietary n-acetylcysteine", "dietary n-acetyl cysteine", "n-acetyl cysteine supplement", "nicotinamide riboside", or "nicotinamide mononucleotide". Resulting accessible trials involved the administration of 7 agents: N-acetylcysteine ($n = 37$); niacin or niacinamide ($n = 4$); nicotinamide ($n = 7$); nicotinamide adenine dinucleotide (NAD; $n = 4$); nicotinamide mononucleotide ($n = 4$); nicotinamide riboside ($n = 11$); or nicotinic acid ($n = 1$). The vast majority of studies did not report the time of day of antioxidant administration (51 of 68 studies; Supplementary Table S2). Five studies involved only daytime or morning administration; no trials have employed exclusively evening dosing. Six of the studies incorporated morning and evening administration; five of these entailed morning administration in addition to evening administration in the same subject group. A single study (PMID 35215405) systematically compared the effects of nicotinamide mononucleotide (NMN) supplementation in the morning vs. evening on self-reported, subjective sleep quality and drowsiness. Subjects were instructed to take supplements either in the AM (between wake-up time and 12:00) or PM (between 18:00 and bedtime). Those in the PM NMN group had reduced daytime drowsiness ratings relative to PM placebo controls, whereas those in the AM NMN group did not differ from AM placebo controls. This study, which appears to be unique to the literature in that it systematically assessed time-of-day effects on antioxidant efficacy, is suggestive of a time-of-day-dependent antioxidant response relevant to sleep and sleepiness (Figure 9B). Future studies should further address this possibility and, at a minimum, specify in reports the time of day at which antioxidants were administered.

Of the 68 studies included in this general review of nicotinamide-based antioxidants, only 39 (57%) assessed study outcomes in both sexes, and only 9 (13%) reported statistical analysis related to sex differences (Supplementary Table S2). Out of the studies which reported statistical analysis of sex differences, four studies (6%) found differences between sexes. Two of these studies found that nicotinamide treatments were more effective in males (PMIDs 8951265, 7660324), whereas two found more striking results in females (PMIDs 34238308, 32320006), with several studies mentioning that larger study sizes would be necessary to properly assess the difference in sex. These equivocal results comport with our understanding of sex differences, as observed in Figures 7 and 8, which suggest that sex differences may reveal themselves differentially based on a subject's prior accumulation of tissue-level oxidative stress (which may be roughly assessed by time of day).

A. Antioxidant Trials in Schizophrenia

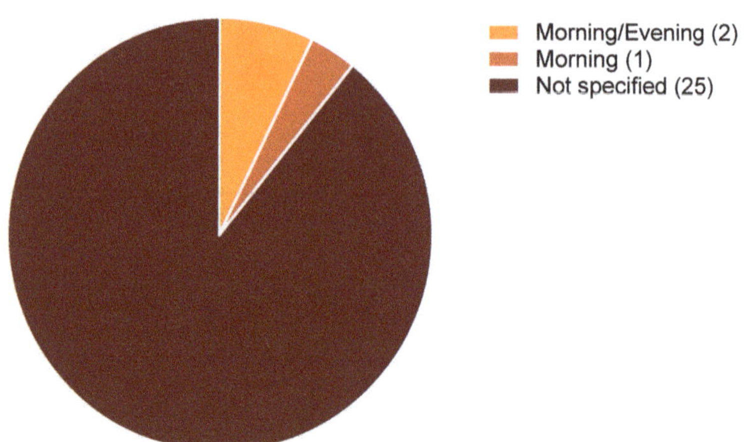

B. Antioxidant Trials Irrespective of Disorder

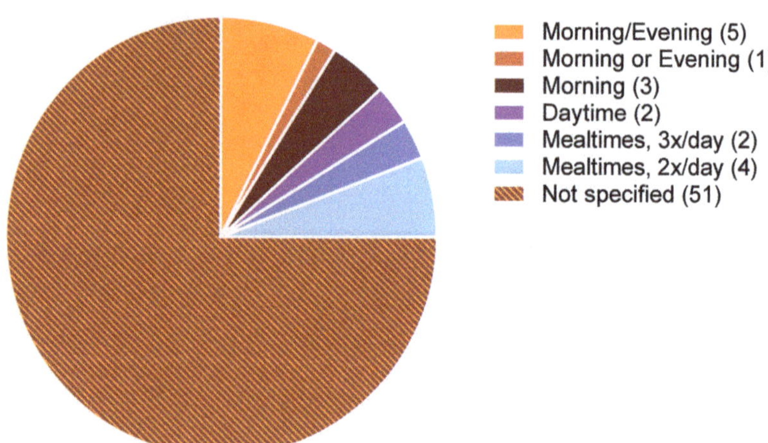

Figure 9. Timing of administration of NAC and related antioxidants in clinical trials. (**A**) Data from 32 clinical trials involving schizophrenia. (**B**) Data from 93 clinical trials reviewed without regard to the disorder targeted. Numbers in parentheses indicate the number of trials included in that category.

Disorders where NAC or other nicotinamide-based treatments are commonly used as adjunctive therapies are often conditions in which sleep is taxed and tissue-level oxidative stress is high. Given these results, we suggest that future clinical studies assess both sexes, as conditions in which oxidative stress is exacerbated will likely reveal sex differences (as suggested by PMID 32320006). Consideration of both time of day and sex together could potentially improve the replicability of study outcomes and the efficacy of antioxidant-based treatments in clinical trials.

5. Conclusions

The studies described here demonstrate that N-acetylcysteine modulates the sleep-wake cycle in mice: reducing the latency to onset of and increasing the amount of time spent in NREMS. The sleep-promoting effect of N-acetylcysteine may contribute to its therapeutic potential in schizophrenia and other neuropsychiatric conditions, provided that it is administered at bedtime. N-acetylcysteine may modulate the electroencephalogram differently in male and female subjects. It is, therefore, recommended that future clinical trials involving N-acetylcysteine incorporate bedtime administration into the design and measure the effects of treatments in both male and female subjects.

Supplementary Materials: The following supporting information can be downloaded at: https://www.mdpi.com/article/10.3390/antiox12051124/s1, Table S1: Antioxidant Trials in Schizophrenia; Table S2: Antioxidant Trials Irrespective of Disorder.

Author Contributions: Conceptualization, P.N.B., M.A.S., B.A.S. and J.P.W.; methodology, P.N.B., M.A.S. and J.P.W.; software, J.P.W.; validation, P.N.B., M.A.S., B.A.S. and J.P.W.; formal analysis, P.N.B., B.A.S. and J.P.W.; investigation, P.N.B., M.A.S., K.M.C. and T.V.; resources, K.M.C. and T.V.; data curation, P.N.B., M.A.S., K.M.C. and T.V.; writing—original draft preparation, P.N.B. and J.P.W.; writing—review and editing, B.A.S. and J.P.W.; visualization, P.N.B., B.A.S. and J.P.W.; supervision, B.A.S. and J.P.W.; project administration, M.A.S., B.A.S. and J.P.W.; funding acquisition, B.A.S. and J.P.W. All authors have read and agreed to the published version of the manuscript.

Funding: Research supported by NIH R01NS078498, NIH R01DA040965 and Good Samaritan Foundation of Legacy Health.

Institutional Review Board Statement: The animal study protocol was approved by the Institutional Animal Care and Use Committee of Washington State University (protocol # 6229).

Data Availability Statement: The data presented in this study are available on request from the corresponding author.

Acknowledgments: We thank Travis Denton and Laken Kruger for technical assistance with the proper handling of research chemicals. We thank Marcos Frank and Christopher Davis for their helpful comments.

Conflicts of Interest: The authors declare no conflict of interest. The funders had no role in the design of the study; in the collection, analyses, or interpretation of data; in the writing of the manuscript; or in the decision to publish the results.

References

1. Borbély, A.A. A two process model of sleep regulation. *Hum. Neurobiol.* **1982**, *1*, 195–204. [PubMed]
2. Harkness, J.H.; Bushana, P.N.; Todd, R.P.; Clegern, W.C.; Sorg, B.A.; Wisor, J.P. Sleep disruption elevates oxidative stress in parvalbumin-positive cells of the rat cerebral cortex. *Sleep* **2019**, *42*, zsy201. [CrossRef] [PubMed]
3. Harkness, J.H.; Gonzalez, A.E.; Bushana, P.N.; Jorgensen, E.T.; Hegarty, D.M.; Di Nardo, A.A.; Prochiantz, A.; Wisor, J.P.; Aicher, S.A.; Brown, T.E.; et al. Diurnal changes in perineuronal nets and parvalbumin neurons in the rat medial prefrontal cortex. *Brain Struct. Funct.* **2021**, *226*, 1135–1153. [CrossRef]
4. Honda, K.; Komoda, Y.; Inoue, S. Oxidized glutathione regulates physiological sleep in unrestrained rats. *Brain Res.* **1994**, *636*, 253–258. [CrossRef]
5. Hill, V.M.; O'Connor, R.M.; Sissoko, G.B.; Irobunda, I.S.; Leong, S.; Canman, J.C.; Stavropoulos, N.; Shirasu-Hiza, M. A bidirectional relationship between sleep and oxidative stress in Drosophila. *PLoS Biol.* **2018**, *16*, e2005206. [CrossRef]
6. Zhang, S.X.; Khalyfa, A.; Wang, Y.; Carreras, A.; Hakim, F.; Neel, B.A.; Brady, M.J.; Qiao, Z.; Hirotsu, C.; Gozal, D. Sleep Fragmentation Promotes NADPH Oxidase 2-Mediated Adipose Tissue Inflammation Leading to Insulin Resistance in Mice. *Int. J. Obes.* **2014**, *38*, 619–624. [CrossRef]
7. Ramanathan, L.; Hu, S.; Frautschy, S.A.; Siegel, J.M. Short-term total sleep deprivation in the rat increases antioxidant responses in multiple brain regions without impairing spontaneous alternation behavior. *Behav. Brain Res.* **2010**, *207*, 305–309. [CrossRef]
8. Ramanathan, L.; Siegel, J.M. Sleep deprivation under sustained hypoxia protects against oxidative stress. *Free Radic. Biol. Med.* **2011**, *51*, 1842–1848. [CrossRef]
9. Winkler, B.S.; DeSantis, N.; Solomon, F. Multiple NADPH-producing pathways control glutathione (GSH) content in retina. *Exp. Eye Res.* **1986**, *43*, 829–847. [CrossRef]

10. Wisor, J.P.; Rempe, M.J.; Schmidt, M.A.; Moore, M.E.; Clegern, W.C. Sleep Slow-Wave Activity Regulates Cerebral Glycolytic Metabolism. *Cereb. Cortex* **2012**, *23*, 1978–1987. [CrossRef]
11. Holdiness, M.R. Clinical pharmacokinetics of N-acetylcysteine. *Clin. Pharmacokinet.* **1991**, *20*, 123–134. [CrossRef] [PubMed]
12. Dickey, D.T.; Muldoon, L.L.; Doolittle, N.D.; Peterson, D.R.; Kraemer, D.F.; Neuwelt, E.A. Effect of N-acetylcysteine route of administration on chemoprotection against cisplatin-induced toxicity in rat models. *Cancer Chemother. Pharmacol.* **2008**, *62*, 235–241. [CrossRef] [PubMed]
13. Sjödin, K.; Nilsson, E.; Hallberg, A.; Tunek, A. Metabolism of N-acetyl-l-cysteine Some structural requirements for the deacetylation and consequences for the oral bioavailability. *Biochem. Pharmacol.* **1989**, *38*, 3981–3985. [CrossRef] [PubMed]
14. Borgström, L.; Kågedal, B.; Paulsen, O. Pharmacokinetics of N-acetylcysteine in man. *Eur. J. Clin. Pharmacol.* **1986**, *31*, 217–222. [CrossRef]
15. Miladinović, Đ.; Muheim, C.; Bauer, S.; Spinnler, A.; Noain, D.; Bandarabadi, M.; Gallusser, B.; Krummenacher, G.; Baumann, C.; Adamantidis, A. SPINDLE: End-to-end learning from EEG/EMG to extrapolate animal sleep scoring across experimental settings, labs and species. *PLoS Comput. Biol.* **2019**, *15*, e1006968. [CrossRef]
16. Grønli, J.; Rempe, M.J.; Clegern, W.C.; Schmidt, M.; Wisor, J.P. Beta EEG reflects sensory processing in active wakefulness and homeostatic sleep drive in quiet wakefulness. *J. Sleep Res.* **2016**, *25*, 257–268. [CrossRef]
17. Marzano, C.; Ferrara, M.; Curcio, G.; Gennaro, L.D. The effects of sleep deprivation in humans: Topographical electroencephalogram changes in non-rapid eye movement (NREM) sleep versus REM sleep. *J. Sleep Res.* **2009**, *19*, 260–268. [CrossRef]
18. Ikeda, M.; Ikeda-Sagara, M.; Okada, T.; Clement, P.; Urade, Y.; Nagai, T.; Sugiyama, T.; Yoshioka, T.; Honda, K.; Inoue, S. Brain oxidation is an initial process in sleep induction. *Neuroscience* **2005**, *130*, 1029–1040. [CrossRef] [PubMed]
19. Kimura, M.; Kapas, L.; Krueger, J.M. Oxidized glutathione promotes sleep in rabbits. *Brain Res. Bull.* **1998**, *45*, 545–548. [CrossRef]
20. Villafuerte, G.; Miguel-Puga, A.; Murillo Rodríguez, E.; Machado, S.; Manjarrez, E.; Arias-Carrión, O. Sleep Deprivation and Oxidative Stress in Animal Models: A Systematic Review. *Oxidative Med. Cell. Longev.* **2015**, *2015*, 234952. [CrossRef] [PubMed]
21. Kerksick, C.; Willoughby, D. The Antioxidant Role of Glutathione and N-Acetyl-Cysteine Supplements and Exercise-Induced Oxidative Stress. *J. Int. Soc. Sport. Nutr.* **2005**, *2*, 38. [CrossRef]
22. Atkuri, K.R.; Mantovani, J.J.; Herzenberg, L.A.; Herzenberg, L.A. N-Acetylcysteine—A safe antidote for cysteine/glutathione deficiency. *Curr. Opin. Pharmacol.* **2007**, *7*, 355–359. [CrossRef]
23. Rushworth, G.F.; Megson, I.L. Existing and potential therapeutic uses for N-acetylcysteine: The need for conversion to intracellular glutathione for antioxidant benefits. *Pharmacol. Ther.* **2014**, *141*, 150–159. [CrossRef]
24. Arent, C.O.; Valvassori, S.S.; Steckert, A.V.; Resende, W.R.; Dal-Pont, G.C.; Lopes-Borges, J.; Amboni, R.T.; Bianchini, G.; Quevedo, J. The effects of n-acetylcysteine and/or deferoxamine on manic-like behavior and brain oxidative damage in mice submitted to the paradoxal sleep deprivation model of mania. *J. Psychiatr. Res.* **2015**, *65*, 71–79. [CrossRef] [PubMed]
25. Komoda, Y.; Honda, K.; Inoue, S. SPS-B, A physiological sleep regulator, from the brainstems of sleep-deprived rats, identified as oxidized glutathione. *Chem. Pharm. Bull.* **1990**, *38*, 2057–2059. [CrossRef] [PubMed]
26. Alzoubi, K.H.; Khabour, O.F.; Salah, H.A.; Abu Rashid, B.E. The combined effect of sleep deprivation and Western diet on spatial learning and memory: Role of BDNF and oxidative stress. *J. Mol. Neurosci.* **2013**, *50*, 124–133. [CrossRef] [PubMed]
27. Kanazawa, L.K.; Vecchia, D.D.; Wendler, E.M.; de AS Hocayen, P.; dos Reis Lívero, F.A.; Stipp, M.C.; Barcaro, I.M.; Acco, A.; Andreatini, R. Quercetin reduces manic-like behavior and brain oxidative stress induced by paradoxical sleep deprivation in mice. *Free Radic. Biol. Med.* **2016**, *99*, 79–86. [CrossRef]
28. Schafer, F.Q.; Buettner, G.R. Redox environment of the cell as viewed through the redox state of the glutathione disulfide/glutathione couple. *Free Radic. Biol. Med.* **2001**, *30*, 1191–1212. [CrossRef]
29. Anwyl, R. Metabotropic glutamate receptors: Electrophysiological properties and role in plasticity. *Brain Res. Rev.* **1999**, *29*, 83–120. [CrossRef]
30. Bridges, R.J.; Natale, N.R.; Patel, S.A. System xc- cystine/glutamate antiporter: An update on molecular pharmacology and roles within the CNS. *Br. J. Pharmacol.* **2012**, *165*, 20–34. [CrossRef]
31. Hameed, M.Q.; Hodgson, N.; Lee, H.H.; Pascual-Leone, A.; MacMullin, P.C.; Jannati, A.; Dhamne, S.C.; Hensch, T.K.; Rotenberg, A. N-acetylcysteine treatment mitigates loss of cortical parvalbumin-positive interneuron and perineuronal net integrity resulting from persistent oxidative stress in a rat TBI model. *Cereb. Cortex* **2023**, *33*, 4070–4084. [CrossRef]
32. Zada, D.; Sela, Y.; Matosevich, N.; Monsonego, A.; Lerer-Goldshtein, T.; Nir, Y.; Appelbaum, L. Parp1 promotes sleep, which enhances DNA repair in neurons. *Mol. Cell* **2021**, *81*, 4979–4993.e7. [CrossRef] [PubMed]
33. De Flora, S.; Izzotti, A.; D'Agostini, F.; Balansky, R.M. Mechanisms of N-acetylcysteine in the prevention of DNA damage and cancer, with special reference to smoking-related end-points. *Carcinogenesis* **2001**, *22*, 999–1013. [CrossRef] [PubMed]
34. Halasi, M.; Wang, M.; Tanmay; Gaponenko, V.; Hay, N.; Andrei. ROS inhibitor N-acetyl-L-cysteine antagonizes the activity of proteasome inhibitors. *Biochem. J.* **2013**, *454*, 201–208. [CrossRef] [PubMed]
35. Fries, G.R.; Kapczinski, F. N-acetylcysteine as a mitochondrial enhancer: A new class of psychoactive drugs? *Rev. Bras. De Psiquiatr.* **2011**, *33*, 321–322. [CrossRef]

36. Jiao, Y.; Ma, S.; Wang, Y.; Li, J.; Shan, L.; Liu, Q.; Liu, Y.; Song, Q.; Yu, F.; Yu, H.; et al. N-Acetyl Cysteine Depletes Reactive Oxygen Species and Prevents Dental Monomer-Induced Intrinsic Mitochondrial Apoptosis In Vitro in Human Dental Pulp Cells. *PLoS ONE* **2016**, *11*, e0147858. [CrossRef]
37. Rossignol, D.A. The use of N-acetylcysteine as a chelator for metal toxicity. In *The Therapeutic Use of N-Acetylcysteine (NAC) in Medicine*; Springer: Berlin/Heidelberg, Germany, 2019; pp. 169–179.
38. Silaidos, C.; Pilatus, U.; Grewal, R.; Matura, S.; Lienerth, B.; Pantel, J.; Eckert, G.P. Sex-associated differences in mitochondrial function in human peripheral blood mononuclear cells (PBMCs) and brain. *Biol. Sex Differ.* **2018**, *9*, 34. [CrossRef]
39. Zhang, J.; Zhu, Y.; Zhan, G.; Fenik, P.; Panossian, L.; Wang, M.M.; Reid, S.; Lai, D.; Davis, J.G.; Baur, J.A. Extended wakefulness: Compromised metabolics in and degeneration of locus ceruleus neurons. *J. Neurosci.* **2014**, *34*, 4418–4431. [CrossRef]
40. Ronchi, J.A.; Figueira, T.R.; Ravagnani, F.G.; Oliveira, H.C.; Vercesi, A.E.; Castilho, R.F. A spontaneous mutation in the nicotinamide nucleotide transhydrogenase gene of C57BL/6J mice results in mitochondrial redox abnormalities. *Free Radic. Biol. Med.* **2013**, *63*, 446–456. [CrossRef]
41. Clark, J.; Clore, E.L.; Zheng, K.; Adame, A.; Masliah, E.; Simon, D.K. Oral N-Acetyl-Cysteine Attenuates Loss of Dopaminergic Terminals in α-Synuclein Overexpressing Mice. *PLoS ONE* **2010**, *5*, e12333. [CrossRef]
42. Durieux, A.M.; Fernandes, C.; Murphy, D.; Labouesse, M.A.; Giovanoli, S.; Meyer, U.; Li, Q.; So, P.-W.; McAlonan, G. Targeting glia with N-acetylcysteine modulates brain glutamate and behaviors relevant to neurodevelopmental disorders in C57BL/6J mice. *Front. Behav. Neurosci.* **2015**, *9*, 343. [CrossRef]
43. Ercal, N.; Treeratphan, P.; Hammond, T.C.; Matthews, R.H.; Grannemann, N.H.; Spitz, D.R. In vivo indices of oxidative stress in lead-exposed C57BL/6 mice are reduced by treatment with meso-2, 3-dimercaptosuccinic acid or N-acetylcysteine. *Free Radic. Biol. Med.* **1996**, *21*, 157–161. [CrossRef]
44. Simon, M.M.; Greenaway, S.; White, J.K.; Fuchs, H.; Gailus-Durner, V.; Wells, S.; Sorg, T.; Wong, K.; Bedu, E.; Cartwright, E.J.; et al. A comparative phenotypic and genomic analysis of C57BL/6J and C57BL/6N mouse strains. *Genome Biol.* **2013**, *14*, R82. [CrossRef] [PubMed]
45. Yolland, C.O.; Phillipou, A.; Castle, D.J.; Neill, E.; Hughes, M.E.; Galletly, C.; Smith, Z.M.; Francis, P.S.; Dean, O.M.; Sarris, J. Improvement of cognitive function in schizophrenia with N-acetylcysteine: A theoretical review. *Nutr. Neurosci.* **2020**, *23*, 139–148. [CrossRef]
46. Steullet, P.; Cabungcal, J.; Monin, A.; Dwir, D.; O'donnell, P.; Cuenod, M.; Do, K. Redox dysregulation, neuroinflammation, and NMDA receptor hypofunction: A "central hub" in schizophrenia pathophysiology? *Schizophr. Res.* **2016**, *176*, 41–51. [CrossRef] [PubMed]
47. Magalhães, P.V.; Dean, O.; Andreazza, A.C.; Berk, M.; Kapczinski, F. Antioxidant treatments for schizophrenia. *Cochrane Database Syst. Rev.* **2016**, *2*, CD008919. [CrossRef]
48. Slattery, J.; Kumar, N.; Delhey, L.; Berk, M.; Dean, O.; Spielholz, C.; Frye, R. Clinical trials of N-acetylcysteine in psychiatry and neurology: A systematic review. *Neurosci. Biobehav. Rev.* **2015**, *55*, 294–321.
49. Ooi, S.L.; Green, R.; Pak, S.C. N-Acetylcysteine for the treatment of psychiatric disorders: A review of current evidence. *BioMed Res. Int.* **2018**, *2018*, 2469486. [CrossRef]
50. Zheng, W.; Zhang, Q.E.; Cai, D.B.; Yang, X.H.; Qiu, Y.; Ungvari, G.; Ng, C.; Berk, M.; Ning, Y.P.; Xiang, Y.T. N-acetylcysteine for major mental disorders: A systematic review and meta-analysis of randomized controlled trials. *Acta Psychiatr. Scand.* **2018**, *137*, 391–400. [CrossRef]
51. Poljsak, B.; Kovač, V.; Milisav, I. Healthy lifestyle recommendations: Do the beneficial effects originate from NAD+ amount at the cellular level? *Oxidative Med. Cell. Longev.* **2020**, *2020*, 8819627. [CrossRef]
52. Maiese, K. Moving to the rhythm with clock (circadian) genes, autophagy, mTOR, and SIRT1 in degenerative disease and cancer. *Curr. Neurovascular Res.* **2017**, *14*, 299–304. [CrossRef] [PubMed]
53. Oblong, J.E. The evolving role of the NAD+/nicotinamide metabolome in skin homeostasis, cellular bioenergetics, and aging. *DNA Repair* **2014**, *23*, 59–63. [CrossRef]
54. Morris, G.; Stubbs, B.; Köhler, C.A.; Walder, K.; Slyepchenko, A.; Berk, M.; Carvalho, A.F. The putative role of oxidative stress and inflammation in the pathophysiology of sleep dysfunction across neuropsychiatric disorders: Focus on chronic fatigue syndrome, bipolar disorder and multiple sclerosis. *Sleep Med. Rev.* **2018**, *41*, 255–265. [CrossRef]
55. Atrooz, F.; Salim, S. Sleep deprivation, oxidative stress and inflammation. *Adv. Protein Chem. Struct. Biol.* **2020**, *119*, 309–336.
56. Palagini, L.; Geoffroy, P.A.; Miniati, M.; Perugi, G.; Biggio, G.; Marazziti, D.; Riemann, D. Insomnia, sleep loss, and circadian sleep disturbances in mood disorders: A pathway toward neurodegeneration and neuroprogression? A theoretical review. *CNS Spectr.* **2021**, *27*, 298–308. [CrossRef] [PubMed]
57. Luu, B.; Rodway, G.W.; Rice, E. Should we be targeting sleep architecture to more effectively treat schizophrenia? *JAAPA* **2018**, *31*, 52–54. [CrossRef]
58. Ettinger, U.; Kumari, V. Effects of sleep deprivation on inhibitory biomarkers of schizophrenia: Implications for drug development. *Lancet Psychiatry* **2015**, *2*, 1028–1035. [CrossRef]

59. Griffith, J.M.; Waldo, M.; Adler, L.E.; Freedman, R. Normalization of auditory sensory gating in schizophrenic patients after a brief period for sleep. *Psychiatry Res.* **1993**, *49*, 29–39. [CrossRef]
60. Gumenyuk, V.; Korzyukov, O.; Roth, T.; Bowyer, S.M.; Drake, C.L. Sleep extension normalizes ERP of waking auditory sensory gating in healthy habitually short sleeping individuals. *PLoS ONE* **2013**, *8*, e59007. [CrossRef] [PubMed]

antioxidants

Systematic Review

Sleep Deprivation-Induced Oxidative Stress in Rat Models: A Scoping Systematic Review

Vlad Sever Neculicioiu [1,*], Ioana Alina Colosi [1], Carmen Costache [1], Dan Alexandru Toc [1], Alexandra Sevastre-Berghian [2], Horațiu Alexandru Colosi [3,*,†] and Simona Clichici [2,†]

[1] Department of Microbiology, "Iuliu Hatieganu" University of Medicine and Pharmacy, 400349 Cluj-Napoca, Romania
[2] Department of Physiology, "Iuliu Hatieganu" University of Medicine and Pharmacy, 400006 Cluj-Napoca, Romania
[3] Division of Medical Informatics and Biostatistics, Department of Medical Education, "Iuliu Hatieganu" University of Medicine and Pharmacy, 400349 Cluj-Napoca, Romania
* Correspondence: neculicioiu.vlad.sever@elearn.umfcluj.ro (V.S.N.); hcolosi@umfcluj.ro (H.A.C.)
† These authors contributed equally to this work.

Citation: Neculicioiu, V.S.; Colosi, I.A.; Costache, C.; Toc, D.A.; Sevastre-Berghian, A.; Colosi, H.A.; Clichici, S. Sleep Deprivation-Induced Oxidative Stress in Rat Models: A Scoping Systematic Review. *Antioxidants* **2023**, *12*, 1600. https://doi.org/10.3390/antiox12081600

Academic Editor: Waldo Cerpa

Received: 15 July 2023
Revised: 1 August 2023
Accepted: 8 August 2023
Published: 11 August 2023

Abstract: Sleep deprivation is highly prevalent in the modern world, possibly reaching epidemic proportions. While multiple theories regarding the roles of sleep exist (inactivity, energy conservation, restoration, brain plasticity and antioxidant), multiple unknowns still remain regarding the proposed antioxidant roles of sleep. The existing experimental evidence is often contradicting, with studies pointing both toward and against the presence of oxidative stress after sleep deprivation. The main goals of this review were to analyze the existing experimental data regarding the relationship between sleep deprivation and oxidative stress, to attempt to further clarify multiple aspects surrounding this relationship and to identify current knowledge gaps. Systematic searches were conducted in three major online databases for experimental studies performed on rat models with oxidative stress measurements, published between 2015 and 2022. A total of 54 studies were included in the review. Most results seem to point to changes in oxidative stress parameters after sleep deprivation, further suggesting an antioxidant role of sleep. Alterations in these parameters were observed in both paradoxical and total sleep deprivation protocols and in multiple rat strains. Furthermore, the effects of sleep deprivation seem to extend beyond the central nervous system, affecting multiple other body sites in the periphery. Sleep recovery seems to be characterized by an increased variability, with the presence of both normalizations in some parameters and long-lasting changes after sleep deprivation. Surprisingly, most studies revealed the presence of a stress response following sleep deprivation. However, the origin and the impact of the stress response during sleep deprivation remain somewhat unclear. While a definitive exclusion of the influence of the sleep deprivation protocol on the stress response is not possible, the available data seem to suggest that the observed stress response may be determined by sleep deprivation itself as opposed to the experimental conditions. Due to this fact, the observed oxidative changes could be attributed directly to sleep deprivation.

Keywords: sleep deprivation; sleep; oxidative stress; stress; glutathione; GSH; GSSG; catalase; CAT; superoxide dismutase; SOD; nitric oxide; NOx; lipid peroxidation; MDA; rat

1. Introduction

Sleep is a highly conserved fundamental aspect observed across the animal kingdom, playing a critical role in maintaining homeostasis. Sleep deprivation (SD) refers mainly to quantitative or qualitative alterations of normal sleep. SD seems to be intrinsically linked to the modern technological world and lifestyle. Decreases in sleep duration or quality can be seen in all age groups, with significant variations determined by a multitude of factors [1]. While SD is certainly common, there is no current agreement on the extent of this phenomenon to epidemic proportions [2]. In humans, SD determines a multitude of negative health effects ranging from daytime sleepiness to an association with multiple chronic conditions such as obesity, diabetes, mental health disorders, cardiovascular disease and neurodegenerative disorders [2–4].

Sleep is a universal phenomenon in most animals. Even though the first experimental evidence regarding the importance of sleep dates back to the 19th century [5], multiple unknowns still persist regarding the nature of sleep, particularly regarding its underlying functions. While several theories attempt to explain the roles of sleep (such as the inactivity theory, energy conservation theory, restoration theory and brain plasticity theory, among others), it is commonly agreed that the roles of sleep are best explained through an integration of these theories [6]. Furthermore, an emerging perspective describes sleep as a consequence of metaregulation. Through this lens, the beneficial role of sleep can be explained by considering sleep as a state of adaptive inactivity or as a default state of cerebral networks [7].

A further theory regarding the role of sleep was first proposed by Reimund in 1994, hypothesizing that sleep may play an antioxidant role in the brain and peripheric organs [8]. Initially, this theory was disregarded due to the frequently contradictory evidence in the literature regarding animal models of SD and oxidative stress. Nevertheless, recent findings appear to validate the antioxidant functions of sleep and the occurrence of oxidative stress after SD in both paradoxical sleep deprivation (PSD/REM sleep) and total sleep deprivation (TSD) protocols in rodent models [9]. Additionally, a multifaceted reciprocal association between sleep and oxidative stress has been proposed and seems to be supported by recent results: sleep may act as a protective mechanism against oxidative stress, while at the same time, a certain level of oxidative stress is required to initiate sleep [9,10]. While the behavioral effects of sleep deprivation (such as low mood, anxiety, memory alterations etc.) have been largely characterized before, numerous unknowns remain regarding the molecular changes underlying these effects. Furthermore, it is believed that the molecular and immune alterations resulting from SD, such as oxidative stress and inflammation, may serve as driving factors for the development of chronic pathologies linked to insufficient sleep [4,11].

Oxidative stress can be defined as "an imbalance between oxidants and antioxidants in favor of the oxidants, leading to a disruption of redox signaling and control and/or molecular damage" [12]. Reactive oxygen species (ROS) span a spectrum of roles depending on their concentration. At low levels (oxidative eustress), ROS encompass crucial functions in multiple physiological processes such as redox signaling, cell death, and immune function. However, at higher concentrations, ROS can be damage-inducing (oxidative distress) [13,14]. Overall, redox homeostasis is maintained through multiple oxidative stress responses following activation by multiple molecular redox switches [13]. The imbalance between oxidants and antioxidants observed in oxidative stress can occur due to various factors, including increased reactive oxygen species (ROS) production, antioxidant enzyme inactivation or increased antioxidant consumption [14]. The excess ROS formed during oxidative stress (superoxide anion, hydrogen peroxide, hydroxyl radical, peroxyl radicals etc.) is counterbalanced by both enzymatic (Superoxide dismutase, Catalase, Glutathione peroxidase and transferase) and nonenzymatic antioxidants (Glutathione, vitamins, etc.). The roles and determination methods of the major antioxidants are relatively well-known and have been previously reviewed [12,14,15]. The unbalanced increased oxidant load can lead to an accumulation of damage, activating multiple programs such as autophagy, mitophagy, apoptosis, necroptosis, and ferroptosis [13]. Furthermore, increased ROS production may determine changes in DNA structure, induce alterations in lipids and proteins, trigger the activation of transcription factors, impact the synthesis of inflammatory cytokines and influence signal transduction pathways [15]. Oxidative damage to lipids generates a wide range of toxic compounds, ranging from lipid hydroperoxides to Malonaldehyde (MDA) and 4-Hydroxynonenal (4-HNE) [16]. These products can further induce protein and DNA damage by cross-linking or denaturation [15,17]. Nitric oxide (NO) exhibits intricate physiological and pathological functions ranging from signaling, inflammation, vascular regulation, regulation of sleep–wake cycles and oxidative damage. NO synthesis is mediated by three enzyme isoforms, each possessing specific localizations and distinct roles: neuronal nitric oxide synthase (nNOS), inducible nitric oxide synthase (iNOS) and endothelial nitric oxide synthase (eNOS) [18]. NO plays an intricate role in relation to oxidative stress. While increased ROS can decrease NO availability through NO scavenging, especially in the cardiovascular system [19], dysregulated nitric oxide can be implicated in oxidative/nitrosative stress, which could further induce oxidative damage in proteins, lipids and DNA [20,21].

A previous systematic review on the relationship between sleep deprivation and oxidative stress has been published by Villafuerte et al. [9], presenting evidence for the antioxidant role of sleep in both brain and non-brain areas. However, multiple unknowns remain regarding this relationship. We performed a scoping review in order to systematically identify and evaluate the newly available research conducted on this topic. The main goals of this review were to evaluate and attempt to clarify the relationship between sleep deprivation and oxidative stress in experimental studies performed on rat models, to provide a timeline of oxidative changes determined by sleep deprivation and to identify and address any remaining knowledge gaps regarding this relationship.

2. Materials and Methods

The methods used in this review were designed and based on the 22-item Preferred Reporting Items for Systematic Reviews and Meta-Analyses extension for Scoping Reviews (PRISMA-ScR) [22].

A scoping systematic review of the literature was performed regarding oxidative changes in experimental sleep-deprived rat models. The searches were performed between 21 and 23 February 2023 in three databases PubMed, Web of Science and Cochrane Library, for articles published between 1 January 2015 and 1 October 2022. The following simplified search algorithm was adapted for each database: (sleep OR sleep deprivation) AND (oxidative stress) AND (rat) AND (publication date 1 January 2015–1 October 2022) NOT (sleep apnea OR neurodegenerative disease). Search results were imported into Endnote 20 (IBM–Clarivate) for management. Title and Abstract screening and full-text evaluation were performed by two researchers (V.S.N. and I.A.C.), and discrepancies were resolved by consensus and oversight from C.C., H.A.C. and S.C. A data charting form was initially piloted and developed by two authors (V.S.N. and I.A.C.) and updated as needed. The following data were extracted from each included study: sleep deprivation protocol and duration, type of control group, subject characteristics (rat breed, age and sex), serum cortisol/corticosterone, oxidative stress parameters (Reduced Glutathione–GSH, Oxidized Glutathione–GSSG, GSH/GSSG ratio, Catalase–CAT, Superoxide dismutase–SOD, Nitric Oxide and Nitric Oxide enzymes), Lipid peroxidation. Data extraction was performed independently by two authors (V.S.N. and I.A.C.).

Comparisons were sought between a control group not exposed to sleep deprivation and a group that was sleep deprived without any other intervention or drug administration. We included experimental sleep deprivation studies performed on rats, using any type of sleep deprivation protocol that included at least one of the previously mentioned oxidative stress parameters. All measurement types were included (enzyme activity, ELISA, Western Blot, Immunohistochemistry, Immunofluorescence, gene expression, etc.).

No automation tools were employed during the search, screening, full-text evaluation or data extraction steps of the review. A limited methodological and bias evaluation was performed.

The included studies were grouped by type of SD protocol (PSD or TSD), by anatomical region in which the oxidative stress measurements were performed and by the length of the SD duration. In some cases, studies that performed oxidative stress measurements in multiple body sites were presented in multiple tables according to the previously detailed grouping.

A comprehensive overview of the methods can be found in Supplementary File S1. A flowchart presenting the study inclusion process is presented in Figure 1.

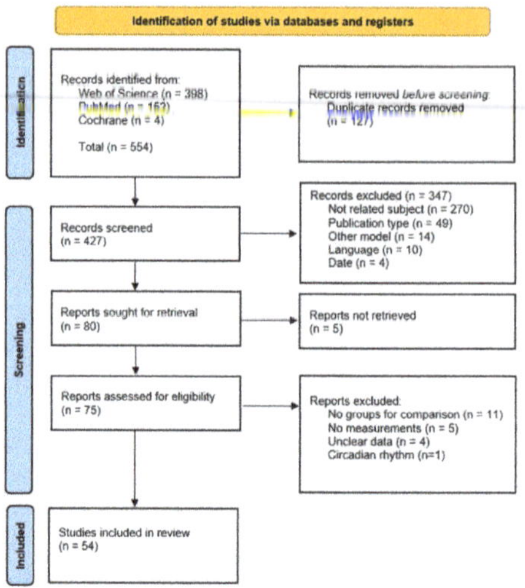

Figure 1. Study inclusion flowchart [23].

3. Results

A total of 54 studies were included in the qualitative analysis. Multiple sleep deprivation protocols were used, such as MSP (multiple small platforms), CP (classical platform/inverted flowerpot), DOW (Disk over water), GH (Gentle handling) and ASD (automated sleep deprivation). Comprehensive descriptions of the mentioned sleep deprivation protocols and respective control groups have been previously reviewed [9,24,25].

Most of the included studies employed paradoxical sleep deprivation (PSD) protocols (87%, $n = 47$), while the others used total sleep deprivation (TSD) protocols (13%, $n = 7$).

The studies that evaluated PSD employed the following SD protocols: Multiple small platforms (68.5%, $n = 37/54$) and Classical Platform/Inverted flowerpot (18.5%, $n = 10/54$).

The studies that evaluated TSD employed the following SD protocols: Disk over water (5.5%, $n = 3$), Gentle handling (3.7%, $n = 2$) and Automated sleep deprivation (3.7%, $n = 2$).

The main results of the review are presented in Tables 1–4 for paradoxical sleep deprivation in the hippocampus, cortex and other brain areas, serum, other non-brain areas, respectively; Tables 5 and 6 for total sleep deprivation in brain and non-brain areas and Table 7 for gene expression in both PSD and TSD protocols. The full-length tables are available in Supplementary File S1.

Overall, only a limited number of studies (11.1%, $n = 6/54$) reported a lack of significant changes in any of the evaluated parameters compared to their respective control group [26–31]. Conversely, the overwhelming majority of results point to either increases or decreases in at least one parameter associated with oxidative stress. Remarkably, a notable pattern seems to emerge across a majority of studies whereby the examined parameters exhibit relatively consistent changes, with either a lack of changes, reductions in some parameters such as antioxidants/antioxidant enzymes (GSH, GSH/GSSG ratio, GPx, SOD, CAT) or increases in others (GSSG and lipid peroxidation). Only three of the examined studies seem to present contradicting results, with increases in the antioxidant GSH [32] or in the antioxidant enzyme SOD [33,34].

The stress response was evaluated in a limited number of studies through the measurement of serum Cortisol/Corticosterone (18.5%, $n = 10/54$), almost exclusively in PSD protocols employing the MSP paradigm. In most of these studies, serum Cortisol/Corticosterone was significantly increased in the sleep deprivation group compared to the control group in both continuous and sleep restriction protocols [34–41]. Only two studies reported no significant changes in this parameter [42,43]. A comprehensive comparison between studies that determined serum stress hormones is available in Table 8.

Table 1. Paradoxical sleep deprivation in the hippocampus.

Reference	SD Protocol	SD Duration	Rat Breed, Sex, Age	Oxidative Stress Marker	Results
[27]	MSP	48 h	Wistar, Male, PND 28	LP	LP (-)
[44]	MSP	48 h	Wistar, Male, Adult	GSH, NOx, LP	GSH ($\downarrow$), NOx ($\uparrow$), LP (-)
[26]	MSP	72 h	Wistar, Male, Adult	GSH, LP	GSH (-), LP (-)
[45]	MSP	72 h	Wistar, Male	LP	LP ($\uparrow$)
[46]	MSP	72 h	Wistar, Male, Adult	GSH, SOD, LP	GSH ($\downarrow$), SOD ($\downarrow$), LP ($\uparrow$)
[47]	MSP	72 h	Wistar, Male, Adult	LP	LP ($\uparrow$)
[31]	MSP	24, 48, 72 h	Wistar, Male, Adult	GSH, GPx, CAT, SOD, LP	GSH (-), GPx (-), CAT (-), SOD (-), LP (-)
[48]	CP	4 days	Wistar, Male, 5 weeks	LP	LP ($\uparrow$) *
[49]	MSP	5 days	Sprague Dawley, Male and Female, 6 weeks	GSH/GSSG ratio, GPX, CAT, SOD, LP	GSH/GSSG ratio ($\downarrow$), GPX (-), CAT ($\downarrow$), SOD ($\downarrow$), LP ($\uparrow$)
[38]	MSP	5 days	Sprague Dawley, Male, 48 weeks	NOx	NOx ($\uparrow$)
[50]	CP	6 days (48 h SD, 48 h Srec, 48 h SD, 48 h Srec, 48 h SD)	Wistar, Male, Adult	CAT, LP	CAT (-), LP ($\uparrow$)
[43]	MSP	21 days (18 h/day)	Sprague Dawley, Male, 12–16 weeks	GSH/GSSG ratio, LP	GSH/GSSG ratio ($\downarrow$), LP ($\uparrow$)
[43]	MSP	21 days (18 h/day) + 5 days Srec	Sprague Dawley, Male, 12–16 weeks	GSH/GSSG ratio, LP	GSH/GSSG ratio ($\downarrow$), LP ($\uparrow$)
[43]	MSP	21 days (18 h/day) + 21 days Srec	Sprague Dawley, Male, 12–16 weeks	GSH/GSSG ratio, LP	GSH/GSSG ratio ($\downarrow$), LP (-)
[51]	MSP	4 weeks (8 h/day)	Wistar, Male, Adult	GSH, GSSG, GSH/GSSG ratio, GPx, CAT	GSH (-), GSSG ($\uparrow$), GSH/GSSG ratio ($\downarrow$), GPx ($\downarrow$), CAT ($\downarrow$)
[52]	MSP	4 weeks (8 h/day)	Wistar, Male, Adult	GSH, GSSG, GSH/GSSG ratio, GPx, CAT	GSH (-), GSSG ($\uparrow$), GSH/GSSG ratio ($\downarrow$), GPx ($\downarrow$), CAT ($\downarrow$)
[53]	MSP	4 weeks (8 h/day)	Wistar, Male, Adult	GSH, GSSG, GSH/GSSG ratio, GPx, CAT, SOD, LP	GSH (-), GSSG ($\uparrow$), GSH/GSSG ratio ($\downarrow$), GPx ($\downarrow$), CAT ($\downarrow$), SOD ($\downarrow$), LP (-)
[54]	MSP	6 weeks (8 h/day)	Wistar, Male, 8–10 weeks	GSH, GSSG, GSH/GSSG ratio, GPx, CAT, SOD, LP	GSH (-), GSSG ($\uparrow$), GSH/GSSG ratio ($\downarrow$), GPx ($\downarrow$), CAT ($\downarrow$), SOD ($\downarrow$), LP (-)
[55]	MSP	6 weeks (8 h/day)	Wistar, Male, Adult	GSH, GSSG, GSH/GSSG ratio, GPx, CAT, SOD, LP	GSH (-), GSSG ($\uparrow$), GSH/GSSG ratio ($\downarrow$), GPx ($\downarrow$), CAT ($\downarrow$), SOD ($\downarrow$), LP (-)
[56]	MSP	6 weeks (8 h/day)	Wistar, Male, Adult	GSH, GSSG, GSH/GSSG ratio, GPx, CAT, SOD	GSH (-), GSSG ($\uparrow$), GSH/GSSG ratio ($\downarrow$), GPx ($\downarrow$), CAT ($\downarrow$), SOD ($\downarrow$)
[57]	MSP	6 weeks (8 h/day)	Wistar, Male, Adult	GSH, GSSG, GSH/GSSG ratio, CAT, SOD, LP	GSH (-), GSSG ($\uparrow$), GSH/GSSG ratio ($\downarrow$), CAT ($\downarrow$), SOD ($\downarrow$), LP (-)
[58]	MSP	8 weeks (8 h/day)	Wistar, Male, Adult	GSH, GSSG, GSH/GSSG ratio, GPx, CAT, SOD, LP	GSH (-), GSSG ($\uparrow$), GSH/GSSG ratio ($\downarrow$), GPx ($\downarrow$), CAT ($\downarrow$), SOD (-), LP (-)
[59]	MSP	8 weeks (8 h/day)	Wistar, Male, Adult	GSH, GSSG, GSH/GSSG ratio, GPx, CAT	GSH (-), GSSG ($\uparrow$), GSH/GSSG ratio ($\downarrow$), GPx ($\downarrow$), CAT ($\downarrow$)

SD: sleep deprivation; MSP: multiple small platforms; CP: classical platform/inverted flowerpot; Srec: sleep recovery; PND: postnatal day; GSH: Reduced Glutathione; GSSG: Oxidized Glutathione; GSH/GSSG ratio: Reduced Glutathione/Oxidized Glutathione ratio; GPx: Glutathione peroxidase; CAT: Catalase; SOD: Superoxide dismutase; NOx: Nitric oxide; LP: Lipid peroxidation; *: dentate gyrus, hippocampus; "$\uparrow$/$\downarrow$": significantly increased/decreased; "-": not significantly increased or decreased.

Table 2. Paradoxical sleep deprivation in cortex and other brain areas.

Reference	SD Protocol	SD Duration	Rat Breed, Sex, Age	Anatomical Site	Oxidative Stress Marker	Results
[28]	MSP	48 h	Wistar, Male, Adult	Cortex	GSH, NOx, LP	GSH (-), NOx (-), LP (-)
[26]	MSP	72 h	Wistar, Male, Adult	Prefrontal cortex	GSH, LP	GSH (-), LP (-)
[45]	MSP	72 h	Wistar, Male	Forebrain cortex	LP	LP (↑)
[47]	MSP	72 h	Wistar, Male	Forebrain cortex	LP	LP (↑)
[48]	CP	4 days	Wistar, Male, 5 weeks	Cortex	LP	LP (↑)
[60]	MSP	7 days	Wistar, Male, 8 weeks	Cortex	GSH, GPx, CAT, SOD, LP	GSH (↓), GPx (↓), CAT (↓), SOD (↓), LP (↑)
[26]	MSP	72 h	Wistar, Male, Adult	Cerebellum, Brainstem	GSH, LP	GSH (-), LP (-)
[36]	MSP	72 h	Sprague Dawley, Adult	Amygdala	SOD, LP	SOD (↓), LP (↑)
[61]	CP	72 h	Sprague Dawley, Male, 8–10 weeks	Thalamus	GSH, CAT, SOD, LP	GSH (↓), CAT (↓), SOD (↓), LP (↑)
[62]	MSP	5 days	Sprague Dawley, Male and Female, Adult	Whole brain	CAT, SOD, LP	CAT (↓), SOD (↓), LP (↑)
[63]	CP	6 days	Wistar, Male, Adult	Locus coeruleus	GSH	GSH (↓)

SD: sleep deprivation; MSP: multiple small platforms; CP: classical platform/inverted flowerpot; GSH: Reduced Glutathione; GPx: Glutathione peroxidase; CAT: Catalase; SOD: Superoxide dismutase; NOx: Nitric oxide; LP: Lipid peroxidation; "↑/↓": significantly increased/decreased; "-": not significantly increased or decreased.

Table 3. Paradoxical sleep deprivation in serum/plasma.

Reference	SD Protocol	SD Duration	Rat Breed, Sex, Age	Oxidative Stress Markers	Results
[39]	MSP	20 h	ns	LP	LP (↑)
[35]	MSP	24 h	Long-Evans, Male, Old	LP	LP (↑)
[64]	CP	24, 36, 48 h	Sprague Dawley, Male, 6 months	SOD, LP	SOD (↓), LP (↑)
[65]	CP	48 h	Sprague Dawley, Male, 6 weeks	SOD, LP	SOD (↓), LP (↑)
[17]	MSP	72 h	Wistar, Male, 10 weeks	NOx	NOx (↓)
[36]	MSP	72 h	Sprague Dawley, Adult	SOD, LP	SOD (↓), LP (↑)
[40]	MSP	5 days	Wistar, Male	LP	LP (↑)
[39]	MSP	5 days (20 h/day) 5 days (20 h/day) + 5 days Srec	ns	LP	LP (↑) LP (-)
[50]	CP	6 days (48 h SD, 48 h Srec, 48 h SD, 48 h Srec, 48 h SD)	Wistar, Male, Adult	CAT, LP	CAT (-), LP (-)
[66]	MSP	7 days (20 h/day) and 7 days continuous	Wistar, Male	CAT, SOD, LP	CAT (-/↓), SOD (-), LP (↑)
[67]	MSP	7 days	Wistar, Male	GPx, SOD	GPx (↓), SOD (↓)
[30]	MSP	21 days (18 h/day)	Wistar, Male, Adult	eNOS	eNOS (-)

SD: sleep deprivation; MSP: multiple small platforms; CP: classical platform/inverted flowerpot; Srec: sleep recovery; GPx: Glutathione peroxidase; CAT: Catalase; SOD: Superoxide dismutase; NOx: Nitric oxide; eNOS: Endothelial Nitric Oxide Synthase; LP: Lipid peroxidation; ns: not specified; "↑/↓": significantly increased/decreased; "-": not significantly increased or decreased.

Table 4. Paradoxical sleep deprivation in other non-brain sites.

Reference	SD Protocol	SD Duration	Rat Breed, Sex, Age	Anatomical Site	Oxidative Stress Markers	Results
[17]	MSP	72 h	Wistar, Male, 10 weeks	Testes, Epididymis	GSH, GPx, CAT, SOD, LP	GSH (↓), GPx (↓), CAT (↓), SOD (↓), LP (↑)
[40]	MSP	5 days	Wistar, Male	Testes	GSH, GPx, LP	GSH (↓), GPx (↓), LP (↑)
[34]	MSP	14 days (20 h/day)	Sprague Dawley, Male, 12 weeks	Testes	GSH, CAT, SOD, LP	GSH (↓), CAT (↓), SOD (↑), LP (↑)
[41]	MSP	21 days (18 h/day)	Wistar, Male, Peripubertal	Testes	GSH, GSSG, GSH/GSSG ratio, LP	GSH (-), GSSG (-), GSH/GSSG ratio (-), LP (↑)
[32]	MSP	21 days (18 h/day)	Wistar, Male, Peripubertal	Epididymis caput, cauda	GSH, LP	GSH (↑), LP (↑)
[33]	MSP	4 days 8 days	Wistar, Male, 3 months	Liver, Pancreas	CAT, SOD, LP	CAT (-), SOD (-/↑), LP (-/↑) CAT (↓/-), SOD (↓/-), LP (↑/-) CAT (-), SOD (↓/-), LP (↑)
		8 days + 20 days Srec	Wistar, Male, 14 months	Liver, Pancreas	CAT, SOD, LP	CAT (↓/-), SOD (-), LP (-) CAT (↓/-), SOD (-/↑), LP (-/↑) CAT (-), SOD (-/↑), LP (-)
[68]	MSP	21 days (14 h/day)	Wistar, Male, Adult	Liver	SOD, LP	SOD (↓), LP (↑)
[69]	CP	21 days (18 h/day) 21 days (22 h/day)	Wistar, Male, Adult	Liver	GPx, SOD, LP	GPx (-), SOD (-), LP (-) GPx (-), SOD (↓), LP (↑)
[70]	CP	72 h 72 h + 72 h Srec	Sprague Dawley, Male, 8–10 weeks	Aorta	SOD, LP	SOD (↓), LP (↑) SOD (-), LP (-)
[71]	CP	5 days	Sprague Dawley, Male, 24 weeks	Aorta	NOx, eNOS, p-eNOS	NOx (↓), eNOS (-), p-eNOS (↓)
[66]	MSP	7 days (20 h/day) and 7 days continuous	Wistar, Male	Saliva Submandibular	CAT, SOD, LP	CAT (-), SOD (-), LP (-) CAT (-), SOD (↓), LP (-/↑)
[72]	CP	21 days (18 h/day)	Wistar, Male	Thyroid	LP	LP (↑)
[45]	MSP	72 h	Wistar, Male	Kidney Erythrocytes	LP	LP (-) LP (↑)
[37]	MSP	4 days	Wistar, Male, 3 months	Plantar muscle Soleus muscle	LP	LP (-) LP (↑)

SD: sleep deprivation; MSP: multiple small platforms; CP: classical platform/inverted flowerpot; Srec: sleep recovery; GSH: Reduced Glutathione; GSSG: Oxidized Glutathione; GPx: Glutathione peroxidase; CAT: Catalase; SOD: Superoxide dismutase; NOx: Nitric oxide; eNOS: Endothelial Nitric Oxide Synthase; p-eNOS: phosphorylated Endothelial Nitric Oxide Synthase; LP: Lipid peroxidation; "↑/↓": significantly increased/decreased; "-": not significantly increased or decreased.

Table 5. Total sleep deprivation in brain areas.

Reference	SD Protocol	SD Duration	Rat Breed, Sex, Age	Anatomical Site	Oxidative Stress Markers	Results
[29]	GH	6 h	Wistar, Male, 10 weeks	Hippocampus	LP	LP (-)
[42]	GH	12 h	Wistar, Female, 13–15 months	Hypothalamus	nNOS	nNOS (↓)
[73]	ASD	14 days	Wistar, Male	Cortex and Hippocampus	GSH, CAT, SOD, LP	GSH (↓), CAT (↓), SOD (↓), LP (↑)
[74]	DOW	5 days SD + 2 days Srec (3 total cycles) + 3 months Srec	Wistar, Male, Weanling	Hippocampus	GPx, CAT, SOD	GPx (↓), CAT (↓), SOD (↓/-)

SD: sleep deprivation; Srec: sleep recovery; GH: Gentle handling; ASD: Automated sleep deprivation; DOW: Disk over water; GSH: Reduced Glutathione; GPx: Glutathione peroxidase; CAT: Catalase; SOD: Superoxide dismutase; nNOS: neuronal Nitric Oxide Synthase; LP: lipid peroxidation; "↑/↓": significantly increased/decreased; "-": not significantly increased or decreased.

Table 6. Total sleep deprivation in non-brain areas.

Reference	SD Protocol	SD Duration	Rat Breed, Sex, Age	Anatomical Site	Oxidative Stress Markers	Results
[75]	DOW	5 days	Wistar, Male	Liver	GPx, CAT, SOD, LP	GPx ($\downarrow$), CAT ($\downarrow$), SOD ($\downarrow$), LP ($\uparrow$)
[76]	DOW	5 days SD + 2 days Srec (3 total cycles)	Wistar, Male, Adult	Liver	GPx, CAT, SOD, LP	GPx ($\downarrow$), CAT ($\downarrow$), SOD ($\downarrow$), LP ($\uparrow$)
[77]	ASD	14 days 6 h/day–1st week 8 h/day–2nd week	Sprague Dawley, Male, PND 19	Plasma	LP	PND 33: LP ($\uparrow$) PND 90: LP (-)

SD: sleep deprivation; Srec: sleep recovery; ASD: Automated sleep deprivation; DOW: Disk over water; PND: Postnatal day; GPx: Glutathione peroxidase; CAT: Catalase; SOD: Superoxide dismutase; LP: Lipid peroxidation; "$\uparrow$/$\downarrow$": significantly increased/decreased; "-": not significantly increased or decreased.

Table 7. Gene expression in paradoxical and total sleep deprivation.

Reference	SD Protocol	SD Duration	Rat Breed, Sex, Age	Anatomical Site	Oxidative Stress Measurements	Results
[78]	PSD–MSP	96 h 21 days (18 h/day)	Wistar-Hannover, Male, Adult	Testes	iNOS, eNOS	iNOS ($\uparrow$), eNOS ($\downarrow$) iNOS ($\uparrow$), eNOS (-)
[77]	TSD–ASD	14 days 6 h/day–1st week 8 h/day–2nd week	Sprague Dawley, Male, PND 19	Prefrontal cortex	GPx, CAT, SOD	PND 33: GPx ($\uparrow$), CAT (-), SOD ($\uparrow$) PND 90: GPx (-), CAT (-), SOD (-)

SD: sleep deprivation; PSD: paradoxical sleep deprivation; TSD: total sleep deprivation; MSP: multiple small platforms; ASD: automated sleep deprivation; PND: postnatal day; GPx: Glutathione peroxidase; CAT: Catalase; SOD: Superoxide dismutase; eNOS: Endothelial Nitric Oxide Synthase; iNOS: Inducible nitric oxide synthase; "$\uparrow$/$\downarrow$": significantly increased/decreased; "-": not significantly increased or decreased.

Table 8. Serum cortisol/corticosterone changes in sleep deprivation.

Reference	SD Type	SD Protocol	SD Duration	Rat Breed, Sex, Age	Cortisol/Corticosterone
[42]	TSD	GH	12 h	Wistar, Female, 13–15 months	Yes -
[35]	PSD	MSP	24 h	Long-Evans, Male, Old	Yes $\uparrow$
[36]	PSD	MSP	72 h	Sprague Dawley, Adult	Yes $\uparrow$
[37]	PSD	MSP	4 days	Wistar, Male, 3 months	Yes $\uparrow$
[38]	PSD	MSP	5 days	Sprague Dawley, Male, 48 weeks	Yes $\uparrow$
[39]	PSD	MSP	20 h 5 days (20 h/day) 5 days (20 h/day) + 5 days Srec	ns	Yes $\uparrow$ Yes $\uparrow$ Yes -
[40]	PSD	MSP	5 days	Wistar, Male	Yes $\uparrow$
[34]	PSD	MSP	14 days (20 h/day)	Sprague Dawley, Male, 12 weeks	Yes $\uparrow$
[41]	PSD	MSP	21 days (18 h/day)	Wistar, Male, Peripubertal	Yes $\uparrow$
[43]	PSD	MSP	21 days (18 h/day) 21 days (18 h/day) + 5 days Srec 21 days (18 h/day) + 21 days Srec	Sprague Dawley, Male, 12–16 weeks	Yes -

SD: Sleep deprivation; TSD: Total sleep deprivation; PSD: Paradoxical sleep deprivation; Srec: Sleep recovery; GH: Gentle handling; MSP: multiple small platforms; ns: not specified; "$\uparrow$": significantly increased; "-": not significantly increased or decreased.

4. Discussion

4.1. Sleep Deprivation Determines Changes in Oxidative Stress Parameters

With a few exceptions [26–31], most studies included in our review revealed changes in at least one parameter associated with oxidative stress in both PSD and TSD protocols. Even when taking only lipid peroxidation into consideration, with some exceptions [26–29,31,44,53–55,57,58], most studies that determined this parameter revealed elevated levels of lipid peroxidation in both PSD and TSD protocols. Moreover, while some of the results were not interpreted in the context of oxidative stress, most of the observed alterations in these parameters were interpreted as or implied to be evidence of oxidative stress by the original authors. The absence of a specific protocol to selectively eliminate short-wave sleep (SWS) limits our ability to determine its potential antioxidant role based on these results. However, these findings provide additional evidence supporting the antioxidant role of paradoxical sleep and sleep as a whole.

As previously observed [9], the evidence from the included studies seems to point to changes in oxidative stress parameters in both brain (whole brain, hippocampus, cortex, thalamus, hypothalamus, amygdala, locus coeruleus) and non-brain areas (serum, testes, epididymis, liver, pancreas, aorta, submandibular glands, thyroid, erythrocytes, soleus muscle). However, some body sites did not show changes in oxidative stress parameters: cerebellum and brainstem [26], saliva [66], kidney [45], and plantar muscle [37]. While the evidence is limited in some cases, these observations might suggest that the antioxidant effects of sleep extend beyond the central nervous system, possibly including most body sites.

Although alterations in oxidative stress parameters have been noted across various body sites, not all sites might exhibit equal resilience to oxidative stress. Interestingly, short SD durations (24–72 h or 48–72 h) might not induce reliable oxidative changes in the hippocampus [26,27,31,44–47] or cortex [26,28,45,47], as opposed to longer SD protocols. In contrast, changes in oxidative stress parameters seem to be reliably encountered in serum in both short and longer SD durations.

The included studies used multiple rat strains, namely Wistar (39), Sprague Dawley (12), Long-Evans (1), and Wistar-Hannover (1). Given that the majority of these studies employed Wistar rats, strain differences might not be readily apparent. While strain differences might exist and account for some degree of variability in the oxidative stress response to SD [9], the current data seem to indicate the presence of an oxidative stress response after SD across all mentioned rat strains and in both males and females in the case of Wistar [42] and Sprague Dawley rats [49,62].

It has been previously proposed that short SD durations might induce an antioxidant compensatory response in multiple body sites [9,11], evidenced by increases in antioxidants or decreases in oxidants. Contrary to these findings, we did not observe this effect in the studies included in our review that employed short (6–24 h) SD protocols [29,31,35,39,42,64]. Based on these data, we cannot exclude the presence of some compensatory mechanisms in short SD protocols. However, even if such mechanisms are present, their overall effect might not be sufficient to counterbalance the global redox state in the examined organs [79].

4.2. Sleep Recovery

Sleep recovery plays a significant role in reversing or at least partially mitigating the effects caused by SD. Somewhat simplified, the sleep recovery time seems to be dependent on the extent of the damage caused by the previous period of SD [43].

Some of the studies included in our review contained sleep recovery periods at the end of the SD period in both PSD [27,33,39,43,70] and TSD [74,76,77] protocols of varying lengths. The presented results are mostly conflicting. While some studies revealed normalizations in multiple parameters related to oxidative stress [39,70], others reported partial recoveries depending on the examined parameter [33] or on the length of the recovery period [43,77]. Lasting changes after sleep recovery were also reported [74,76], as well as the absence of changes after sleep recovery [27].

A comparison can somewhat be drawn between the studies that revealed normalizations in oxidative stress parameters [39,70] and studies that reported lasting changes after SD [74,76]. While the available evidence is limited, the results seem to suggest that longer SD periods would induce longer-lasting changes as compared to shorter SD periods. This interpretation would be consistent with the idea that longer SD periods induce increased amounts of ROS and subsequently increased cellular and tissue damage. While age might play a role in the observed results [74], it may be the case that the examined body sites present different sleep recovery dynamics.

Most of the studies included in our review that measured serum cortisol/corticosterone revealed increases in these parameters, at different SD durations, during both continuous SD and sleep restriction protocols. Of note, most of these studies employed PSD protocols, and only one study used a TSD protocol [42]. Only two of the included studies reported insignificant changes in the levels of serum cortisol/corticosterone [42,43]. Bajaj et al. [42] observed increases in serum cortisol in the sleep-deprived group, although these changes did not reach statistical significance. This fact can be at least partially explained by the short SD duration of 12 h, exclusive inclusion of female animal groups or the gentle handling protocol used. However, it is worth noting that the results presented by Konakanchi et al. [43] are in contradiction to those reported by other authors.

Given the fact that only one of the included studies employed a TSD protocol with a short SD duration [42], we believe that a comparison between the stress-inducing effects of PSD and TSD protocols is not possible based on these findings.

Only two of the included studies evaluated the effects of sleep recovery at the end of the SD period on stress hormones [39,43]. Even though limited, the results seem to indicate that 5 days of sleep recovery are sufficient to normalize serum corticosterone levels after 5 days of SD [39]. Based on previous evidence, it appears that the normalization of the increased corticosterone/ACTH levels induced by SD might occur at an even faster rate after 24 h [93] or even after 4 h [94], depending on the experimental conditions such as the length of the previous sleep deprivation paradigm. Nevertheless, SD may potentially still present longer-lasting effects on multiple hormones and neurotransmitters [93], including the stress response, by influencing the reaction to a subsequent stressor [94,95].

It is currently believed that the stress response observed in rodent models following SD is likely determined by a combination of stressors originating both directly from SD and those induced by the SD protocol itself. While the extent and differences between SD protocols in inducing stress are controversial, multiple studies seem to suggest that the most commonly used SD protocols induce a variable stress response in rodents [96–98]. In certain SD protocols, such as the MSP or CP paradigms, it is possible to at least partially assess the source and possible influence of the stress response by introducing a second control group placed on wide platforms or a steel grid that allow for undisturbed sleep. Although the utilization of a large platform group is known to have certain limitations, such as disrupted normal sleep patterns [99] and even a potential stress response [96], it provides additional insights for understanding the stress response during SD. In our review, less than half of the studies that used the MSP or CP paradigms employed the use of a second wide platform/grid control group (40.4%, $n = 19/47$) [26,36,38,43,48,51–61,69,70,72]. Interestingly, with the exception of one study [60], all studies consistently indicated the absence of significant differences in the assessed oxidative stress parameters between the cage control and the wide platform/grid control group. Furthermore, in three of the aforementioned studies, serum cortisol/corticosterone levels were also assessed, indicating no statistically significant differences between the control groups [36,38,43].

Overall, the presence of a stress response after SD is well-known (see reviews [11,98]). Despite the elevated levels of stress hormones observed in the majority of studies included in the analysis, the precise underlying cause of this response still remains somewhat unclear. Previous research suggests that stress determined by SD may exhibit distinct characteristics when compared to other forms of stress. These include a gradual rise in corticosterone, a lack of habituation in the HPA axis, a diminished response in ACTH and the absence of increases in adrenal gland size [94].

Most of the results presented in our review indicate elevated corticosterone levels after SD, coupled with a lack of notable differences between multiple control groups in oxidative stress parameters and stress hormones. While limited to the presented data, these results might suggest that sleep deprivation may itself act as the primary determinant of the observed stress response rather than the specific experimental protocol. If this is the case, the observed changes in oxidative stress parameters are most likely a direct consequence of sleep deprivation and not determined by the stress of the experimental protocol.

To further complicate matters, stress might play a physiological role in the induction of sleep, potentially facilitating the occurrence of sleep rebound subsequent to exposure to a stressor. This process is presumed to play an evolutionary role in mitigating the effects associated with stress exposure [100]. Furthermore, the stress response determined by SD may be physiological in nature, playing a crucial role in facilitating REM sleep rebound after SD [101].

The timing of measurement represents another crucial aspect to consider when examining the stress response during SD. As previously reviewed [98], the measurement of stress hormones at the end of the SD protocol may introduce a potential source of bias, as it could mask transient increases in stress hormone levels that occur at the beginning of the procedure.

4.5. Mechanisms of Sleep Deprivation-Induced Oxidative Stress

The literature on experimental SD frequently exhibits ambiguity, as multiple studies demonstrate the absence of oxidative stress following sleep deprivation [26–31,102,103]. This fact may likely be attributed primarily to the overall variability in experimental conditions. However, most evidence from a previous systematic review [9] and from the current review seems to point towards the fact that sleep deprivation represents an oxidative challenge.

The precise mechanism through which sleep deprivation induces oxidative stress remains incompletely understood. However, from a conceptual standpoint, heightened levels of ROS could arise as a result of increased production, reduced clearance, or a combination of both mechanisms. Moreover, sleep deprivation might give rise to distinct adverse conditions that determine the accumulation of ROS [104]. If oxidative stress is indeed an outcome of sleep deprivation, it is likely attributable to a complex interplay of various site- and possibly species-specific mechanisms, including but not limited to: increased glucose consumption during prolonged waking, stress hormones induced oxidative stress, increased metabolism and mitochondrial dysfunction, endoplasmic reticulum dysfunction, and gut dysbiosis linked to Nox enzyme (NADPH oxidase) induced ROS hyperproduction in the gut.

Physiological sleep is distinguished by a notable reduction in whole-body energy expenditure, with slow-wave sleep, in particular, exhibiting a decrease of approximately 15–35% [105]. Moreover, the diminished neuron activity and metabolism observed during sleep contribute to a reduction in brain glucose utilization [105–107]. As a result, sleep deprivation and prolonged wakefulness might lead to heightened metabolism, increased glucose consumption and increased production of ROS [43,104].

As previously shown, the activation of the HPA axis and the consequent release of stress hormones may induce oxidative stress in various body sites, including the central nervous system and the periphery [87–92]. Regardless of its specific origin, the observed stress response during sleep deprivation may, to some extent, contribute to the generation of ROS in multiple body sites.

As previously reviewed, mitochondrial dysfunction, endoplasmic reticulum dysfunction and gut dysbiosis might be other mechanisms through which SD generates oxidative stress [104]. The heightened metabolism observed during SD translates into increased mitochondrial oxygen-dependent ATP synthesis, consequently leading to an elevated production of ROS. The accumulation of misfolded or unfolded proteins observed during SD can induce oxidative stress by imposing an increased demand for protein folding processes. Moreover, the observed gut dysbiosis during SD could potentially contribute to the generation of ROS by triggering the hyperactivation of the Nox enzyme, which is responsible for a significant portion of ROS production in the gut.

Additionally, certain data suggest a connection between insufficient sleep duration and heightened susceptibility to acute oxidative stress. This phenomenon could potentially be attributed to an elevated baseline accumulation of ROS, leading to an increased sensitization to acute oxidative stress [10].

4.6. Negative Health Outcomes Associated with Sleep Deprivation

There are still several unresolved questions concerning the relationship between SD and various pathologies. SD has been linked to a diverse array of both acute and chronic adverse health consequences, encompassing daytime sleepiness, mental health disorders, neurodegenerative disorders, metabolic disorders such as diabetes and obesity, as well as cardiovascular disease (CV), among other conditions [2–4,108].

From a conceptual standpoint, SD can be viewed as a stressor that induces an increase in the allostatic load, thereby contributing to dysfunctions across multiple organ systems. The increase in allostatic load is likely mediated through the multiple consequences of SD and circadian disruption, such as oxidative stress, inflammation and multiple hormone dysregulations (Cortisol, Insulin, etc.) [11].

Multiple mechanistic links have been proposed in order to explain the previously mentioned SD-induced pathologies. In the CV system, SD is reported to induce endothelial dysfunction (through chronobiological disruption, oxidative stress, inflammation, and autonomic dysregulation) [108] and hypertension [109,110]. Furthermore, short sleep durations have been associated with an increased risk of metabolic syndrome [111], obesity and diabetes [112], further increasing CV risk factors. Multiple mechanisms have been proposed to explain the effects of sleep deprivation on metabolism, such as changes in energy intake (increased hunger and appetite) and possibly expenditure, hormonal changes (decreased Leptin and glucagon-like peptide-1, increased Ghrelin, altered Cortisol rhythms), inflammation and oxidative stress, sympathetic predominance (either from reduced vagal tone or increased sympathetic activity). Overall, these mechanisms may lead to Insulin resistance and altered energy metabolism (reviewed in [112]). Previous studies have shown that SD leads to an elevated food intake in both human subjects and rodent models. However, it is noteworthy that differential effects on overall weight have been observed, with rodents experiencing weight loss and humans exhibiting weight gain [113]. SD has been shown to affect various biochemical and molecular pathways found in multiple neurological or neurodegenerative disorders such as Alzheimer's disease (AD), Parkinson's disease, Multiple sclerosis, Huntington's disease and stroke [10,114]. Overall, the proposed main mechanisms through which SD determines these changes might be summarised as: decreased neuronal clearance of misfolded proteins (α-synuclein, amyloid-β, tau), impairment of the glymphatic system, neuroinflammation and oxidative stress (reviewed in [114]). Furthermore, it is important to note that oxidative stress can induce protein misfolding and aggregation, genomic instability and DNA damage. Consequently, the neuronal accumulation of ROS might be seen as a contributing element in the pathogenesis of neurodegenerative diseases [10,33]. SD has also been shown to determine the activation of astrocytes and microglia in the central nervous system, leading to neuronal injury through increased levels of proinflammatory factors [115]. SD determines alterations in multiple pathways that may lead to the pathology observed in AD, such as neuroinflammation and oxidative stress, endothelial damage, impaired glial pathway, inhibition of neurogenesis and cholinergic neurons, impairment of spatial and working memory, impairment of long-term potentiation and synaptic plasticity, decreased amyloid-β clearance and subsequent accumulation of amyloid-β and tau [114]. In addition, SD has been shown to directly negatively influence memory. While the exact mechanisms through which SD determines memory impairment are not completely known, multiple pathways have been proposed, such as oxidative stress, neuroinflammation, neuronal damage, decreased synaptic plasticity, changes in neurotransmitters and gene expression (reviewed in [116]).

5. Limitations

Our review presents several limitations determined by the design of the study, by the original studies from which the data were extracted and by the existing experimental sleep deprivation protocols.

5.1. Limitations Inherent to the Review Methodology and Results

The first limitation of our review refers to the search algorithm. While we employed systematic search strategies, the searches were performed in three major indexing databases and included studies published between 2015 and 2022. Studies published before 2015 were not included in order to avoid duplications with the previous review written by Villafuerte et al. [9]. Most of the included studies utilized PSD protocols through the MSP paradigm. All included studies utilized a limited number of rat strains as the SD model, with Wistar rats being the most frequently used. As a consequence of these factors, the findings presented in our review cannot be widely generalised to other rodent models, such as mice. Even though strain differences are not evident from these data, they have been suggested before in the context of SD and may account for at least some degree of variability in the observed results [9].

5.2. Limitations Inherent to the Original Studies

We encountered several challenges associated with data extraction due to the often limited and unclear aspects regarding the methodology of the included studies. These include the general timeline of the experiments (mainly the presence of sleep recovery periods after SD and until euthanasia), stress reduction procedures such as maintaining social hierarchy in the case of the MSP paradigm or limited human interaction and the description of the control group.

The majority of the studies included in the analysis utilized behavioral tests. However, it was not consistently indicated whether these tests were conducted during or at the end of the SD protocol. To address this potential bias, sleep recovery periods were taken into consideration only when explicitly described as such by the original studies or when presented on a clear experimental timeline.

Furthermore, we observed an increased variability in the implementation of the specific SD protocols among the included studies.

5.3. Limitations Determined by the Existing Sleep Deprivation Protocols

The studies included in our review employed a multitude of sleep deprivation protocols, such as MSP, CP, DOW, GH and ASD. Sleep deprivation research has been carried out since the 1950s, leading to the development of multiple protocols aimed at eliminating sleep. An ideal SD protocol would be able to completely abolish a desired sleep phase and would be free from any inherent bias factors associated with the procedure itself. Additionally, such a protocol should include a comparable control group not exposed to SD. It is well-known that all the currently available SD protocols suffer from some drawbacks [9,117]. Previous research has shown that the most commonly used PSD protocols could effectively eliminate or nearly eliminate REM sleep [99,118]. However, while some data seem to point out an additional partial reduction in SWS (37% SWS reduction in socially stable rats in the case of MSP and 31% SWS reduction in the case of CP) [99], other studies have reported minimal reductions in SWS through the CP paradigm [118]. Furthermore, as stated before, most SD protocols might be stress-inducing [96–98].

6. Quality and Bias Evaluation of Included Studies

A limited quality and bias evaluation was performed for all included studies by examining the following criteria: control group description, randomisation, total number of experimental animals and social hierarchy maintenance in the case of studies that employed the MSP paradigm. Such an evaluation was mainly motivated by the frequent variations in the employed SD protocols. Overall, the studies included in our analysis were of good methodological quality. Consequently, no studies were excluded based on this criterion.

All included studies contained at least one control group that was not exposed to sleep deprivation. More than two-thirds of the included studies (72.2%, $n = 39/54$) presented a detailed description of the control group, while the remaining studies specified the existence of a control group without providing a detailed description. Moreover, some studies that employed the MSP or CP paradigm included two separate control groups: cage control and wide platform/grid control (40.4%, $n = 19/47$).

More than half of the included studies specified the use of randomisation in the creation of the groups (66.6%, $n = 36/54$), but only one study offered a detailed explanation of the randomisation method.

In the majority of the included studies, the total count of experimental animals was explicitly provided or could be inferred from the sample sizes (77.7%, $n = 42/54$), whereas in a limited number of studies, it remained unclear (22.2%, $n = 12/54$).

The evaluation of social hierarchy maintenance was motivated by two factors: the frequent use of the MSP SD protocol in the included studies and the effectiveness of this procedure to further mitigate stress within the MSP paradigm [119]. Of the included studies that employed the MSP sleep deprivation protocol, only a minority (13.5%, $n = 5/37$) provided, at the very least, partial insights regarding the maintenance of social hierarchy throughout the experimental period. The remaining studies either lacked any specific information regarding the maintenance of social hierarchy or provided unclear data.

7. Conclusions

The inherent variability in sleep deprivation protocols coupled with factors such as possible sex and differences frequently adds a layer of complexity to the interpretation of data regarding sleep deprivation. As such, definitive conclusions are difficult to draw regarding the multifaceted relationship between sleep deprivation and oxidative stress.

The currently available data seem to further suggest that both paradoxical and total sleep deprivation can determine alterations in oxidative stress parameters. These changes seem to be relatively consistent and can be seen in both brain (whole brain, hippocampus, cortex, thalamus, hypothalamus, amygdala, locus coeruleus) and non-brain areas (serum, testes, epididymis, liver, pancreas, aorta, submandibular glands, thyroid, erythrocytes, soleus muscle), in multiple rat strains (Wistar, Sprague Dawley, Long-Evans, Wistar-Hannover) and in both males and females in the case of Wistar and Sprague Dawley rats.

Sleep recovery seems to be characterized by extensive variability determined by a multitude of factors ranging from the duration of sleep deprivation to the duration of sleep recovery, among others. Furthermore, short daily sleep deprivation seems to be at least partially compensated even when performed for longer time periods, at least when considering lipid peroxidation.

Most available data seem to suggest the presence of a stress response after sleep deprivation. Stress has traditionally been considered a significant confounding variable in studies investigating sleep deprivation. However, the origin of the stress response and the overall effects of stress in SD studies remain somewhat unclear. The findings outlined in this review seem to provide evidence that supports the hypothesis of sleep deprivation being a stressor in itself. Consequently, the oxidative changes observed during sleep deprivation are most likely a direct result of sleep deprivation rather than an indirect effect of the experimental conditions.

8. Future Directions

Multiple unknowns remain regarding the relationship between oxidative stress and sleep deprivation, requiring to be addressed in future studies: sex differences in response to sleep deprivation and sleep recovery, dynamics of sleep recovery, the existence of an antioxidant compensatory mechanism in short SD durations and the molecular mechanisms through which SD determines oxidative stress.

Supplementary Materials: The following supporting information can be downloaded at: https://www.mdpi.com/article/10.3390/antiox12081600/s1, Supplementary file S1.

Author Contributions: Conceptualization, V.S.N., S.C. and C.C.; methodology, V.S.N., H.A.C., I.A.C. and A.S.-B.; literature search algorithm, H.A.C.; literature search, V.S.N. and D.A.T.; screening, inclusion, data extraction, V.S.N. and I.A.C.; writing, original draft preparation, V.S.N.; writing, review and editing, V.S.N., H.A.C., I.A.C. and D.A.T.; visualization, D.A.T. and A.S.-B.; supervision, H.A.C., C.C. and S.C.; project administration, S.C., H.A.C. and C.C. All authors have read and agreed to the published version of the manuscript.

Funding: This work was granted by project PDI-PFE-CDI 2021, entitled Increasing the Performance of Scientific Research, Supporting Excellence in Medical Research and Innovation, PROGRES, no. 40PFE/30.12.2021.

Institutional Review Board Statement: Not applicable.

Informed Consent Statement: Not applicable.

Data Availability Statement: All data are contained within the article and Supplementary file.

Conflicts of Interest: The authors declare no conflict of interest.

References

1. Kocevska, D.; Lysen, T.S.; Dotinga, A.; Koopman-Verhoeff, M.E.; Luijk, M.P.C.M.; Antypa, N.; Biermasz, N.R.; Blokstra, A.; Brug, J.; Burk, W.J.; et al. Sleep Characteristics across the Lifespan in 1.1 Million People from the Netherlands, United Kingdom and United States: A Systematic Review and Meta-Analysis. *Nat. Hum. Behav.* **2021**, *5*, 113–122. [CrossRef] [PubMed]
2. Neculicioiu, V.S.; Colosi, I.A.; Costache, C.; Sevastre-Berghian, A.; Clichici, S. Time to Sleep?—A Review of the Impact of the COVID-19 Pandemic on Sleep and Mental Health. *Int. J. Environ. Res. Public Health* **2022**, *19*, 3497. [CrossRef]
3. Colten, H.R.; Altevogt, B.M. Institute of Medicine (US) Committee on Sleep Medicine and Research. In *Extent and Health Consequences of Chronic Sleep Loss and Sleep Disorders*; National Academies Press (US): Washington, DC, USA, 2006.
4. Garbarino, S.; Lanteri, P.; Bragazzi, N.L.; Magnavita, N.; Scoditti, E. Role of Sleep Deprivation in Immune-Related Disease Risk and Outcomes. *Commun. Biol.* **2021**, *4*, 1304. [CrossRef] [PubMed]
5. Andersen, M.L. A Brief Report on Early Sleep Studies. *Sleep Sci.* **2020**, *13*, 1–2. [PubMed]
6. Brinkman, J.E.; Reddy, V.; Sharma, S. Physiology of Sleep. In *StatPearls [Internet]*; StatPearls Publishing: St. Petersburg, FL, USA, 2022.
7. Vyazovskiy, V.V. Sleep, Recovery, and Metaregulation: Explaining the Benefits of Sleep. *Nat. Sci. Sleep* **2015**, *7*, 171–184. [CrossRef] [PubMed]
8. Reimund, E. The Free Radical Flux Theory of Sleep. *Med. Hypotheses* **1994**, *43*, 231–233. [CrossRef]
9. Villafuerte, G.; Miguel-Puga, A.; Rodríguez, E.M.; Machado, S.; Manjarrez, E.; Arias-Carrión, O. Sleep Deprivation and Oxidative Stress in Animal Models: A Systematic Review. *Oxid. Med. Cell. Longev.* **2015**, *2015*, 234952. [CrossRef]
10. Hill, V.M.; O'Connor, R.M.; Sissoko, G.B.; Irobunda, I.S.; Leong, S.; Canman, J.C.; Stavropoulos, N.; Shirasu-Hiza, M. A Bidirectional Relationship between Sleep and Oxidative Stress in Drosophila. *PLoS Biol.* **2018**, *16*, e2005206. [CrossRef]
11. Atrooz, F.; Salim, S. Sleep Deprivation, Oxidative Stress and Inflammation. *Adv. Protein Chem. Struct. Biol.* **2020**, *119*, 309–336.
12. Mobbs, C. Oxidative Stress and Acidosis, Molecular Responses To. In *Encyclopedia of Stress*; Fink, G., Ed.; Elsevier: Amsterdam, The Netherlands, 2007; Volume 3, p. 45.
13. Sies, H. Chapter 13—Oxidative Stress: Eustress and Distress in Redox Homeostasis. In *Stress: Physiology, Biochemistry, and Pathology*; Fink, G., Ed.; Academic Press: Cambridge, MA, USA, 2019; pp. 153–163.
14. Katerji, M.; Filippova, M.; Duerksen-Hughes, P. Approaches and Methods to Measure Oxidative Stress in Clinical Samples: Research Applications in the Cancer Field. *Oxid. Med. Cell. Longev.* **2019**, *2019*, 1279250. [CrossRef]
15. Birben, E.; Sahiner, U.M.; Sackesen, C.; Erzurum, S.; Kalayci, O. Oxidative Stress and Antioxidant Defense. *World Allergy Organ. J.* **2012**, *5*, 9–19. [CrossRef] [PubMed]
16. Ayala, A.; Muñoz, M.F.; Argüelles, S. Lipid Peroxidation: Production, Metabolism, and Signaling Mechanisms of Malondialdehyde and 4-Hydroxy-2-Nonenal. *Oxid. Med. Cell. Longev.* **2014**, *2014*, 360438. [CrossRef] [PubMed]
17. Hamed, M.A.; Akhigbe, T.M.; Akhigbe, R.E.; Aremu, A.O.; Oyedokun, P.A.; Gbadamosi, J.A.; Anifowose, P.E.; Adewole, M.A.; Aboyeji, O.O.; Yisau, H.O.; et al. Glutamine Restores Testicular Glutathione-Dependent Antioxidant Defense and Upregulates NO/CGMP Signaling in Sleep Deprivation-Induced Reproductive Dysfunction in Rats. *Biomed. Pharmacother.* **2022**, *148*, 112765. [CrossRef] [PubMed]
18. Goshi, E.; Zhou, G.; He, Q. Nitric Oxide Detection Methods in Vitro and in Vivo. *Med. Gas Res.* **2019**, *9*, 192–207. [PubMed]
19. Pierini, D.; Bryan, N. Nitric Oxide Availability as a Marker of Oxidative Stress. In *Advanced Protocols in Oxidative Stress III*; Armstrong, D., Ed.; Springer: New York, NY, USA, 2015; Volume 3.
20. Wang, F.; Yuan, Q.; Chen, F.; Pang, J.; Pan, C.; Xu, F.; Chen, Y. Fundamental Mechanisms of the Cell Death Caused by Nitrosative Stress. *Front. Cell Dev. Biol.* **2021**, *9*, 742483. [CrossRef]
21. Cialoni, D.; Brizzolari, A.; Samaja, M.; Bosco, G.; Paganini, M.; Pieri, M.; Lancellotti, V.; Marroni, A. Nitric Oxide and Oxidative Stress Changes at Depth in Breath-Hold Diving. *Front. Physiol.* **2020**, *11*, 609642. [CrossRef] [PubMed]
22. Tricco, A.C.; Lillie, E.; Zarin, W.; O'Brien, K.K.; Colquhoun, H.; Levac, D.; Moher, D.; Peters, M.D.J.; Horsley, T.; Weeks, L.; et al. PRISMA Extension for Scoping Reviews (PRISMA-ScR): Checklist and Explanation. *Ann. Intern. Med.* **2018**, *169*, 467–473. [CrossRef] [PubMed]

23. Page, M.J.; McKenzie, J.E.; Bossuyt, P.M.; Boutron, I.; Hoffmann, T.C.; Mulrow, C.D.; Shamseer, L.; Tetzlaff, J.M.; Akl, E.A.; Brennan, S.E.; et al. The PRISMA 2020 Statement: An Updated Guideline for Reporting Systematic Reviews. *BMJ* **2021**, *372*, n71. [CrossRef]

24. Colavito, V.; Fabene, P.F.; Grassi-Zucconi, G.; Pifferi, F.; Lamberty, Y.; Bentivoglio, M.; Bertini, G. Experimental Sleep Deprivation as a Tool to Test Memory Deficits in Rodents. *Front. Syst. Neurosci.* **2013**, *7*, 106. [CrossRef]

25. Wright, C.J.; Milosavljevic, S.; Pocivavsek, A. The Stress of Losing Sleep: Sex-Specific Neurobiological Outcomes. *Neurobiol. Stress* **2023**, *24*, 100543. [CrossRef]

26. Turan, I.; Sayan Ozacmak, H.; Ozacmak, V.H.; Ergenc, M.; Bayraktaroğlu, T. The Effects of Glucagon-like Peptide 1 Receptor Agonist (Exenatide) on Memory Impairment, and Anxiety- and Depression-like Behavior Induced by REM Sleep Deprivation. *Brain Res. Bull.* **2021**, *174*, 194–202. [CrossRef]

27. Sahin, L.; Cevik, O.S.; Cevik, K.; Guven, C.; Taskin, E.; Kocahan, S. Mild Regular Treadmill Exercise Ameliorated the Detrimental Effects of Acute Sleep Deprivation on Spatial Memory. *Brain Res.* **2021**, *1759*, 147367. [CrossRef] [PubMed]

28. Mohammed, H.S.; Khadrawy, Y.A. Electrophysiological and Neurochemical Evaluation of the Adverse Effects of REM Sleep Deprivation and Epileptic Seizures on Rat's Brain. *Life Sci.* **2021**, *273*, 119303. [CrossRef]

29. Forouzanfar, F.; Gholami, J.; Foroughnia, M.; Payvar, B.; Nemati, S.; Khodadadegan, M.A.; Saheb, M.; Hajali, V. The Beneficial Effects of Green Tea on Sleep Deprivation-Induced Cognitive Deficits in Rats: The Involvement of Hippocampal Antioxidant Defense. *Heliyon* **2021**, *7*, e08336. [CrossRef] [PubMed]

30. Yildirim, G.; Ozcan, K.M.; Keskin, O.; Tekeli, F.; Kaymaz, A.A. Effects of Chronic REM Sleep Deprivation on Lipocalin-2, Nitric Oxide Synthase-3, Interleukin-6 and Cardiotrophin-1 Levels: An Experimental Rat Model. *Sleep Biol. Rhythm.* **2019**, *17*, 305–310. [CrossRef]

31. Nabaee, E.; Kesmati, M.; Shahriari, A.; Khajehpour, L.; Torabi, M. Cognitive and Hippocampus Biochemical Changes Following Sleep Deprivation in the Adult Male Rat. *Biomed. Pharmacother.* **2018**, *104*, 69–76. [CrossRef] [PubMed]

32. Siervo, G.E.M.L.; Ogo, F.M.; Valério, A.D.; Silva, T.N.X.; Staurengo-Ferrari, L.; Alvarenga, T.A.; Cecchini, R.; Verri, W.A.; Guarnier, F.A.; Andersen, M.L.; et al. Sleep Restriction in Wistar Rats Impairs Epididymal Postnatal Development and Sperm Motility in Association with Oxidative Stress. *Reprod. Fertil. Dev.* **2017**, *29*, 1813–1820. [CrossRef]

33. Hernández Santiago, K.; López-López, A.L.; Sánchez-Muñoz, F.; Cortés Altamirano, J.L.; Alfaro-Rodríguez, A.; Bonilla-Jaime, H. Sleep Deprivation Induces Oxidative Stress in the Liver and Pancreas in Young and Aging Rats. *Heliyon* **2021**, *7*, e06466. [CrossRef]

34. Medubi, L.J.; Nwosu, N.C.; Medubi, O.O.; Lawal, O.R.; Ama, C.; Kusemiju, T.O.; Osinubi, A.A. Increased de Novo Glutathione Production Enhances Sexual Dysfunctions in Rats Subjected to Paradoxical Sleep Deprivation. *JBRA Assist. Reprod.* **2021**, *25*, 215–222. [CrossRef]

35. Solanki, N.; Atrooz, F.; Asghar, S.; Salim, S. Tempol Protects Sleep-Deprivation Induced Behavioral Deficits in Aggressive Male Long-Evans Rats. *Neurosci. Lett.* **2016**, *612*, 245–250. [CrossRef]

36. Zhang, L.; Guo, H.-L.; Zhang, H.-Q.; Xu, T.-Q.; He, B.; Wang, Z.-H.; Yang, Y.-P.; Tang, X.-D.; Zhang, P.; Liu, F.-E. Melatonin Prevents Sleep Deprivation-Associated Anxiety-like Behavior in Rats: Role of Oxidative Stress and Balance between GABAergic and Glutamatergic Transmission. *Am. J. Transl. Res.* **2017**, *9*, 2231–2242. [PubMed]

37. Mônico-Neto, M.; Lee, K.S.; da Luz, M.H.M.; Pino, J.M.V.; Ribeiro, D.A.; Cardoso, C.M.; Sueur-Maluf, L.L.; Tufik, S.; Antunes, H.K.M. Histopathological Changes and Oxidative Damage in Type I and Type II Muscle Fibers in Rats Undergoing Paradoxical Sleep Deprivation. *Cell. Signal.* **2021**, *81*, 109939. [CrossRef] [PubMed]

38. Gou, X.-J.; Cen, F.; Fan, Z.-Q.; Xu, Y.; Shen, H.-Y.; Zhou, M.-M. Serum and Brain Metabolomic Variations Reveal Perturbation of Sleep Deprivation on Rats and Ameliorate Effect of Total Ginsenoside Treatment. *Int. J. Genom. Proteom.* **2017**, *2017*, 5179271. [CrossRef] [PubMed]

39. Olayaki, L.A.; Sulaiman, S.O.; Anoba, N.B. Vitamin C Prevents Sleep Deprivation-Induced Elevation in Cortisol and Lipid Peroxidation in the Rat Plasma. *Niger. J. Physiol. Sci.* **2015**, *30*, 5–9. [PubMed]

40. Rizk, N.I.; Rizk, M.S.; Mohamed, A.S.; Naguib, Y.M. Attenuation of Sleep Deprivation Dependent Deterioration in Male Fertility Parameters by Vitamin C. *Reprod. Biol. Endocrinol.* **2020**, *18*, 2. [CrossRef] [PubMed]

41. Siervo, G.E.M.L.; Ogo, F.M.; Staurengo-Ferrari, L.; Anselmo-Franci, J.A.; Cunha, F.Q.; Cecchini, R.; Guarnier, F.A.; Verri, W.A., Jr.; Fernandes, G.S.A. Sleep Restriction during Peripuberty Unbalances Sexual Hormones and Testicular Cytokines in Rats. *Biol. Reprod.* **2019**, *100*, 112–122. [CrossRef] [PubMed]

42. Bajaj, P.; Kaur, G. Acute Sleep Deprivation-Induced Anxiety and Disruption of Hypothalamic Cell Survival and Plasticity: A Mechanistic Study of Protection by Butanol Extract of Tinospora Cordifolia. *Neurochem. Res.* **2022**, *47*, 1692–1706. [CrossRef]

43. Konakanchi, S.; Raavi, V.; Ml, H.K.; Shankar Ms, V. Effect of Chronic Sleep Deprivation and Sleep Recovery on Hippocampal CA3 Neurons, Spatial Memory and Anxiety-like Behavior in Rats. *Neurobiol. Learn. Mem.* **2022**, *187*, 107559. [CrossRef]

44. Aboul Ezz, H.S.; Noor, A.E.; Mourad, I.M.; Fahmy, H.; Khadrawy, Y.A. Neurochemical Effects of Sleep Deprivation in the Hippocampus of the Pilocarpine-Induced Rat Model of Epilepsy. *Iran. J. Basic Med. Sci.* **2021**, *24*, 85–91.

45. Roubalová, L.; Vošahlíková, M.; Slaninová, J.; Kaufman, J.; Alda, M.; Svoboda, P. Tissue-Specific Protective Properties of Lithium: Comparison of Rat Kidney, Erythrocytes and Brain. *Naunyn. Schmiedebergs. Arch. Pharmacol.* **2021**, *394*, 955–965. [CrossRef]

46. Zuo, J.-X.; Li, M.; Jiang, L.; Lan, F.; Tang, Y.-Y.; Kang, X.; Zou, W.; Wang, C.-Y.; Zhang, P.; Tang, X.-Q. Hydrogen Sulfide Prevents Sleep Deprivation-Induced Hippocampal Damage by Upregulation of Sirt1 in the Hippocampus. *Front. Neurosci.* **2020**, *14*, 169. [CrossRef]

47. Vosahlikova, M.; Roubalova, L.; Cechova, K.; Kaufman, J.; Musil, S.; Miksik, I.; Alda, M.; Svoboda, P. Na$^+$/K$^+$-ATPase and Lipid Peroxidation in Forebrain Cortex and Hippocampus of Sleep-Deprived Rats Treated with Therapeutic Lithium Concentration for Different Periods of Time. *Prog. Neuropsychopharmacol. Biol. Psychiatry* **2020**, *102*, 109953. [CrossRef]

48. Kwon, K.J.; Lee, E.J.; Kim, M.K.; Jeon, S.J.; Choi, Y.Y.; Shin, C.Y.; Han, S.-H. The Potential Role of Melatonin on Sleep Deprivation-Induced Cognitive Impairments: Implication of FMRP on Cognitive Function. *Neuroscience* **2015**, *301*, 403–414. [CrossRef] [PubMed]

49. Zhang, Q.; Su, G.; Zhao, T.; Wang, S.; Sun, B.; Zheng, L.; Zhao, M. The Memory Improving Effects of Round Scad (Decapterus Maruadsi) Hydrolysates on Sleep Deprivation-Induced Memory Deficits in Rats via Antioxidant and Neurotrophic Pathways. *Food Funct.* **2019**, *10*, 7733–7744. [CrossRef] [PubMed]

50. Andrabi, M.; Andrabi, M.M.; Kunjunni, R.; Sriwastva, M.K.; Bose, S.; Sagar, R.; Srivastava, A.K.; Mathur, R.; Jain, S.; Subbiah, V. Lithium Acts to Modulate Abnormalities at Behavioral, Cellular, and Molecular Levels in Sleep Deprivation-Induced Mania-like Behavior. *Bipolar Disord.* **2020**, *22*, 266–280. [CrossRef] [PubMed]

51. Alzoubi, K.H.; Al-Jamal, F.F.; Mahasneh, A.F. Cerebrolysin Prevents Sleep Deprivation Induced Memory Impairment and Oxidative Stress. *Physiol. Behav.* **2020**, *217*, 112823. [CrossRef]

52. Alzoubi, K.H.; Al Mosabih, H.S.; Mahasneh, A.F. The Protective Effect of Edaravone on Memory Impairment Induced by Chronic Sleep Deprivation. *Psychiatry Res.* **2019**, *281*, 112577. [CrossRef] [PubMed]

53. Alzoubi, K.H.; Mayyas, F.A.; Khabour, O.F.; Bani Salama, F.M.; Alhashimi, F.H.; Mhaidat, N.M. Chronic Melatonin Treatment Prevents Memory Impairment Induced by Chronic Sleep Deprivation. *Mol. Neurobiol.* **2016**, *53*, 3439–3447. [CrossRef]

54. Massadeh, A.M.; Alzoubi, K.H.; Milhem, A.M.; Rababa'h, A.M.; Khabour, O.F. Evaluating the Effect of Selenium on Spatial Memory Impairment Induced by Sleep Deprivation. *Physiol. Behav.* **2022**, *244*, 113669. [CrossRef]

55. Alzoubi, K.H.; Rababa'h, A.M.; Owaisi, A.; Khabour, O.F. L-Carnitine Prevents Memory Impairment Induced by Chronic REM-Sleep Deprivation. *Brain Res. Bull.* **2017**, *131*, 176–182. [CrossRef]

56. Alzoubi, K.H.; Khabour, O.F.; Albawaana, A.S.; Alhashimi, F.H.; Athamneh, R.Y. Tempol Prevents Chronic Sleep-Deprivation Induced Memory Impairment. *Brain Res. Bull.* **2016**, *120*, 144–150. [CrossRef] [PubMed]

57. Mhaidat, N.M.; Alzoubi, K.H.; Khabour, O.F.; Tashtoush, N.H.; Banihani, S.A.; Abdul-razzak, K.K. Exploring the Effect of Vitamin C on Sleep Deprivation Induced Memory Impairment. *Brain Res. Bull.* **2015**, *113*, 41–47. [CrossRef] [PubMed]

58. Alzoubi, K.H.; Mayyas, F.; Abu Zamzam, H.I. Omega-3 Fatty Acids Protects against Chronic Sleep-Deprivation Induced Memory Impairment. *Life Sci.* **2019**, *227*, 1–7. [CrossRef] [PubMed]

59. Alzoubi, K.H.; Malkawi, B.S.; Khabour, O.F.; El-Elimat, T.; Alali, F.Q. *Arbutus andrachne* L. Reverses Sleep Deprivation-Induced Memory Impairments in Rats. *Mol. Neurobiol.* **2018**, *55*, 1150–1156. [CrossRef]

60. Suganya, K.; Kayalvizhi, E.; Yuvaraj, R.; Chandrasekar, M.; Kavitha, U.; Konakanchi Suresh, K. Effect of Withania Somnifera on the Antioxidant and Neurotransmitter Status in Sleep Deprivation Induced Wistar Rats. *Bioinformation* **2020**, *16*, 631–637. [CrossRef]

61. Anis Syahirah, M.S.; Che Badariah, A.A.; Idris, L.; Rosfaiizah, S.; Liza, N. Impact of Rapid Eye Movement Sleep Deprivation on Pain Behaviour and Oxidative Stress in the Thalamus: Role of Tualang Honey Supplementation. *Malays. J. Med. Sci.* **2022**, *29*, 69–79.

62. Ahmad, L.; Mujahid, M.; Mishra, A.; Rahman, M.A. Protective Role of Hydroalcoholic Extract of Cajanus Cajan Linn Leaves against Memory Impairment in Sleep Deprived Experimental Rats. *J. Ayurveda Integr. Med.* **2020**, *11*, 471–477. [CrossRef]

63. Jameie, S.B.; Mesgar, S.; Aliaghaei, A.; Raoofi, A.; Amini, M.; Khodagholi, F.; Danyali, S.; Sorraya, M.; Sadeghi, Y. Neuroprotective Effect of Exogenous Melatonin on the Noradrenergic Neurons of Adult Male Rats' Locus Coeruleus Nucleus Following REM Sleep Deprivation. *J. Chem. Neuroanat.* **2019**, *100*, 101656. [CrossRef]

64. Hao, L.; Yang, R.; Xu, H. Effects of Exercise of Equal Intensity on Working Memory and BDNF Protein Expression in the Prefrontal Cortex in Rats with Different Degrees of Sleep Deprivation. *Int. J. Clin. Exp. Med.* **2019**, *12*, 11490–11496.

65. Zhang, L.; Tang, M.; Wang, F.; Zhao, Y.; Liu, Z. The Impact of Different Exercise Intensities on Working Memory and BDNF Protein Expression in Prefrontal Cortex of Sleep Deprivation Rat. *Int. J. Clin. Exp. Med.* **2017**, *10*, 9265–9270.

66. Lasisi, T.J.; Shittu, S.-T.T.; Abeje, J.I.; Ogunremi, K.J.; Shittu, S.A. Paradoxical Sleep Deprivation Induces Oxidative Stress in the Submandibular Glands of Wistar Rats. *J. Basic Clin. Physiol. Pharmacol.* **2021**, *33*, 399–408. [CrossRef] [PubMed]

67. Gu, X.; Zhang, S.; Ma, W.; Wang, Q.; Li, Y.; Xia, C.; Xu, Y.; Zhang, T.; Yang, L.; Zhou, M. The Impact of Instant Coffee and Decaffeinated Coffee on the Gut Microbiota and Depression-Like Behaviors of Sleep-Deprived Rats. *Front. Microbiol.* **2022**, *13*, 778512. [CrossRef]

68. Han, C.; Li, F.; Liu, Y.; Ma, J.; Yu, X.; Wu, X.; Zhang, W.; Li, D.; Chen, D.; Dai, N.; et al. Modified Si-Ni-San Decoction Ameliorates Central Fatigue by Improving Mitochondrial Biogenesis in the Rat Hippocampus. *Evid. Based. Complement. Alternat. Med.* **2018**, *2018*, 9452127. [CrossRef] [PubMed]

69. Li, Y.; Zhang, Y.; Ji, G.; Shen, Y.; Zhao, N.; Liang, Y.; Wang, Z.; Liu, M.; Lin, L. Autophagy Triggered by Oxidative Stress Appears to Be Mediated by the AKT/MTOR Signaling Pathway in the Liver of Sleep-Deprived Rats. *Oxid. Med. Cell. Longev.* **2020**, *2020*, 6181630. [CrossRef] [PubMed]

70. Nawi, A.; Eu, K.L.; Faris, A.N.A.; Wan Ahmad, W.A.N.; Noordin, L. Lipid Peroxidation in the Descending Thoracic Aorta of Rats Deprived of REM Sleep Using the Inverted Flowerpot Technique. *Exp. Physiol.* **2020**, *105*, 1223–1231. [CrossRef] [PubMed]

71. Jiang, J.; Gan, Z.; Li, Y.; Zhao, W.; Li, H.; Zheng, J.-P.; Ke, Y. REM Sleep Deprivation Induces Endothelial Dysfunction and Hypertension in Middle-Aged Rats: Roles of the ENOS/NO/CGMP Pathway and Supplementation with L-Arginine. *PLoS ONE* **2017**, *12*, e0182746. [CrossRef] [PubMed]

72. Li, Y.; Zhang, W.; Liu, M.; Zhang, Q.; Lin, Z.; Jia, M.; Liu, D.; Lin, L. Imbalance of Autophagy and Apoptosis Induced by Oxidative Stress May Be Involved in Thyroid Damage Caused by Sleep Deprivation in Rats. *Oxid. Med. Cell. Longev.* **2021**, *2021*, 5645090. [CrossRef] [PubMed]

73. Tang, H.; Li, K.; Dou, X.; Zhao, Y.; Huang, C.; Shu, F. The Neuroprotective Effect of Osthole against Chronic Sleep Deprivation (CSD)-Induced Memory Impairment in Rats. *Life Sci.* **2020**, *263*, 118524. [CrossRef]

74. Chang, H.-M.; Lin, H.-C.; Cheng, H.-L.; Liao, C.-K.; Tseng, T.-J.; Renn, T.-Y.; Lan, C.-T.; Chen, L.-Y. Melatonin Successfully Rescues the Hippocampal Molecular Machinery and Enhances Anti-Oxidative Activity Following Early-Life Sleep Deprivation Injury. *Antioxidants* **2021**, *10*, 774. [CrossRef]

75. Renn, T.-Y.; Yang, C.-P.; Wu, U.-I.; Chen, L.-Y.; Mai, F.-D.; Tikhonova, M.A.; Amstislavskaya, T.G.; Liao, W.-C.; Lin, C.-T.; Liu, Y.-C.; et al. Water Composed of Reduced Hydrogen Bonds Activated by Localized Surface Plasmon Resonance Effectively Enhances Anti-Viral and Anti-Oxidative Activities of Melatonin. *Chem. Eng. J.* **2022**, *427*, 131626. [CrossRef]

76. Chen, H.-C.; Cheng, C.-Y.; Chen, L.-Y.; Chang, C.-C.; Yang, C.-P.; Mai, F.-D.; Liao, W.-C.; Chang, H.-M.; Liu, Y.-C. Plasmon-Activated Water Effectively Relieves Hepatic Oxidative Damage Resulting from Chronic Sleep Deprivation. *RSC Adv.* **2018**, *8*, 9618–9626. [CrossRef] [PubMed]

77. Atrooz, F.; Liu, H.; Kochi, C.; Salim, S. Early Life Sleep Deprivation: Role of Oxido-Inflammatory Processes. *Neuroscience* **2019**, *406*, 22–37. [CrossRef] [PubMed]

78. Alvarenga, T.A.; Hirotsu, C.; Mazaro-Costa, R.; Tufik, S.; Andersen, M.L. Impairment of Male Reproductive Function after Sleep Deprivation. *Fertil. Steril.* **2015**, *103*, 1355–1362.e1. [CrossRef]

79. Vollert, C.; Zagaar, M.; Hovatta, I.; Taneja, M.; Vu, A.; Dao, A.; Levine, A.; Alkadhi, K.; Salim, S. Exercise Prevents Sleep Deprivation-Associated Anxiety-like Behavior in Rats: Potential Role of Oxidative Stress Mechanisms. *Behav. Brain Res.* **2011**, *224*, 233–240. [CrossRef] [PubMed]

80. Martínez de Toda, I.; González-Sánchez, M.; Díaz-Del Cerro, E.; Valera, G.; Carracedo, J.; Guerra-Pérez, N. Sex Differences in Markers of Oxidation and Inflammation. Implications for Ageing. *Mech. Ageing Dev.* **2023**, *211*, 111797. [CrossRef] [PubMed]

81. Hajali, V.; Andersen, M.L.; Negah, S.S.; Sheibani, V. Sex Differences in Sleep and Sleep Loss-Induced Cognitive Deficits: The Influence of Gonadal Hormones. *Horm. Behav.* **2019**, *108*, 50–61. [CrossRef] [PubMed]

82. Dib, R.; Gervais, N.J.; Mongrain, V. A Review of the Current State of Knowledge on Sex Differences in Sleep and Circadian Phenotypes in Rodents. *Neurobiol. Sleep Circadian Rhythm.* **2021**, *11*, 100068. [CrossRef]

83. Wang, S.; Su, G.; Zhang, Q.; Zhao, T.; Liu, Y.; Zheng, L.; Zhao, M. Walnut (Juglans Regia) Peptides Reverse Sleep Deprivation-Induced Memory Impairment in Rat via Alleviating Oxidative Stress. *J. Agric. Food Chem.* **2018**, *66*, 10617–10627. [CrossRef]

84. Hsu, J.-C.; Lee, Y.-S.; Chang, C.-N.; Chuang, H.-L.; Ling, E.-A.; Lan, C.-T. Sleep Deprivation Inhibits Expression of NADPH-d and NOS While Activating Microglia and Astroglia in the Rat Hippocampus. *Cells Tissues Organs* **2003**, *173*, 242–254. [CrossRef]

85. Hirotsu, C.; Tufik, S.; Andersen, M.L. Interactions between Sleep, Stress, and Metabolism: From Physiological to Pathological Conditions. *Sleep Sci.* **2015**, *8*, 143–152. [CrossRef]

86. Gong, S.; Miao, Y.-L.; Jiao, G.-Z.; Sun, M.-J.; Li, H.; Lin, J.; Luo, M.-J.; Tan, J.-H. Dynamics and Correlation of Serum Cortisol and Corticosterone under Different Physiological or Stressful Conditions in Mice. *PLoS ONE* **2015**, *10*, e0117503. [CrossRef] [PubMed]

87. Sahin, E.; Gümüşlü, S. Immobilization Stress in Rat Tissues: Alterations in Protein Oxidation, Lipid Peroxidation and Antioxidant Defense System. *Comp. Biochem. Physiol. C. Toxicol. Pharmacol.* **2007**, *144*, 342–347. [CrossRef]

88. Zafir, A.; Banu, N. Induction of Oxidative Stress by Restraint Stress and Corticosterone Treatments in Rats. *Indian J. Biochem. Biophys.* **2009**, *46*, 53–58. [PubMed]

89. Samarghandian, S.; Azimi-Nezhad, M.; Farkhondeh, T.; Samini, F. Anti-Oxidative Effects of Curcumin on Immobilization-Induced Oxidative Stress in Rat Brain, Liver and Kidney. *Biomed. Pharmacother.* **2017**, *87*, 223–229. [CrossRef]

90. Ali, F.F.; Rifaai, R.A. Preventive Effect of Omega-3 Fatty Acids in a Rat Model of Stress-Induced Liver Injury. *J. Cell. Physiol.* **2019**, *234*, 11960–11968. [CrossRef] [PubMed]

91. Nirupama, M.; Devaki, M.; Nirupama, R.; Yajurvedi, H.N. Chronic Intermittent Stress-Induced Alterations in the Spermatogenesis and Antioxidant Status of the Testis Are Irreversible in Albino Rat. *J. Physiol. Biochem.* **2013**, *69*, 59–68. [CrossRef]

92. García-Díaz, E.C.; Gómez-Quiroz, L.E.; Arenas-Ríos, E.; Aragón-Martínez, A.; Ibarra-Arias, J.A.; del Socorro I Retana-Márquez, M. Oxidative Status in Testis and Epididymal Sperm Parameters after Acute and Chronic Stress by Cold-Water Immersion in the Adult Rat. *Syst. Biol. Reprod. Med.* **2015**, *61*, 150–160. [CrossRef]

93. Andersen, M.L.; Martins, P.J.F.; D'Almeida, V.; Bignotto, M.; Tufik, S. Endocrinological and Catecholaminergic Alterations during Sleep Deprivation and Recovery in Male Rats. *J. Sleep Res.* **2005**, *14*, 83–90. [CrossRef]

94. Meerlo, P.; Koehl, M.; van der Borght, K.; Turek, F.W. Sleep Restriction Alters the Hypothalamic-Pituitary-Adrenal Response to Stress. *J. Neuroendocrinol.* **2002**, *14*, 397–402. [CrossRef]

95. Suchecki, D.; Tiba, P.A.; Tufik, S. Paradoxical Sleep Deprivation Facilitates Subsequent Corticosterone Response to a Mild Stressor in Rats. *Neurosci. Lett.* **2002**, *320*, 45–48. [CrossRef]

96. Suchecki, D.; Lobo, L.L.; Hipólide, D.C.; Tufik, S. Increased ACTH and Corticosterone Secretion Induced by Different Methods of Paradoxical Sleep Deprivation. *J. Sleep Res.* **1998**, *7*, 276–281. [CrossRef] [PubMed]

97. Coenen, A.M.; van Luijtelaar, E.L. Stress Induced by Three Procedures of Deprivation of Paradoxical Sleep. *Physiol. Behav.* **1985**, *35*, 501–504. [CrossRef] [PubMed]

98. Nollet, M.; Wisden, W.; Franks, N.P. Sleep Deprivation and Stress: A Reciprocal Relationship. *Interface Focus* **2020**, *10*, 20190092. [CrossRef]

99. Machado, R.B.; Hipólide, D.C.; Benedito-Silva, A.A.; Tufik, S. Sleep Deprivation Induced by the Modified Multiple Platform Technique: Quantification of Sleep Loss and Recovery. *Brain Res.* **2004**, *1004*, 45–51. [CrossRef] [PubMed]

100. Suchecki, D.; Machado, R.B.; Tiba, P.A. Stress-Induced Sleep Rebound: Adaptive Behavior and Possible Mechanisms. *Sleep Sci.* **2009**, *2*, 151–160.

101. Machado, R.B.; Tufik, S.; Suchecki, D. Role of Corticosterone on Sleep Homeostasis Induced by REM Sleep Deprivation in Rats. *PLoS ONE* **2013**, *8*, e63520. [CrossRef] [PubMed]

102. Melgarejo-Gutiérrez, M.; Acosta-Peña, E.; Venebra-Muñoz, A.; Escobar, C.; Santiago-García, J.; Garcia-Garcia, F. Sleep Deprivation Reduces Neuroglobin Immunoreactivity in the Rat Brain. *Neuroreport* **2013**, *24*, 120–125. [CrossRef]

103. Gopalakrishnan, A.; Ji, L.L.; Cirelli, C. Sleep Deprivation and Cellular Responses to Oxidative Stress. *Sleep* **2004**, *27*, 27–35. [CrossRef]

104. Vaccaro, A.; Kaplan Dor, Y.; Nambara, K.; Pollina, E.A.; Lin, C.; Greenberg, M.E.; Rogulja, D. Sleep Loss Can Cause Death through Accumulation of Reactive Oxygen Species in the Gut. *Cell* **2020**, *181*, 1307–1328.e15. [CrossRef]

105. Briançon-Marjollet, A.; Weiszenstein, M.; Henri, M.; Thomas, A.; Godin-Ribuot, D.; Polak, J. The Impact of Sleep Disorders on Glucose Metabolism: Endocrine and Molecular Mechanisms. *Diabetol. Metab. Syndr.* **2015**, *7*, 25. [CrossRef]

106. Morselli, L.; Leproult, R.; Balbo, M.; Spiegel, K. Role of Sleep Duration in the Regulation of Glucose Metabolism and Appetite. *Best Pract. Res. Clin. Endocrinol. Metab.* **2010**, *24*, 687–702. [CrossRef] [PubMed]

107. Boyle, P.J.; Scott, J.C.; Krentz, A.J.; Nagy, R.J.; Comstock, E.; Hoffman, C. Diminished Brain Glucose Metabolism Is a Significant Determinant for Falling Rates of Systemic Glucose Utilization during Sleep in Normal Humans. *J. Clin. Investig.* **1994**, *93*, 529–535. [CrossRef] [PubMed]

108. Holmer, B.J.; Lapierre, S.S.; Jake-Schoffman, D.E.; Christou, D.D. Effects of Sleep Deprivation on Endothelial Function in Adult Humans: A Systematic Review. *Geroscience* **2021**, *43*, 137–158. [CrossRef] [PubMed]

109. Palagini, L.; Bruno, R.M.; Gemignani, A.; Baglioni, C.; Ghiadoni, L.; Riemann, D. Sleep Loss and Hypertension: A Systematic Review. *Curr. Pharm. Des.* **2013**, *19*, 2409–2419. [CrossRef]

110. Sá Gomes E Farias, A.V.; de Lima Cavalcanti, M.P.; de Passos Junior, M.A.; Vechio Koike, B.D. The Association between Sleep Deprivation and Arterial Pressure Variations: A Systematic Literature Review. *Sleep Med. X* **2022**, *4*, 100042. [CrossRef] [PubMed]

111. Xi, B.; He, D.; Zhang, M.; Xue, J.; Zhou, D. Short Sleep Duration Predicts Risk of Metabolic Syndrome: A Systematic Review and Meta-Analysis. *Sleep Med. Rev.* **2014**, *18*, 293–297. [CrossRef] [PubMed]

112. Antza, C.; Kostopoulos, G.; Mostafa, S.; Nirantharakumar, K.; Tahrani, A. The Links between Sleep Duration, Obesity and Type 2 Diabetes Mellitus. *J. Endocrinol.* **2021**, *252*, 125–141. [CrossRef]

113. Knutson, K.L.; Spiegel, K.; Penev, P.; Van Cauter, E. The Metabolic Consequences of Sleep Deprivation. *Sleep Med. Rev.* **2007**, *11*, 163–178. [CrossRef]

114. Bishir, M.; Bhat, A.; Essa, M.M.; Ekpo, O.; Ihunwo, A.O.; Veeraraghavan, V.P.; Mohan, S.K.; Mahalakshmi, A.M.; Ray, B.; Tuladhar, S.; et al. Sleep Deprivation and Neurological Disorders. *Biomed Res. Int.* **2020**, *2020*, 5764017. [CrossRef]

115. Xue, R.; Wan, Y.; Sun, X.; Zhang, X.; Gao, W.; Wu, W. Nicotinic Mitigation of Neuroinflammation and Oxidative Stress After Chronic Sleep Deprivation. *Front. Immunol.* **2019**, *10*, 2546. [CrossRef]

116. Chen, P.; Ban, W.; Wang, W.; You, Y.; Yang, Z. The Devastating Effects of Sleep Deprivation on Memory: Lessons from Rodent Models. *Clocks Sleep* **2023**, *5*, 276–294. [CrossRef]

117. Tufik, S.; Andersen, M.L.; Bittencourt, L.R.A.; de Mello, M.T. Paradoxical Sleep Deprivation: Neurochemical, Hormonal and Behavioral Alterations. Evidence from 30 Years of Research. *An. Acad. Bras. Cienc.* **2009**, *81*, 521–538. [CrossRef]

118. Maloney, K.J.; Mainville, L.; Jones, B.E. Differential C-Fos Expression in Cholinergic, Monoaminergic, and GABAergic Cell Groups of the Pontomesencephalic Tegmentum after Paradoxical Sleep Deprivation and Recovery. *J. Neurosci.* **1999**, *19*, 3057–3072. [CrossRef]
119. Suchecki, D.; Tufik, S. Social Stability Attenuates the Stress in the Modified Multiple Platform Method for Paradoxical Sleep Deprivation in the Rat. *Physiol. Behav.* **2000**, *68*, 309–316. [CrossRef]

 *antioxidants*

Review

The Unfolded Protein Response: A Double-Edged Sword for Brain Health

Magdalena Gebert [1], Jakub Sławski [2], Leszek Kalinowski [1,3], James F. Collawn [4] and Rafal Bartoszewski [2,*]

[1] Department of Medical Laboratory Diagnostics—Fahrenheit Biobank BBMRI.pl, Medical University of Gdansk, 80-134 Gdansk, Poland
[2] Department of Biophysics, Faculty of Biotechnology, University of Wroclaw, F. Joliot-Curie 14a Street, 50-383 Wroclaw, Poland
[3] BioTechMed Centre, Department of Mechanics of Materials and Structures, Gdansk University of Technology, 11/12 Narutowicza Street, 80-233 Gdansk, Poland
[4] Department of Cell, Developmental, and Integrative Biology, University of Alabama at Birmingham, Birmingham, AL 35233, USA
* Correspondence: rafal.bartoszewski@uwr.edu.pl

Abstract: Efficient brain function requires as much as 20% of the total oxygen intake to support normal neuronal cell function. This level of oxygen usage, however, leads to the generation of free radicals, and thus can lead to oxidative stress and potentially to age-related cognitive decay and even neurodegenerative diseases. The regulation of this system requires a complex monitoring network to maintain proper oxygen homeostasis. Furthermore, the high content of mitochondria in the brain has elevated glucose demands, and thus requires a normal redox balance. Maintaining this is mediated by adaptive stress response pathways that permit cells to survive oxidative stress and to minimize cellular damage. These stress pathways rely on the proper function of the endoplasmic reticulum (ER) and the activation of the unfolded protein response (UPR), a cellular pathway responsible for normal ER function and cell survival. Interestingly, the UPR has two opposing signaling pathways, one that promotes cell survival and one that induces apoptosis. In this narrative review, we discuss the opposing roles of the UPR signaling pathways and how a better understanding of these stress pathways could potentially allow for the development of effective strategies to prevent age-related cognitive decay as well as treat neurodegenerative diseases.

Keywords: endoplasmic reticulum stress; mitochondria unfolded protein response; oxidative stress; neurodegeneration; proteostasis; calcium; brain; nitrosative stress; oxygen homeostasis

Citation: Gebert, M.; Sławski, J.; Kalinowski, L.; Collawn, J.F.; Bartoszewski, R. The Unfolded Protein Response: A Double-Edged Sword for Brain Health. *Antioxidants* **2023**, *12*, 1648. https://doi.org/10.3390/antiox12081648

Academic Editor: Waldo Cerpa

Received: 26 July 2023
Revised: 14 August 2023
Accepted: 19 August 2023
Published: 21 August 2023

1. Introduction

Proper oxygen (O_2) homeostasis is essential for human survival, and the human brain consumes about 20% of the total oxygen to support neurons and glia [1–4]. Unmet brain oxygen needs during ischemic stroke limit ATP synthesis [5,6]. Oxygen consumption results in the generation of free radicals and non-radicals including superoxide ($O_2^{\cdot-}$) and hydroxyl anions ($^{\cdot}OH$), and hydrogen peroxide (H_2O_2) [7–10]. Although this is an unavoidable consequence of oxygen-dependent brain activity, if not controlled properly, it leads to oxidative stress and neurodegeneration [11–19]. Thus, maintaining proper oxygen homeostasis in brain tissues requires a balanced level of O_2-derived free radicals and non-radicals [1]. In this review, we discuss how the unfolded protein response (UPR) regulates oxygen homeostasis in the endoplasmic reticulum (ER) and mitochondria to support neuronal cell viability, but also how these stress pathways can promote cognitive decline and potentially neuronal diseases.

Given that maintaining the redox balance is necessary for cell survival, it is surprising that the brain is so susceptible to oxidative stress and oxidative damage [1]. This vulnerability to brain oxygen damage is believed to be a compromise between brain function and the biochemical organization that is required for survival [20]. This organization includes a high content of mitochondria, an increased glucose demand, and a high influx of neuronal Ca^{2+}. Furthermore, there is increased microglia activity, as well as increased neuronal nitric oxide synthase (nNOS) and nicotinamide adenine dinucleotide phosphate (NAPDH) oxidase (NOX) signaling, along with the presence of autoxidizable neurotransmitters. This metabolism machinery generates hydrogen peroxide, high concentrations of peroxidable lipids, elevated levels of cytochrome P_{450}, and the enrichment of brain tissues in redox-active transition metals such as Fe^{2+} and Cu^+ [1,11–19,21,22]. All of this leads to potential stress that needs to be properly and safely regulated.

In this complex system, brain cells have to efficiently modulate their signaling pathways to maintain their redox balance and utilize universal adaptive stress responses in order to survive periods of elevated oxidation levels and minimize cellular damage. These stress pathways depend on the proper function of the endoplasmic reticulum (ER) and activation of the unfolded protein response (UPR), a set of complex molecular pathways that regulate proper ER function required for cell survival, or in the case of unmitigated cell stress, lead to cell death. In this review, we discuss the Janus faces of this complex signaling pathway in the context of managing the "oxidant burden" of the brain [23,24].

2. Role of the ER in Maintaining Neuron Cell Homeostasis

2.1. Calcium Regulation and Signaling

Connecting synaptic activity with the biochemical signals of neurons occurs through utilizing calcium ions (Ca^{2+}) as the main second messenger to regulate activity-dependent signaling [25,26]. Brain calcium fluxes lead to high ATP demands that restore the ion levels after calcium influx through the plasma membrane receptor. When impaired, intracellular calcium homeostasis leads to increased generation of mitochondrial reactive oxygen species (ROS) [27]. The ER, the main cellular calcium storage compartment, remains a critical system responsible for the calcium balance in neurons [28]. ER calcium release in response to small increases in its cytosolic levels is termed calcium-induced calcium release (CICR), whereas the reduction in calcium concentration in ER lumen is referred to as storage-operated calcium entry (SOCE) [28]. Both of these mechanisms amplify cytosolic calcium levels and allow the ER, at least in theory, to generate calcium transients independently of any plasma membrane depolarization [29]. Furthermore, ER calcium release and uptake in neurons relies on the membrane potential and contributes to its modulation by accelerating increases and decreases in the calcium cytosolic levels.

The excessive influx of calcium into neurons mainly occurs through the activation of N-methyl-D-aspartate (NMDA) receptors by glutamate, and results in CICR [28]. Although the influx of calcium through NMDA receptors is the underlying basis of neurodegeneration caused by excitotoxicity, calcium stores within the endoplasmic reticulum (ER) can also be released through ryanodine receptors (RyR) and inositol 1,4,5-trisphosphate receptors (IP_3R) under these conditions, and this can amplify the pathological calcium signals [28,29]. As a consequence, the activation of the mitochondrial calcium buffering system can occur and lead to rapid mitochondrial damage due to increased permeability of the transition pore (mPTP) [28,30,31]. Furthermore, the increase in intracellular calcium concentration is accompanied by O_2^- release and the generation of $OH\cdot$ in the Fenton reaction, which is catalyzed by superoxide dismutase (SOD) [32,33].

ER calcium release in the region of mitochondria-associated membranes (MAMs) [34,35] has been shown to support the ATP demand-related mitochondrial uptake of calcium [36,37]. Mitochondrial calcium uptake leads to increases in the activity of the Krebs cycle enzymes [36–39]. Despite multiple pathways that allow mitochondrial calcium release that include both ion exchangers and the transient opening of the mitochondrial permeability transition pore (mPTP) [30,31], mitochondria remain prone to calcium overload. This unfortunately leads to reduced ATP synthesis, increased ROS formation [40,41], and eventually cell death [42]. This highlights the importance of the cooperation between mitochondria and ER in regulating intracellular calcium levels and neuronal cell viability.

2.2. The ER and Proteostasis

The spatial organization of the brain dependence on this complex neuronal structure is maintained by the continuous protein profile-related remodeling of synapses [43–45]. Their proper function relies on the biogenesis of plasma membranes that are enriched with specific proteins, including cell adhesion molecules, ion channels, receptors, and transporters [46]. The ER is a central compartment for the secretory protein pathway, which is important for membrane protein maturation and lipid biosynthesis, and this pathway remains critical both during and after brain development [47,48]. Proper ER functions are crucial for both synapse formation and plasticity as well for cognitive functions [47–51].

The ER also contains enzymes and chaperones that assist in various protein folding scenarios and mediates their posttranslational maturation [52]. This protein maturation machinery includes chaperone immunoglobulin binding protein (BiP; also known as *HSPA5* or Grp78) [53], different oxidoreductases of the protein disulfide isomerase (PDI) family [54], and the peptidyl prolyl cis-trans isomerases (PPIs) [55]. Protein quality control of the ER-maturating glycosylated proteins is ensured by the calnexin–calreticulin system [56], whereas terminally misfolded peptides are exported from the ER and degraded either by the proteasome (ER-associated degradation (ERAD)) or the lysosome (ER-to-lysosome-associated degradation (ERLAD)) [57,58]. Random oxidation of mRNA is one of the consequences of the brain oxygen burden [59], and this can increase translational errors [60], reduce the successful protein folding in ER [61–63], and provide challenges for the ER-associated degradation system. Furthermore, impaired efficiency of ER-related protein maturation can result in deregulation of brain redox homeostasis and lead to oxidative damage. Oxidative stress can also impair ER proteostasis and ER-associated degradation, leading to accumulation and aggregation of misfolded proteins, as is observed during neurodegeneration [64,65].

2.3. The ER Lipid Biosynthesis

ER-localized enzymes are also responsible for the synthesis of the majority of cellular lipids that are another key component of the brain. These membrane lipids allow the brain cells to grow, proliferate, differentiate, and modulate neurons and glia cell function, including neurotransmission [66–68]. Interestingly, the brain is enriched in long-chain polyunsaturated fatty acids that are sensitive to oxidation, but neurons do not store energy in the form of glycogen or lipid droplets. Therefore, fatty acid oxidation primarily occurs in astrocytes that transfer the related metabolites to neurons [69]. Furthermore, stressed neurons release peroxidated fatty acids to be endocytosed and stored in lipid droplets by neighboring astrocytes that utilize this storage to support the stimulated neuron energy requirements [70]. This lipid crosstalk between the neurons and astrocytes ensures proper brain function, while minimizing the risk of oxidative stress [69]. This cooperation between the neurons and astrocytes prevents a buildup of peroxidated fatty acids in neurons during periods of prolonged stimulation [70].

Cholesterol, on the other hand, is enriched in synaptic membranes and serves as a regulator of neurotransmissions. It is synthetized de novo in both neurons and astrocytes [71,72]. The cholesterol synthesis pathway is dependent upon the ER-associated sterol regulatory element-binding protein (SREBP) system that is activated by low cholesterol levels in ER membranes and is very sensitive to the alterations in ER homeostasis [73,74].

In summary, ER homeostasis (as presented in Figure 1) remains one of the key factors for brain development and function, including the redox balance. ER homeostasis is stabilized by the presence of the UPR. The UPR promotes cellular survival by reducing ER damage during stress, or alternatively promotes cell death during prolonged or unmitigated stress [75]. This negative scenario is a common characteristic of neurodegenerative diseases caused by aggregates of mutant proteins or through loss of function of genes responsible for proteostasis [75–78]. Thus, the ability of UPR to determine cell fate is a crucial element of brain aging and potential neurodegeneration.

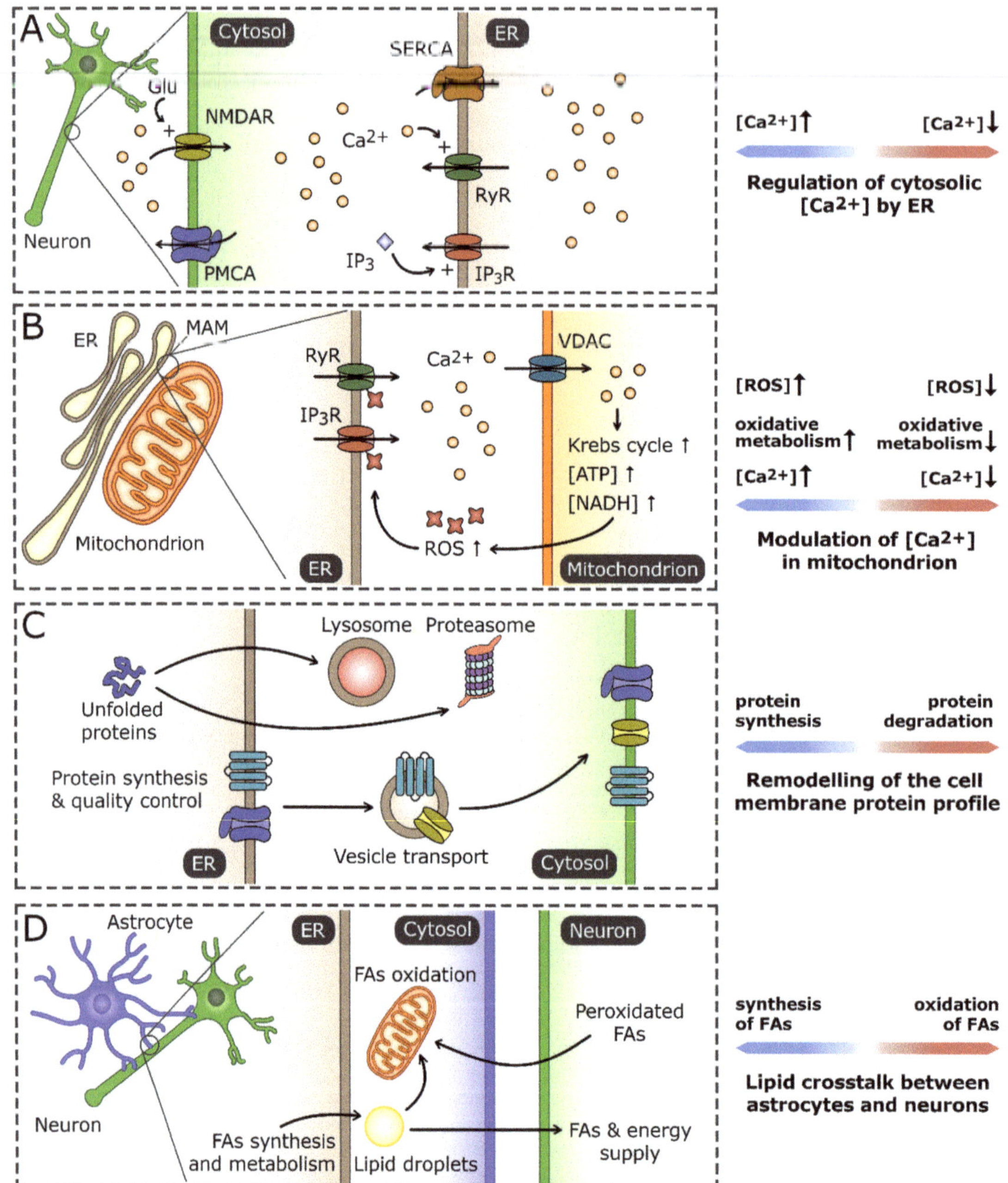

Figure 1. The role of the endoplasmic reticulum (ER) in maintaining neuron cell homeostasis. (**A**) As the main Ca^{2+} reservoir, the ER is crucial for the regulation of cytosolic Ca^{2+} concentration using pumps and channels localized in ER membrane. Those include sarco/endoplasmic reticulum Ca^{2+} ATPase (SERCA), Ca^{2+}-activated ryanodine receptors (RyRs), and inositol-1,4,5-trisphosphate (IP_3)-gated IP_3 receptors (IP_3Rs). They cooperate with the cell membrane Ca^{2+} transporters that regulate the influx of extracellular Ca^{2+}, exemplified by plasma membrane Ca^{2+} ATPase (PMCA) and N-methyl-D-aspartate receptor (NMDAR). (**B**) Ca^{2+} homeostasis processes in the ER and mitochondrion

are tightly interconnected, primarily by virtue of the regions of mitochondria-associated membranes (MAMs). An increase in Ca^{2+} concentration in MAM promotes its influx into the mitochondrion, mainly through voltage-dependent anion channel (VDAC). High Ca^{2+} concentration stimulates the activity of the oxidative processes in the mitochondrion, leading to the increased production of reactive oxygen species (ROS). In turn, ROS-dependent modifications of ER Ca^{2+} channels increase their permeability for Ca^{2+} and the efflux of Ca^{2+} from ER, which closes the positive-feedback loop. (**C**) The ER is a central cell compartment where the synthesis and quality control of secretory and membrane proteins takes place. The properly folded proteins are directed through secretory pathway to the cell membrane, whereas irreversibly unfolded/misfolded proteins are exported and eventually degraded either in lysosomes or proteasomes. (**D**) ER-based lipid crosstalk between neurons and astrocytes. Fatty acids (FAs) and the products of their oxidation synthesized in astrocytes are delivered to neurons to support their demand for energy and membrane building components. In turn, nonfunctional peroxidated FAs released by neurons are endocytosed by astrocytes and stored in lipid droplets or catabolized by the mitochondrial FA oxidation pathway.

3. The Unfolded Protein Response Pathway

The proper ratio between folded and unfolded proteins in the ER is an essential component of ER homeostasis [79]. Nevertheless, numerous cellular and environmental and physiological insults, including gene mutations, prion transmission, virial infections and ROS, promote ER stress. This results in the extensive accumulation of misfolded or incompletely folded proteins in the lumen of this organelle [75–78,80–87]. This type of disturbance of proteostasis calls for reductions in the protein synthetic load and increases in the availability of ER chaperones such as BiP [88]. Consequently, the pool of BiP associated with the ER UPR transmembrane proteins is released into the ER lumen to facilitate folding while simultaneously activating the UPR proteins (Figure 2A). These UPR proteins include protein kinase RNA (PKR)-like ER kinase (PERK), inositol-requiring transmembrane kinase/endoribonuclease (IRE1α), and activating transcription factor 6 (ATF6)) [89]. After BiP release, both IRE1 and PERK self-associate and undergo trans-autophosphorylation to become functional [88–91], whereas ATF6 translocates to the Golgi, where it is subjected to intermembrane proteolysis by site 1 and 2 proteases, yielding the nuclear-targeted transcription factor ATF6f (p50) [92–95].

PERK phosphorylates an alpha subunit of the eukaryotic initiation factor 2 (eIF2α), yielding P-eIF2α [96,97]. This in turn reduces the global rates of protein synthesis by inhibiting the activity of its own guanine nucleotide exchange factor [98]. The PERK-mediated reduction in cellular protein synthesis, referred to as the integrated stress response (ISR), reduces the ER peptide influx and allows correction of the degradation of misfolded proteins [99–102]. Nevertheless, the ISR-related translational blockage does not apply to the translation of a limited number of specific genes, including the growth arrest and DNA damage-inducible protein (GADD34), proapoptotic CCAAT/enhancer binding homologous protein (CHOP), and activating transcription factor 4 (ATF4) [89,103–106]. ATF4 enhances expression of antiapoptotic factors as well as—along with nuclear factor erythroid 2–related factor 2 (NRF2)—modulates glutathione (GSH) synthesis and the response to oxidative stress [107,108]. If the ER stress is diminished, GADD34 dephosphorylates P-eIF2α and thus reverses the translational blockage when the stress response is resolved [109].

Upon trans-autophosphorylation, IRE1's endoribonuclease (RNase) activity is initiated, which allows it to degrade a subset of mRNAs to reduce the ER load of newly translated proteins in a process called IRE1-dependent decay (RIDD) [106,110]. Secondly, IRE1 splices the mRNA transcript of the X-box binding protein 1 (XBP1) transcription factor into an mRNA that encodes a transcriptionally active isoform of this protein (XBP1s) [111].

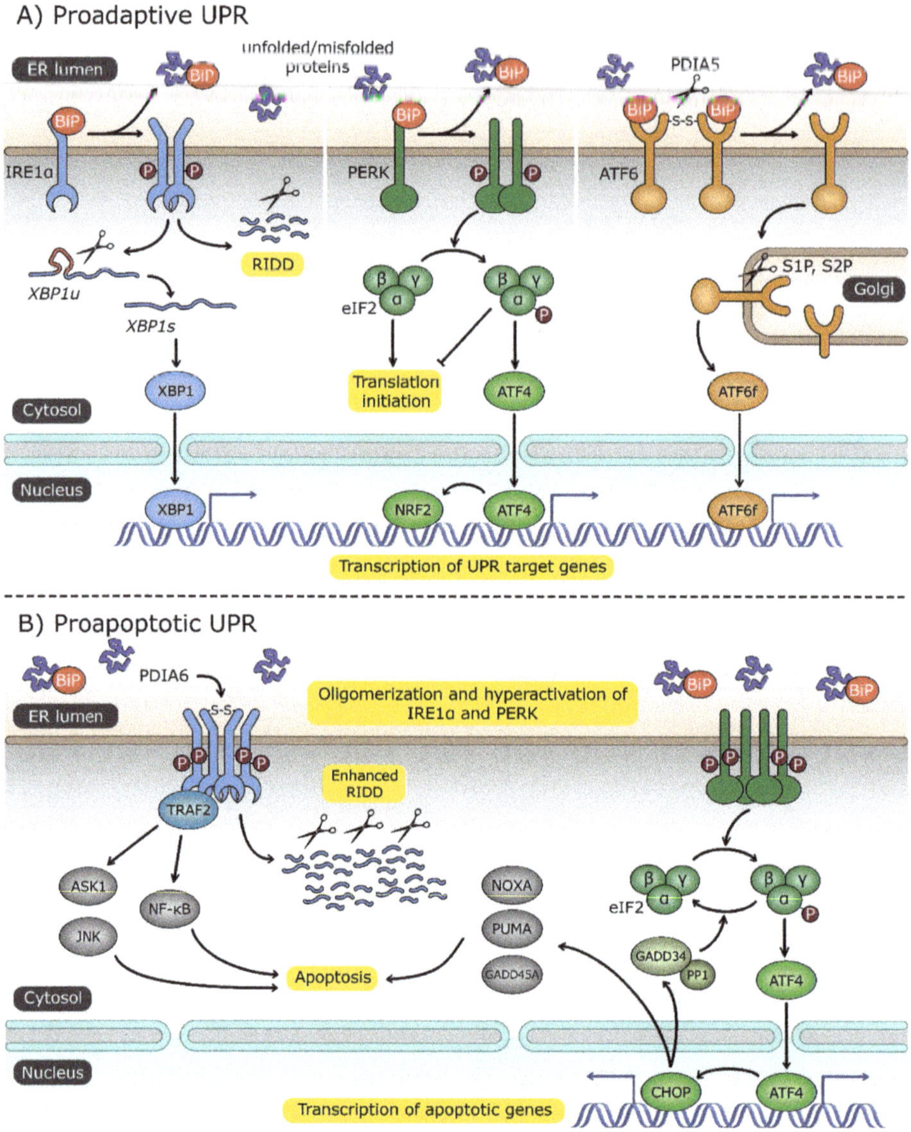

A) Proadaptive UPR

B) Proapoptotic UPR

Figure 2. The unfolded protein response (UPR) pathway. (**A**) Three UPR sensors—inositol-requiring protein 1α (IRE1α), protein kinase RNA (PKR)-like endoplasmic reticulum kinase (PERK) and activating transcription factor 6 (ATF6)—are localized in endoplasmic reticulum (ER) membrane and share a common activation signal: the dissociation of binding immunoglobulin protein (BiP) chaperone in response to increased level of unfolded/misfolded proteins. Dimerization of IRE1α, followed by its trans-autophosphorylation, activates its RNase domain. The primary target of IRE1α is the unspliced X box-binding protein 1 (XBP1u) transcript. Spliced XBP1 mRNA (XBP1s) encodes transcription factor XBP1s, which activates UPR-associated genes. IRE1α also degrades certain mRNAs through the regulated IRE1-dependent decay (RIDD) process. Upon dimerization and trans-autophosphorylation, PERK phosphorylates eukaryotic translation initiator factor 2α (eIF2α) to attenuate general protein translation. Phosphorylated eIF2α promotes expression of activating transcription factor 4 (ATF4) and nuclear factor erythroid 2-related factor 2 (NRF2), which are involved in the response to ER and oxidative stress, respectively. ER stress triggers the cleavage of disulfide bonds, stabilizing ATF6 oligomers by protein disulfide isomerase family A member 5 (PDIA5), and this is followed by its transport to the Golgi apparatus where it is processed by site 1 and site 2 proteases (S1P, S2P). Cytosolic ATF6 fragment (ATF6f) is released and imported to the nucleus, where it plays the role of an active transcription factor. (**B**) Under extensive and persistent ER stress, the UPR switches from proadaptive to a proapoptotic character. Oligomerized IRE1α, stabilized by the disulfide bonds formed by protein disulfide isomerase family A member 6 (PDIA6), recruits tumor necrosis factor (TNF) receptor-associated factor 2 (TRAF2), which in turn activates the proapoptotic signal-regulating kinase 1/Janus N-terminal kinase (ASK1/JNK) and the nuclear factor kappa-light-chain-enhancer of activated B cells (NF-κB) signaling pathways. ATF4 promotes the expression of CCAAT/enhancer-binding protein homologous protein (CHOP) and transcription factor targeting apoptotic genes, including growth arrest and DNA damage-inducible 45 alpha (GADD45A), p53 upregulated modulator of apoptosis (PUMA), phorbol-12-myristate-13-acetate-induced protein 1 (NOXA), and growth arrest and DNA damage-inducible 34 (GADD34). GADD34 forms a complex with protein phosphatase 1 (PP1) to dephosphorylate eIF2α and reverse the inhibition of translation.

Both ATF6f and the XBP1s mediate a wide transcriptional reprogramming of stressed ER cells. These transcription factors work both cooperatively and independently to reduce ER peptide influx, increase folding processes in ER, and improve misfolded protein removal [82,112–114]. Furthermore, both ATF6f and XBP1s stimulate ER lipid membrane biosynthesis and chaperone transcription to increase the volume and folding capacity of the ER. They also promote the expression of the genes responsible for ERAD, including synoviolin 1 (HRD1), which is XBP1-induced, and the suppressor/enhancer of lin-12-like (SEL1L), which is induced by both ATF6f and XBP1s [115–117] and N-glycosylation [82,98,118–120]. Notably, ATF6f and XBP1s transcriptional targets include prosurvival transcripts [111,114,118,121,122]. Although the ER requires increased production of membrane lipids in order to increase the ER volume during the UPR, this approach remains the most straightforward mechanism for the cell to resolve the stress and improve protein folding [123]. Despite the fact that all of the UPR branches stimulate lipid biogenesis [120,124–126], XBP1s remain the most critical for efficient increasing the ER volume [127–129].

The UPR can also realign its three signaling branches towards cell death programs (Figure 2B). The UPR-related cell death shifts the balance away from the proadaptive signals in cases where the cellular damage is too severe or the adaptative response fails [114,130,131]. Both PERK and ATF6f continuously stimulate expression of CHOP, whereas IRE1 leads to the activation of the Janus N-terminal kinase (JNK) [130,132–134]. The RIDD allows for the accumulation of proapoptotic factors by degrading their specific miRNAs that target these factors [135,136]. Furthermore, upon eventual hyperactivation of IRE1, in addition to RIDD, this RNAse forms a scaffold for the activation of proinflammatory and apoptotic ASK1-JNK and NF-kβ pathways [137,138]. IRE1-ASK1-JNK signaling leads to the inhibition of mitochondrial respiration and enhanced ROS production [139]. Interestingly, IRE1 activation can also prevent the proapoptotic activity of ATF6f [140].

The UPR cell death decision is also supported by changes in levels of other apoptotic factors such as growth arrest and DNA damage-inducible alpha (GADD45A), p53 upregulated modulator of apoptosis (PUMA), and phorbol-12-myristate-13-acetate-induced protein 1 (PMAIP1, also known as NOXA) [82,130,131,141–144]. Notably, PUMA and NOXA provide the link between UPR-induced cell death and mitochondrial apoptosis [145]. Since these two proteins contribute to the outer mitochondrial membrane permeabilization, their accumulation during ER stress can result in enhanced ROS efflux from mitochondria and accelerated oxidative stress [146]. Furthermore, if cells are exposed to strong and chronic ER insults, potent activation of PERK signals will result in a rapid decline in ATP levels accompanied by an intensive release of ER-stored calcium that leads to necroptosis [147–152]. Notably, necroptosis is also often associated with increased ROS levels [153–156]. It is also worth mentioning that both proadaptive and apoptotic aspects of the UPR are modulated at the posttranscriptional levels by the accompanying ER stress specific changes in noncoding RNAs, especially microRNAs [114,121,131,136,157–165].

4. The Mitochondrial UPR

Since mitochondria play a central role in terms of ROS-produced oxidative stress in brain, the impairment of ATP production and deregulation of mitochondrial function may also deregulate protein import and homeostasis in these organelles, and result in the induction of the mitochondrial UPR (UPRmt) [166–169]. In order to respond to such an insult, the mitochondrial UPR pathway has to adjust both mitochondria and nuclear encoded genes in order to increase the levels of ROS scavengers and mitochondrial chaperones and proteases. Chronic stress can lead to apoptosis [166–170].

It has been suggested that the mitochondrial UPR can serve as a protective mechanism against ATP depletion, mitochondrial protein misfolding or loss of mitochondrial inner membrane potential [168,171]. For example, the activation of UPRmt favors glycolysis [170,172], while at the same time it stimulates mitochondrial ROS removal [168]. The UPRmt has also been associated with a number of human diseases, including cancers, cardiac pathophysiology, neurodegeneration and Alzheimer's disease [168,171,173–175].

While significant progress on deciphering the UPRmt mechanisms was achieved initially in *C. elegans*, it is only recently that the human UPRmt has become better characterized [176]. It has been shown, for example, that the UPRmt can result in the activation of the PERK axis of the UPR and thus increase levels of ATF4, ATF5 and CHOP as well as participate in ISR [166–169,177–180]. Mitochondrial disfunction has also been shown to lead to eIF2 phosphorylation, and this promotes the translation of ATF4, CHOP and activating transcription factor 5 (ATF5). These factors stimulate the transcription of the genes responsible for the recovery from mitochondrial insults including the mitochondrial chaperones [166,168,169,178–180]. ATF4 induces the transcription of the supercomplex assembly factor 1 (*SCAF1*) that supports OXPHOS metabolic reprograming [181]. Furthermore, ATF5 serves as sensor of mitochondrial homeostasis since its activity is inhibited when the protein import into healthy mitochondria is restored [182]. Since ATF5 contains both a mitochondrial translocation signal and a nuclear localization signal. During non-stress conditions, it is selectively imported into mitochondria for subsequent degradation by resident proteases [182].

Depending on the cause of mitochondrial dysfunction, different kinases can phosphorylate eIF2 [176]. Besides the ER stress and oxidative stress-related PERK kinase, eIF2 can be also phosphorylated by ribosome-associated general control nonderepressible 2 (GCN2) during stalled translation [183,184], whereas in the absence of heme or with the binding of the death ligand signal enhancer (DELE1), a mitochondrial protein that is exported to cytosol during stress, the eIF2 heme-regulated inhibitor (HRI) is activated [176,185,186]. Furthermore, eIF2 can also be phosphorylated by protein kinase R (PKR) activated by mitochondrial matrix-generated dsRNA [187]. Interestingly, the ISR-related translational blockage includes blocking the synthesis of the mitochondrial subunits of the channels responsible for protein import to attenuate mitochondrial stress [176,188,189]. Given the importance of mitochondrial homeostasis in the brain, understanding the crosstalk between the mitochondrial and the ER UPR pathways will require further study [190].

5. Oxidative Insults Can Cause ER Stress

Increased cellular oxidation can disrupt ER homeostasis and trigger UPR activation and eventually lead to cell death. These oxidative insult-related ER stressors include deregulation of ER calcium homeostasis, nitrosative stress, and mitochondrially generated ROS, as well as ischemic events, discussed below [191–196]. Calcium homeostasis is a critical component here (Figure 3). Calcium influx to the ER is mediated by pumps from the sarco/endoplasmic reticulum calcium transport ATPase (SERCA) family, whereas the efflux occurs via the inositol 1,4,5-trisphosphate (IP$_3$) receptors (IP$_3$R) channels, the ryanodine receptor (RyR) channels, and a heterogeneous collection of calcium leak pores [28,197,198].

Importantly, although sulfoxidation of cysteine 674 in SERCA will prevent calcium influx to ER, the nitric oxide-mediated glutathionylation of this cysteine residue has an opposite effect [191,199,200]. These independent reports stress the importance of maintaining proper redox homeostasis in terms of ER calcium storage. Furthermore, ROS-dependent posttranslational modifications of IP_3R and RyR channels enhance calcium efflux from ER and consequently impair the calcium-dependent protein folding machinery (calnexin and calreticulin) and lead to the activation of UPR [89,201,202].

A) Deregulation of ER Ca²⁺ homeostasis by ROS

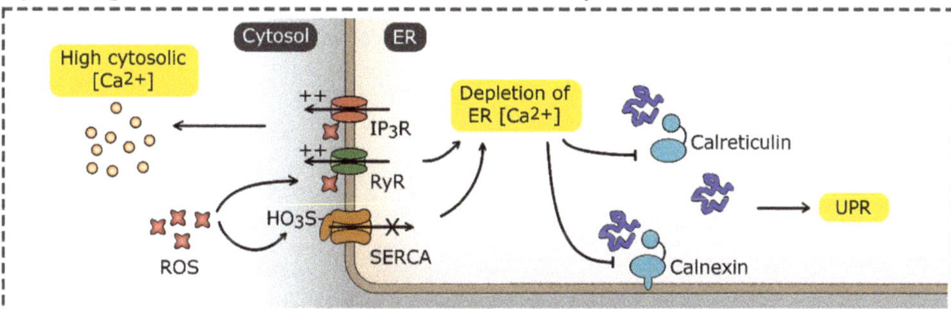

B) Ca²⁺ crosstalk between ER and mitochondrion

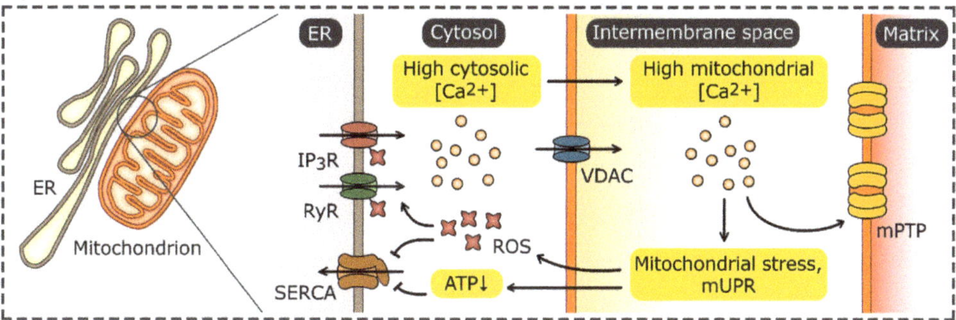

C) Induction of UPR by nitroso-oxidative stress

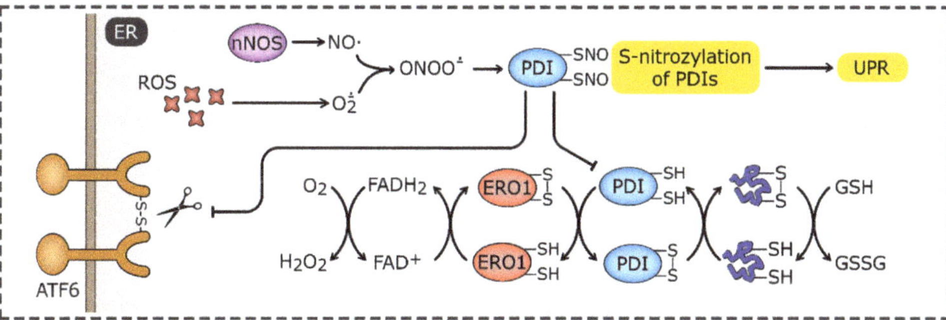

Figure 3. Induction of UPR by oxidative stress. (**A**) Elevated reactive oxygen species (ROS) levels may cause the oxidation of endoplasmic reticulum (ER) calcium transporters, most notably, ryanodine receptors (RyRs), and inositol-1,4,5-trisphosphate (IP_3) receptors (IP_3Rs). Elevated ROS levels also promote sulfoxidation of Cys674 of sarco/endoplasmic reticulum Ca^{2+} ATPase (SERCA). These modifications lead to efflux of Ca^{2+} from ER and impairment of Ca^{2+}-dependent chaperons, calnexin

and calreticulin. (**B**) The disturbance of ER Ca^{2+} homeostasis may spread through mitochondria-associated membranes and target the mitochondrion, causing the Ca^{2+} influx through the voltage-dependent anion channel (VDAC). High Ca^{2+} concentrations induce mitochondrial stress, which leads to activation of the mitochondrial unfolded protein response (mUPR) and formation of mitochondrial permeability transition pores (mPTP). Increased leakage of ROS from electron transport chains and depletion of ATP enhances further ER stress and deregulation of Ca^{2+} homeostasis. (**C**) Increased ROS concentrations combined with the production of NO by nNOS (neuronal nitric oxide synthase) leads to the formation of peroxynitrate (ONOO⁻) which reacts with thiol group of proteins. S-nitrosylation inhibits the activity of modified proteins, including protein disulfide isomerases (PDIs). PDIs, accompanied by ER oxidoreductin 1 (ERO1), catalyze the formation and cleavage of disulfide bonds, and are one of the crucial components of the ER proteostasis system. The reduced-to-oxidized ratio of glutathione (GSH/GSSG), which plays a role analogous to PDIs, may also be increased by the oxidative environment in ER. PDIs also directly affect the UPR sensors and activate transcription factor 6 (ATF6) and inositol-requiring protein 1α (IRE1α).

Calcium depletion of ER can also be attributed to the crosstalk between the ER and mitochondria and the fact that efficient calcium influx to the ER requires ATP. Hence, oxidative stress-related alterations of the mitochondrial calcium pool and function may impair ER calcium balance and activate the UPR (Figure 3B). Mitochondrial associated membrane (MAM) regions of the ER are known to amplify calcium release and signaling [36,203]. Furthermore, the increased release of mitochondrial H_2O_2 also stimulates ER calcium release via the oxidation of IP_3 receptors [201]. Disturbed MAM signaling has been associated with both Alzheimer's disease (AD) and amyotrophic lateral sclerosis (ALS), neurodegenerative diseases that are associated with ER stress [204,205]. Additionally, IP_3R channels are regulated by the ER membrane presenilins that are also considered ER calcium leak channels [206,207], and mutations in the presenilins are associated with AD [208–211]. Although the role of presenilins in maintaining ER calcium homeostasis requires further study, some of the mutations in these proteins were shown to disturb UPR signaling [212].

In neurons, oxidative stress-related damage results in reduced ATP and NADH synthesis and eventually impairment of complex I that leads to increased levels of $O_2^{\cdot-}$ [213]. This leads to ER stress and activation of the apoptotic branch of the UPR, including the ER-stress associated caspase 12 [193,214–216]. Furthermore, the increase in mitochondrial ROS (both $O_2^{\cdot-}$ and H_2O_2) along with the NO synthesized by nNOS can result in formation of peroxynitrite (ONOO⁻) [217] and leads to the formation of S-nitrosylated proteins [218]. Notably, PDIs that facilitate proper disulfide bond formation and rearrangements in ER can be S-nitrosylated, and if so, their activity is inhibited and leads to the accumulation of misfolded polyubiquitinated proteins in ER and activation of the UPR [219,220]. Since increases in PDI activity serve as a neuroprotective mechanism preventing accumulation of immature and misfolded proteins upon ischemia and during neurodegenerative disorders, the oxidative stress-related impairment of these ER resident chaperones can dramatically influence neurodegeneration [221].

Ischemic events in the brain affect mitochondrial function and result in elevated ROS levels and limit ATP production. This would therefore inhibit energy-dependent cellular functions including the maintenance of ion homeostasis and the redox potential [222–227]. Notably, the ischemic ATP level reduction is accompanied by the accumulation of NADH and acyl esters of coenzyme A and carnitine, and these acyl esters were shown to impair both mitochondrial function and structure [228,229]. These changes would impair protein and lipid synthesis, as well as protein folding in ER, and therefore activate the UPR and UPRmt [88,166–169]. An unmet oxygen cellular demand results in increased levels of BiP as well as PERK activation [95,230–242]. This suggests that reduced ATP production due to hypoxia or mitochondrial dysfunction can be at least partially counteracted by reducing global translation by an integrated stress response, whereas the related ATF4 signaling restores the mitochondrial and ER balance [166–169,177].

Although mild and short-lived ischemic events are well controlled by hypoxia-inducible factors (HIFs) that allow both adaptation and survival of neural cells and prevent extensive ROS formation [243–248], the rapid reestablishment of normal oxygen levels is often accompanied by overproduction of ROS and cellular damage that is referred to as ischemia–reperfusion injury [243,244,249–254]. This damage is accompanied by hyperoxidation of NADH in some neurons and consequently enhanced generation of $O_2^{\cdot-}$ and acute oxidative stress [255–257]. Not surprisingly, ischemia–reperfusion injury has been also associated with the rapid depletion of ER calcium and extensive activation of UPR and UPRmt [258–282].

ROS may also react and change properties of other ER-important molecules such as lipids, proteins and nucleic acids and thus impair ER function. For example, mRNA oxidation that has been observed in neurodegenerative diseases, including AD and ALS [61–63], can result in ribosome stalking and disturbances of cotranslational folding that could eventually contribute to ER stress [59]. Furthermore, ROS-related lipid oxidation can alter ER membrane composition that may also activate the UPR via IRE1 or PERK [283–286]. Furthermore, since cholesterol autoxidation is proportional to ROS levels, the oxidative stress can result in increased generation of non-enzymatically produced oxysterol [287] that can also disrupt ER membranes and lead to activation of the UPR [288–290].

6. ER Stress Contributions to Oxidative Stress

Disulfide bond generation in the ER is an oxidative process that utilizes O_2 and H_2O_2 as the electron acceptors [291,292]. Oxygen is required by oxidases such as ER oxidoreductin 1 (ERO1) [293], whereas H_2O_2 is generated by the glutathione peroxidases 7 or 8 (GPX7, GPX8) and peroxiredoxin IV (PRDX4) [292,294–296]. These two types of enzymes are involved in disulfide bond generation complement and control each other since ERO1 catalysis results in H_2O_2 formation that has to be reduced by GPX7 and GPX8 [297]. Notably, PRDX4 reactions rely on other sources of H_2O_2 in ER [297]. PDIs mediate oxidation of cysteine residues in the proteins that require oxidative folding in ER [294]. Although, this oxidative protein folding system is well maintained during normal physiological conditions, during prolonged stress, disulfide bond formation in ER may contribute to oxidative stress through the PERK branch of the UPR [298–300]. During chronic stress, the PERK signals switch from the integrated stress response to the propagation of proapoptotic CHOP signaling. The increased expression of some CHOP target genes such as *ERO1* may contribute to enhanced ROS generation in ER. Upon ER stress, the expression of GPX8 peroxidase increases as well [297], and thus the importance of CHOP-ERO1 axis in inducing oxidative stress in vivo remains unclear. Other studies, however, have indicated that increased ERO1 levels can result in increased efflux of ER calcium through IP$_3$R channels [301,302], and these in turn activate the JNK pathway and stimulate ROS production by the oxidases NOX2 and NOX4 [303,304]. Consequently, ERO1-mediated efflux of ER calcium leads to oxidative stress and amplifies CHOP signaling [303,304]. Furthermore, the ER-stress related increase in H_2O_2 generation leads to elevated oxidized GSH levels and thus further reduces the cellular ROS buffering capacities [299,305].

More importantly, chronic or exaggerated ER stress results in dramatic ER calcium efflux as well as activation of UPR apoptotic signaling that can support mitochondrial ROS release and lead to oxidative stress [146]. As mentioned, UPR-induced intrinsic apoptosis relies on B-cell lymphoma 2 (BCL2) repression and induction of BH3-only proteins, including the BCL-2 interacting mediator of cell death (BIM), NOXA, PUMA, death receptor 5 (DR5), and proto-oncogene c (CRK) [82,306–311]. Such a programmed increase in mitochondrial outer membrane permeability allows the release of cytochrome c, changing the gating of mPTPs, and the balance between ER and mitochondrial calcium pools, all of which leads to mitochondrial dysfunction and ROS generation [207,312,313]. Furthermore, ER stress-related increases in cytosolic calcium may stimulate phospholipase A$_2$ activity and consequently enhance peroxidation of unsaturated lipids and contribute to oxidative stress [314,315].

Taken together, depending on the pathological situation, the chronic or exacerbated activity of this pathway caused by accumulation of mutated misfolded proteins in neurodegenerative diseases such as AD and ALS can also induce ROS production (Figure 4) [64,316].

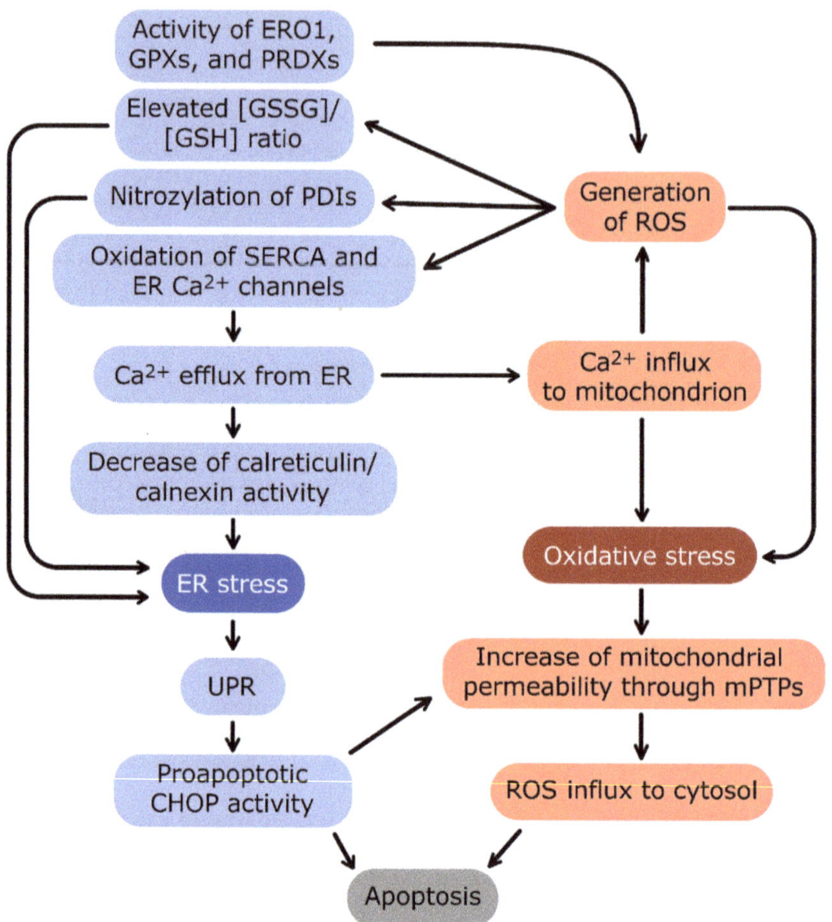

Figure 4. The crosstalk between ER stress and oxidative stress. The main linkage between endoplasmic reticulum (ER) and mitochondrion homeostasis is the Ca^{2+} concentration interdependence.

Ca^{2+} efflux from ER and influx into mitochondria are connected by a positive-feedback loop: oxidative stress and reactive oxygen species (ROS) generation induce the release of Ca^{2+} from ER, and in turn, high Ca^{2+} stimulates the oxidative stress. The important source of ROS in ER is the activity of enzymes catalyzing redox reactions: protein disulfide isomerases (PDIs), ER oxidoreductin 1 (ERO1), glutathione peroxidases (GPXs), and peroxiredoxins (PRDXs). ER stress induces activation of proadaptive unfolded protein response (UPR). In the case of prolonged and excessive stress, UPR activates apoptotic transcription factor CCAAT/enhancer-binding protein homologous protein (CHOP) and severe oxidative stress leads to formation of mitochondrial permeability transition pores (mPTPs). Both pathways trigger eventual apoptosis of cells during unmitigated cellular stress conditions.

7. Discussion

Given the complexity of the processes described above and challenges that the brain cells experience while maintaining oxygen homeostasis, it is important to understand molecular mechanisms that assure their proper functioning and survival, as well as the role that the ER plays in this regulation.

Although it seems obvious that oxidative stress accompanies brain pathologies and aging, the role of proadaptive stage of UPR pathway remains underappreciated both in research and clinical approaches. The majority of current approaches focus on the elimination of death-related signals during chronic ER stress, and that is understandable given the pathomechanisms of many of the neurodegenerative diseases, including ALS, AD, Parkinson's disease (PD), and prion diseases [64,316–324]. In these cases, the chronic ER stress will have devastating effects on cell survival. Notably, some studies have shown the benefits of supporting adaptive UPR activity in these disease models. For example, the neuroprotective effects of the transgenic increased levels of XBP1s in a PD mice model [325] and the use of chemical chaperones such as 4-phenyl butyric acid (4-*PBA*) to reduce stress [326]. Furthermore, the forced activation of ATF6 in forebrain neurons improved functional recovery in a mouse model of stroke and Huntington's disease [327,328].

Alternatively, UPR-inhibiting approaches have also been tested. The PERK pathway inhibitor ISRIB [329] was able to attenuate amyloid β-induced neuronal cell death in AD [330], and was also shown to be promising for therapies targeting ALS [331] and traumatic brain injury (TBI) [332]. Furthermore, the "free radical theory of aging" proposes that the long-term accumulation of oxidative stress incidents will eventually manifest itself by impairing the cellular abilities of maintaining homeostasis, including mitochondrial and ER function [333,334]. Although ROS scavengers seem like a straightforward strategy to cope with neurodegeneration, successful approaches to improve aging-related declines in cognitive function in humans with antioxidants are rarely successful [335,336]. Furthermore, similar limitations were observed during clinical trials using antioxidant strategies in stroke or cardiac ischemia [28,198,337–341]. The main challenges of antioxidant therapies are related to the short half-life of ROS, and this requires scavenger molecules to be extremely efficient, lipid-permeable, and usually used at very high concentrations [28,342].

Thus, development of effective strategies against neurodegeneration and aging requires extension of therapeutic strategies towards other mechanisms that regulate brain cell homeostasis, including the UPR. Notably, a recent study showed the importance of proper balance between the proadaptive and proapoptotic activity of IRE1 in aging brain by demonstrating that XBP1 expression alleviated many of the age-related functional changes [343]. Furthermore, activation of PERK signaling may also have neuroprotective effects [344].

8. Conclusions

Here, we have discussed how regulation of the UPR in the ER and mitochondria deals with oxidative stress, how the two collaborate to regulate redox homeostasis, and how things can go wrong with the high oxygen demands of neuronal cells. More insight, however, is needed to understand how these pathways can be manipulated to control the key translations between the survival and death pathways. Both sides of the UPR pathways need to be considered. The findings discussed here emphasize the role of the adaptive ER stress responses for preserving proper brain cell homeostasis. This suggests that reprograming the UPR pathways in order to increase the cellular survival pathways rather than the apoptotic pathways should be tested. Only a precise understanding of mechanisms governing both brain cell redox homeostasis and its crosstalk with UPR_{mt} and the ER UPR will lead to effective therapies for age-related cognitive decay and neurodegenerative diseases.

Author Contributions: All authors wrote, read, and revised the final version of the manuscript. All authors have read and agreed to the published version of the manuscript. The figures were prepared by J.S.

Funding: This work has been supported by National Science Center "OPUS" 2020/37/B/NZ3/00861 Program to R.B.

Conflicts of Interest: The authors declare no conflict of interest.

References

1. Cobley, J.N.; Fiorello, M.L.; Bailey, D.M. 13 reasons why the brain is susceptible to oxidative stress. *Redox Biol.* **2018**, *15*, 490–503. [CrossRef] [PubMed]
2. Mink, J.W.; Blumenschine, R.J.; Adams, D.B. Ratio of central nervous system to body metabolism in vertebrates: Its constancy and functional basis. *Am. J. Physiol.* **1981**, *241*, R203–R212. [CrossRef]
3. Goyal, M.S.; Hawrylycz, M.; Miller, J.A.; Snyder, A.Z.; Raichle, M.E. Aerobic glycolysis in the human brain is associated with development and neotenous gene expression. *Cell Metab.* **2014**, *19*, 49–57. [CrossRef] [PubMed]
4. Nedergaard, M.; Ransom, B.; Goldman, S.A. New roles for astrocytes: Redefining the functional architecture of the brain. *Trends Neurosci.* **2003**, *26*, 523–530. [CrossRef] [PubMed]
5. Desai, S.; Rocha, M.; Jovin, T.; Jadhav, A. High Variability in Neuronal Loss: Time is Brain, Re-quantified. *Stroke* **2019**, *50*, 34–37. [CrossRef] [PubMed]
6. Bailey, D.M.; Bartsch, P.; Knauth, M.; Baumgartner, R.W. Emerging concepts in acute mountain sickness and high-altitude cerebral edema: From the molecular to the morphological. *Cell Mol. Life Sci.* **2009**, *66*, 3583–3594. [CrossRef]
7. Carter, R. Oxygen: The Molecule that made the World. *J. R. Soc. Med.* **2003**, *96*, 46–47. [CrossRef]
8. Sawyer, D.T.; Valentine, J.S. How Super Is Superoxide. *Acc. Chem. Res.* **1981**, *14*, 393–400. [CrossRef]
9. Winterbourn, C.C. Reconciling the chemistry and biology of reactive oxygen species. *Nat. Chem. Biol.* **2008**, *4*, 278–286. [CrossRef]
10. Sies, H. Biochemistry of Oxidative Stress. *Eur. J. Cancer Clin.* **1987**, *23*, 1798. [CrossRef]
11. Campese, V.M.; Ye, S.H.; Zhong, H.Q. Reactive oxygen species (ROS) and central regulation of the sympathetic nervous system (SNS) activity. *Hypertension* **2002**, *40*, 382.
12. Halliwell, B. Reactive oxygen species and the central nervous system. *J. Neurochem.* **1992**, *59*, 1609–1623. [CrossRef] [PubMed]
13. Gutowicz, M. The influence of reactive oxygen species on the central nervous system. *Postep. Hig. Med. Dosw.* **2011**, *65*, 104–113. [CrossRef]
14. Halliwell, B. Oxidative stress and neurodegeneration: Where are we now? *J. Neurochem.* **2006**, *97*, 1634–1658. [CrossRef] [PubMed]
15. Michalska, P.; Leon, R. When It Comes to an End: Oxidative Stress Crosstalk with Protein Aggregation and Neuroinflammation Induce Neurodegeneration. *Antioxidants* **2020**, *9*, 740. [CrossRef] [PubMed]
16. Quinn, P.M.J.; Ambrosio, A.F.; Alves, C.H. Oxidative Stress, Neuroinflammation and Neurodegeneration: The Chicken, the Egg and the Dinosaur. *Antioxidants* **2022**, *11*, 1554. [CrossRef]
17. Picca, A.; Calvani, R.; Coelho-Junior, H.J.; Landi, F.; Bernabei, R.; Marzetti, E. Mitochondrial Dysfunction, Oxidative Stress, and Neuroinflammation: Intertwined Roads to Neurodegeneration. *Antioxidants* **2020**, *9*, 647. [CrossRef]
18. Zeevalk, G.D.; Bernard, L.P.; Song, C.; Gluck, M.; Ehrhart, J. Mitochondrial inhibition and oxidative stress: Reciprocating players in neurodegeneration. *Antioxid. Redox Signal.* **2005**, *7*, 1117–1139. [CrossRef]
19. Andersen, J.K. Oxidative stress in neurodegeneration: Cause or consequence? *Nat. Med.* **2004**, *10* (Suppl. S7), S18–S25. [CrossRef]
20. Salim, S. Oxidative Stress and the Central Nervous System. *J. Pharmacol. Exp. Ther.* **2017**, *360*, 201–205. [CrossRef]
21. Chance, B.; Sies, H.; Boveris, A. Hydroperoxide metabolism in mammalian organs. *Physiol. Rev.* **1979**, *59*, 527–605. [CrossRef]
22. Bailey, D.M.; Willie, C.K.; Hoiland, R.L.; Bain, A.R.; MacLeod, D.B.; Santoro, M.A.; DeMasi, D.K.; Andrijanic, A.; Mijacika, T.; Barak, O.F.; et al. Surviving Without Oxygen: How Low Can the Human Brain Go? *High Alt. Med. Biol.* **2017**, *18*, 73–79. [CrossRef]
23. Bailey, D.M. Radical dioxygen: From gas to (unpaired!) electrons. *Adv. Exp. Med. Biol.* **2003**, *543*, 201–221.
24. Pryor, W.A.; Houk, K.N.; Foote, C.S.; Fukuto, J.M.; Ignarro, L.J.; Squadrito, G.L.; Davies, K.J. Free radical biology and medicine: It's a gas, man! *Am. J. Physiol. Regul. Integr. Comp. Physiol.* **2006**, *291*, R491–R511. [CrossRef] [PubMed]
25. Zucker, R.S. Calcium- and activity-dependent synaptic plasticity. *Curr. Opin. NeuroBiol.* **1999**, *9*, 305–313. [CrossRef]
26. Wheeler, D.B.; Randall, A.; Tsien, R.W. Roles of N-type and Q-type Ca2+ channels in supporting hippocampal synaptic transmission. *Science* **1994**, *264*, 107–111. [CrossRef]
27. Gorlach, A.; Bertram, K.; Hudecova, S.; Krizanova, O. Calcium and ROS: A mutual interplay. *Redox Biol.* **2015**, *6*, 260–271. [CrossRef] [PubMed]
28. Gleichmann, M.; Mattson, M.P. Neuronal calcium homeostasis and dysregulation. *Antioxid. Redox Signal.* **2011**, *14*, 1261–1273. [CrossRef]
29. Collin, T.; Franconville, R.; Ehrlich, B.E.; Llano, I. Activation of metabotropic glutamate receptors induces periodic burst firing and concomitant cytosolic Ca2+ oscillations in cerebellar interneurons. *J. Neurosci.* **2009**, *29*, 9281–9291. [CrossRef]
30. Huser, J.; Blatter, L.A. Fluctuations in mitochondrial membrane potential caused by repetitive gating of the permeability transition pore. *Biochem. J.* **1999**, *343 Pt 2*, 311–317. [CrossRef] [PubMed]
31. Bernardi, P.; Krauskopf, A.; Basso, E.; Petronilli, V.; Blachly-Dyson, E.; Di Lisa, F.; Forte, M.A. The mitochondrial permeability transition from in vitro artifact to disease target. *FEBS J.* **2006**, *273*, 2077–2099. [CrossRef]
32. Kelley, E.E.; Khoo, N.K.; Hundley, N.J.; Malik, U.Z.; Freeman, B.A.; Tarpey, M.M. Hydrogen peroxide is the major oxidant product of xanthine oxidase. *Free Radic. Biol. Med.* **2010**, *48*, 493–498. [CrossRef]

33. Nishino, T.; Okamoto, K.; Kawaguchi, Y.; Hori, H.; Matsumura, T.; Eger, B.T.; Pai, E.F.; Nishino, T. Mechanism of the conversion of xanthine dehydrogenase to xanthine oxidase: Identification of the two cysteine disulfide bonds and crystal structure of a non-convertible rat liver xanthine dehydrogenase mutant. *J. Biol. Chem.* **2005**, *280*, 24888–24894. [CrossRef]

34. Rowland, A.A.; Voeltz, G.K. Endoplasmic reticulum-mitochondria contacts: Function of the junction. *Nat. Rev. Mol. Cell Biol.* **2012**, *13*, 607–625. [CrossRef] [PubMed]

35. Rizzuto, R.; Pinton, P.; Carrington, W.; Fay, F.S.; Fogarty, K.E.; Lifshitz, L.M.; Tuft, R.A.; Pozzan, T. Close contacts with the endoplasmic reticulum as determinants of mitochondrial Ca^{2+} responses. *Science* **1998**, *280*, 1763–1766. [CrossRef] [PubMed]

36. Booth, D.M.; Enyedi, B.; Geiszt, M.; Varnai, P.; Hajnoczky, G. Redox Nanodomains Are Induced by and Control Calcium Signaling at the ER-Mitochondrial Interface. *Mol. Cell* **2016**, *63*, 240–248. [CrossRef] [PubMed]

37. Rizzuto, R.; Bernardi, P.; Pozzan, T. Mitochondria as all-round players of the calcium game. *J. Physiol.* **2000**, *529 Pt 1*, 37–47. [CrossRef]

38. Nichols, B.J.; Denton, R.M. Towards the molecular basis for the regulation of mitochondrial dehydrogenases by calcium ions. *Mol. Cell Biochem.* **1995**, *149*, 203–212. [CrossRef]

39. Denton, R.M. Regulation of mitochondrial dehydrogenases by calcium ions. *Biochim. Biophys Acta* **2009**, *1787*, 1309–1316. [CrossRef] [PubMed]

40. Murphy, M.P. How mitochondria produce reactive oxygen species. *Biochem. J.* **2009**, *417*, 1–13. [CrossRef] [PubMed]

41. Pryde, K.R.; Hirst, J. Superoxide is produced by the reduced flavin in mitochondrial complex I: A single, unified mechanism that applies during both forward and reverse electron transfer. *J. Biol. Chem.* **2011**, *286*, 18056–18065. [CrossRef]

42. Breckwoldt, M.O.; Pfister, F.M.; Bradley, P.M.; Marinkovic, P.; Williams, P.R.; Brill, M.S.; Plomer, B.; Schmalz, A.; St Clair, D.K.; Naumann, R.; et al. Multiparametric optical analysis of mitochondrial redox signals during neuronal physiology and pathology in vivo. *Nat. Med.* **2014**, *20*, 555–560. [CrossRef] [PubMed]

43. Vasquez, G.E.; Medinas, D.B.; Urra, H.; Hetz, C. Emerging roles of endoplasmic reticulum proteostasis in brain development. *Cells Dev.* **2022**, *170*, 203781. [CrossRef]

44. Cohen-Cory, S. The developing synapse: Construction and modulation of synaptic structures and circuits. *Science* **2002**, *298*, 770–776. [CrossRef]

45. Zeng, H.; Sanes, J.R. Neuronal cell-type classification: Challenges, opportunities and the path forward. *Nat. Rev. Neurosci.* **2017**, *18*, 530–546. [CrossRef]

46. Martinez, G.; Khatiwada, S.; Costa-Mattioli, M.; Hetz, C. ER Proteostasis Control of Neuronal Physiology and Synaptic Function. *Trends Neurosci.* **2018**, *41*, 610–624. [CrossRef] [PubMed]

47. Schwarz, D.S.; Blower, M.D. The endoplasmic reticulum: Structure, function and response to cellular signaling. *Cell Mol. Life Sci.* **2016**, *73*, 79–94. [CrossRef]

48. Kennedy, M.J.; Hanus, C. Architecture and Dynamics of the Neuronal Secretory Network. *Annu. Rev. Cell Dev. Biol.* **2019**, *35*, 543–566. [CrossRef]

49. Balch, W.E.; Morimoto, R.I.; Dillin, A.; Kelly, J.W. Adapting proteostasis for disease intervention. *Science* **2008**, *319*, 916–919. [CrossRef]

50. Hetz, C. Adapting the proteostasis capacity to sustain brain healthspan. *Cell* **2021**, *184*, 1545–1560. [CrossRef] [PubMed]

51. Sossin, W.S.; Costa-Mattioli, M. Translational Control in the Brain in Health and Disease. *Cold Spring Harb. Perspect. Biol.* **2019**, *11*, a032912. [CrossRef]

52. Gidalevitz, T.; Stevens, F.; Argon, Y. Orchestration of secretory protein folding by ER chaperones. *Biochim. Biophys Acta* **2013**, *1833*, 2410–2424. [CrossRef]

53. Pobre, K.F.R.; Poet, G.J.; Hendershot, L.M. The endoplasmic reticulum (ER) chaperone BiP is a master regulator of ER functions: Getting by with a little help from ERdj friends. *J. Biol. Chem.* **2019**, *294*, 2098–2108. [CrossRef]

54. Ellgaard, L.; Ruddock, L.W. The human protein disulphide isomerase family: Substrate interactions and functional properties. *Embo. Rep.* **2005**, *6*, 28–32. [CrossRef] [PubMed]

55. Ninagawa, S.; George, G.; Mori, K. Mechanisms of productive folding and endoplasmic reticulum-associated degradation of glycoproteins and non-glycoproteins. *Biochim. Biophys. Acta Gen. Subj.* **2021**, *1865*, 129812. [CrossRef] [PubMed]

56. Kozlov, G.; Gehring, K. Calnexin cycle-structural features of the ER chaperone system. *FEBS J.* **2020**, *287*, 4322–4340. [CrossRef] [PubMed]

57. Smith, M.H.; Ploegh, H.L.; Weissman, J.S. Road to ruin: Targeting proteins for degradation in the endoplasmic reticulum. *Science* **2011**, *334*, 1086–1090. [CrossRef]

58. Fregno, I.; Molinari, M. Proteasomal and lysosomal clearance of faulty secretory proteins: ER-associated degradation (ERAD) and ER-to-lysosome-associated degradation (ERLAD) pathways. *Crit. Rev. Biochem. Mol. Biol.* **2019**, *54*, 153–163. [CrossRef]

59. Poulsen, H.E.; Specht, E.; Broedbaek, K.; Henriksen, T.; Ellervik, C.; Mandrup-Poulsen, T.; Tonnesen, M.; Nielsen, P.E.; Andersen, H.U.; Weimann, A. RNA modifications by oxidation: A novel disease mechanism? *Free Radic. Biol. Med.* **2012**, *52*, 1353–1361. [CrossRef]

60. Tanaka, M.; Chock, P.B.; Stadtman, E.R. Oxidized messenger RNA induces translation errors. *Proc. Natl. Acad. Sci. USA* **2007**, *104*, 66–71. [CrossRef]

61. Nunomura, A.; Perry, G.; Pappolla, M.A.; Wade, R.; Hirai, K.; Chiba, S.; Smith, M.A. RNA oxidation is a prominent feature of vulnerable neurons in Alzheimer's disease. *J. Neurosci.* **1999**, *19*, 1959–1964. [CrossRef] [PubMed]

62. Ding, Q.; Dimayuga, E.; Keller, J.N. Oxidative stress alters neuronal RNA- and protein-synthesis: Implications for neural viability. *Free Radic. Res.* **2007**, *41*, 903–910. [CrossRef] [PubMed]

63. Chang, Y.; Kong, Q.; Shan, X.; Tian, G.; Ilieva, H.; Cleveland, D.W.; Rothstein, J.D.; Borchelt, D.R.; Wong, P.C.; Lin, C.L. Messenger RNA oxidation occurs early in disease pathogenesis and promotes motor neuron degeneration in ALS. *PLoS ONE* **2008**, *3*, e2849. [CrossRef] [PubMed]

64. Hetz, C.; Saxena, S. ER stress and the unfolded protein response in neurodegeneration. *Nat. Rev. Neurol.* **2017**, *13*, 477–491. [CrossRef] [PubMed]

65. Freeman, O.J.; Mallucci, G.R. The UPR and synaptic dysfunction in neurodegeneration. *Brain Res.* **2016**, *1648*, 530–537. [CrossRef]

66. O'Brien, J.S.; Sampson, E.L. Lipid composition of the normal human brain: Gray matter, white matter, and myelin. *J. Lipid Res.* **1965**, *6*, 537–544. [CrossRef]

67. Rouser, G.; Galli, C.; Kritchevsky, G. Lipid Class Composition of Normal Human Brain and Variations in Metachromatic Leucodystrophy, Tay-Sachs, Niemann-Pick, Chronic Gaucher's and Alzheimer's Diseases. *J. Am. Oil. Chem. Soc.* **1965**, *42*, 404–410. [CrossRef]

68. Puchkov, D.; Haucke, V. Greasing the synaptic vesicle cycle by membrane lipids. *Trends Cell. Biol.* **2013**, *23*, 493–503. [CrossRef]

69. Barber, C.N.; Raben, D.M. Lipid Metabolism Crosstalk in the Brain: Glia and Neurons. *Front. Cell Neurosci.* **2019**, *13*, 212. [CrossRef]

70. Ioannou, M.S.; Jackson, J.; Sheu, S.H.; Chang, C.L.; Weigel, A.V.; Liu, H.; Pasolli, H.A.; Xu, C.S.; Pang, S.; Matthies, D.; et al. Neuron-Astrocyte Metabolic Coupling Protects against Activity-Induced Fatty Acid Toxicity. *Cell* **2019**, *177*, 1522–1535 e1514. [CrossRef]

71. Camargo, N.; Brouwers, J.F.; Loos, M.; Gutmann, D.H.; Smit, A.B.; Verheijen, M.H. High-fat diet ameliorates neurological deficits caused by defective astrocyte lipid metabolism. *FASEB J. Off. Publ. Fed. Am. Soc. Exp. Biol.* **2012**, *26*, 4302–4315. [CrossRef] [PubMed]

72. Chen, J.; Zhang, X.; Kusumo, H.; Costa, L.G.; Guizzetti, M. Cholesterol efflux is differentially regulated in neurons and astrocytes: Implications for brain cholesterol homeostasis. *Biochim. Biophys. Acta* **2013**, *1831*, 263–275. [CrossRef] [PubMed]

73. Jacquemyn, J.; Cascalho, A.; Goodchild, R.E. The ins and outs of endoplasmic reticulum-controlled lipid biosynthesis. *EMBO Rep.* **2017**, *18*, 1905–1921. [CrossRef]

74. Shimano, H. Sterol regulatory element-binding proteins (SREBPs): Transcriptional regulators of lipid synthetic genes. *Prog. Lipid Res.* **2001**, *40*, 439–452. [CrossRef] [PubMed]

75. Hetz, C.; Mollereau, B. Disturbance of endoplasmic reticulum proteostasis in neurodegenerative diseases. *Nat. Rev. Neurosci.* **2014**, *15*, 233–249. [CrossRef]

76. Mallucci, G.R.; Klenerman, D.; Rubinsztein, D.C. Developing Therapies for Neurodegenerative Disorders: Insights from Protein Aggregation and Cellular Stress Responses. *Annu. Rev. Cell Dev. Biol.* **2020**, *36*, 165–189. [CrossRef]

77. Ogen-Shtern, N.; Ben David, T.; Lederkremer, G.Z. Protein aggregation and ER stress. *Brain Res.* **2016**, *1648*, 658–666. [CrossRef]

78. Hamdan, N.; Kritsiligkou, P.; Grant, C.M. ER stress causes widespread protein aggregation and prion formation. *J. Cell Biol.* **2017**, *216*, 2295–2304. [CrossRef]

79. Bravo, R.; Parra, V.; Gatica, D.; Rodriguez, A.E.; Torrealba, N.; Paredes, F.; Wang, Z.V.; Zorzano, A.; Hill, J.A.; Jaimovich, E.; et al. Endoplasmic reticulum and the unfolded protein response: Dynamics and metabolic integration. *Int. Rev. Cell Mol. Biol.* **2013**, *301*, 215–290. [CrossRef]

80. Bartoszewska, S.; Collawn, J.F.; Bartoszewski, R. The Role of the Hypoxia-Related Unfolded Protein Response (UPR) in the Tumor Microenvironment. *Cancers* **2022**, *14*, 4870. [CrossRef]

81. Bartoszewska, S.; Collawn, J.F. Unfolded protein response (UPR) integrated signaling networks determine cell fate during hypoxia. *Cell Mol. Biol. Lett.* **2020**, *25*, 18. [CrossRef] [PubMed]

82. Gebert, M.; Sobolewska, A.; Bartoszewska, S.; Cabaj, A.; Crossman, D.K.; Kroliczewski, J.; Madanecki, P.; Dabrowski, M.; Collawn, J.F.; Bartoszewski, R. Genome-wide mRNA profiling identifies X-box-binding protein 1 (XBP1) as an IRE1 and PUMA repressor. *Cell Mol. Life Sci.* **2021**, *78*, 7061–7080. [CrossRef] [PubMed]

83. Moszynska, A.; Collawn, J.F.; Bartoszewski, R. IRE1 Endoribonuclease Activity Modulates Hypoxic HIF-1alpha Signaling in Human Endothelial Cells. *Biomolecules* **2020**, *10*, 895. [CrossRef]

84. Fu, L.; Rab, A.; Tang, L.; Bebok, Z.; Rowe, S.M.; Bartoszewski, R.; Collawn, J.F. DeltaF508 CFTR surface stability is regulated by DAB2 and CHIP-mediated ubiquitination in post-endocytic compartments. *PLoS ONE* **2015**, *10*, e0123131. [CrossRef]

85. Bartoszewski, R.; Rab, A.; Fu, L.; Bartoszewska, S.; Collawn, J.; Bebok, Z. CFTR expression regulation by the unfolded protein response. *Methods Enzym.* **2011**, *491*, 3–24. [CrossRef]

86. Bartoszewski, R.; Rab, A.; Jurkuvenaite, A.; Mazur, M.; Wakefield, J.; Collawn, J.F.; Bebok, Z. Activation of the unfolded protein response by deltaF508 CFTR. *Am. J. Respir Cell Mol. Biol.* **2008**, *39*, 448–457. [CrossRef]

87. Bartoszewski, R.; Rab, A.; Twitty, G.; Stevenson, L.; Fortenberry, J.; Piotrowski, A.; Dumanski, J.P.; Bebok, Z. The mechanism of cystic fibrosis transmembrane conductance regulator transcriptional repression during the unfolded protein response. *J. Biol. Chem.* **2008**, *283*, 12154–12165. [CrossRef]

88. Almanza, A.; Carlesso, A.; Chintha, C.; Creedican, S.; Doultsinos, D.; Leuzzi, B.; Luis, A.; McCarthy, N.; Montibeller, L.; More, S.; et al. Endoplasmic reticulum stress signalling—From basic mechanisms to clinical applications. *FEBS J.* **2019**, *286*, 241–278. [CrossRef]

89. Hetz, C. The unfolded protein response: Controlling cell fate decisions under ER stress and beyond. *Nat. Rev. Mol. Cell Biol.* **2012**, *13*, 89–102. [CrossRef] [PubMed]

90. Zhou, J.; Liu, C.Y.; Back, S.H.; Clark, R.L.; Peisach, D.; Xu, Z.; Kaufman, R.J. The crystal structure of human IRE1 luminal domain reveals a conserved dimerization interface required for activation of the unfolded protein response. *Proc. Natl. Acad. Sci. USA* **2006**, *103*, 14343–14348. [CrossRef] [PubMed]

91. Carrara, M.; Prischi, F.; Nowak, P.R.; Ali, M.M. Crystal structures reveal transient PERK luminal domain tetramerization in endoplasmic reticulum stress signaling. *EMBO J.* **2015**, *34*, 1589–1600. [CrossRef] [PubMed]

92. Schroder, M.; Kaufman, R.J. The mammalian unfolded protein response. *Annu. Rev. Biochem.* **2005**, *74*, 739–789. [CrossRef]

93. Ye, J.; Rawson, R.B.; Komuro, R.; Chen, X.; Dave, U.P.; Prywes, R.; Brown, M.S.; Goldstein, J.L. ER stress induces cleavage of membrane-bound ATF6 by the same proteases that process SREBPs. *Mol. Cell* **2000**, *6*, 1355–1364. [CrossRef] [PubMed]

94. Haze, K.; Yoshida, H.; Yanagi, H.; Yura, T.; Mori, K. Mammalian transcription factor ATF6 is synthesized as a transmembrane protein and activated by proteolysis in response to endoplasmic reticulum stress. *Mol. Biol. Cell* **1999**, *10*, 3787–3799. [CrossRef] [PubMed]

95. Ye, J.; Koumenis, C. ATF4, an ER stress and hypoxia-inducible transcription factor and its potential role in hypoxia tolerance and tumorigenesis. *Curr. Mol. Med.* **2009**, *9*, 411–416. [CrossRef]

96. Lavoie, H.; Li, J.J.; Thevakumaran, N.; Therrien, M.; Sicheri, F. Dimerization-induced allostery in protein kinase regulation. *Trends Biochem. Sci.* **2014**, *39*, 475–486. [CrossRef]

97. Liu, Z.; Lv, Y.; Zhao, N.; Guan, G.; Wang, J. Protein kinase R-like ER kinase and its role in endoplasmic reticulum stress-decided cell fate. *Cell Death Dis.* **2015**, *6*, e1822. [CrossRef]

98. Baird, T.D.; Wek, R.C. Eukaryotic initiation factor 2 phosphorylation and translational control in metabolism. *Adv. Nutr.* **2012**, *3*, 307–321. [CrossRef]

99. Calabrese, E.J.; Bachmann, K.A.; Bailer, A.J.; Bolger, P.M.; Borak, J.; Cai, L.; Cedergreen, N.; Cherian, M.G.; Chiueh, C.C.; Clarkson, T.W.; et al. Biological stress response terminology: Integrating the concepts of adaptive response and preconditioning stress within a hormetic dose-response framework. *Toxicol. Appl. Pharmacol.* **2007**, *222*, 122–128. [CrossRef]

100. Rzymski, T.; Harris, A.L. The unfolded protein response and integrated stress response to anoxia. *Clin. Cancer Res.* **2007**, *13*, 2537–2540. [CrossRef]

101. Blais, J.; Bell, J.C. Novel therapeutic target: The PERKs of inhibiting the integrated stress response. *Cell Cycle* **2006**, *5*, 2874–2877. [CrossRef] [PubMed]

102. Herman, J.P. Integrated circuits controlling the stress response. *Neurosci. Res.* **2006**, *55*, S25.

103. Rutkowski, D.T.; Kaufman, R.J. All roads lead to ATF4. *Dev. Cell* **2003**, *4*, 442–444. [CrossRef]

104. Wortel, I.M.N.; van der Meer, L.T.; Kilberg, M.S.; van Leeuwen, F.N. Surviving Stress: Modulation of ATF4-Mediated Stress Responses in Normal and Malignant Cells. *Trends Endocrinol. Metab.* **2017**, *28*, 794–806. [CrossRef]

105. Gonen, N.; Sabath, N.; Burge, C.B.; Shalgi, R. Widespread PERK-dependent repression of ER targets in response to ER stress. *Sci. Rep.* **2019**, *9*, 4330. [CrossRef]

106. Han, J.; Back, S.H.; Hur, J.; Lin, Y.H.; Gildersleeve, R.; Shan, J.; Yuan, C.L.; Krokowski, D.; Wang, S.; Hatzoglou, M.; et al. ER-stress-induced transcriptional regulation increases protein synthesis leading to cell death. *Nat. Cell Biol.* **2013**, *15*, 481–490. [CrossRef]

107. He, F.; Zhang, P.; Liu, J.; Wang, R.; Kaufman, R.J.; Yaden, B.C.; Karin, M. ATF4 suppresses hepatocarcinogenesis by inducing SLC7A11 (xCT) to block stress-related ferroptosis. *J. Hepatol.* **2023**, *79*, 362–377. [CrossRef] [PubMed]

108. Lange, P.S.; Chavez, J.C.; Pinto, J.T.; Coppola, G.; Sun, C.W.; Townes, T.M.; Geschwind, D.H.; Ratan, R.R. ATF4 is an oxidative stress-inducible, prodeath transcription factor in neurons in vitro and in vivo. *J. Exp. Med.* **2008**, *205*, 1227–1242. [CrossRef] [PubMed]

109. Novoa, I.; Zeng, H.; Harding, H.P.; Ron, D. Feedback inhibition of the unfolded protein response by GADD34-mediated dephosphorylation of eIF2alpha. *J. Cell Biol.* **2001**, *153*, 1011–1022. [CrossRef]

110. Maurel, M.; Chevet, E.; Tavernier, J.; Gerlo, S. Getting RIDD of RNA: IRE1 in cell fate regulation. *Trends Biochem. Sci.* **2014**, *39*, 245–254. [CrossRef] [PubMed]

111. Yoshida, H.; Matsui, T.; Yamamoto, A.; Okada, T.; Mori, K. XBP1 mRNA is induced by ATF6 and spliced by IRE1 in response to ER stress to produce a highly active transcription factor. *Cell* **2001**, *107*, 881–891. [CrossRef] [PubMed]

112. Ruggiano, A.; Foresti, O.; Carvalho, P. Quality control: ER-associated degradation: Protein quality control and beyond. *J. Cell Biol.* **2014**, *204*, 869–879. [CrossRef] [PubMed]

113. Bartoszewski, R.; Kroliczewski, J.; Piotrowski, A.; Jasiecka, A.J.; Bartoszewska, S.; Vecchio-Pagan, B.; Fu, L.; Sobolewska, A.; Matalon, S.; Cutting, G.R.; et al. Codon bias and the folding dynamics of the cystic fibrosis transmembrane conductance regulator. *Cell Mol. Biol. Lett.* **2016**, *21*, 23. [CrossRef]

114. Bartoszewska, S.; Cabaj, A.; Dabrowski, M.; Collawn, J.F.; Bartoszewski, R. miR-34c-5p modulates X-box-binding protein 1 (XBP1) expression during the adaptive phase of the unfolded protein response. *FASEB J.* **2019**, *33*, 11541–11554. [CrossRef]

115. Kaneko, M.; Yasui, S.; Niinuma, Y.; Arai, K.; Omura, T.; Okuma, Y.; Nomura, Y. A different pathway in the endoplasmic reticulum stress-induced expression of human HRD1 and SEL1 genes. *FEBS Lett.* **2007**, *581*, 5355–5360. [CrossRef]

116. Yamamoto, K.; Suzuki, N.; Wada, T.; Okada, T.; Yoshida, H.; Kaufman, R.J.; Mori, K. Human HRD1 promoter carries a functional unfolded protein response element to which XBP1 but not ATF6 directly binds. *J. Biochem.* **2008**, *144*, 477–486. [CrossRef]

117. Dibdiakova, K.; Saksonova, S.; Pilchova, I.; Klacanova, K.; Tatarkova, Z.; Racay, P. Both thapsigargin- and tunicamycin-induced endoplasmic reticulum stress increases expression of Hrd1 in IRE1-dependent fashion. *Neurol. Res.* **2019**, *41*, 177–188. [CrossRef]
118. Zhang, K.; Kaufman, R.J. Signaling the unfolded protein response from the endoplasmic reticulum. *J. Biol. Chem.* **2004**, *279*, 25935–25938. [CrossRef]
119. Mori, K.; Kawahara, T.; Yoshida, H.; Yanagi, H.; Yura, T. Signalling from endoplasmic reticulum to nucleus: Transcription factor with a basic-leucine zipper motif is required for the unfolded protein-response pathway. *Genes Cells* **1996**, *1*, 803–817. [CrossRef]
120. Yoshida, H.; Okada, T.; Haze, K.; Yanagi, H.; Yura, T.; Negishi, M.; Mori, K. ATF6 activated by proteolysis binds in the presence of NF-Y (CBF) directly to the cis-acting element responsible for the mammalian unfolded protein response. *Mol. Cell Biol.* **2000**, *20*, 6755–6767. [CrossRef] [PubMed]
121. Bartoszewski, R.; Brewer, J.W.; Rab, A.; Crossman, D.K.; Bartoszewska, S.; Kapoor, N.; Fuller, C.; Collawn, J.F.; Bebok, Z. The unfolded protein response (UPR)-activated transcription factor X-box-binding protein 1 (XBP1) induces microRNA-346 expression that targets the human antigen peptide transporter 1 (TAP1) mRNA and governs immune regulatory genes. *J. Biol. Chem.* **2011**, *286*, 41862–41870. [CrossRef]
122. Li, M.; Baumeister, P.; Roy, B.; Phan, T.; Foti, D.; Luo, S.; Lee, A.S. ATF6 as a transcription activator of the endoplasmic reticulum stress element: Thapsigargin stress-induced changes and synergistic interactions with NF-Y and YY1. *Mol. Cell Biol.* **2000**, *20*, 5096–5106. [CrossRef]
123. Schuck, S.; Prinz, W.A.; Thorn, K.S.; Voss, C.; Walter, P. Correction: Membrane expansion alleviates endoplasmic reticulum stress independently of the unfolded protein response. *J. Cell Biol.* **2021**, *220*, jcb.20090707402092021c, Erratum in *J. Cell Biol.* **2009**, *187*, 525–536. [CrossRef] [PubMed]
124. Tsuru, A.; Imai, Y.; Saito, M.; Kohno, K. Novel mechanism of enhancing IRE1alpha-XBP1 signalling via the PERK-ATF4 pathway. *Sci. Rep.* **2016**, *6*, 24217. [CrossRef] [PubMed]
125. Bommiasamy, H.; Back, S.H.; Fagone, P.; Lee, K.; Meshinchi, S.; Vink, E.; Sriburi, R.; Frank, M.; Jackowski, S.; Kaufman, R.J.; et al. ATF6alpha induces XBP1-independent expansion of the endoplasmic reticulum. *J. Cell Sci.* **2009**, *122*, 1626–1636. [CrossRef]
126. Maiuolo, J.; Bulotta, S.; Verderio, C.; Benfante, R.; Borgese, N. Selective activation of the transcription factor ATF6 mediates endoplasmic reticulum proliferation triggered by a membrane protein. *Proc. Natl. Acad. Sci. USA* **2011**, *108*, 7832–7837. [CrossRef]
127. Sriburi, R.; Bommiasamy, H.; Buldak, G.L.; Robbins, G.R.; Frank, M.; Jackowski, S.; Brewer, J.W. Coordinate regulation of phospholipid biosynthesis and secretory pathway gene expression in XBP-1(S)-induced endoplasmic reticulum biogenesis. *J. Biol. Chem.* **2007**, *282*, 7024–7034. [CrossRef] [PubMed]
128. Sriburi, R.; Jackowski, S.; Mori, K.; Brewer, J.W. XBP1: A link between the unfolded protein response, lipid biosynthesis, and biogenesis of the endoplasmic reticulum. *J. Cell Biol.* **2004**, *167*, 35–41. [CrossRef]
129. Lee, A.H.; Chu, G.C.; Iwakoshi, N.N.; Glimcher, L.H. XBP-1 is required for biogenesis of cellular secretory machinery of exocrine glands. *EMBO J.* **2005**, *24*, 4368–4380. [CrossRef]
130. Bartoszewski, R.; Gebert, M.; Janaszak-Jasiecka, A.; Cabaj, A.; Kroliczewski, J.; Bartoszewska, S.; Sobolewska, A.; Crossman, D.K.; Ochocka, R.; Kamysz, W.; et al. Genome-wide mRNA profiling identifies RCAN1 and GADD45A as regulators of the transitional switch from survival to apoptosis during ER stress. *FEBS J* **2020**, *287*, 2923–2947. [CrossRef]
131. Gebert, M.; Bartoszewska, S.; Janaszak-Jasiecka, A.; Moszynska, A.; Cabaj, A.; Kroliczewski, J.; Madanecki, P.; Ochocka, R.J.; Crossman, D.K.; Collawn, J.F.; et al. PIWI proteins contribute to apoptosis during the UPR in human airway epithelial cells. *Sci. Rep.* **2018**, *8*, 16431. [CrossRef]
132. Han, D.; Lerner, A.G.; Vande Walle, L.; Upton, J.P.; Xu, W.; Hagen, A.; Backes, B.J.; Oakes, S.A.; Papa, F.R. IRE1alpha kinase activation modes control alternate endoribonuclease outputs to determine divergent cell fates. *Cell* **2009**, *138*, 562–575. [CrossRef]
133. Urano, F.; Wang, X.Z.; Bertolotti, A.; Zhang, Y.H.; Chung, P.; Harding, H.P.; Ron, D. Coupling of stress in the ER to activation of JNK protein kinases by transmembrane protein kinase IRE1. *Science* **2000**, *287*, 664–666. [CrossRef] [PubMed]
134. Iurlaro, R.; Munoz-Pinedo, C. Cell death induced by endoplasmic reticulum stress. *FEBS J* **2016**, *283*, 2640–2652. [CrossRef] [PubMed]
135. Upton, J.P.; Wang, L.; Han, D.; Wang, E.S.; Huskey, N.E.; Lim, L.; Truitt, M.; McManus, M.T.; Ruggero, D.; Goga, A.; et al. IRE1alpha cleaves select microRNAs during ER stress to derepress translation of proapoptotic Caspase-2. *Science* **2012**, *338*, 818–822. [CrossRef]
136. Gebert, M.; Bartoszewska, S.; Opalinski, L.; Collawn, J.F.; Bartoszewski, R. IRE1-mediated degradation of pre-miR-301a promotes apoptosis through upregulation of GADD45A. *bioRxiv* **2023**. [CrossRef]
137. Chen, L.; Xu, S.; Liu, L.; Wen, X.; Xu, Y.; Chen, J.; Teng, J. Cab45S inhibits the ER stress-induced IRE1-JNK pathway and apoptosis via GRP78/BiP. *Cell Death Dis.* **2014**, *5*, e1219. [CrossRef]
138. Adams, C.J.; Kopp, M.C.; Larburu, N.; Nowak, P.R.; Ali, M.M.U. Structure and Molecular Mechanism of ER Stress Signaling by the Unfolded Protein Response Signal Activator IRE1. *Front. Mol. Biosci.* **2019**, *6*, 11. [CrossRef]
139. Win, S.; Than, T.A.; Fernandez-Checa, J.C.; Kaplowitz, N. JNK interaction with Sab mediates ER stress induced inhibition of mitochondrial respiration and cell death. *Cell Death Dis.* **2014**, *5*, e989. [CrossRef] [PubMed]
140. Walter, F.; O'Brien, A.; Concannon, C.G.; Dussmann, H.; Prehn, J.H.M. ER stress signaling has an activating transcription factor 6alpha (ATF6)-dependent "off-switch". *J. Biol. Chem.* **2018**, *293*, 18270–18284. [CrossRef] [PubMed]

141. Reimertz, C.; Kogel, D.; Rami, A.; Chittenden, T.; Prehn, J.H. Gene expression during ER stress-induced apoptosis in neurons: Induction of the BH3-only protein Bbc3/PUMA and activation of the mitochondrial apoptosis pathway. *J. Cell Biol.* **2003**, *162*, 587–597. [CrossRef]
142. Gupta, S.; Giricz, Z.; Natoni, A.; Donnelly, N.; Deegan, S.; Szegezdi, E.; Samali, A. NOXA contributes to the sensitivity of PERK-deficient cells to ER stress. *FEBS Lett.* **2012**, *586*, 4023–4030. [CrossRef]
143. Wang, Q.; Mora-Jensen, H.; Weniger, M.A.; Perez-Galan, P.; Wolford, C.; Hai, T.; Ron, D.; Chen, W.; Trenkle, W.; Wiestner, A.; et al. ERAD inhibitors integrate ER stress with an epigenetic mechanism to activate BH3-only protein NOXA in cancer cells. *Proc. Natl. Acad. Sci. USA* **2009**, *106*, 2200–2205. [CrossRef] [PubMed]
144. Rosebeck, S.; Sudini, K.; Chen, T.; Leaman, D.W. Involvement of Noxa in mediating cellular ER stress responses to lytic virus infection. *Virology* **2011**, *417*, 293–303. [CrossRef]
145. Shibue, T.; Suzuki, S.; Okamoto, H.; Yoshida, H.; Ohba, Y.; Takaoka, A.; Taniguchi, T. Differential contribution of Puma and Noxa in dual regulation of p53-mediated apoptotic pathways. *EMBO J.* **2006**, *25*, 4952–4962. [CrossRef]
146. Figueira, T.R.; Barros, M.H.; Camargo, A.A.; Castilho, R.F.; Ferreira, J.C.; Kowaltowski, A.J.; Sluse, F.E.; Souza-Pinto, N.C.; Vercesi, A.E. Mitochondria as a source of reactive oxygen and nitrogen species: From molecular mechanisms to human health. *Antioxid. Redox Signal.* **2013**, *18*, 2029–2074. [CrossRef] [PubMed]
147. Saveljeva, S.; Mc Laughlin, S.L.; Vandenabeele, P.; Samali, A.; Bertrand, M.J. Endoplasmic reticulum stress induces ligand-independent TNFR1-mediated necroptosis in L929 cells. *Cell Death Dis.* **2015**, *6*, e1587. [CrossRef] [PubMed]
148. Livezey, M.; Huang, R.; Hergenrother, P.J.; Shapiro, D.J. Strong and sustained activation of the anticipatory unfolded protein response induces necrotic cell death. *Cell Death Differ.* **2018**, *25*, 1796–1807. [CrossRef]
149. Shirjang, S.; Mansoori, B.; Asghari, S.; Duijf, P.H.G.; Mohammadi, A.; Gjerstorff, M.; Baradaran, B. MicroRNAs in cancer cell death pathways: Apoptosis and necroptosis. *Free Radic. Biol. Med.* **2019**, *139*, 1–15. [CrossRef]
150. Kishino, A.; Hayashi, K.; Maeda, M.; Jike, T.; Hidai, C.; Nomura, Y.; Oshima, T. Caspase-8 Regulates Endoplasmic Reticulum Stress-Induced Necroptosis Independent of the Apoptosis Pathway in Auditory Cells. *Int. J. Mol. Sci.* **2019**, *20*, 5896. [CrossRef]
151. Ding, B.; Parmigiani, A.; Divakaruni, A.S.; Archer, K.; Murphy, A.N.; Budanov, A.V. Sestrin2 is induced by glucose starvation via the unfolded protein response and protects cells from non-canonical necroptotic cell death. *Sci. Rep.* **2016**, *6*, 22538. [CrossRef]
152. Cheng, S.B.; Nakashima, A.; Huber, W.J.; Davis, S.; Banerjee, S.; Huang, Z.; Saito, S.; Sadovsky, Y.; Sharma, S. Pyroptosis is a critical inflammatory pathway in the placenta from early onset preeclampsia and in human trophoblasts exposed to hypoxia and endoplasmic reticulum stressors. *Cell Death Dis.* **2019**, *10*, 927. [CrossRef]
153. Yang, Z.; Wang, Y.; Zhang, Y.; He, X.; Zhong, C.Q.; Ni, H.; Chen, X.; Liang, Y.; Wu, J.; Zhao, S.; et al. RIP3 targets pyruvate dehydrogenase complex to increase aerobic respiration in TNF-induced necroptosis. *Nat. Cell Biol.* **2018**, *20*, 186–197. [CrossRef]
154. Qiu, X.; Zhang, Y.; Han, J. RIP3 is an upregulator of aerobic metabolism and the enhanced respiration by necrosomal RIP3 feeds back on necrosome to promote necroptosis. *Cell Death Differ.* **2018**, *25*, 821–824. [CrossRef]
155. Fulda, S. Alternative cell death pathways and cell metabolism. *Int. J. Cell Biol.* **2013**, *2013*, 463637. [CrossRef]
156. Gong, Y.; Fan, Z.; Luo, G.; Yang, C.; Huang, Q.; Fan, K.; Cheng, H.; Jin, K.; Ni, Q.; Yu, X.; et al. The role of necroptosis in cancer biology and therapy. *Mol. Cancer* **2019**, *18*, 100. [CrossRef] [PubMed]
157. Byrd, A.E.; Brewer, J.W. Micro(RNA)managing endoplasmic reticulum stress. *IUBMB Life* **2013**, *65*, 373–381. [CrossRef] [PubMed]
158. Kim, T.; Croce, C.M. MicroRNA and ER stress in cancer. *Semin Cancer Biol.* **2021**, *75*, 3–14. [CrossRef] [PubMed]
159. Mukherji, S.; Ebert, M.S.; Zheng, G.X.; Tsang, J.S.; Sharp, P.A.; van Oudenaarden, A. MicroRNAs can generate thresholds in target gene expression. *Nat. Genet.* **2011**, *43*, 854–859. [CrossRef] [PubMed]
160. Byrd, A.; Brewer, J. MicroRNA-mediated repression of XBP1: A novel mechanism for regulation of a UPR transcriptional activator. *J. Immunol.* **2011**, *186*. [CrossRef]
161. Maurel, M.; Chevet, E. Endoplasmic reticulum stress signaling: The microRNA connection. *Am. J. Physiol. Cell Physiol.* **2013**, *304*, C1117–C1126. [CrossRef] [PubMed]
162. Cheung, O.; Mirshahi, F.; Min, H.; Zhou, H.; Fuchs, M.; Sanyal, A.J. Silencing Microrna Mir-34a and 451 Promotes Recovery from Unfolded Protein Response (Upr) and Reverses Nonalcoholic Fatty Liver Disease (Nafld). *Hepatology* **2008**, *48*, 366a–367a.
163. Bartoszewska, S.; Kamysz, W.; Jakiela, B.; Sanak, M.; Kroliczewski, J.; Bebok, Z.; Bartoszewski, R.; Collawn, J.F. miR-200b downregulates CFTR during hypoxia in human lung epithelial cells. *Cell Mol. Biol. Lett.* **2017**, *22*, 23. [CrossRef]
164. Bartoszewska, S.; Kochan, K.; Madanecki, P.; Piotrowski, A.; Ochocka, R.; Collawn, J.F.; Bartoszewski, R. Regulation of the unfolded protein response by microRNAs. *Cell Mol. Biol. Lett.* **2013**, *18*, 555–578. [CrossRef]
165. Bartoszewska, S.; Slawski, J.; Collawn, J.F.; Bartoszewski, R. HIF-1-Induced hsa-miR-429: Understanding Its Direct Targets as the Key to Developing Cancer Diagnostics and Therapies. *Cancers* **2023**, *15*, 2903. [CrossRef]
166. Melber, A.; Haynes, C.M. UPR(mt) regulation and output: A stress response mediated by mitochondrial-nuclear communication. *Cell Res.* **2018**, *28*, 281–295. [CrossRef]
167. Kueh, H.Y.; Niethammer, P.; Mitchison, T.J. Maintenance of mitochondrial oxygen homeostasis by cosubstrate compensation. *Biophys. J.* **2013**, *104*, 1338–1348. [CrossRef] [PubMed]
168. Shpilka, T.; Haynes, C.M. The mitochondrial UPR: Mechanisms, physiological functions and implications in ageing. *Nat. Rev. Mol. Cell Biol.* **2018**, *19*, 109–120. [CrossRef]
169. Munch, C. The different axes of the mammalian mitochondrial unfolded protein response. *BMC Biol.* **2018**, *16*, 81. [CrossRef]

170. Nargund, A.M.; Pellegrino, M.W.; Fiorese, C.J.; Baker, B.M.; Haynes, C.M. Mitochondrial import efficiency of ATFS-1 regulates mitochondrial UPR activation. *Science* **2012**, *337*, 587–590. [CrossRef] [PubMed]

171. Quiles, J.M.; Gustafsson, A.B. Mitochondrial Quality Control and Cellular Proteostasis: Two Sides of the Same Coin. *Front. Physiol.* **2020**, *11*, 515. [CrossRef]

172. Nargund, A.M.; Fiorese, C.J.; Pellegrino, M.W.; Deng, P.; Haynes, C.M. Mitochondrial and nuclear accumulation of the transcription factor ATFS-1 promotes OXPHOS recovery during the UPR(mt). *Mol. Cell* **2015**, *58*, 123–133. [CrossRef] [PubMed]

173. Sorrentino, V.; Romani, M.; Mouchiroud, L.; Beck, J.S.; Zhang, H.; D'Amico, D.; Moullan, N.; Potenza, F.; Schmid, A.W.; Rietsch, S.; et al. Enhancing mitochondrial proteostasis reduces amyloid-beta proteotoxicity. *Nature* **2017**, *552*, 187–193. [CrossRef]

174. Roth, K.G.; Mambetsariev, I.; Kulkarni, P.; Salgia, R. The Mitochondrion as an Emerging Therapeutic Target in Cancer. *Trends Mol. Med.* **2020**, *26*, 119–134. [CrossRef] [PubMed]

175. Wilkins, H.M.; Weidling, I.W.; Ji, Y.; Swerdlow, R.H. Mitochondria-Derived Damage-Associated Molecular Patterns in Neurodegeneration. *Front. Immunol.* **2017**, *8*, 508. [CrossRef] [PubMed]

176. Anderson, N.S.; Haynes, C.M. Folding the Mitochondrial UPR into the Integrated Stress Response. *Trends Cell Biol.* **2020**, *30*, 428–439. [CrossRef]

177. Baker, B.M.; Nargund, A.M.; Sun, T.; Haynes, C.M. Protective coupling of mitochondrial function and protein synthesis via the eIF2alpha kinase GCN-2. *PLoS Genet.* **2012**, *8*, e1002760. [CrossRef]

178. Aldridge, J.E.; Horibe, T.; Hoogenraad, N.J. Discovery of genes activated by the mitochondrial unfolded protein response (mtUPR) and cognate promoter elements. *PLoS ONE* **2007**, *2*, e874. [CrossRef] [PubMed]

179. Munch, C.; Harper, J.W. Mitochondrial unfolded protein response controls matrix pre-RNA processing and translation. *Nature* **2016**, *534*, 710–713. [CrossRef]

180. Quiros, P.M.; Prado, M.A.; Zamboni, N.; D'Amico, D.; Williams, R.W.; Finley, D.; Gygi, S.P.; Auwerx, J. Multi-omics analysis identifies ATF4 as a key regulator of the mitochondrial stress response in mammals. *J. Cell Biol.* **2017**, *216*, 2027–2045. [CrossRef]

181. Balsa, E.; Soustek, M.S.; Thomas, A.; Cogliati, S.; Garcia-Poyatos, C.; Martin-Garcia, E.; Jedrychowski, M.; Gygi, S.P.; Enriquez, J.A.; Puigserver, P. ER and Nutrient Stress Promote Assembly of Respiratory Chain Supercomplexes through the PERK-eIF2alpha Axis. *Mol. Cell* **2019**, *74*, 877–890 e876. [CrossRef]

182. Fiorese, C.J.; Schulz, A.M.; Lin, Y.F.; Rosin, N.; Pellegrino, M.W.; Haynes, C.M. The Transcription Factor ATF5 Mediates a Mammalian Mitochondrial UPR. *Curr. Biol.* **2016**, *26*, 2037–2043. [CrossRef] [PubMed]

183. Inglis, A.J.; Masson, G.R.; Shao, S.; Perisic, O.; McLaughlin, S.H.; Hegde, R.S.; Williams, R.L. Activation of GCN2 by the ribosomal P-stalk. *Proc. Natl. Acad. Sci. USA* **2019**, *116*, 4946–4954. [CrossRef] [PubMed]

184. Ishimura, R.; Nagy, G.; Dotu, I.; Chuang, J.H.; Ackerman, S.L. Activation of GCN2 kinase by ribosome stalling links translation elongation with translation initiation. *Elife* **2016**, *5*, e14295. [CrossRef] [PubMed]

185. Rafie-Kolpin, M.; Chefalo, P.J.; Hussain, Z.; Hahn, J.; Uma, S.; Matts, R.L.; Chen, J.J. Two heme-binding domains of heme-regulated eukaryotic initiation factor-2alpha kinase. N terminus and kinase insertion. *J. Biol. Chem.* **2000**, *275*, 5171–5178. [CrossRef]

186. Guo, X.; Aviles, G.; Liu, Y.; Tian, R.; Unger, B.A.; Lin, Y.T.; Wiita, A.P.; Xu, K.; Correia, M.A.; Kampmann, M. Mitochondrial stress is relayed to the cytosol by an OMA1-DELE1-HRI pathway. *Nature* **2020**, *579*, 427–432. [CrossRef]

187. Kim, Y.; Park, J.; Kim, S.; Kim, M.; Kang, M.G.; Kwak, C.; Kang, M.; Kim, B.; Rhee, H.W.; Kim, V.N. PKR Senses Nuclear and Mitochondrial Signals by Interacting with Endogenous Double-Stranded RNAs. *Mol. Cell* **2018**, *71*, 1051–1063 e1056. [CrossRef]

188. Rainbolt, T.K.; Atanassova, N.; Genereux, J.C.; Wiseman, R.L. Stress-regulated translational attenuation adapts mitochondrial protein import through Tim17A degradation. *Cell Metab.* **2013**, *18*, 908–919. [CrossRef]

189. Wang, X.; Zuo, X.; Kucejova, B.; Chen, X.J. Reduced cytosolic protein synthesis suppresses mitochondrial degeneration. *Nat. Cell Biol.* **2008**, *10*, 1090–1097. [CrossRef]

190. Hu, D.; Liu, Z.; Qi, X. UPR^{mt} activation protects against MPP$^+$-induced toxicity in a cell culture model of Parkinson's disease. *Biochem. Biophys. Res. Commun.* **2021**, *569*, 17–22. [CrossRef]

191. Eletto, D.; Chevet, E.; Argon, Y.; Appenzeller-Herzog, C. Redox controls UPR to control redox. *J. Cell Sci.* **2014**, *127*, 3649–3658. [CrossRef]

192. Dejeans, N.; Tajeddine, N.; Beck, R.; Verrax, J.; Taper, H.; Gailly, P.; Calderon, P.B. Endoplasmic reticulum calcium release potentiates the ER stress and cell death caused by an oxidative stress in MCF-7 cells. *Biochem. Pharmacol.* **2010**, *79*, 1221–1230. [CrossRef]

193. Inoue, T.; Suzuki-Karasaki, Y. Mitochondrial superoxide mediates mitochondrial and endoplasmic reticulum dysfunctions in TRAIL-induced apoptosis in Jurkat cells. *Free Radic. Biol. Med.* **2013**, *61*, 273–284. [CrossRef] [PubMed]

194. Santos, C.X.; Tanaka, L.Y.; Wosniak, J.; Laurindo, F.R. Mechanisms and implications of reactive oxygen species generation during the unfolded protein response: Roles of endoplasmic reticulum oxidoreductases, mitochondrial electron transport, and NADPH oxidase. *Antioxid. Redox Signal.* **2009**, *11*, 2409–2427. [CrossRef] [PubMed]

195. Uehara, T.; Nakamura, T.; Yao, D.; Shi, Z.Q.; Gu, Z.; Ma, Y.; Masliah, E.; Nomura, Y.; Lipton, S.A. S-nitrosylated protein-disulphide isomerase links protein misfolding to neurodegeneration. *Nature* **2006**, *441*, 513–517. [CrossRef]

196. Yang, W.; Paschen, W. Unfolded protein response in brain ischemia: A timely update. *J. Cereb. Blood Flow. Metab.* **2016**, *36*, 2044–2050. [CrossRef]

197. Berridge, M.J.; Bootman, M.D.; Roderick, H.L. Calcium signalling: Dynamics, homeostasis and remodelling. *Nat. Rev. Mol. Cell Biol.* **2003**, *4*, 517–529. [CrossRef]

198. Popugaeva, E.; Pchitskaya, E.; Bezprozvanny, I. Dysregulation of neuronal calcium homeostasis in Alzheimer's disease—A therapeutic opportunity? *Biochem. Biophys. Res. Commun.* **2017**, *483*, 998–1004. [CrossRef]

199. Raturi, A.; Ortiz-Sandoval, C.; Simmen, T. Redox dependence of endoplasmic reticulum (ER) Ca^{2+} signaling. *Histol. Histopathol.* **2014**, *29*, 543–552. [CrossRef]

200. Tong, X.; Evangelista, A.; Cohen, R.A. Targeting the redox regulation of SERCA in vascular physiology and disease. *Curr. Opin Pharmacol.* **2010**, *10*, 133–138. [CrossRef]

201. Bansaghi, S.; Golenar, T.; Madesh, M.; Csordas, G.; RamachandraRao, S.; Sharma, K.; Yule, D.I.; Joseph, S.K.; Hajnoczky, G. Isoform- and species-specific control of inositol 1,4,5-trisphosphate (IP3) receptors by reactive oxygen species. *J. Biol. Chem.* **2014**, *289*, 8170–8181. [CrossRef]

202. Csordas, G.; Hajnoczky, G. SR/ER-mitochondrial local communication: Calcium and ROS. *Biochim. Biophys. Acta* **2009**, *1787*, 1352–1362. [CrossRef] [PubMed]

203. Sies, H. Hydrogen peroxide as a central redox signaling molecule in physiological oxidative stress: Oxidative eustress. *Redox Biol.* **2017**, *11*, 613–619. [CrossRef] [PubMed]

204. Paillusson, S.; Stoica, R.; Gomez-Suaga, P.; Lau, D.H.W.; Mueller, S.; Miller, T.; Miller, C.C.J. There's Something Wrong with my MAM; the ER-Mitochondria Axis and Neurodegenerative Diseases. *Trends Neurosci.* **2016**, *39*, 146–157. [CrossRef] [PubMed]

205. Stoica, R.; De Vos, K.J.; Paillusson, S.; Mueller, S.; Sancho, R.M.; Lau, K.F.; Vizcay-Barrena, G.; Lin, W.L.; Xu, Y.F.; Lewis, J.; et al. ER-mitochondria associations are regulated by the VAPB-PTPIP51 interaction and are disrupted by ALS/FTD-associated TDP-43. *Nat. Commun.* **2014**, *5*, 3996. [CrossRef] [PubMed]

206. Tu, H.; Nelson, O.; Bezprozvanny, A.; Wang, Z.; Lee, S.F.; Hao, Y.H.; Serneels, L.; De Strooper, B.; Yu, G.; Bezprozvanny, I. Presenilins form ER Ca2+ leak channels, a function disrupted by familial Alzheimer's disease-linked mutations. *Cell* **2006**, *126*, 981–993. [CrossRef]

207. Kiviluoto, S.; Vervliet, T.; Ivanova, H.; Decuypere, J.P.; De Smedt, H.; Missiaen, L.; Bultynck, G.; Parys, J.B. Regulation of inositol 1,4,5-trisphosphate receptors during endoplasmic reticulum stress. *Biochim. Biophys. Acta* **2013**, *1833*, 1612–1624. [CrossRef]

208. Yang, Y.; Bagyinszky, E.; An, S.S.A. Presenilin-1 (PSEN1) Mutations: Clinical Phenotypes beyond Alzheimer's Disease. *Int. J. Mol. Sci.* **2023**, *24*, 8417. [CrossRef]

209. Hernandez-Sapiens, M.A.; Reza-Zaldivar, E.E.; Marquez-Aguirre, A.L.; Gomez-Pinedo, U.; Matias-Guiu, J.; Cevallos, R.R.; Mateos-Diaz, J.C.; Sanchez-Gonzalez, V.J.; Canales-Aguirre, A.A. Presenilin mutations and their impact on neuronal differentiation in Alzheimer's disease. *Neural. Regen. Res.* **2022**, *17*, 31–37. [CrossRef]

210. Cheng, Z.; Shang, Y.; Xu, X.; Dong, Z.; Zhang, Y.; Du, Z.; Lu, X.; Zhang, T. Presenilin 1 mutation likely contributes to U1 small nuclear RNA dysregulation and Alzheimer's disease-like symptoms. *NeuroBiol. Aging* **2021**, *100*, 1–10. [CrossRef]

211. Sutovsky, S.; Smolek, T.; Turcani, P.; Petrovic, R.; Brandoburova, P.; Jadhav, S.; Novak, P.; Attems, J.; Zilka, N. Neuropathology and biochemistry of early onset familial Alzheimer's disease caused by presenilin-1 missense mutation Thr116Asn. *J. Neural. Transm.* **2018**, *125*, 965–976. [CrossRef] [PubMed]

212. Katayama, T.; Imaizumi, K.; Honda, A.; Yoneda, T.; Kudo, T.; Takeda, M.; Mori, K.; Rozmahel, R.; Fraser, P.; George-Hyslop, P.S.; et al. Disturbed activation of endoplasmic reticulum stress transducers by familial Alzheimer's disease-linked presenilin-1 mutations. *J. Biol. Chem.* **2001**, *276*, 43446–43454. [CrossRef]

213. Lieberman, D.N.; Mody, I. Regulation of NMDA channel function by endogenous Ca^{2+}-dependent phosphatase. *Nature* **1994**, *369*, 235–239. [CrossRef]

214. Nakagawa, T.; Yuan, J. Cross-talk between two cysteine protease families. Activation of caspase-12 by calpain in apoptosis. *J. Cell Biol.* **2000**, *150*, 887–894. [CrossRef] [PubMed]

215. Szegezdi, E.; Fitzgerald, U.; Samali, A. Caspase-12 and ER-stress-mediated apoptosis: The story so far. *Ann. N. Y. Acad. Sci.* **2003**, *1010*, 186–194. [CrossRef]

216. Momoi, T. Caspases involved in ER stress-mediated cell death. *J. Chem. Neuroanat.* **2004**, *28*, 101–105. [CrossRef]

217. Carballal, S.; Bartesaghi, S.; Radi, R. Kinetic and mechanistic considerations to assess the biological fate of peroxynitrite. *Biochim. Biophys. Acta* **2014**, *1840*, 768–780. [CrossRef] [PubMed]

218. Hlaing, K.H.; Clement, M.V. Formation of protein S-nitrosylation by reactive oxygen species. *Free Radic. Res.* **2014**, *48*, 996–1010. [CrossRef] [PubMed]

219. Oka, O.B.; Bulleid, N.J. Forming disulfides in the endoplasmic reticulum. *Biochim. Biophys. Acta* **2013**, *1833*, 2425–2429. [CrossRef]

220. Nadanaka, S.; Okada, T.; Yoshida, H.; Mori, K. Role of disulfide bridges formed in the luminal domain of ATF6 in sensing endoplasmic reticulum stress. *Mol. Cell Biol.* **2007**, *27*, 1027–1043. [CrossRef]

221. Nakamura, T.; Lipton, S.A. Redox modulation by S-nitrosylation contributes to protein misfolding, mitochondrial dynamics, and neuronal synaptic damage in neurodegenerative diseases. *Cell Death Differ.* **2011**, *18*, 1478–1486. [CrossRef]

222. Thiele, R.H. Subcellular Energetics and Metabolism: A Cross-Species Framework. *Anesth. Analg.* **2017**, *124*, 1857–1871. [CrossRef]

223. Koritzinsky, M.; Levitin, F.; van den Beucken, T.; Rumantir, R.A.; Harding, N.J.; Chu, K.C.; Boutros, P.C.; Braakman, I.; Wouters, B.G. Two phases of disulfide bond formation have differing requirements for oxygen. *J. Cell Biol.* **2013**, *203*, 615–627. [CrossRef]

224. May, D.; Itin, A.; Gal, O.; Kalinski, H.; Feinstein, E.; Keshet, E. Ero1-L alpha plays a key role in a HIF-1-mediated pathway to improve disulfide bond formation and VEGF secretion under hypoxia: Implication for cancer. *Oncogene* **2005**, *24*, 1011–1020. [CrossRef]

225. Arnould, T.; Michiels, C.; Alexandre, I.; Remacle, J. Effect of hypoxia upon intracellular calcium concentration of human endothelial cells. *J. Cell Physiol.* **1992**, *152*, 215–221. [CrossRef]
226. Dorner, A.J.; Wasley, L.C.; Kaufman, R.J. Protein dissociation from GRP78 and secretion are blocked by depletion of cellular ATP levels. *Proc. Natl. Acad. Sci. USA* **1990**, *87*, 7429–7432. [CrossRef] [PubMed]
227. Bartoszewski, R.; Matalon, S.; Collawn, J.F. Ion channels of the lung and their role in disease pathogenesis. *Am. J. Physiol. Lung Cell. Mol. Physiol.* **2017**, *313*, L859–L872. [CrossRef]
228. Ford, D.A.; Han, X.; Horner, C.C.; Gross, R.W. Accumulation of unsaturated acylcarnitine molecular species during acute myocardial ischemia: Metabolic compartmentalization of products of fatty acyl chain elongation in the acylcarnitine pool. *Biochemistry* **1996**, *35*, 7903–7909. [CrossRef] [PubMed]
229. Neely, J.R.; Feuvray, D. Metabolic products and myocardial ischemia. *Am. J. Pathol.* **1981**, *102*, 282–291.
230. Binet, F.; Sapieha, P. ER Stress and Angiogenesis. *Cell Metab.* **2015**, *22*, 560–575. [CrossRef] [PubMed]
231. Kalinowski, L.; Janaszak-Jasiecka, A.; Siekierzycka, A.; Bartoszewska, S.; Wozniak, M.; Lejnowski, D.; Collawn, J.F.; Bartoszewski, R. Posttranscriptional and transcriptional regulation of endothelial nitric-oxide synthase during hypoxia: The role of microRNAs. *Cell Mol. Biol. Lett.* **2016**, *21*, 16. [CrossRef]
232. Sun, L.L.; Chen, C.M.; Zhang, J.; Wang, J.; Yang, C.Z.; Lin, L.Z. Glucose-Regulated Protein 78 Signaling Regulates Hypoxia-Induced Epithelial-Mesenchymal Transition in A549 Cells. *Front. Oncol.* **2019**, *9*, 137. [CrossRef] [PubMed]
233. Song, M.S.; Park, Y.K.; Lee, J.H.; Park, K. Induction of glucose-regulated protein 78 by chronic hypoxia in human gastric tumor cells through a protein kinase C-epsilon/ERK/AP-1 signaling cascade. *Cancer Res.* **2001**, *61*, 8322–8330. [PubMed]
234. Koong, A.C.; Auger, E.A.; Chen, E.Y.; Giaccia, A.J. The Regulation of Grp78 and Messenger-Rna Levels by Hypoxia Is Modulated by Protein-Kinase-C Activators and Inhibitors. *Radiat. Res.* **1994**, *138*, S60–S63. [CrossRef] [PubMed]
235. Raiter, A.; Weiss, C.; Bechor, Z.; Ben-Dor, I.; Battler, A.; Kaplan, B.; Hardy, B. Activation of GRP78 on endothelial cell membranes by an ADAM15-derived peptide induces angiogenesis. *J. Vasc. Res.* **2010**, *47*, 399–411. [CrossRef]
236. Koong, A.C.; Chen, E.Y.; Lee, A.S.; Brown, J.M.; Giaccia, A.J. Increased cytotoxicity of chronic hypoxic cells by molecular inhibition of GRP78 induction. *Int. J. Radiat. Oncol. Biol. Phys.* **1994**, *28*, 661–666. [CrossRef]
237. Koumenis, C.; Naczki, C.; Koritzinsky, M.; Rastani, S.; Diehl, A.; Sonenberg, N.; Koromilas, A.; Wouters, B.G. Regulation of protein synthesis by hypoxia via activation of the endoplasmic reticulum kinase PERK and phosphorylation of the translation initiation factor eIF2alpha. *Mol. Cell Biol.* **2002**, *22*, 7405–7416. [CrossRef] [PubMed]
238. Blais, J.D.; Filipenko, V.; Bi, M.; Harding, H.P.; Ron, D.; Koumenis, C.; Wouters, B.G.; Bell, J.C. Activating transcription factor 4 is translationally regulated by hypoxic stress. *Mol. Cell Biol.* **2004**, *24*, 7469–7482. [CrossRef]
239. Scheuner, D.; Song, B.; McEwen, E.; Liu, C.; Laybutt, R.; Gillespie, P.; Saunders, T.; Bonner-Weir, S.; Kaufman, R.J. Translational control is required for the unfolded protein response and in vivo glucose homeostasis. *Mol. Cell* **2001**, *7*, 1165–1176. [CrossRef]
240. Liu, L.; Cash, T.P.; Jones, R.G.; Keith, B.; Thompson, C.B.; Simon, M.C. Hypoxia-induced energy stress regulates mRNA translation and cell growth. *Mol. Cell* **2006**, *21*, 521–531. [CrossRef]
241. Wang, Y.; Alam, G.N.; Ning, Y.; Visioli, F.; Dong, Z.; Nor, J.E.; Polverini, P.J. The unfolded protein response induces the angiogenic switch in human tumor cells through the PERK/ATF4 pathway. *Cancer Res.* **2012**, *72*, 5396–5406. [CrossRef] [PubMed]
242. Bensellam, M.; Maxwell, E.; Jonas, J.C.; Chan, J.; Laybutt, D.R. Hypoxia induces beta cell death by inhibiting the adaptive UPR. *Diabetologia* **2015**, *58*, S235.
243. Semenza, G.L. Hypoxia-inducible factor 1: Regulator of mitochondrial metabolism and mediator of ischemic preconditioning. *Biochim. Biophys. Acta* **2011**, *1813*, 1263–1268. [CrossRef] [PubMed]
244. Chandel, N.S.; Maltepe, E.; Goldwasser, E.; Mathieu, C.E.; Simon, M.C.; Schumacker, P.T. Mitochondrial reactive oxygen species trigger hypoxia-induced transcription. *Proc. Natl. Acad. Sci. USA* **1998**, *95*, 11715–11720. [CrossRef]
245. Sanjuan-Pla, A.; Cervera, A.M.; Apostolova, N.; Garcia-Bou, R.; Victor, V.M.; Murphy, M.P.; McCreath, K.J. A targeted antioxidant reveals the importance of mitochondrial reactive oxygen species in the hypoxic signaling of HIF-1alpha. *FEBS Lett.* **2005**, *579*, 2669–2674. [CrossRef]
246. Kim, J.W.; Tchernyshyov, I.; Semenza, G.L.; Dang, C.V. HIF-1-mediated expression of pyruvate dehydrogenase kinase: A metabolic switch required for cellular adaptation to hypoxia. *Cell Metab.* **2006**, *3*, 177–185. [CrossRef]
247. Guzy, R.D.; Hoyos, B.; Robin, E.; Chen, H.; Liu, L.; Mansfield, K.D.; Simon, M.C.; Hammerling, U.; Schumacker, P.T. Mitochondrial complex III is required for hypoxia-induced ROS production and cellular oxygen sensing. *Cell Metab.* **2005**, *1*, 401–408. [CrossRef]
248. Brunelle, J.K.; Bell, E.L.; Quesada, N.M.; Vercauteren, K.; Tiranti, V.; Zeviani, M.; Scarpulla, R.C.; Chandel, N.S. Oxygen sensing requires mitochondrial ROS but not oxidative phosphorylation. *Cell Metab.* **2005**, *1*, 409–414. [CrossRef]
249. Zweier, J.L. Measurement of superoxide-derived free radicals in the reperfused heart. Evidence for a free radical mechanism of reperfusion injury. *J. Biol. Chem.* **1988**, *263*, 1353–1357. [CrossRef]
250. Ambrosio, G.; Zweier, J.L.; Duilio, C.; Kuppusamy, P.; Santoro, G.; Elia, P.P.; Tritto, I.; Cirillo, P.; Condorelli, M.; Chiariello, M.; et al. Evidence That Mitochondrial Respiration Is a Source of Potentially Toxic Oxygen-Free Radicals in Intact Rabbit Hearts Subjected to Ischemia and Reflow. *J. Biol. Chem.* **1993**, *268*, 18532–18541. [CrossRef] [PubMed]
251. Jaskiewicz, M.; Moszynska, A.; Gebert, M.; Collawn, J.F.; Bartoszewski, R. EPAS1 resistance to miRNA-based regulation contributes to prolonged expression of HIF-2 during hypoxia in human endothelial cells. *Gene* **2023**, *868*, 147376. [CrossRef] [PubMed]

252. Jaskiewicz, M.; Moszynska, A.; Kroliczewski, J.; Cabaj, A.; Bartoszewska, S.; Charzynska, A.; Gebert, M.; Dabrowski, M.; Collawn, J.F.; Bartoszewski, R. The transition from HIF-1 to HIF-2 during prolonged hypoxia results from reactivation of PHDs and HIF1A mRNA instability. *Cell Mol. Biol. Lett.* **2022**, *27*, 109. [CrossRef]

253. Moszynska, A.; Jaskiewicz, M.; Serocki, M.; Cabaj, A.; Crossman, D.K.; Bartoszewska, S.; Gebert, M.; Dabrowski, M.; Collawn, J.F.; Bartoszewski, R. The hypoxia-induced changes in miRNA-mRNA in RNA-induced silencing complexes and HIF-2 induced miRNAs in human endothelial cells. *FASEB J. Off. Publ. Fed. Am. Soc. Exp. Biol.* **2022**, *36*, e22412. [CrossRef]

254. Jaskiewicz, M.; Moszynska, A.; Serocki, M.; Kroliczewski, J.; Bartoszewska, S.; Collawn, J.F.; Bartoszewski, R. Hypoxia-inducible factor (HIF)-3a2 serves as an endothelial cell fate executor during chronic hypoxia. *EXCLI J.* **2022**, *21*, 454–469. [CrossRef] [PubMed]

255. Foster, K.A.; Margraf, R.R.; Turner, D.A. NADH hyperoxidation correlates with enhanced susceptibility of aged rats to hypoxia. *NeuroBiol. Aging* **2008**, *29*, 598–613. [CrossRef] [PubMed]

256. Kogure, K.; Busto, R.; Schwartzman, R.J.; Scheinberg, P. The dissociation of cerebral blood flow, metabolism, and function in the early stages of developing cerebral infarction. *Ann. Neurol.* **1980**, *8*, 278–290. [CrossRef]

257. Hoek, J.B.; Rydstrom, J. Physiological roles of nicotinamide nucleotide transhydrogenase. *Biochem. J.* **1988**, *254*, 1–10. [CrossRef]

258. Imarisio, C.; Alchera, E.; Bangalore Revanna, C.; Valente, G.; Follenzi, A.; Trisolini, E.; Boldorini, R.; Carini, R. Oxidative and ER stress-dependent ASK1 activation in steatotic hepatocytes and Kupffer cells sensitizes mice fatty liver to ischemia/reperfusion injury. *Free Radic. Biol. Med.* **2017**, *112*, 141–148. [CrossRef]

259. Van Kooten, C.; Pacchiarotta, T.; van der Pol, P.; de Fijter, J.; Schlagwein, N.; van Gijlswijk, D.; Mayboroda, O. ER Stress and Loss of GRP78 Expression Provides a Link Between Renal Ischemia/Reperfusion Injury and the Urinary Metabolome. *Am. J. Transpl.* **2016**, *16*, 638. [CrossRef]

260. Rao, J.; Yue, S.; Fu, Y.; Zhu, J.; Wang, X.; Busuttil, R.W.; Kupiec-Weglinski, J.W.; Lu, L.; Zhai, Y. ATF6 mediates a pro-inflammatory synergy between ER stress and TLR activation in the pathogenesis of liver ischemia-reperfusion injury. *Am. J. Transpl.* **2014**, *14*, 1552–1561. [CrossRef]

261. Gao, F.; Shen, X.; Lu, T.; Liu, J.; Busuttil, R.W.; Kupiec-Weglinski, J.W.; Zhai, Y. IL-23 in Liver Ischemia/Reperfusion Injury (IRI): A Synergy between ER Stress and TLR4 Activation. *Am. J. Transpl.* **2012**, *12*, 223.

262. Balachandran, P.; Dubray, B.J.; Upadhya, G.A.; Jia, J.; Anderson, C.; Chapman, W.D. ER Stress Is an Important Mediator of Ischemia Reperfusion Injury in Hepatocytes Isolated from Steatotic Livers. *Am. J. Transpl.* **2011**, *11*, 503.

263. Kaser, A.; Tomczak, M.; Blumberg, R.S. "ER stress(ed out)!": Paneth cells and ischemia-reperfusion injury of the small intestine. *Gastroenterology* **2011**, *140*, 393–396. [CrossRef]

264. Ren, F.; Liu, J.; Gao, F.; Shen, X.D.; Busuttil, R.W.; Kupiec-Weglinski, J.W.; Zhai, Y. Endoplasmic Reticulum (ER) Stress Modulates Tissue Inflammatory Responses and Its Implication in Liver Ischemia/Reperfusion Injury (IRI). *Liver Transpl.* **2010**, *16*, S100.

265. Vilatoba, M.; Eckstein, C.; Ringland, S.; Bilbao, G.; Thompson, A.; Eckhoff, D.E.; Contreras, J.L. Sodium 4-phenylbutyrate (PBA) protects against liver ischemia reperfusion injury (I/R-injury) by inhibition of endoplasmic reticulum (ER)-stress mediated apoptosis. *Am. J. Transpl.* **2005**, *5*, 536.

266. Ricca, L.; Lecorche, E.; Hamelin, J.; Balducci, G.; Azoulay, D.; Lemoine, A. The Unfolded Protein Response (Upr) Can Participate to the Liver Ischemic Postconditioning Protection against Ischemia/Reperfusion (I/R) Injury Via The Modulation of Nf-Kb/Chop/Il-1 Beta Signaling Pathway. *Transpl. Int.* **2014**, *27*, 15.

267. Zhang, C.; He, S.; Li, Y.; Li, F.; Liu, Z.; Liu, J.; Gong, J. Bisoprolol protects myocardium cells against ischemia/reperfusion injury by attenuating unfolded protein response in rats. *Sci. Rep.* **2017**, *7*, 11859. [CrossRef] [PubMed]

268. Le Pape, S.; Dimitrova, E.; Hannaert, P.; Konovalov, A.; Volmer, R.; Ron, D.; Thuillier, R.; Hauet, T. Polynomial algebra reveals diverging roles of the unfolded protein response in endothelial cells during ischemia-reperfusion injury. *FEBS Lett.* **2014**, *588*, 3062–3067. [CrossRef]

269. Kim, H.; Zhao, J.; Lee, D.; Bai, X.; Cypel, M.; Keshavjee, S.; Liu, M. Protein Kinase C delta-Mediated Unfolded Protein Response and Necrotic Cell Death Contributes to Ischemia-Reperfusion Induced Injury in Lung Transplantation. *J. Heart Lung Transpl.* **2014**, *33*, S83. [CrossRef]

270. Wang, Z.V.; Deng, Y.F.; Gao, N.G.; Pedrozo, Z.; Li, D.L.; Tan, W.; Liang, N.; Lehrman, M.A.; Rothermel, B.A.; Lee, A.H.; et al. The Unfolded Protein Response Directly Activates the Hexosamine Biosynthetic Pathway to Protect the Heart from Ischemia/Reperfusion Injury. *Circulation* **2013**, *128*, A11565.

271. Li, Y.P.; Wang, S.L.; Liu, B.; Tang, L.; Kuang, R.R.; Wang, X.B.; Zhao, C.; Song, X.D.; Cao, X.M.; Wu, X.; et al. Sulforaphane prevents rat cardiomyocytes from hypoxia/reoxygenation injury in vitro via activating SIRT1 and subsequently inhibiting ER stress. *Acta Pharmacol. Sin.* **2016**, *37*, 344–353. [CrossRef] [PubMed]

272. Xu, J.; Hu, H.; Chen, B.; Yue, R.; Zhou, Z.; Liu, Y.; Zhang, S.; Xu, L.; Wang, H.; Yu, Z. Lycopene Protects against Hypoxia/Reoxygenation Injury by Alleviating ER Stress Induced Apoptosis in Neonatal Mouse Cardiomyocytes. *PLoS ONE* **2015**, *10*, e0136443. [CrossRef] [PubMed]

273. Guan, G.; Yang, L.; Huang, W.; Zhang, J.; Zhang, P.; Yu, H.; Liu, S.; Gu, X. Mechanism of interactions between endoplasmic reticulum stress and autophagy in hypoxia/reoxygenation-induced injury of H9c2 cardiomyocytes. *Mol. Med. Rep.* **2019**, *20*, 350–358. [CrossRef] [PubMed]

274. Xing, J.; Xu, H.; Liu, C.; Wei, Z.; Wang, Z.; Zhao, L.; Ren, L. Melatonin ameliorates endoplasmic reticulum stress in N2a neuroblastoma cell hypoxia-reoxygenation injury by activating the AMPK-Pak2 pathway. *Cell Stress Chaperones* **2019**, *24*, 621–633. [CrossRef] [PubMed]
275. Li, T.; Chen, L.; Yu, Y.; Yang, B.; Li, P.; Tan, X.Q. Resveratrol alleviates hypoxia/reoxygenation injury-induced mitochondrial oxidative stress in cardiomyocytes. *Mol. Med. Rep.* **2019**, *19*, 2774–2780. [CrossRef]
276. Deng, T.; Wang, Y.; Wang, C.; Yan, H. FABP4 silencing ameliorates hypoxia reoxygenation injury through the attenuation of endoplasmic reticulum stress-mediated apoptosis by activating PI3K/Akt pathway. *Life Sci.* **2019**, *224*, 149–156. [CrossRef]
277. Sun, M.Y.; Ma, D.S.; Zhao, S.; Wang, L.; Ma, C.Y.; Bai, Y. Salidroside mitigates hypoxia/reoxygenation injury by alleviating endoplasmic reticulum stress-induced apoptosis in H9c2 cardiomyocytes. *Mol. Med. Rep.* **2018**, *18*, 3760–3768. [CrossRef]
278. Xu, Y.; Wang, W.; Jin, K.; Zhu, Q.; Lin, H.; Xie, M.; Wang, D. Perillyl alcohol protects human renal tubular epithelial cells from hypoxia/reoxygenation injury via inhibition of ROS, endoplasmic reticulum stress and activation of PI3K/Akt/eNOS pathway. *Biomed. Pharmacother* **2017**, *95*, 662–669. [CrossRef]
279. Lei, X.; Zhang, S.; Hu, H.X.; Yue, R.C.; Wang, H.; Chen, H.Y.; Tan, C.Y.; Li, K. Lycopene protects cardiomyocytes from hypoxia/reoxygenation injury via attenuating endoplasmic reticulum stress. *J. Am. Coll. Cardiol.* **2014**, *64*, C88–C89. [CrossRef]
280. Wu, X.D.; Zhang, Z.Y.; Sun, S.; Li, Y.Z.; Wang, X.R.; Zhu, X.Q.; Li, W.H.; Liu, X.H. Hypoxic preconditioning protects microvascular endothelial cells against hypoxia/reoxygenation injury by attenuating endoplasmic reticulum stress. *Apoptosis* **2013**, *18*, 85–98. [CrossRef]
281. Samarasinghe, D.A.; Tapner, M.; Farrell, G.C. Role of oxidative stress in hypoxia-reoxygenation injury to cultured rat hepatic sinusoidal endothelial cells. *Hepatology* **2000**, *31*, 160–165. [CrossRef]
282. Samarasinghe, D.A.; Farrell, G.C. Role of redox stress in hypoxia-reoxygenation injury to hepatic sinusoidal endothelial cells. *Hepatology* **1996**, *24*, 444. [CrossRef]
283. Hou, N.S.; Gutschmidt, A.; Choi, D.Y.; Pather, K.; Shi, X.; Watts, J.L.; Hoppe, T.; Taubert, S. Activation of the endoplasmic reticulum unfolded protein response by lipid disequilibrium without disturbed proteostasis in vivo. *Proc. Natl. Acad. Sci. USA* **2014**, *111*, E2271–E2280. [CrossRef] [PubMed]
284. Pineau, L.; Colas, J.; Dupont, S.; Beney, L.; Fleurat-Lessard, P.; Berjeaud, J.M.; Berges, T.; Ferreira, T. Lipid-induced ER stress: Synergistic effects of sterols and saturated fatty acids. *Traffic* **2009**, *10*, 673–690. [CrossRef]
285. Ariyama, H.; Kono, N.; Matsuda, S.; Inoue, T.; Arai, H. Decrease in membrane phospholipid unsaturation induces unfolded protein response. *J. Biol. Chem.* **2010**, *285*, 22027–22035. [CrossRef]
286. Volmer, R.; van der Ploeg, K.; Ron, D. Membrane lipid saturation activates endoplasmic reticulum unfolded protein response transducers through their transmembrane domains. *Proc. Natl. Acad. Sci. USA* **2013**, *110*, 4628–4633. [CrossRef] [PubMed]
287. Testa, G.; Staurenghi, E.; Zerbinati, C.; Gargiulo, S.; Iuliano, L.; Giaccone, G.; Fanto, F.; Poli, G.; Leonarduzzi, G.; Gamba, P. Changes in brain oxysterols at different stages of Alzheimer's disease: Their involvement in neuroinflammation. *Redox. Biol.* **2016**, *10*, 24–33. [CrossRef]
288. Park, S.H.; Shin, D.; Lim, S.S.; Lee, J.Y.; Kang, Y.H. Purple perilla extracts allay ER stress in lipid-laden macrophages. *PLoS ONE* **2014**, *9*, e110581. [CrossRef]
289. Li, J.; Zheng, X.; Lou, N.; Zhong, W.; Yan, D. Oxysterol binding protein-related protein 8 mediates the cytotoxicity of 25-hydroxycholesterol. *J. Lipid Res.* **2016**, *57*, 1845–1853. [CrossRef]
290. Weigel, T.K.; Kulas, J.A.; Ferris, H.A. Oxidized cholesterol species as signaling molecules in the brain: Diabetes and Alzheimer's disease. *Neuronal Signal.* **2019**, *3*, NS20190068. [CrossRef]
291. Bulleid, N.J.; Ellgaard, L. Multiple ways to make disulfides. *Trends Biochem. Sci.* **2011**, *36*, 485–492. [CrossRef]
292. Appenzeller-Herzog, C. Glutathione- and non-glutathione-based oxidant control in the endoplasmic reticulum. *J. Cell Sci.* **2011**, *124*, 847–855. [CrossRef]
293. Ramming, T.; Appenzeller-Herzog, C. The physiological functions of mammalian endoplasmic oxidoreductin 1: On disulfides and more. *Antioxid. Redox Signal.* **2012**, *16*, 1109–1118. [CrossRef]
294. Appenzeller-Herzog, C.; Ellgaard, L. The human PDI family: Versatility packed into a single fold. *Biochim. Biophys. Acta* **2008**, *1783*, 535–548. [CrossRef] [PubMed]
295. Ramming, T.; Appenzeller-Herzog, C. Destroy and exploit: Catalyzed removal of hydroperoxides from the endoplasmic reticulum. *Int. J. Cell Biol.* **2013**, *2013*, 180906. [CrossRef] [PubMed]
296. Araki, K.; Iemura, S.; Kamiya, Y.; Ron, D.; Kato, K.; Natsume, T.; Nagata, K. Ero1-alpha and PDIs constitute a hierarchical electron transfer network of endoplasmic reticulum oxidoreductases. *J. Cell Biol.* **2013**, *202*, 861–874. [CrossRef]
297. Ramming, T.; Hansen, H.G.; Nagata, K.; Ellgaard, L.; Appenzeller-Herzog, C. GPx8 peroxidase prevents leakage of H2O2 from the endoplasmic reticulum. *Free Radic. Biol. Med.* **2014**, *70*, 106–116. [CrossRef]
298. Harding, H.P.; Zhang, Y.; Zeng, H.; Novoa, I.; Lu, P.D.; Calfon, M.; Sadri, N.; Yun, C.; Popko, B.; Paules, R.; et al. An integrated stress response regulates amino acid metabolism and resistance to oxidative stress. *Mol. Cell* **2003**, *11*, 619–633. [CrossRef]
299. Haynes, C.M.; Titus, E.A.; Cooper, A.A. Degradation of misfolded proteins prevents ER-derived oxidative stress and cell death. *Mol. Cell* **2004**, *15*, 767–776. [CrossRef]
300. Marciniak, S.J.; Yun, C.Y.; Oyadomari, S.; Novoa, I.; Zhang, Y.; Jungreis, R.; Nagata, K.; Harding, H.P.; Ron, D. CHOP induces death by promoting protein synthesis and oxidation in the stressed endoplasmic reticulum. *Genes Dev.* **2004**, *18*, 3066–3077. [CrossRef] [PubMed]

301. Anelli, T.; Bergamelli, L.; Margittai, E.; Rimessi, A.; Fagioli, C.; Malgaroli, A.; Pinton, P.; Ripamonti, M.; Rizzuto, R.; Sitia, R. Ero1alpha regulates Ca^{2+} fluxes at the endoplasmic reticulum-mitochondria interface (MAM). *Antioxid. Redox Signal.* **2012**, *16*, 1077–1087. [CrossRef] [PubMed]
302. Li, G.; Mongillo, M.; Chin, K.T.; Harding, H.; Ron, D.; Marks, A.R.; Tabas, I. Role of ERO1-alpha-mediated stimulation of inositol 1,4,5-triphosphate receptor activity in endoplasmic reticulum stress-induced apoptosis. *J. Cell Biol.* **2009**, *186*, 783–792. [CrossRef] [PubMed]
303. Li, G.; Scull, C.; Ozcan, L.; Tabas, I. NADPH oxidase links endoplasmic reticulum stress, oxidative stress, and PKR activation to induce apoptosis. *J. Cell Biol.* **2010**, *191*, 1113–1125. [CrossRef]
304. Pedruzzi, E.; Guichard, C.; Ollivier, V.; Driss, F.; Fay, M.; Prunet, C.; Marie, J.C.; Pouzet, C.; Samadi, M.; Elbim, C.; et al. NAD(P)H oxidase Nox-4 mediates 7-ketocholesterol-induced endoplasmic reticulum stress and apoptosis in human aortic smooth muscle cells. *Mol. Cell Biol.* **2004**, *24*, 10703–10717. [CrossRef] [PubMed]
305. Malhotra, J.D.; Miao, H.; Zhang, K.; Wolfson, A.; Pennathur, S.; Pipe, S.W.; Kaufman, R.J. Antioxidants reduce endoplasmic reticulum stress and improve protein secretion. *Proc. Natl. Acad. Sci. USA* **2008**, *105*, 18525–18530. [CrossRef]
306. McCullough, K.D.; Martindale, J.L.; Klotz, L.O.; Aw, T.Y.; Holbrook, N.J. Gadd153 sensitizes cells to endoplasmic reticulum stress by down-regulating Bcl2 and perturbing the cellular redox state. *Mol. Cell Biol.* **2001**, *21*, 1249–1259. [CrossRef]
307. Austgen, K.; Johnson, E.T.; Park, T.J.; Curran, T.; Oakes, S.A. The adaptor protein CRK is a pro-apoptotic transducer of endoplasmic reticulum stress. *Nat Cell Biol.* **2011**, *14*, 87–92. [CrossRef]
308. Cunha, D.A.; Igoillo-Esteve, M.; Gurzov, E.N.; Germano, C.M.; Naamane, N.; Marhfour, I.; Fukaya, M.; Vanderwinden, J.M.; Gysemans, C.; Mathieu, C.; et al. Death protein 5 and p53-upregulated modulator of apoptosis mediate the endoplasmic reticulum stress-mitochondrial dialog triggering lipotoxic rodent and human beta-cell apoptosis. *Diabetes* **2012**, *61*, 2763–2775. [CrossRef]
309. Wali, J.A.; Rondas, D.; McKenzie, M.D.; Zhao, Y.; Elkerbout, L.; Fynch, S.; Gurzov, E.N.; Akira, S.; Mathieu, C.; Kay, T.W.; et al. The proapoptotic BH3-only proteins Bim and Puma are downstream of endoplasmic reticulum and mitochondrial oxidative stress in pancreatic islets in response to glucotoxicity. *Cell Death Dis.* **2014**, *5*, e1124. [CrossRef] [PubMed]
310. Puthalakath, H.; O'Reilly, L.A.; Gunn, P.; Lee, L.; Kelly, P.N.; Huntington, N.D.; Hughes, P.D.; Michalak, E.M.; McKimm-Breschkin, J.; Motoyama, N.; et al. ER stress triggers apoptosis by activating BH3-only protein Bim. *Cell* **2007**, *129*, 1337–1349. [CrossRef]
311. Namba, T.; Tian, F.; Chu, K.; Hwang, S.Y.; Yoon, K.W.; Byun, S.; Hiraki, M.; Mandinova, A.; Lee, S.W. CDIP1-BAP31 complex transduces apoptotic signals from endoplasmic reticulum to mitochondria under endoplasmic reticulum stress. *Cell Rep.* **2013**, *5*, 331–339. [CrossRef] [PubMed]
312. Rizzuto, R.; De Stefani, D.; Raffaello, A.; Mammucari, C. Mitochondria as sensors and regulators of calcium signalling. *Nat. Rev. Mol. Cell Biol.* **2012**, *13*, 566–578. [CrossRef] [PubMed]
313. Boehning, D.; Patterson, R.L.; Sedaghat, L.; Glebova, N.O.; Kurosaki, T.; Snyder, S.H. Cytochrome c binds to inositol (1,4,5) trisphosphate receptors, amplifying calcium-dependent apoptosis. *Nat. Cell Biol.* **2003**, *5*, 1051–1061. [CrossRef]
314. Carafoli, E.; Krebs, J. Why Calcium? How Calcium Became the Best Communicator. *J. Biol. Chem.* **2016**, *291*, 20849–20857. [CrossRef] [PubMed]
315. Reis, A.; Spickett, C.M. Chemistry of phospholipid oxidation. *Biochim. Biophys. Acta* **2012**, *1818*, 2374–2387. [CrossRef] [PubMed]
316. Esmaeili, Y.; Yarjanli, Z.; Pakniya, F.; Bidram, E.; Los, M.J.; Eshraghi, M.; Klionsky, D.J.; Ghavami, S.; Zarrabi, A. Targeting autophagy, oxidative stress, and ER stress for neurodegenerative disease treatment. *J. Control Release* **2022**, *345*, 147–175. [CrossRef]
317. Brown, D.R. Neurodegeneration and oxidative stress: Prion disease results from loss of antioxidant defence. *Folia Neuropathol.* **2005**, *43*, 229–243.
318. Banerjee, R.; Kaidery, N.A.; Thomas, B. Oxidative Stress in Parkinson's Disease: Role in Neurodegeneration and Targets for Therapeutics. *ACS Symp. Ser.* **2015**, *1200*, 147–176. [CrossRef]
319. Wang, L.; Colodner, K.J.; Feany, M.B. Protein misfolding and oxidative stress promote glial-mediated neurodegeneration in an Alexander disease model. *J. Neurosci.* **2011**, *31*, 2868–2877. [CrossRef]
320. Guyon, A.; Rousseau, J.; Lamothe, G.; Tremblay, J.P. The protective mutation A673T in amyloid precursor protein gene decreases Abeta peptides production for 14 forms of Familial Alzheimer's Disease in SH-SY5Y cells. *PLoS ONE* **2020**, *15*, e0237122. [CrossRef]
321. De Strooper, B.; Voet, T. Alzheimer's disease: A protective mutation. *Nature* **2012**, *488*, 38–39. [CrossRef] [PubMed]
322. Credle, J.J.; Forcelli, P.A.; Delannoy, M.; Oaks, A.W.; Permaul, E.; Berry, D.L.; Duka, V.; Wills, J.; Sidhu, A. alpha-Synuclein-mediated inhibition of ATF6 processing into COPII vesicles disrupts UPR signaling in Parkinson's disease. *NeuroBiol. Dis.* **2015**, *76*, 112–125. [CrossRef] [PubMed]
323. Hetz, C.; Russelakis-Carneiro, M.; Maundrell, K.; Castilla, J.; Soto, C. Caspase-12 and endoplasmic reticulum stress mediate neurotoxicity of pathological prion protein. *EMBO J.* **2003**, *22*, 5435–5445. [CrossRef]
324. Zoghbi, H.Y.; Orr, H.T. Glutamine repeats and neurodegeneration. *Annu. Rev. Neurosci.* **2000**, *23*, 217–247. [CrossRef] [PubMed]
325. Valdes, P.; Mercado, G.; Vidal, R.L.; Molina, C.; Parsons, G.; Court, F.A.; Martinez, A.; Galleguillos, D.; Armentano, D.; Schneider, B.L.; et al. Control of dopaminergic neuron survival by the unfolded protein response transcription factor XBP1. *Proc. Natl. Acad. Sci. USA* **2014**, *111*, 6804–6809. [CrossRef]
326. Kaneko, M. Molecular pharmacological studies on the protection mechanism against endoplasmic reticulum stress-induced neurodegenerative disease. *Yakugaku Zasshi* **2012**, *132*, 1437–1442. [CrossRef]

327. Naranjo, J.R.; Zhang, H.; Villar, D.; Gonzalez, P.; Dopazo, X.M.; Moron-Oset, J.; Higueras, E.; Oliveros, J.C.; Arrabal, M.D.; Prieto, A.; et al. Activating transcription factor 6 derepression mediates neuroprotection in Huntington disease. *J. Clin. Investig.* **2016**, *126*, 627–638. [CrossRef]

328. Yu, Z.; Sheng, H.; Liu, S.; Zhao, S.; Glembotski, C.C.; Warner, D.S.; Paschen, W.; Yang, W. Activation of the ATF6 branch of the unfolded protein response in neurons improves stroke outcome. *J. Cereb. Blood Flow. Metab.* **2017**, *37*, 1069–1079. [CrossRef]

329. Rabouw, H.H.; Langereis, M.A.; Anand, A.A.; Visser, L.J.; de Groot, R.J.; Walter, P.; van Kuppeveld, F.J.M. Small molecule ISRIB suppresses the integrated stress response within a defined window of activation. *Proc. Natl. Acad. Sci. USA* **2019**, *116*, 2097–2102. [CrossRef]

330. Hosoi, T.; Kakimoto, M.; Tanaka, K.; Nomura, J.; Ozawa, K. Unique pharmacological property of ISRIB in inhibition of Abeta-induced neuronal cell death. *J. Pharmacol. Sci.* **2016**, *131*, 292–295. [CrossRef]

331. Bugallo, R.; Marlin, E.; Baltanas, A.; Toledo, E.; Ferrero, R.; Vinueza-Gavilanes, R.; Larrea, L.; Arrasate, M.; Aragon, T. Fine tuning of the unfolded protein response by ISRIB improves neuronal survival in a model of amyotrophic lateral sclerosis. *Cell Death Dis.* **2020**, *11*, 397. [CrossRef] [PubMed]

332. Frias, E.S.; Hoseini, M.S.; Krukowski, K.; Paladini, M.S.; Grue, K.; Ureta, G.; Rienecker, K.D.A.; Walter, P.; Stryker, M.P.; Rosi, S. Aberrant cortical spine dynamics after concussive injury are reversed by integrated stress response inhibition. *Proc. Natl. Acad. Sci. USA* **2022**, *119*, e2209427119. [CrossRef] [PubMed]

333. Beckman, K.B.; Ames, B.N. The free radical theory of aging matures. *Physiol. Rev.* **1998**, *78*, 547–581. [CrossRef] [PubMed]

334. Salmon, A.B.; Richardson, A.; Perez, V.I. Update on the oxidative stress theory of aging: Does oxidative stress play a role in aging or healthy aging? *Free Radic Biol. Med.* **2010**, *48*, 642–655. [CrossRef]

335. Pratico, D. Evidence of oxidative stress in Alzheimer's disease brain and antioxidant therapy: Lights and shadows. *Ann. New York Acad. Sci.* **2008**, *1147*, 70–78. [CrossRef]

336. Quick, K.L.; Ali, S.S.; Arch, R.; Xiong, C.; Wozniak, D.; Dugan, L.L. A carboxyfullerene SOD mimetic improves cognition and extends the lifespan of mice. *NeuroBiol. Aging* **2008**, *29*, 117–128. [CrossRef]

337. Dirksen, M.T.; Laarman, G.J.; Simoons, M.L.; Duncker, D.J. Reperfusion injury in humans: A review of clinical trials on reperfusion injury inhibitory strategies. *Cardiovasc. Res.* **2007**, *74*, 343–355. [CrossRef]

338. Pearce, K.A.; Boosalis, M.G.; Yeager, B. Update on vitamin supplements for the prevention of coronary disease and stroke. *Am. Fam. Physician.* **2000**, *62*, 1359–1366.

339. Papadakis, M.; Nagel, S.; Buchan, A.M. Development and efficacy of NXY-059 for the treatment of acute ischemic stroke. *Futur. Neurol.* **2008**, *3*, 229–240. [CrossRef]

340. Feuerstein, G.Z.; Zaleska, M.M.; Krams, M.; Wang, X.; Day, M.; Rutkowski, J.L.; Finklestein, S.P.; Pangalos, M.N.; Poole, M.; Stiles, G.L.; et al. Missing steps in the STAIR case: A Translational Medicine perspective on the development of NXY-059 for treatment of acute ischemic stroke. *J. Cereb Blood Flow Metab.* **2008**, *28*, 217–219. [CrossRef]

341. Willcox, B.J.; Curb, J.D.; Rodriguez, B.L. Antioxidants in cardiovascular health and disease: Key lessons from epidemiologic studies. *Am. J. Cardiol.* **2008**, *101*, 75D–86D. [CrossRef] [PubMed]

342. An, Z.; Yan, J.; Zhang, Y.; Pei, R. Applications of nanomaterials for scavenging reactive oxygen species in the treatment of central nervous system diseases. *J. Mater. Chem. B* **2020**, *8*, 8748–8767. [CrossRef] [PubMed]

343. Cabral-Miranda, F.; Tamburini, G.; Martinez, G.; Ardiles, A.O.; Medinas, D.B.; Gerakis, Y.; Hung, M.D.; Vidal, R.; Fuentealba, M.; Miedema, T.; et al. Unfolded protein response IRE1/XBP1 signaling is required for healthy mammalian brain aging. *EMBO J.* **2022**, *41*, e111952. [CrossRef]

344. Shacham, T.; Patel, C.; Lederkremer, G.Z. PERK Pathway and Neurodegenerative Disease: To Inhibit or to Activate? *Biomolecules* **2021**, *11*, 354. [CrossRef] [PubMed]

 *antioxidants*

Article

Redox Regulation of Microglial Inflammatory Response: Fine Control of NLRP3 Inflammasome through Nrf2 and NOX4

Alejandra Palomino-Antolín [1,†], Céline Decouty-Pérez [1,†], Víctor Farré-Alins [1], Paloma Narros-Fernández [1], Ana Belen Lopez-Rodriguez [1], María Álvarez-Rubal [1], Inés Valencia [1], Francisco López-Muñoz [2,3], Eva Ramos [4], Antonio Cuadrado [5], Ana I. Casas [6,7,8,9], Alejandro Romero [4,*] and Javier Egea [1,*]

[1] Unidad de Investigación, Hospital Santa Cristina, Instituto de Investigación Sanitaria Princesa (IIS-IP), 28006 Madrid, Spain; alejandra.palominoantolin@gmail.com (A.P.-A.); celine.decouty@estudiante.uam.es (C.D.-P.); victorfarre@hotmail.com (V.F.-A.); narrosfp@tcd.ie (P.N.-F.); anabelen.lopez.externo@salud.madrid.org (A.B.L.-R.)

[2] Faculty of Health, Camilo José Cela University of Madrid (UCJC), 28692 Madrid, Spain

[3] Neuropsychopharmacology Unit, Hospital 12 de Octubre Research Institute, 28041 Madrid, Spain

[4] Department of Pharmacology and Toxicology, Faculty of Veterinary Medicine, Complutense University of Madrid, 28040 Madrid, Spain; eva.ramos@ucm.es

[5] Instituto de Investigaciones Biomédicas "Alberto Sols" UAM-CSIC, Centro de Investigación Biomédica en Red Sobre Enfermedades Neurodegenerativas (CIBERNED), Instituto de Salud Carlos III (ISCIII), 28031 Madrid, Spain; antonio.cuadrado@uam.es

[6] Pharmacology and Personalised Medicine, Maastricht University, 6211 LK Maastricht, The Netherlands

[7] Neurology Clinic, University Hospital Essen, 45147 Essen, Germany

[8] Center for Translational Neuro- and Behavioral Sciences (C-TNBS), 45147 Essen, Germany

[9] Department of Neurology, University Hospital Essen, 45147 Essen, Germany

[*] Correspondence: manarome@ucm.es (A.R.); javier.egea@inv.uam.es (J.E.); Tel.: +34-913943970 (A.R.); +34-915574402 (J.E.)

[†] These authors contributed equally to the work.

Citation: Palomino-Antolín, A.; Decouty-Pérez, C.; Farré-Alins, V.; Narros-Fernández, P.; Lopez-Rodriguez, A.B.; Álvarez-Rubal, M.; Valencia, I.; López-Muñoz, F.; Ramos, E.; Cuadrado, A.; et al. Redox Regulation of Microglial Inflammatory Response: Fine Control of NLRP3 Inflammasome through Nrf2 and NOX4. *Antioxidants* **2023**, *12*, 1729. https://doi.org/10.3390/antiox12091729

Academic Editors: Ferdinando Nicoletti and Waldo Cerpa

Received: 3 July 2023
Revised: 21 August 2023
Accepted: 5 September 2023
Published: 7 September 2023

Abstract: The role of inflammation and immunity in the pathomechanism of neurodegenerative diseases has become increasingly relevant within the past few years. In this context, the NOD-like receptor protein 3 (NLRP3) inflammasome plays a crucial role in the activation of inflammatory responses by promoting the maturation and secretion of pro-inflammatory cytokines such as interleukin-1β and interleukin-18. We hypothesized that the interplay between nuclear factor erythroid 2-related factor 2 (Nrf2) and NADPH oxidase 4 (NOX4) may play a critical role in the activation of the NLRP3 inflammasome and subsequent inflammatory responses. After priming mixed glial cultures with lipopolysaccharide (LPS), cells were stimulated with ATP, showing a significant reduction of IL1-β release in NOX4 and Nrf2 KO mice. Importantly, NOX4 inhibition using GKT136901 also reduced IL-1β release, as in NOX4 KO mixed glial cultures. Moreover, we measured NOX4 and NLRP3 expression in wild-type mixed glial cultures following LPS treatment, observing that both increased after TLR4 activation, while 24 h treatment with tert-butylhydroquinone, a potent Nrf2 inducer, significantly reduced NLRP3 expression. LPS administration resulted in significant cognitive impairment compared to the control group. Indeed, LPS also modified the expression of NLRP3 and NOX4 in mouse hippocampus. However, mice treated with GKT136901 after LPS impairment showed a significantly improved discrimination index and recovered the expression of inflammatory genes to normal levels compared with wild-type animals. Hence, we here validate NOX4 as a key player in NLRP3 inflammasome activation, suggesting NOX4 pharmacological inhibition as a potent therapeutic approach in neurodegenerative diseases.

Keywords: inflammation and immunity; NLRP3 inflammasome; Nrf2; NOX4; glial cultures; KO mice; neurodegenerative diseases

1. Introduction

Inflammation is a protective response to infection or injury characterized by the recruitment and activation of immune cells to fight infection and remove debris. The inflammatory response eliminates harmful stimuli and restores tissue homeostasis through tissue and wound repair. Excessive inflammation contributes to chronic inflammatory disease, which can also lead to sepsis and subsequent multiorgan failure [1]. NOD-like receptor protein 3 (NLRP3) inflammasome is a cytoplasmic multiprotein complex that plays a crucial role in regulating the innate immune response [2]. Activation of the NLRP3 inflammasome requires two steps: priming and activation. The priming step involves the transcriptional upregulation of NLRP3 and pro-interleukin-1β (IL-1β) genes by nuclear factor-kappa B (NF-κB) in response to pathogen-associated molecular patterns (PAMPs) or danger-associated molecular patterns (DAMPs) [3]. The second step involves the assembly and activation of NLRP3 inflammasome complex, which activates caspase-1. This caspase-1 activation leads to the maturation and secretion of proinflammatory cytokines, such as IL-1β and interleukin-18 (IL-18). Additionally, GSDM is also cleaved by caspase-1, forming pores in the membrane and, consequently, pyroptosis [4,5]. Dysregulation of NLRP3 inflammasome activation has been implicated in the pathogenesis of various diseases, including neurodegenerative [6], diabetes, and autoimmune diseases [7].

Reactive oxygen species (ROS) are highly reactive molecules that are produced during normal cellular metabolism. They play an essential role in cellular signaling and regulation of various physiological processes such as cell proliferation, differentiation, cellular senescence, and apoptosis [8]. However, the excessive production of ROS can cause oxidative stress, which can lead to cellular damage, aging, and various diseases [9]. Therefore, it is crucial to maintain a balance between ROS production and scavenging to prevent the deleterious effects of oxidative stress. The production and scavenging of ROS are tightly regulated by a complex network of enzymes and transcription factors.

NADPH oxidases (NOX) are considered an important enzymatic sources of ROS production [10]. Of these, NOX4 appears to be the most suitable therapeutic target because it is induced in various cells and tissues under ischemia or hypoxia [11]. It catalyzes the transfer of electrons from NADPH to molecular oxygen to produce a superoxide anion ($O_2^{\bullet}$) and hydrogen peroxide (H_2O_2), which are the primary ROS produced by NOX4. In this regard, NOX4-derived ROS have been shown to play a critical role in the priming and activation of the NLRP3 inflammasome [12] by activating NF-κB signaling, leading to the upregulation of NLRP3 and pro-IL-1β genes [12]. Moreover, NOX4-derived ROS have been shown to activate the NLRP3 inflammasome directly by inducing potassium efflux and calcium influx, leading to the activation of caspase-1 and the subsequent release of IL-1β and IL-18. On the other hand, nuclear factor erythroid 2-related factor 2 (Nrf2) is a transcription factor that is found imprisoned in the cytoplasm, but under stress conditions such as the formation of ROS, its translocation to the nucleus is induced, leading to the transcription of a battery of genes encoding antioxidant and detoxification enzymes [13]. In this context, Nrf2 plays a critical role in protecting cells from oxidative stress by upregulating the expression of various antioxidant genes, including glutathione peroxidase, superoxide dismutase, and catalase [13]. Additionally, Nrf2 has been shown to be a negative regulator of NLRP3 inflammasome activation by reducing ROS production and modulating the activity of NF-κB [14]. Therefore, the interplay between NOX4 and Nrf2 in the regulation of NLRP3 inflammasome ROS homeostasis is a complex and dynamic process. Moreover, NOX4-derived ROS can contribute to NLRP3 inflammasome activation, while Nrf2 activation can modify ROS levels and mitigate NLRP3 inflammasome activation.

In this manuscript, we aimed to contribute to the understanding of the role of NOX4 and Nrf2 in the activation of the NLRP3 inflammasome. Our data show that (i) both NOX4 and Nrf2 participate in NLRP3 inflammasome activation; (ii) in NOX4 and Nrf2 KO animals, there was a reduction in the release of IL-1β in response to LPS plus ATP; (iii) NOX4 inhibition using GKT136901 also reduced IL-1β release as well as NOX4 KO mixed glial cultures. In addition, we measured NOX4 and NLRP3 expression in WT mixed glial cultures following LPS treatment, observing that both increased after TLR4 activation, while 24 h treatment with tert-butylhydroquinone (tBHQ), a potent Nrf2 inducer, significantly reduced NLRP3 expression. In vivo LPS administration resulted in significant cognitive impairment compared to the control group. However, mice treated with GKT136901 after LPS impairment showed a significantly improved discrimination index and recovery of the expression of inflammatory genes to normal levels. This shows that NOX4 is a key player in NLRP3 inflammasome activation.

2. Materials and Methods

2.1. Mixed Glial Cultures

Mixed glial cultures were prepared from the cerebral cortices of postnatal day 3 C57BL/6N wild-type NOX4 knockout and Nrf2 knockout mice, as previously described [15,16]. Briefly, after removing the meninges and blood vessels, the forebrain was carefully dissociated in DMEM/F12 medium by repeated pipetting. After mechanical dissociation, cells were seeded in DMEM/F12 containing 20% FBS at a density of 300,000 cells/mL and incubated at 37 °C in humidified 5% CO_2 and 95% air. After 5 days in vitro (DIV), the medium was replaced with DMEM/F12 and 10% FBS. Cultures were used at confluency reached after 10–12 DIV. Cultures were treated with LPS (1 µg/mL), ATP (5 mM) (Sigma-Aldrich, Madrid, Spain), tBHQ (10 µM), and GKT136901 (1 µM).

2.2. Determination of IL-1β Levels in the Culture Medium

After the different drug treatments, IL-1β levels were measured by using a specific ELISA kit. Supernatant samples were collected at the indicated time points and subjected to the ELISA analysis according to the supplier's recommendations (DY401, R&D Systems, Minneapolis, MN, USA).

2.3. Animals

All animal experimentation was performed under the license PROEX 013/18 granted by the Ethics Committee of Universidad Autónoma de Madrid (Madrid, Spain) and in compliance with the Cruelty to Animals Act, 1876, and the European Community Directive, 86/609/EEC. Every effort was made to minimize stress to the animals. Animals were housed under controlled conditions (22 ± 1 °C, 55–65% humidity, 12 h light-dark cycle), and have been given free access to water and standard laboratory chow. Experiments were performed on 3–4-month male C57BL/6N wild-type mice, NOX4 KO mice, C57BL/6J, and Nrf2 KO mice. NOX4 KO mice were generated in 2010 by deleting the NADPH binding pockets located in exons 14 and 15 to directly assess NOX4 function in vivo without modifying its expression [11]. In this work, Kleinschnitz C. et al. verified that NOX4 in these animals had no activity. For our experiment, animals were randomly divided into the different experimental groups (vehicle, LPS 250 µg/kg, LPS 250 µg/kg + GKT136901 10 mg/kg). A mixture of DMSO/water in a ratio of 1/99 (10 mL/kg) was used as a vehicle. All the treatments were injected intraperitoneally (i.p.). For the analyses of protein and transcriptional changes, animals were terminally anesthetized with a mix of ketamin:xylacin 1:2 (Ketolar 50 mg/mL, Pfizer, Madrid, Spain; Xilagesic 20 mg/mL, Calier Labs, Barcelona, Spain) and transcardially perfused with heparinized saline. Hippocampi were gently removed and stored at −80 °C until use.

2.4. Novel Object Recognition Test in Mice

The novel object recognition (NOR) test is a behavioral test commonly used to assess recognition memory in mice [17]. For 3 consecutive days, the animals were placed on a field (40 × 40 × 40 cm, made of polyvinyl chloride) for 10 min. On the first day (T0), mice explored an empty box. On day 2 (T1), the animal was placed in the field with two identical objects (cylindrical glass bottles, too heavy for the mouse to move, 22 cm high, 9 cm in diameter) and allowed to explore for 10 min. On the third day (T2), a new object (new object, square object) was placed in place of one of the old objects (familiar object) and the other was left as is. Object exploration was measured using a stopwatch with the animal sniffing, wiping, or looking at the object from a distance of up to 2 cm. All object positions were balanced between groups, and objects and NOR fields were washed with 0.1% acetic acid between trials to balance olfactory cues. The amount of time spent examining new or familiar objects was videotaped for 10 min and scored by a blind observer. NOR is based on the premise that rodents have an innate preference for new objects, so mice that remember familiar objects spend more time exploring new objects. Discrimination index in the T2 was estimated as follows: Discrimination Index (%) = [Time exploring novel object − Time exploring familiar object)/(Time exploring novel object + Time exploring familiar object)] × 100. LPS (250 μg/kg) was injected intraperitoneally (i.p.) immediately after the end of the T1 phase. GKT136901 was dissolved in a mixture of DMSO/water in a ratio of 1/99. GKT136901 5 (10 mg/kg i.p.) was administered immediately after the end of the T0 and T1 phases.

2.5. Quantitative Real-Time PCR

Total RNA was extracted from mixed glial cultures and hippocampi using the TRIzol method (10296-028, Invitrogen, Carlsbad, CA, USA). cDNA was synthesized using the iS-cript cDNA synthesis kit (1708891, Biorad, Hercules, CA, USA) according to the manufacturer's instructions. Quantitative polymerase chain reaction (qPCR) was performed using the 7900HT Fast Real-Time PCR System (Applied Biosystems, Waltham, MA, USA) in 384-well format with Power SYBR Green PCR Master Mix (Thermo Fisher, Waltham, MA, USA). Data were normalized to the expression of the housekeeping gene B2M (NM_009735). Specific primers were designed using the NCBI nucleotide data base and Primer 3 software (Version 4.1.0) (http://biotools.umassmed.edu/bioapps/primer3_www.cgi (accessed on 3 February 2023). Primer sequences were checked with BLAST (http://blast.ncbi.nlm.nih.gov/Blast.cgi (accessed on 3 February 2023)) and purchased from Sigma or IDT. The comparative CT method (or the $2^{\Delta CT}$ method) [18,19] was used to determine differences in gene expression between B2M and control samples. The target genes and the specific primers were the following: Nlrp3 (NM_145827) (forward, 5′-TTCAATCTGTTGTTCAGCTC-3′; reverse, 5′-GTCTAATTCCAGCCATCTGTAG-3′), NOX4 (NM_001285835) (forward, 5′-GGGACATTAAACGATTAAACAAGAATCC-3′); reverse, 5′-GGAAGTATTGGCTTCTTATTGG-3′), Il1b (NM_008361) (forward, 5′-GAAGAGCCCATCCTCTGTGA-3′; reverse, 5′-TTCATCTCGGAGCCTGTAG-3′).

2.6. Immunoblotting and Image Analysis

After the different treatments, mixed glial cultures and mouse hippocampi were lysed in ice-cold lysis buffer (1% Nonidet P-40, 10% glycerol, 137 mM NaCl, 20 mM Tris–HCl, pH 7.5, 1 g/mL leupeptin, 1 mM PMSF, 20 mM NaF, 1 mM sodium pyrophosphate, and 1 mM Na_3VO_4). Proteins (30 μg) from cell lysates were separated via SDS-PAGE and transferred to Immobilon-P membranes (Millipore Corp., Billerica, MA, USA). Membranes were incubated with anti-NLRP3 (1:1000; AdipoGen, San Diego, CA, USA), anti-NOX4 (1:1000; R&D Systems, Minneapolis, MN, USA), anti-NQO1 (1:1000; Cell Signaling Technology, Danvers, MA, USA) or anti-βactin (1:50,000; Sigma-Aldrich, Madrid, Spain). Appropriate peroxidase-conjugated secondary antibodies (1:5000; Santa Cruz Biotechnology, Dallas, TX, USA) were used for protein detection via enhanced chemiluminescence on the ImageQuant LAS 4000 min. Different band intensities corresponding to immunoblot detection of protein samples were quantified using the Scion Image program. Immunoblots corresponded to a representative experiment that was repeated three or four times with similar results.

2.7. Immunocytochemistry

For the immunofluorescence experiments, primary glial cells were always seeded in 24-well plates, and each well required a coverslip (Ø 13 mm; VWR, Leicestershire, UK), in which the cells grew. After the different treatments were applied, it was necessary to wash three times with PBS 1X and then fix the cells with 4% paraformaldehyde in PBS 1X. The next step was to permeabilize the cells with PBS 0.3% Triton for 10 min at room temperature, then apply a blocking solution containing PBS 1X plus 10% bovine serum albumin (BSA), for 1 h. Primary antibody was diluted in PBS 1X with 1% BSA and incubated overnight at 4 °C: NRF2 (1:100; ref). The next day, we applied the secondary antibody donkey anti-rabbit Alexa Fluor 488 (1:800, ThermoFisher Cat# A21206) for 1 h at room temperature. After incubation in the secondary antibody, the sections were washed four times for 15 min each in PBS, mounted with a small drop of VECTASHIELD mounting medium containing DAPI (Vector Laboratories Inc., Newark, CA, USA, Cat. No. H-1000), and the slides were cover-slipped. All immunostained brain images were acquired with a Leica SP5 confocal microscope. Images were processed with the program ImageJ 1.52e.

2.8. Statistical Analysis

For multiple comparisons, one- or two-way analysis of variance (ANOVA) was used, with the factors being genotype (wild type (WT) or NOX4 or Nrf2 knockout (KO)) and treatment (vehicle or GKT136901 treatment). Subsequent post hoc comparisons (Bonferroni's test) were performed with a level of significance set at $p < 0.05$. Data are presented as mean $\pm$ standard error of the mean (SEM). Symbols in the graphs denote post hoc tests. Statistical analyses were carried out with the SPSS 22.0 software package (SPSS, Inc., Chicago, IL, USA).

3. Results

3.1. Genetic and Pharmacological Deficiency of NOX4 Reduces NLRP3 Inflammasome Components and IL-1β Release in Primary Mixed Glial Cultures

In order to investigate the relationship between NLRP3 inflammasome and NOX4, we used a well-characterized model of NLRP3 inflammasome activation (LPS 4 h + ATP 30 min) in WT and in NOX4 deficient mice, and we also used the pharmacological inhibitor of NOX4 (GKT136901). We measured IL-1β release in mixed glial cultures using the protocol shown in Figure 1A. We used LPS for 4 h to increase the expression of NLRP3 inflammasome components, a well described process called "priming". In the last 30 min, we added ATP to allow NLRP3 oligomerization and activation, and, finally, the release of IL-1β to the culture medium. Stimulation of WT cultures with LPS (1 µg/mL, 4 h) plus ATP (5 mM, 30 min) produced a significant release of IL-1β (Figure 1A). However, in NOX4 KO cultures we observed a significant reduction of IL-1β release compared to WT (60% reduction). Moreover, in wild-type cultures treated with the same stimuli, we observed the same effect using GKT136901 (1 µM), a selective pharmacological inhibitor of NOX4 (Figure 1A). To better explore the mechanism of NLRP3 inflammasome inhibition, we measured NLRP3 inflammasome components (Nlrp3 and Il1b) with RT-PCR using the same protocol. As illustrated in Figure 1B, LPS produced the increase in Nlrp3 and Il1b mRNA both in WT and NOX4 KO mice, and GKT136901 significantly reduced Nlrp3 and Il1b mRNA levels only in WT mice.

Next, we checked the levels of NLRP3 using WB to confirm the data obtained by RT-PCR. Figure 2 shows the Western blot of NLRP3 in WT and NOX4 KO mice in the different conditions. Incubation of cells with LPS for 2 and 4 h significantly increased the production of NLRP3. Co-treatment with GKT136901 partially reduced NLRP3 protein levels in WT mice but not in NOX4 KO mice, confirming the results obtained in mRNA. Together, these data indicate that NOX4 inhibition is important for NLRP3 inflammasome activation in mixed glia cultures.

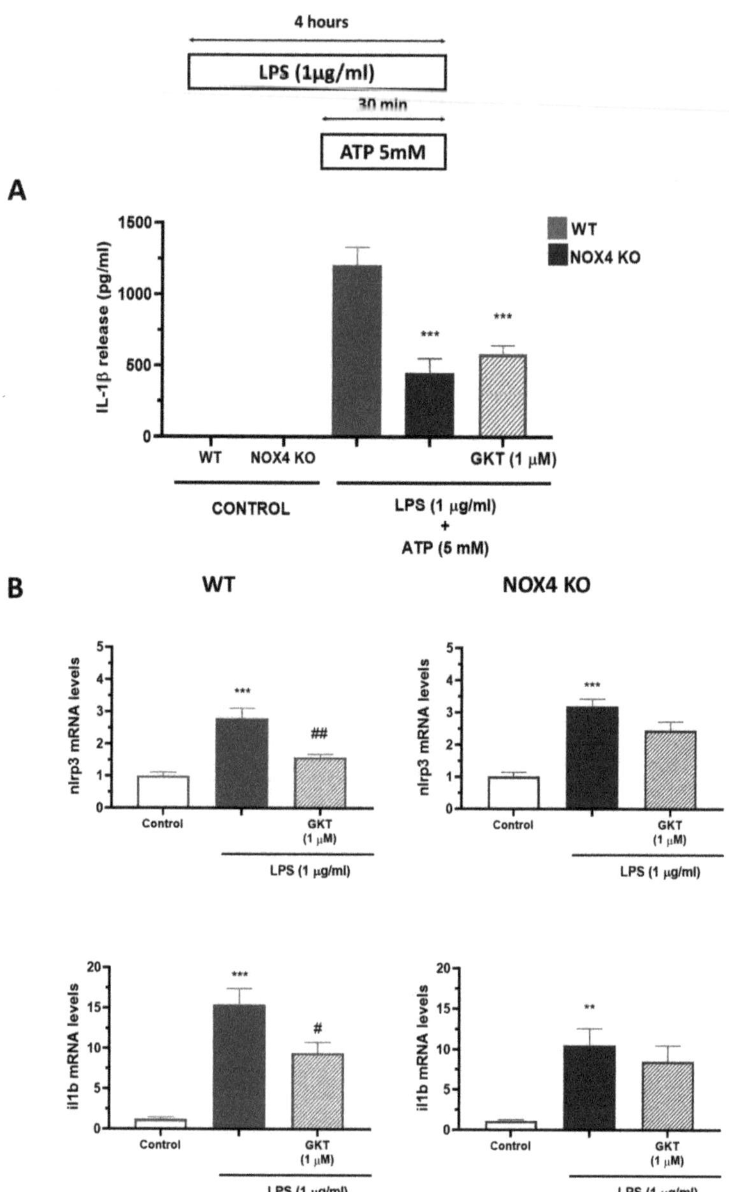

Figure 1. Nox4 activity is required for NLRP3 inflammasome activation. (**A**) Mixed glial cultures of WT and NOX4 KO mice were stimulated with LPS (1 µg/mL) for 4 h plus ATP (5 mM) during the last 30 min, in the absence or presence of GKT136901 (1 µM) following the protocol at the top of the figure. Inflammasome activation was analyzed via ELISA measurement of IL-1β in the supernatant. Mean ± SEM ($n = 6$). Two-way ANOVA followed by Bonferroni's tests *** $p < 0.001$ vs. WT LPS. (**B**) At the end of the experiment, cells were harvested and analyzed for the expression of NLRP3 inflammasome components Nlrp3 and Il1b via RT-PCR. Mean ± SEM ($n = 5$). One-way ANOVA with Tukey's multiple comparisons test. *** $p < 0.001$ and ** $p < 0.01$ vs. control; ## $p < 0.01$ and # $p < 0.05$ vs. LPS-treated cells.

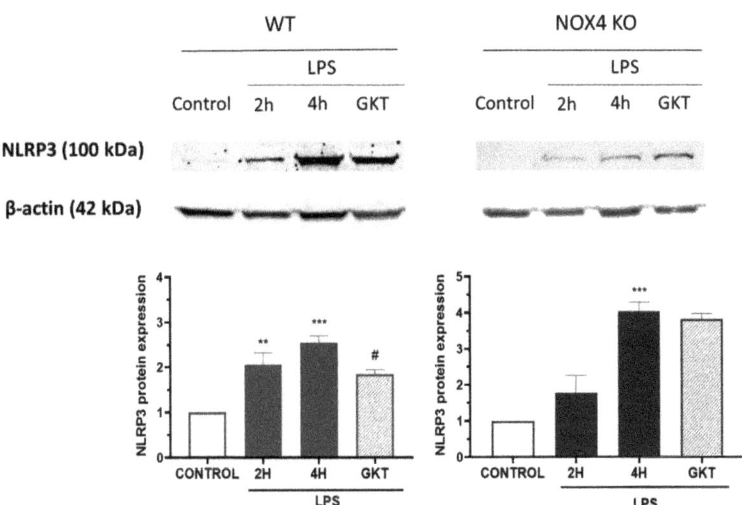

Figure 2. NLRP3 levels in mixed glial cultures of WT and Nox4 KO mice after LPS treatment. Mixed glial cultures of WT and NOX4 KO mice were stimulated with LPS (1 µg/mL) for 2 and 4 h, in the absence or presence of GKT136901 (1 µM) following the protocol at the top of Figure 1. Changes in NLRP3 protein amounts were determined via Western blot. A representative Western blot image of NLRP3 protein and actin is shown (top). Mean ± SEM (n = 5). One-way ANOVA with Tukey's multiple comparisons test. *** $p < 0.001$ and ** $p < 0.01$ vs. control; # $p < 0.05$ vs. 4 h LPS-treated cells.

3.2. NOX4 Activity Resulted in Nrf2 Translocation to the Nucleus That Is Necessary for NLRP3 Inflammasome Activation

As stated in the Introduction, NOX are one of the main sources of ROS and, importantly, the only known enzyme family that has ROS formation as its sole known function [20]. Of these, NOX4 appears to be the most promising target for various diseases like neurodegenerative diseases or brain ischemia [21]. On the other hand, Nrf2 is known as a master regulator of antioxidant and anti-inflammatory responses, and in the presence of oxidative stress, Nrf2 translocates into the nucleus to induce phase II antioxidant response, a set of key proteins that detoxify xenobiotics. Hence, we wanted to explore whether NOX4-derived ROS produced by LPS treatment can induce the translocation of Nrf2 to the nucleus. To confirm this hypothesis, we evaluated the nuclear translocation of Nrf2 induced by 4 h treatment with LPS. Mixed glial cultures from WT and NOX4 KO mice were treated with LPS for 4 h, then were fixed and double-stained with anti-Nrf2 and DAPI. As shown in Figure 3A,B, in control conditions, Nrf2 was predominantly present in the cytosol; however, in the presence of LPS 4 h Nrf2 was predominantly located in the nucleus in WT cultures (Figure 3A) but not in NOX4 KO animals (Figure 3B). We used tBHQ as positive control of Nrf2 translocation. tBHQ treatment resulted in the translocation of Nrf2 to the nucleus both in WT and in NOX4 KO cultures. To corroborate the induction of phase II enzymes by LPS treatment, we analyzed the protein levels of NAD(P)H quinone oxidoreductase 1 (NQO1) (Figure 3C), one of the phase II enzymes induced by Nrf2. LPS treatment significantly increased NQO protein levels by 2-fold and co-incubation with GKT136901 partially reduced the increase in NQO1. Together these results suggest that Nrf2 participates in NOX4-derived ROS activation of NLRP3 inflammasome.

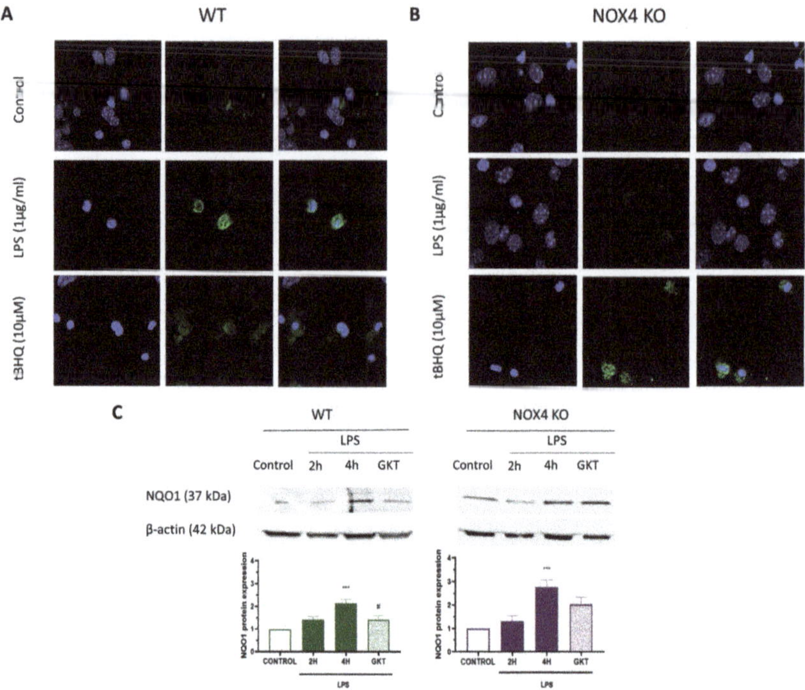

Figure 3. LPS treatment for 4 h induces Nrf2 translocation to the nucleus and NQO1 protein levels. High magnification confocal images of mixed glial cultures from WT (**A**) and NOX4 KO (**B**) immunostained with Nrf2 (green) and counterstained with DAPI (blue) to illustrate nuclei. Mixed glial cultures were untreated (Control), treated with LPS (1 μg/mL) for 4 h or treated with tBHQ (10 μM) for 24 h. (**C**) NQO1 protein levels in WT and NOX4 KO mice. Animals were treated with LPS (1 μg/mL) for 2 and 4 h, in the absence or presence of GKT136901 (10 μM). A representative Western blot image of NQO1 is shown at the top. Mean ± SEM (n = 5). One-way ANOVA with Tukey's multiple comparisons test. *** $p < 0.001$ vs. control; # $p < 0.05$ vs. LPS-treated animals.

3.3. Nrf2 Is Required for NLRP3 Inflammasome Activation

Next, we used the protocol shown on top of Figure 4 to determine the participation of Nrf2 in NLRP3 inflammasome activation. Stimulation of WT cultures with LPS (1 μg/mL, 4 h) plus ATP (5 mM, 30 min) produced a significant release of IL-1β (Figure 4A). However, in Nrf2 KO cultures, we did not observe any IL-1β release. Furthermore, we observed the same blocking effect on IL-1β release using a 24 h pre-treatment with tBHQ (10 μM) after LPS plus ATP (Figure 4A). To explore further in the mechanism of NLRP3 inflammasome, we measured NLRP3 inflammasome components (Nlrp3 and Il1b) by RT-PCR using the same protocol. As illustrated in Figure 4B, LPS resulted in an increase in Nlrp3 and Il1b mRNA both in WT and NOX4 KO mice, and 24 h pre-treatment with tBHQ significantly reduced Nlrp3 and Il1b mRNA levels in both WT and NOX4 KO mice. Together, these data indicate that Nrf2 is necessary for NLRP3 inflammasome activation in mixed glia cultures.

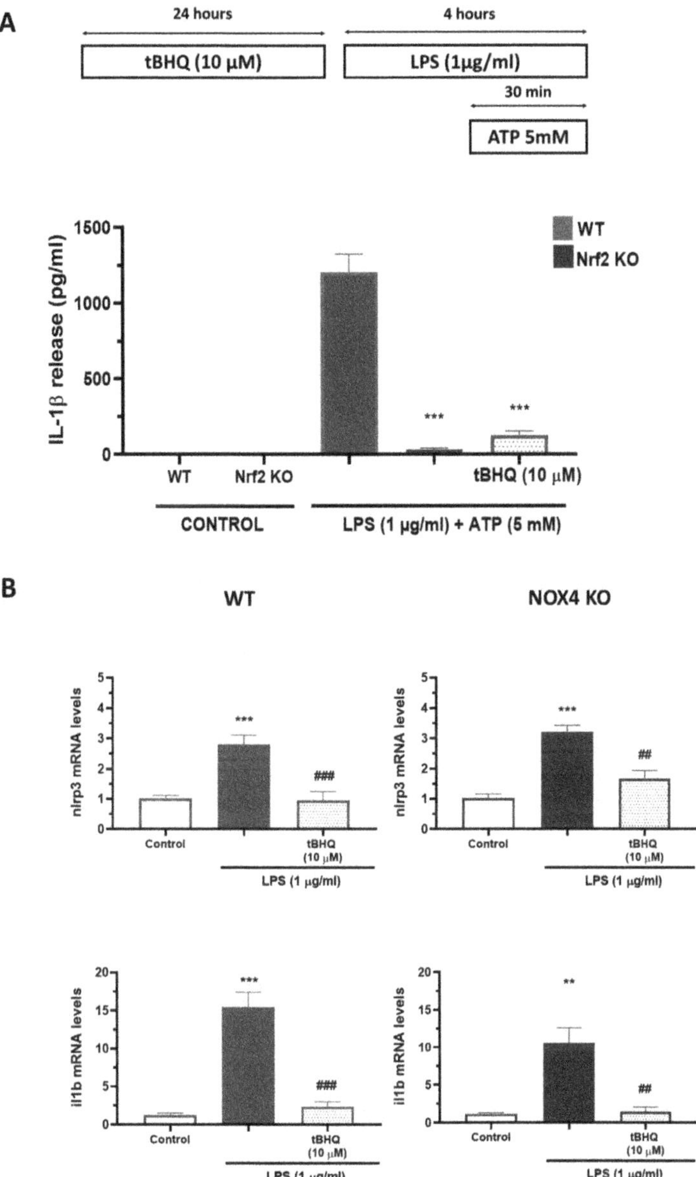

Figure 4. Nrf2 activity is required for NLRP3 inflammasome activation. (**A**) Mixed glial cultures of WT and Nrf2 KO mice were treated with tBHQ (10 µM) for 24 h followed by stimulation LPS (1 µg/mL) for 4 h plus ATP (5 mM) during the last 30 min, following the protocol at the top of the figure. ELISA measurements were performed for quantification of IL-1β secretion. Mean ± SEM ($n = 5$). Two-way ANOVA followed by Bonferroni's tests *** $p < 0.001$ vs. WT LPS. (**B**) Mixed glial cultures of WT and NOX4 KO mice were treated with tBHQ (10 µM) for 24 h followed by the stimulation with LPS (1 µg/mL) for 4 h. At the end of the experiment, cells were harvested and analyzed for the expression of NLRP3 inflammasome components Nlrp3 and Il1b by RT-PCR. Mean ± SEM ($n = 5$). One-way ANOVA with Tukey's multiple comparisons test. *** $p < 0.001$ and ** $p < 0.01$ vs. control; ### $p < 0.001$ and ## $p < 0.01$ vs. LPS-treated cells.

3.4. NOX4 Genetic Deletion and Inhibition with GKT136901 Results in Memory Impairment Induced by LPS Administration In Vivo

Neurodegenerative diseases, such as Alzheimer's and Parkinson's disease, are characterized by several pathological features, including aggregation of specific proteins, selective neuronal loss, and chronic inflammation. To evaluate how NOX4 could be controlling neuroinflammation in vivo, we used the LPS-model that drives a transient sickness behavior response characterized by weight loss and memory loss without affecting locomotor activity [22]. We evaluated the memory loss induced by LPS (250 µg/kg) using WT and NOX4 KO mice and pharmacological inhibition of NOX4 with GKT136901 using the novel object recognition test (NOR) in mice, following the protocol shown in Figure 5A. In these conditions, administration of LPS significantly impaired NOR performance in WT animals (Figure 5B). Memory impairment produced by LPS injection was partially reversed with pharmacological treatment using the inhibitor of NOX4 GKT136901. Furthermore, in NOX4 KO animals, injection of LPS did not produce any memory impairment compared with NOX4 KO vehicle mice (Figure 5B). Taken together, the absence of NOX4 and its pharmacological inhibition improves memory impairment induced by LPS challenge in mice.

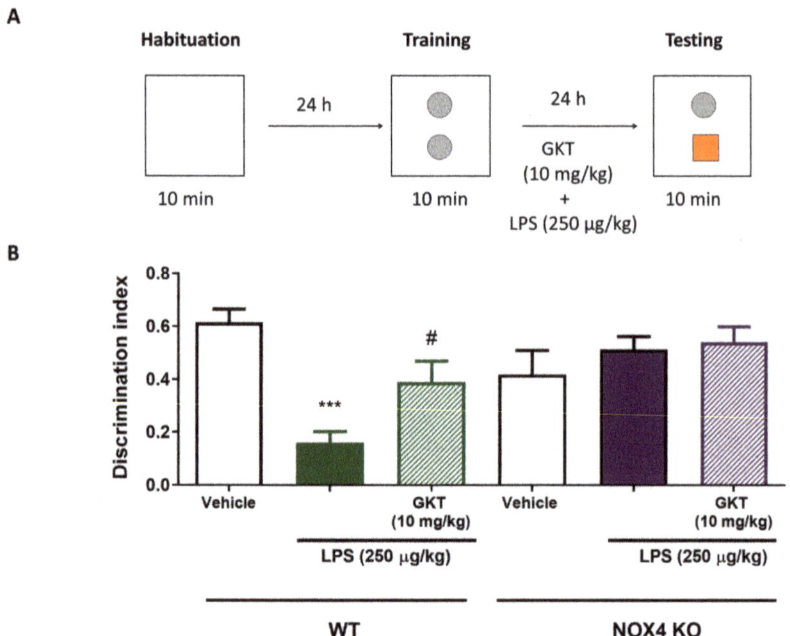

Figure 5. Genetic ablation and pharmacological inhibition of Nox4 improves LPS-induced memory impairment. (**A**) Illustration of the protocol used to evaluate novel object recognition (NOR) test. WT and Nox4 KO mice were treated with LPS (250 µg/kg, i.p.) alone or in combination with GKT (10 mg/kg, i.p.) and after 24 h were subjected to the test. (**B**) Discrimination index in WT and Nox4 KO mice in saline-treated, LPS-treated and GKT-treated groups. Mean ± SEM ($n = 7$). Two-way ANOVA followed by Bonferroni's tests *** $p < 0.001$ vs. WT vehicle and # $p < 0.05$ vs. WT LPS.

3.5. NOX4 Genetic and Pharmacological Inhibition Modulate Inflammasome Component Expression In Vivo

The LPS model of transient inflammation is characterized by an acute inflammatory response during the first few hours and a high increase in proinflammatory cytokine expression, which remains elevated for up to 24 h. Therefore, following the experimental protocol of Figure 5, we evaluated NLRP3 inflammasome components (Nlrp3 and Il-1b) with RT-PCR in mouse hippocampus at 24 h post-LPS administration. At 24 h post-LPS injection, there was a significant increase in Nlrp3 and Il-1b mRNA levels in WT animals (Figure 6A,B). GKT136901 co-administration with LPS significantly reduced the expression of both Nrlp3 and Il-1b genes. Additionally, there was a significant increase in the mRNA levels of NOX4 in LPS-treated mice compared to vehicle in WT mice but not in NOX4 KO mice. Co-administration of GKT136901 reversed NOX4 mRNA increase near to vehicle levels (Figure 6C).

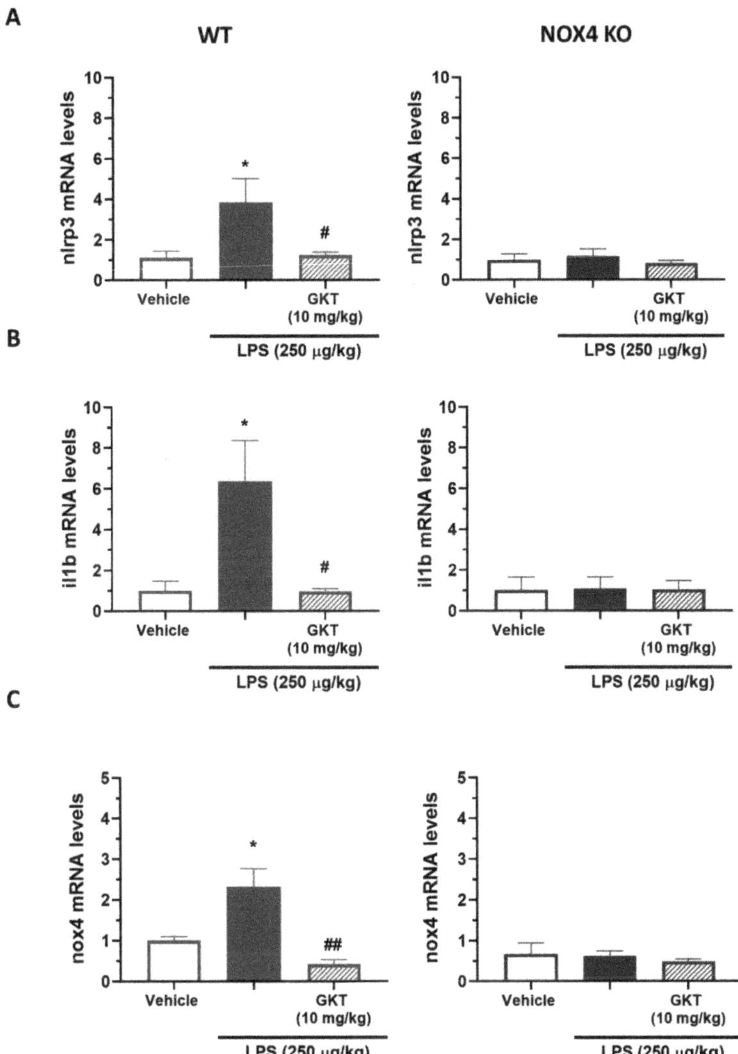

Figure 6. NOX4 activity is necessary for NLRP3 inflammasome components and Nox4 expression in hippocampus of LPS-treated mice. Expression of genes of the inflammasome NLRP3 complex and NOX4 in the hippocampus of WT and Nox4 KO mice after 24 h LPS and/or GKT treatment. mRNA levels of inflammasome components Nlrp3 (**A**), Il1b (**B**), and Nox4 (**C**) in the hippocampus of WT ($n = 5$) and Nox4 KO ($n = 5$) mice treated with LPS (250 µg/kg) and/or GKT (10 mg/kg) for 24 h. Mean ± SEM ($n = 5$). One-way ANOVA with Tukey's multiple comparisons test. * $p < 0.05$ vs. vehicle; ## $p < 0.01$ and # $p < 0.05$ vs. LPS-treated animals.

Finally, we wanted to confirm these results via Western blot in the hippocampus of animals 24 h post-LPS administration. At the protein level, we observed that there was an increase in NLRP3 and NOX4 amounts in LPS-treated WT animals (Figure 7). These increases were not observed in LPS-treated NOX4 KO animals or in WT animals treated with GKT136901, corroborating the results observed in mRNA. These results confirm that NOX4 is a key protein for NLRP3 inflammasome component expression and activation, and that its pharmacological inhibition can be a good pharmacological treatment to reduce inflammatory levels and improve memory impairment.

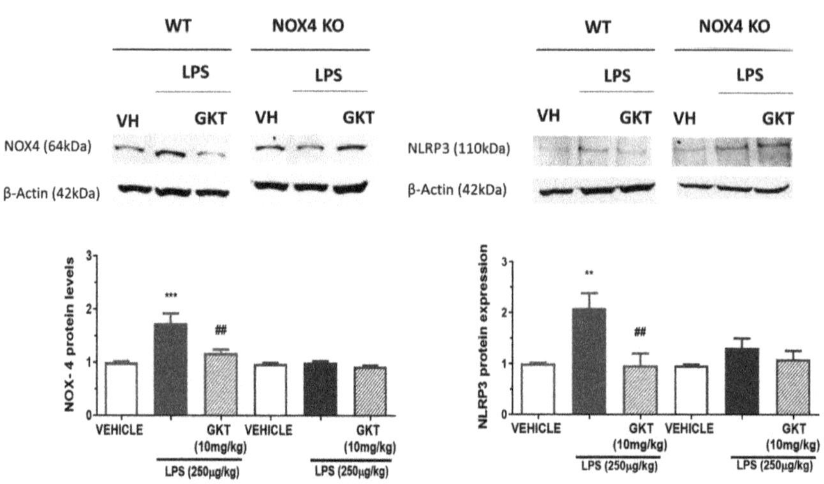

Figure 7. NOX4 and NLRP3 protein levels in the hippocampus of WT and NOX4 KO mice treated with LPS for 24 h. Changes in Nox4 (**left**) and NLRP3 (**right**) levels in WT and NOX4 KO mice. Animals were treated with LPS (250 μg/kg) for 24 h, in the absence or presence of GKT136901 (10 mg/kg). Representative Western blot image of Nox4, NLRP3, and actin is shown on top. Mean ± SEM (n = 5). One-way ANOVA with Tukey's multiple comparisons test. *** $p < 0.001$ and ** $p < 0.01$ vs. control; ## $p < 0.01$ vs. LPS-treated animals.

4. Discussion

We here validate NOX4 as a key player in NLRP3 inflammasome activation, suggesting NOX4 pharmacological inhibition as a potent therapeutic approach in neurodegenerative diseases. In this research, we contribute to the understanding of the role of NOX4 and Nrf2 in the activation of the NLRP3 inflammasome. Our data show that both NOX4 and Nrf2 participate in NLRP3 inflammasome activation, since (i) we showed a reduction in IL-1β release in response to LPS plus ATP in NOX4 and Nrf2 KO animals; (ii) pharmacological inhibition of NOX4 using GKT136901 and Nrf2 activation using tBHQ also reduced IL-1β release; (iii) we measured NOX4 and NLRP3 expression in WT mixed glial cultures following LPS treatment, observing that both increased after TLR4 activation, while 24 h treatment with tert-butylhydroquinone, a potent Nrf2 inducer, significantly reduced NLRP3 expression. In vivo LPS administration resulted in significant cognitive impairment compared to the control group. However, mice treated with GKT136901 after LPS impairment showed a significantly improved discrimination index in the NOR test and recovery of the expression of inflammatory genes to normal levels. Nevertheless, we should consider that our study has some limitations. In this study, to directly assess NOX4 function in vivo without altering its expression, we selected NOX4 KO mice generated in 2010 by deleting the NADPH binding pockets located in exons 14 and 15. In in vitro experiments, other NOX isoforms are present that may be involved in LPS-dependent ROS generation. Therefore, it would be of interest in future experiments to use selective inhibitors of the remaining isoforms to determine their involvement.

Aging is characterized by a gradual, cumulative deterioration in physiological functions over time which results in increased susceptibility to diseases, in particular neurodegenerative diseases [23]. Microglial cells are key mediators of age-related neuroinflammation, and it has been demonstrated that an increase in microglial activity can be an early event that leads to oxidative damage and cell degeneration [24]. Reactive microgliosis resulted in an increase in inflammatory cytokines [25] and increased microglial NADPH-derived ROS accumulation, which are considered to be key events in the central nervous system pathogenesis [26]. Here, we have shown that NOX4 activity is necessary for a proper inflammatory response since genetic ablation and pharmacological inhibition of NOX4 reduced NLRP3 inflammasome activation by 60% (Figure 1). Moreover, NOX4 activation is important for the expression of NLRP3 inflammasome components induced by TLR4 activation (Figures 1 and 2). It has been demonstrated by different authors that there is a link between NOX4 activity and NLRP3 inflammasome activation. In Kupffer cells, NOX4-derived ROS promoted NLRP3 activation and significantly increased the expression of inflammasome components both in vitro and in vivo, aggravating liver inflammatory injury [27]. This mechanism has also been observed in various models of inflammation, such as acute pancreatitis [28], in high-glucose-induced endothelial dysfunction [29], in osteoarthritis [30], and in myocardial ischemia-reperfusion injury [31]. We also observed an in vivo anti-inflammatory effect in our model of LPS-induced memory impairment (Figure 5). In fact, genetic ablation, or pharmacological inhibition of NOX4, improved animal behavior and restored normal levels of NLRP3 inflammasome components and NOX4 (Figures 6 and 7), confirming the results obtained in vitro.

Nrf2 plays an important role in regulating oxidative stress, inflammation, mitochondrial function, and in autophagy [28]. Under normal conditions, Keap1 binds to Nrf2 in the cytoplasm and promotes its ubiquitination and proteasomal degradation [31,32]. Intracellular ROS or electrophiles alter the conformation of the Nrf2/Keap1 complex, thereby inhibiting Nrf2 ubiquitination. Thus, Nrf2 translocates to the cell nucleus, binds to the regulatory enhancer sequence ARE, and promotes the expression of antioxidant and anti-inflammatory genes [30]. On the other hand, NLRP3 also detects cellular stressors, like pathogen- and danger-associated molecular patterns (PAMPs and DAMPs) that induce inflammasome activation. The main link between Nrf2 and NLRP3 are intracellular ROS, that could come from both mitochondrial ROS and NOX activation, and can activate both proteins; however, its actions are opposite. In fact, it has been demonstrated that Nrf2 activation before NLRP3 inflammasome activation reduces inflammation in various in vitro and in vivo models of central and peripheral inflammation [33–36]. Consistent with all these studies, we have shown that 24 h treatment with tBHQ prior to NLRP3 inflammasome activation reduces NLRP3 inflammasome components both in WT and NOX4 KO mice (Figure 4). However, it is unexpected that both Nrf2 activation and Nrf2 ablation have the same consequence on the NLRP3 inflammasome activation (Figure 4). It has been recently proposed that a physical interaction of the Nrf2 complex with caspase-1 contributes to the requirement of Nrf2 expression for inflammasome activation [37,38]. Here, we showed that Nrf2 translocates to the nucleus upon LPS exposure to cells and that NOX4 activity is critical for this translocation, as Nrf2 does not translocate to the nucleus in NOX4 KO animals (Figure 3). These experiments point to NOX4 as the crucial protein for both Nrf2 translocation to the nucleus and for NLRP3 activation. Probably, other sources of ROS, like mitochondrial ROS, could be participating in this process, since NLRP3 inflammasome activation is only partially blocked in NOX4 KO mice. Hence, it is possible that the early production of NOX4-derived ROS could be a crucial step for Nrf2-NLRP3 rupture of physical interaction, but this hypothesis needs to be further investigated.

5. Conclusions

We present evidence for a complex crosstalk between NOX4, Nrf2, and NLRP3 inflammasome pathways. We contribute to the understanding of the role of NOX4 and Nrf2 in the activation of the NLRP3 inflammasome. We demonstrate that NOX4 is a key player in

NLRP3 inflammasome activation suggesting NOX4 pharmacological inhibition as a potent therapeutic approach in neurodegenerative diseases.

Author Contributions: Conceptualization, A.J.C., A.R. and J.E.; Data curation, A.P.-A., C.D.-P., A.R. and J.E.; Formal analysis, A.P.-A., C.D.-P., V.F.-A., P.N.-F., A.B.L.-R., M.A.-R. and I.V.; Funding acquisition, A.R. and J.E.; Investigation, A.P.-A., C.D.-P., V.F.-A., A.B.L.-R., P.N.-F., F.L.-M. and E.R.; Methodology, A.P.-A., C.D.-P., V.F.-A., A.B.L.-R. and P.N.-F.; Project administration, J.E.; Supervision, A.R. and J.E.; Writing—original draft, A.R. and J.E.; Writing—review and editing, A.P.-A., C.D.-P., A.C., A.R. and J.E. All authors have read and agreed to the published version of the manuscript.

Funding: This work was supported by grants from Fundación Mutua Madrileña and Fondo de Investigaciones Sanitarias (FIS) (ISCIII/FEDER) (Programa Miguel Servet CPII19/00005; PI16/00735; PI19/00082 and PI22/00362) to J.E. This research has been funded by Instituto de Salud Carlos III (RICORS-RD21/0006/0009) and co-financed with FEDER Funds and/or from the European funds of the Recovery, Transformation and Resilience Plan and by NextGenerationEU. UCJC INFLAMAMEL 2022-07 project to A.R. and Instituto de Salud Carlos III (Sara Borrell CD19/00092) to A.B.L.-R. and (Sara Borrell CD22/00101) to I.V.

Institutional Review Board Statement: The animal study protocol was approved by the Ethics Committee of Universidad Autónoma de Madrid (Madrid, Spain), in compliance with the Cruelty to Animals Act, 1876, and the European Community Directive, 86/609/EEC. All animal experimentation was performed under the license PROEX 013/18.

Informed Consent Statement: Not applicable.

Data Availability Statement: The data presented in this study are available on request from the corresponding author.

Acknowledgments: We thank Instituto/Fundación Teófilo Hernando de I + D del Medicamento for its continued support.

Conflicts of Interest: The authors declare no conflict of interest.

References

1. Weavers, H.; Martin, P. The cell biology of inflammation: From common traits to remarkable immunological adaptations. *J. Cell Biol.* **2020**, *219*, e202004003. [CrossRef] [PubMed]
2. Wang, Z.; Zhang, S.; Xiao, Y.; Zhang, W.; Wu, S.; Qin, T.; Yue, Y.; Qian, W.; Li, L. NLRP3 Inflammasome and Inflammatory Diseases. *Oxidative Med. Cell. Longev.* **2020**, *2020*, 4063562. [CrossRef]
3. Bauernfeind, F.G.; Horvath, G.; Stutz, A.; Alnemri, E.S.; MacDonald, K.; Speert, D.; Fernandes-Alnemri, T.; Wu, J.; Monks, B.G.; Fitzgerald, K.A.; et al. Cutting Edge: NF-κB Activating Pattern Recognition and Cytokine Receptors License NLRP3 Inflammasome Activation by Regulating NLRP3 Expression. *J. Immunol.* **2009**, *183*, 787–791. [CrossRef] [PubMed]
4. Schroder, K.; Tschopp, J. The Inflammasomes. *Cell* **2010**, *140*, 821–832. [CrossRef]
5. Wen, H.; Miao, E.A.; Ting, J.P.Y. Mechanisms of NOD-like receptor-associated inflammasome activation. *Immunity* **2013**, *39*, 432–441. [CrossRef]
6. Heneka, M.T.; Kummer, M.P.; Stutz, A.; Delekate, A.; Schwartz, S.; Vieira-Saecker, A.; Griep, A.; Axt, D.; Remus, A.; Tzeng, T.C.; et al. NLRP3 is activated in Alzheimer's disease and contributes to pathology in APP/PS1 mice. *Nature* **2013**, *493*, 674–678. [CrossRef]
7. Hoffman, H.M.; Mueller, J.L.; Broide, D.H.; Wanderer, A.A.; Kolodner, R.D. Mutation of a new gene encoding a putative pyrin-like protein causes familial cold autoinflammatory syndrome and Muckle-Wells syndrome. *Nat. Genet.* **2001**, *29*, 301–305. [CrossRef]
8. Dröge, W. Free radicals in the physiological control of cell function. *Physiol. Rev.* **2002**, *82*, 47–95. [CrossRef]
9. Gandhi, S.; Abramov, A.Y. Mechanism of oxidative stress in neurodegeneration. *Oxidative Med. Cell. Longev.* **2012**, *2012*, 428010. [CrossRef] [PubMed]
10. Knaus, U.G.; Leto, T.L. *NADPH Oxidases Methods and Protocols Methods in Molecular Biology 1982*; Humana Press: Totowa, NJ, USA, 2019.
11. Kleinschnitz, C.; Grund, H.; Wingler, K.; Armitage, M.E.; Jones, E.; Mittal, M.; Barit, D.; Schwarz, T.; Geis, C.; Kraft, P.; et al. Post-stroke inhibition of induced NADPH Oxidase type 4 prevents oxidative stress and neurodegeneration. *PLoS Biol.* **2010**, *8*, e1000479. [CrossRef]

12. Moon, J.S.; Nakahira, K.; Chung, K.P.; DeNicola, G.M.; Koo, M.J.; Pabón, M.A.; Rooney, K.T.; Yoon, J.H.; Ryter, S.W.; Stout-Delgado, H.; et al. NOX4-dependent fatty acid oxidation promotes NLRP3 inflammasome activation in macrophages. *Nat. Med.* **2016**, *22*, 1002–1012. [CrossRef]
13. Kensler, T.W.; Wakabayashi, N.; Biswal, S. Cell survival responses to environmental stresses via the Keap1-Nrf2-ARE pathway. *Annu. Rev. Pharmacol. Toxicol.* **2007**, *47*, 89–116. [CrossRef]
14. Zhao, C.; Gillette, D.D.; Li, X.; Zhang, Z.; Wen, H. Nuclear factor E2-related factor-2 (Nrf2) is required for NLRP3 and AIM2 inflammasome activation. *J. Biol. Chem.* **2014**, *289*, 17020–17029. [CrossRef]
15. Parada, E.; Buendia, I.; Navarro, E.; Avendaño, C.; Egea, J.; López, M.G. Microglial HO-1 induction by curcumin provides antioxidant, antineuroinflammatory, and glioprotective effects. *Mol. Nutr. Food Res.* **2015**, *59*, 1690–1700. [CrossRef] [PubMed]
16. Farré-Alins, V.; Narros-Fernández, P.; Palomino-Antolín, A.; Decouty-Pérez, C.; Lopez-Rodriguez, A.B.; Parada, E.; Muñoz-Montero, A.; Gómez-Rangel, V.; López-Muñoz, F.; Ramos, E.; et al. Melatonin reduces NLRP3 inflammasome activation by increasing α7 nAChR-mediated autophagic flux. *Antioxidants* **2020**, *9*, 1299. [CrossRef]
17. Lueptow, L.M. Novel object recognition test for the investigation of learning and memory in mice. *J. Vis. Exp.* **2017**, *2017*, e55718.
18. Livak, K.J.; Schmittgen, T.D. Analysis of relative gene expression data using real-time quantitative PCR and the 2(-Delta Delta C(T)) Method. *Methods* **2001**, *25*, 402–408. [CrossRef] [PubMed]
19. Schmittgen, T.; Livak, K. Analyzing real-time PCR data by the comparative C_T method. *Nat. Protoc.* **2008**, *3*, 1101–1108. [CrossRef]
20. Casas, A.I.; Dao, V.T.V.; Daiber, A.; Maghzal, G.J.; Di Lisa, F.; Kaludercic, N.; Leach, S.; Cuadrado, A.; Jaquet, V.; Seredenina, T.; et al. Reactive Oxygen-Related Diseases: Therapeutic Targets and Emerging Clinical Indications. *Antioxid. Redox Signal.* **2015**, *23*, 1171–1185. [CrossRef]
21. Ma, M.W.; Wang, J.; Zhang, Q.; Wang, R.; Dhandapani, K.M.; Vadlamudi, R.K.; Brann, D.W. NADPH oxidase in brain injury and neurodegenerative disorders. *Mol. Neurodegener.* **2017**, *12*, 7. [CrossRef] [PubMed]
22. Berg, B.M.; Godbout, J.P.; Kelley, K.W.; Johnson, R.W. α-Tocopherol attenuates lipopolysaccharide-induced sickness behavior in mice. *Brain Behav. Immun.* **2004**, *18*, 149–157. [CrossRef]
23. Tarafdar, A.; Pula, G. The role of NADPH oxidases and oxidative stress in neurodegenerative disorders. *Int. J. Mol. Sci.* **2018**, *19*, 3824. [CrossRef]
24. von Bernhardi, R.; Eugenín-von Bernhardi, L.; Eugenín, J. Microglial cell dysregulation in brain aging and neurodegeneration. *Front. Aging Neurosci.* **2015**, *7*, 124. [CrossRef] [PubMed]
25. Ardura-Fabregat, A.; Boddeke, E.W.G.M.; Boza-Serrano, A.; Brioschi, S.; Castro-Gomez, S.; Ceyzériat, K.; Dansokho, C.; Dierkes, T.; Gelders, G.; Heneka, M.T.; et al. Targeting Neuroinflammation to Treat Alzheimer's Disease. *CNS Drugs* **2017**, *31*, 1057–1082. [CrossRef]
26. Zuroff, L.; Daley, D.; Black, K.L.; Koronyo-Hamaoui, M. *Clearance of Cerebral Aβ in Alzheimer's Disease: Reassessing the Role of Microglia and Monocytes*; Springer International Publishing: Berlin/Heidelberg, Germany, 2017; Volume 74.
27. Zhai, L.; Pei, H.; Yang, Y.; Zhu, Y.; Ruan, S. NOX4 promotes Kupffer cell inflammatory response via ROS-NLRP3 to aggravate liver inflammatory injury in acute liver injury. *Aging* **2022**, *14*, 6905–6916. [CrossRef]
28. Jin, H.-Z.; Yang, X.-J.; Zhao, K.-L.; Mei, F.-C.; Zhou, Y.; You, Y.-D.; Wang, W.-X. Apocynin alleviates lung injury by suppressing NLRP3 inflammasome activation and NF-κB signaling in acute pancreatitis. *Int. Immunopharmacol.* **2019**, *75*, 105821. [CrossRef] [PubMed]
29. Luo, X.; Hu, Y.; He, S.; Ye, Q.; Lv, Z.; Liu, J.; Chen, X. Dulaglutide inhibits high glucose-induced endothelial dysfunction and NLRP3 inflammasome activation. *Arch. Biochem. Biophys.* **2019**, *671*, 203–209. [CrossRef] [PubMed]
30. Mourmoura, E.; Papathanasiou, I.; Trachana, V.; Konteles, V.; Tsoumpou, A.; Goutas, A.; Papageorgiou, A.A.; Stefanou, N.; Tsezou, A. Leptin-depended NLRP3 inflammasome activation in osteoarthritic chondrocytes is mediated by ROS. *Mech. Ageing Dev.* **2022**, *208*, 111730. [CrossRef]
31. Wang, F.; Wang, H.; Liu, X.; Yu, H.; Huang, X.; Huang, W.; Wang, G. Neuregulin-1 alleviate oxidative stress and mitigate inflammation by suppressing NOX4 and NLRP3/caspase-1 in myocardial ischaemia-reperfusion injury. *J. Cell. Mol. Med.* **2021**, *25*, 1783–1795. [CrossRef] [PubMed]
32. Taguchi, K.; Motohashi, H.; Yamamoto, M. Molecular mechanisms of the Keap1-Nrf2 pathway in stress response and cancer evolution. *Genes Cells* **2011**, *16*, 123–140. [CrossRef] [PubMed]
33. Ma, Q. Role of Nrf2 in oxidative stress and toxicity. *Annu. Rev. Pharmacol. Toxicol.* **2013**, *53*, 401–426. [CrossRef]
34. Shi, Q.; Qian, Y.; Wang, B.; Liu, L.; Chen, Y.; Chen, C.; Feng, L.; Chen, J.; Dong, N. Glycyrrhizin protects against particulate matter-induced lung injury via regulation of endoplasmic reticulum stress and NLRP3 inflammasome-mediated pyroptosis through Nrf2/HO-1/NQO1 signaling pathway. *Int. Immunopharmacol.* **2023**, *120*, 110371. [CrossRef] [PubMed]
35. Gao, Z.; Zhan, H.; Zong, W.; Sun, M.; Linghu, L.; Wang, G.; Meng, F.; Chen, M. Salidroside alleviates acetaminophen-induced hepatotoxicity via Sirt1-mediated activation of Akt/Nrf2 pathway and suppression of NF-κB/NLRP3 inflammasome axis. *Life Sci.* **2023**, *327*, 121793. [CrossRef]
36. Chen, X.; Tian, C.; Zhang, Z.; Qin, Y.; Meng, R.; Dai, X.; Zhong, Y.; Wei, X.; Zhang, J.; Shen, C. Astragaloside IV Inhibits NLRP3 Inflammasome-Mediated Pyroptosis via Activation of Nrf-2/HO-1 Signaling Pathway and Protects against Doxorubicin-Induced Cardiac Dysfunction. *Front. Biosci.* **2023**, *28*, 45. [CrossRef] [PubMed]

37. Liu, Z.; Tu, K.; Zou, P.; Liao, C.; Ding, R.; Huang, Z.; Huang, Z.; Yao, X.; Chen, J.; Zhang, Z. Hesperetin ameliorates spinal cord injury by inhibiting NLRP3 inflammasome activation and pyroptosis through enhancing Nrf2 signaling. *Int. Immunopharmacol.* **2023**, *118*, 1101033. [CrossRef] [PubMed]

38. Garstkiewicz, M.; Strittmatter, G.E.; Grossi, S.; Sand, J.; Fenini, G.; Werner, S.; French, L.E.; Beer, H.D. Opposing effects of Nrf2 and Nrf2-activating compounds on the NLRP3 inflammasome independent of Nrf2-mediated gene expression. *Eur. J. Immunol.* **2017**, *47*, 806–817. [CrossRef] [PubMed]

 *antioxidants*

Article

c-Abl Phosphorylates MFN2 to Regulate Mitochondrial Morphology in Cells under Endoplasmic Reticulum and Oxidative Stress, Impacting Cell Survival and Neurodegeneration

Alexis Martinez [1,2,†], Cristian M. Lamaizon [1,3,†], Cristian Valls [1], Fabien Llambi [4], Nancy Leal [1], Patrick Fitzgerald [4], Cliff Guy [4], Marcin M. Kamiński [4], Nibaldo C. Inestrosa [2,5], Brigitte van Zundert [2,6,7], Gonzalo I. Cancino [8], Andrés E. Dulcey [9], Silvana Zanlungo [10], Juan J. Marugan [9], Claudio Hetz [11,12,13,14], Douglas R. Green [4] and Alejandra R. Alvarez [1,2,3,*]

1   Cell Signaling Laboratory, Department of Cell and Molecular Biology, Biological Sciences Faculty, Pontificia Universidad Católica de Chile, Santiago 8331150, Chile
2   Basal Center for Aging and Regeneration, Pontificia Universidad Católica de Chile (CARE UC), Santiago 8331150, Chile
3   Millennium Institute on Immunology and Immunotherapy, Biological Sciences Faculty, Pontificia Universidad Católica de Chile, Santiago 8331150, Chile
4   Department of Immunology, St. Jude Children's Research Hospital, Memphis, TN 38105, USA
5   Center of Excellence in Biomedicine of Magallanes (CEBIMA), University of Magallanes, Punta Arenas 6210427, Chile
6   Institute of Biomedical Sciences, Faculty of Medicine & Faculty of Life Sciences, Universidad Andres Bello, Santiago 8370146, Chile
7   Department of Neurology, University of Massachusetts Chan Medical School (UMMS), Worcester, MA 01655, USA
8   Laboratory of Neurobiology, Department of Cell and Molecular Biology, Biological Sciences Faculty, Pontificia Universidad Católica de Chile, Santiago 8331150, Chile
9   Early Translation Branch, National Center for Advancing Translational Sciences (NCATS), NIH, 9800 Medical Center Drive, Rockville, MD 20850, USA
10 Department of Gastroenterology, Faculty of Medicine, Pontificia Universidad Católica de Chile, Av. Libertador Bernardo O'Higgins 340, Santiago 8331150, Chile; szanlungo@uc.cl
11 Biomedical Neuroscience Institute (BNI), Faculty of Medicine, University of Chile, Santiago 8330015, Chile
12 Center for Geroscience, Brain Health and Metabolism (GERO), Santiago 8380453, Chile
13 Program of Cellular and Molecular Biology, Institute of Biomedical Sciences, University of Chile, Santiago 8330015, Chile
14 The Buck Institute for Research in Aging, Novato, CA 94945, USA
*   Correspondence: aalvarez@bio.puc.cl
†   These authors contributed equally to this work.

Citation: Martinez, A.; Lamaizon, C.M.; Valls, C.; Llambi, F.; Leal, N.; Fitzgerald, P.; Guy, C.; Kamiński, M.M.; Inestrosa, N.C.; van Zundert, B.; et al. c-Abl Phosphorylates MFN2 to Regulate Mitochondrial Morphology in Cells under Endoplasmic Reticulum and Oxidative Stress, Impacting Cell Survival and Neurodegeneration. *Antioxidants* **2023**, *12*, 2007. https://doi.org/10.3390/antiox12112007

Academic Editor: Dmitry B. Zorov

Received: 14 September 2023
Revised: 17 October 2023
Accepted: 13 November 2023
Published: 16 November 2023

Abstract: The endoplasmic reticulum is a subcellular organelle key in the control of synthesis, folding, and sorting of proteins. Under endoplasmic reticulum stress, an adaptive unfolded protein response is activated; however, if this activation is prolonged, cells can undergo cell death, in part due to oxidative stress and mitochondrial fragmentation. Here, we report that endoplasmic reticulum stress activates c-Abl tyrosine kinase, inducing its translocation to mitochondria. We found that endoplasmic reticulum stress-activated c-Abl interacts with and phosphorylates the mitochondrial fusion protein MFN2, resulting in mitochondrial fragmentation and apoptosis. Moreover, the pharmacological or genetic inhibition of c-Abl prevents MFN2 phosphorylation, mitochondrial fragmentation, and apoptosis in cells under endoplasmic reticulum stress. Finally, in the amyotrophic lateral sclerosis mouse model, where endoplasmic reticulum and oxidative stress has been linked to neuronal cell death, we demonstrated that the administration of c-Abl inhibitor neurotinib delays the onset of symptoms. Our results uncovered a function of c-Abl in the crosstalk between endoplasmic reticulum stress and mitochondrial dynamics via MFN2 phosphorylation.

Keywords: c-Abl; mitofusin 2; apoptosis; mitochondrial fusion; amyotrophic lateral sclerosis; endoplasmic reticulum stress

1. Introduction

The endoplasmic reticulum (ER) is a membranous organelle key in the control of synthesis, folding, and sorting of proteins. The folding capacity of the ER is constantly challenged by physiological demands and disease states. To adjunt proteostasis, cells engage a dynamic intracellular signaling pathway known as the unfolded protein response (UPR), enforcing adaptive programs that improve central aspects of the entire secretory pathway, whereas uncompensated ER stress results in oxidative stress and apoptosis [1]. The execution of cell death by ER-damaging insults largely depends on the intrinsic mitochondrial apoptosis pathway. Thus, a crosstalk between the ER and mitochondria is essential to determine cell fate under ER stress [2]. Interestingly, early studies suggested that c-Abl kinase operates as a signaling interphase between the ER stress and oxidative stress and mitochondria, mediated by its translocation to the mitochondria and the engagement of apoptosis programs [3]. However, the mechanisms and molecular targets of c-Abl at the mitochondria are not completely explored.

c-Abl is a non-receptor tyrosine kinase with functions in neurulation, cytoskeleton dynamics, synapsis, and apoptosis in the central nervous system [4–14]. Emerging evidence suggests that the localization of c-Abl to the cytosol has a significant impact on cytoskeleton dynamics, in addition to influencing cells under stress. For example, c-Abl is activated by oxidative stress impacting the activation of the canonical mitochondrial apoptosis pathway [15–18]. Also, c-Abl activation under cellular stress leads to the activation of the pro-apoptotic transcription factors p73 [19] and MST1 [20], and thus promoting the expression of proapoptotic genes. Several studies using the pharmacological targeting of c-Abl indicate that its activation has a pathogenic role in different human diseases linked to abnormal protein aggregation and oxidative stress. c-Abl exhibits a proapoptotic function in neurons exposed to amyloid β [21,22], and it is involved in neurodegeneration in Parkinson's disease [23–31] and Alzheimer's disease (AD) [32–36], in addition to lysosomal storage disorders [37,38] and amyotrophic lateral sclerosis (ALS) [39–41]. However, how c-Abl regulates mitochondrial dynamics and cell fate in cells under ER stress is still unknown.

Here, we report that c-Abl regulates mitochondrial morphology in response to ER stress. In cells under ER stress, c-Abl translocates to the mitochondria and triggers mitochondrial fragmentation. Interestingly, when c-Abl is pharmacologically or genetically inhibited in cells under ER stress, mitochondrial fragmentation and apoptosis are rescued. Then, we demonstrate that c-Abl phosphorylates mitofusin 2 (MFN2), an essential regulator of the mitochondria fusion machinery, specifically at Y269, which alters the MFN2 GTP binding affinity, influencing mitochondrial dynamics and cell survival under ER stress. Interestingly, the regulation of mitochondrial morphology by c-Abl-MFN2 has therapeutical potential, as demonstrated by its pharmacological inhibition using cell culture models for ALS. Furthermore, the treatment with neurotinib, a novel allosteric c-Abl inhibitor with high brain penetrance in ALS transgenic mice mutant for superoxide 1 (SOD1), results in a delayed disease onset. Overall, our results uncover that c-Abl modulates mitochondrial dynamics and apoptosis in cells under ER stress via MFN2 phosphorylation.

2. Materials and Methods

2.1. Primary Culture of Rat Hippocampal Neurons

Rat hippocampal cultures were prepared as described previously with some modifications [22,42]. Hippocampi from Sprague Dawley rats on embryonic day 18 were removed, dissected free of meninges in Ca^{2+}/Mg^{2+}-free HBSS, and rinsed twice with HBSS by allowing the tissue to settle to the bottom of the tube. After the second wash, the tissue was resuspended in HBSS containing 0.25% trypsin, and incubated for 15 min at 37 °C. After three rinses with HBSS, the tissue was mechanically dissociated in plating medium (DMEM; Invitrogen, Waltham, MA, USA), supplemented with 10% horse serum (Invitrogen), 100 U/mL penicillin, and 100 μg/mL streptomycin by gentle passage through Pasteur pipettes. The dissociated hippocampal cells were seeded onto poly-L-lysine-coated six-well culture plates at a density of 7×10^5 cells per well in plating medium. The cultures were maintained at 37 °C in 5% CO_2 for 2 h before the plating medium was replaced with Neurobasal growth medium (Invitrogen) supplemented with B27 (Invitrogen), 2 mM L-glutamine, 100 U/mL penicillin, and 100 µg/mL streptomycin. On day 2, the cultured neurons were treated with AraC 2 µM for 24 h; this method resulted in cultures highly enriched in neurons (~5% glia). For hippocampal cultures from *ABL1* conditional knockout mice (*ABL1-cKO*), homozygous c-Abl-floxed mice were kindly donated by Dr. AJ Koleske (Yale School of Medicine, USA) and bred in our animal facility. *ABL1-cKO* mice were bred from *ABL1^{loxP}/ABL1^{loxP}* and Nestin-Cre$^+$, obtained from Jackson Labs. These mice have loxP sites upstream and downstream of exon 5 of the *Abl1* gene. This strain was originated and maintained on a mixed B6.129S4, C57BL/6 background, and did not display any gross physical or behavioral abnormalities. Genotyping was performed using a PCR-based screening to evaluate c-Abl ablation [11]. Male and female mice were housed on a 12/12 h light/dark cycle at 24 °C with ad libitum access to food and water.

2.2. Immunofluorescence

The hippocampal neurons or MEF cells were seeded onto poly-L-lysine-coated coverslips in 24-well culture plates at a density of 2.5×10^4 cells per well. The cells were rinsed twice in ice-cold PBS, fixed with a freshly prepared 4% paraformaldehyde/4% sucrose in PBS for 20 min, and permeabilized for 5 min with 0.2% Triton X-100 in PBS. After several rinses in ice-cold PBS, the cells were incubated in 3% BSA in PBS (blocking solution) for 60 min at room temperature, followed by an overnight incubation at 4 °C with primary antibodies. The cells were extensively washed with PBS and then incubated with Alexa-conjugated secondary antibodies (Invitrogen) for 60 min at room temperature. The cells were mounted in mounting medium and analyzed using confocal microscopy. The primary antibodies used were as follows: mouse anti-c-Abl, mouse anti-GAPDH, rabbit anti-actin, rabbit anti-βIII tubulin, and rabbit anti-TOM20 (SCBT, Dallas, TX, USA); rabbit anti-phospho-c-Abl Y412 (Sigma-Adrich, St. Louis, MO, USA); and cytochrome c (BD Biosciences, Franklin Lakes, NJ, USA). The mitochondrial marker Mitotracker Deep Red was obtained from Invitrogen.

2.3. Immunoblot Analysis

The treated cells were washed with ice-cold PBS and immediately lysed with radioimmunoprecipitation assay (RIPA) buffer (containing 50 mM Tris, 150 mM NaCl, 1 mM EGTA, 1 mM EDTA, 0.5% deoxycholate, 1% NP-40, and 0.1% SDS) supplemented with protease inhibitors (1 mM PMSF, 1 µg/mL aprotinin, 10 µg/mL leupeptin, 1 mM Na_3VO_4, and 50 mM NaF). The homogenates were maintained on ice for 30 min and then were centrifuged at $10,000 \times g$ for 5 min. The supernatant was recovered, and the protein concentration was determined using the BCA protein assay kit (Pierce, Appleton, WI, USA). The proteins were resolved in SDS-PAGE, transferred to a PVDF membrane, and reacted with primary antibodies. The reactions were followed by incubation with secondary peroxidase-labeled antibodies (Pierce) and developed using the ECL technique (Thermo Scientific, Waltham, MA, USA). The primary antibodies were the same as those used for immunofluorescence, with the addition of rabbit anti-p-CRKIII Y221, anti-caspase3, anti-cleaved-caspase 3, and anti-MFN2 (Cell Signaling Technology, Danvers, MA, USA), and rabbit anti-FLAG (Sigma Aldrich, St. Louis, MO, USA).

2.4. Coimmunoprecipitation Assay

Protein extract was obtained from MEF cells lysed in non-denaturing lysis buffer (20 mM Tris, 137 mM NaCl, 1 mM EDTA, and 1% NP-40) containing a mixture of protease and phosphatase inhibitors. Immunoprecipitations were performed using anti-c-Abl (SCBT), and MFN2 (Cell Signaling Technology) and agarose beads anti-FLAG (Sigma Aldrich). Complexes were isolated using protein G Sepharose. Tissue and cell lysates were separated by SDS-PAGE, transferred to Nitrocellulose membranes (Fisher Thermo

Scientific), and immunoblotted with phosphotyrosine (pTyr) antibody (Millipore Bioscience Research Reagents, Burlington, MA, USA), anti-c-Abl, anti-FLAG, and anti-MFN2 antibody.

2.5. In Vitro Phosphorylation Assay

Kinase assay mixtures contained 25 mM HEPES, pH 7.25, 100 mM NaCl, 5 mM MgCl$_2$, 5% glycerol, 100 ng of bovine serum albumin/μL, 1 mM sodium orthovanadate, and 10 nM c-Abl kinase. c-Abl was purified to >90% purity, as previously described [43]. After a 5 min preincubation at 30 °C, 25 μL reactions were initiated by the addition of immunoprecipitated agarose bead-FLAG-MFN2 or GST–CrkII, 5 μM ATP, and 0.25 μCi of [γ-^{32}P]ATP. c-Abl incubated with 0.5 μCi of [γ-^{32}P]ATP without a substrate was used for normalization. All reaction mixtures were incubated at 30 °C for different periods of time, and the reaction mixtures were washed four times with 0.75% v/v phosphoric acid. Alternatively, the reactions were terminated by the addition of ice-cold SDS sample buffer and were resolved by SDS-PAGE. The gels were dried and exposed for autoradiography, and quantified using a Molecular Dynamics PhosphorImaging system and the ImageQuant LAS 500 equipment.

2.6. Molecular Biology

pLZRS-FLAG-MFN2 was generated by PCR and restriction digest using 5′ EcoRI and 3′ XhoI. The pMSCV-mCherry vector has been described previously [44]. Outer mitochondrial membrane-targeted Cerulean (Cerulean-OMM) was generated by inserting the C-terminal domain of human BCL-xL (a.a. 208-233) downstream of Cerulean by PCR amplification with 5′ EcoRI and 3′ SalI restriction sites and cloning into pMX vector. The MFN2 mutant was generated by PCR using the proofreading Pfu polymerase (Promega, Madison, WI, USA), followed by DpnI digestion of the methylated parental plasmid. The oligonucleotides used were as follows: MFN2-Y269F-REV: 5′- GCACCTCCTCCATGAACTCAGGCTCCGA -3′; MFN2-Y269F-FOR: 5′- CTCGGAGCCTGAGTTCATGGAGGAGGTG -3′ (gene ID: 170731). pcDNA3-c-Abl-full-length, pcDNA3-c-Abl-kinase-dead-ful-length, and pcDNA3-c-Abl- ΔC (lacking the C-terminal domain a.a 1-634) were cloned into pMSCV-IRES-mCherry by PCR and restriction digest using 5′ Eco/kozak and 3′ SalI/STOP/ERT2-3′. All constructs were verified by DNA sequencing.

2.7. Cell Lines

MEF cells were obtained from ATCC. All stable MEF cell lines were generated by retroviral transduction. Briefly, Phoenix amphotropic virus producer cells were transfected with the appropriate plasmid using Lipofectamine 2000 (Thermo Fisher, Waltham, MA, USA) for 48 h. The target cells were infected with virus containing culture medium from the packaging cells supplemented with 5 μg/mL polybrene. Stable transductants were selected following the addition of 200 μg/mL Zeocin (Invitrogen, Waltham, MA, USA) (pLZRS vectors) or were sorted by flow cytometry for Venus-, mCherry-, or Cerulean-positive cells (pMX vectors).

2.8. Microscopy and Cell Death Assay

To assess the mitochondrial morphology, the cells were stained with Mitotracker Deep Red or with anti-TOM20. Images were collected using confocal microscopy (Zeiss LSM510, Thornwood, NY, USA) and analyzed with the NIH ImageJ 1.53 software. Fragmentation was judged based on the mitochondrial distribution of a normal cell, and cells exhibiting over 80% mitochondrial fragmentation were counted as fragmented. For each condition, at least 100 cells were counted. The quantification was carried out by an individual blinded to the conditions. TUNEL staining was performed using an apoptosis detection kit (Roche Molecular Biochemicals, Indianapolis, IN, USA). Briefly, the cells were incubated in 0.1% Triton X-100 in PBS. Then, the sections were immersed in the TUNEL reaction mixture for 60 min at 37 °C and washed twice in PBS (pH 7.4). Then, the cells were labeled with phalloidin-TRITC (red).

The confocal images of neurons were obtained using a Carl Zeiss 633 (numerical aperture 1.4) objective with sequential acquisition settings at the maximal resolution of the confocal (1024 × 1024 pixels), or using a Carl Zeiss Axiovert 200M motorized inverted microscope equipped with a precision motorized XY stage (Carl Zeiss MicroImaging, Thornwood, NY, USA). The confocal microscope settings were kept the same for all scans when fluorescence intensity was compared. All measurements were performed using the NIH ImageJ software.

2.9. Onset and Survival Analysis

Male WT or SOD1 G93A mice were fed ad libitum with a control diet, a diet containing nilotinib, or a diet containing neurotinib (a novel allosteric c-Abl inhibitor with high brain penetrance, patent number WO2019/173761 A1) [35,45]. The diet administration began from the time of weaning until euthanasia. The rodent diet was manufactured by Envigo/Teklad (Madison, WI, USA) with the incorporation of nilotinib at 200 ppm or neurotinib at 67 ppm into the NIH-31Open Formula Mouse/Rat Sterilizabile Diet (7017), followed by irradiation handling of the final product. All experimental protocols followed ethical guidelines. The onset was defined as the first day from birth when animals displayed weakness or coordination defects in their posterior limbs; i.e., when the animal was suspended by the tail, the hindlimb was collapsed, partially collapsed towards the lateral midline, trembled, or retracted. The endpoint was determined when animals were no longer able to feed themselves, or when the animal was suspended by the tail and there was rigid paralysis; or when the animal was allowed to walk and there was no forward motion; or when the animal was placed on its left and right side and it was not able to right itself within 10 s [46]. Euthanasia was performed with CO_2, followed by perfusion with NaCl 0.9%. The lumbar spinal cord tissue was dissected, and half was frozen for Western blot analysis, while the other half was submerged in PFA 4% overnight for IHC analysis. On the next day, the spinal cord was transferred to 30% sucrose. For nerve tissue, sciatic nerves were dissected from each animal and fixed in 2.5% (*v/v*) glutaraldehyde in PBS for TEM analysis.

2.10. Fluorescent Immunohistochemistry

The lumbar spinal cord tissues, harvested from both WT and SOD1 G93A mice fed with either the control diet, nilotinib-containing diet, or neurotinib-containing diet, were frozen in the Tissue-Tek® O.C.T. Compound (Sakura Finetek, Torrance, CA, USA) and subsequently sectioned using a cryostat at a thickness of 25 µm. The sections were collected every 200 µm in a solution of PBS with 0.02% azide.

To prepare the sections for analysis, the free-floating sections underwent permeabilization for 30 min using a 0.2% Triton X-100 solution in PBS, followed by blocking in a 3% BSA solution in PBS for 2 h at room temperature. Next, the sections were incubated overnight at 4 °C with a primary antibody against NeuN (Abcam, Cambridge, UK). On the following day, the sections were thoroughly rinsed with PBS, and then incubated for 2 h at room temperature and protected from light with the corresponding Alexa-conjugated secondary antibody (Invitrogen). This was followed by several washes with PBS. At least 6 sections were mounted on slides and allowed to dry overnight at room temperature, while being protected from light. On the following day, the slides were cover-slipped with glass slides using mounting medium.

Fluorescence images were captured using a Zeiss Axioscope 5 microscope. All images were consistently acquired using the same settings, and subsequently quantified using the NIH ImageJ software.

2.11. TEM and Morphological Analysis

The nerve tissue was prepared and fixed in 2.5% (*v/v*) glutaraldehyde in PBS. Ultra-thin sections were mounted in a 300 mesh Formvar/carbon copper grids (Tedpella INC, Redding, CA, USA) and contrasted with uranyl acetate. Images were captured at different magnifications using a Philips Tecnai 12 transmission electron microscopy (Eindhoven, The Netherlands) at 80 kV equipped with a SIS CDD Megaview G2 camera and the iTEM Olympus Soft Imaging Solutions software (Windows NT 6.1).

The morphological analysis of TEM images was performed with the NIH ImageJ software. The images of individual mitochondria were generated from TEM images of sciatic nerve axons, and each mitochondrion was categorized as normal or swollen based on a disrupted external mitochondrial membrane with fragmented or swollen cristae/matrix, as described before [47].

2.12. Statistical Analysis

The mean and SEM values and the number of experiments are indicated in each figure. Statistical analysis was performed using one-way ANOVA, followed by Student's *t* test using GraphPad Prism (version 5.0). For onset and survival analysis, a log-rank (Mantel–Cox) test was performed between curves. For IHC analysis, a one-way ANOVA with Dunnett's T3 multiple comparison test was performed between groups. The results are presented as mean ± SEM.

3. Results

3.1. c-Abl Is Required to Induce Mitochondrial Fragmentation under ER Stress

To investigated whether c-Abl is involved in the morphological changes affecting mitochondria in response to ER stress in the nervous system, we evaluated if c-Abl was activated in neuronal cells exposed to the *N*-glycosylation inhibitor tunicamycin (1 µg/mL) or thapsigargin 1 µM. The primary hippocampal neurons exposed to tunicamycin or thapsigargin showed an increase in phospho-c-Abl signal on immunofluorescence assays (Figure 1A). This increase in phospho-c-Abl was also detected as early as 1 h after tunicamycin treatments, prior to and during ER stress, as shown by the accumulation of UPR markers CHOP and Bip by Western blot assays (Figure 1B). Then, we asked whether c-Abl tyrosine kinase is important for ER stress-induced mitochondrial fragmentation. The primary neuronal cultures under ER stress were treated with Imatinib (5 µM), a pharmacological c-Abl inhibitor. Immunofluorescence for βIII-tubulin (neuronal marker) and TOM20 (mitochondrial marker) showed that shortened mitochondria induced by ER stress were rescued by imatinib (Figure 1C,D), suggesting that activated c-Abl influences mitochondrial morphology. To confirm these results, we exposed MEF cells with tunicamycin and then we stained mitochondria with MitoTracker Deep Red FM to quantify mitochondrial sphericity in the cellular space using the IMARIS version 7.6.5. software. Higher values for sphericity are associated with more fragmented and/or rounded mitochondria. While the control and imatinib-treated cells exhibited a tubular mitochondrial network, tunicamycin treatment induced evident mitochondrial fragmentation, which was significantly prevented by imatinib (Figure 1E,F). These results suggest that the activation of c-Abl is required for the alterations of mitochondrial morphology elicited during the ER stress response. Importantly, the treatment of cells with imatinib protected against ER stress-induced cell death as demonstrated using the MTT (Figure 1G) and Sytox (Figure 1H) assays in the MEF cells. These effects correlated with the translocation of cytochrome c to the cytosol (Figure 1A) and the upregulation of ER stress-proapoptotic marker CHOP (Figure 1B).

Although imatinib is a well-known c-Abl inhibitor, other targets have been described such as PDGFR and c-KIT [48]. Thus, we employed a genetic approach to ablate c-Abl expression in the nervous system. We conditionally ablated c-Abl expression in the neurons by intercrossing *ABL1$^{flox/flox}$* mice with Nestin-Cre transgenic mice to generate *ABL1-cKO*. First, we confirmed the successful deletion of c-Abl in embryonic brain extracts by Western blot (Figure 2A). We then generated primary neuronal cultures from *ABL1-cKO* mouse embryos and littermate control animals and evaluated mitochondrial length in neurites by staining for the mitochondrial marker TOM20 (Figure 2B). At basal level, *ABL1-cKO* neurons presented unaltered mitochondrial length (Figure 2C). Then, we analyzed the mitochondrial morphology of WT and *ABL1-cKO* neurons exposed to tunicamycin for

8 h. Although WT neurons exhibited a 50% reduction in mitochondrial length, *ABL1-cKO* neurons displayed mitochondrial lengths similar to control and untreated WT neurons (Figure 2C). These results indicate that c-Abl contributes to mitochondrial fragmentation under ER stress.

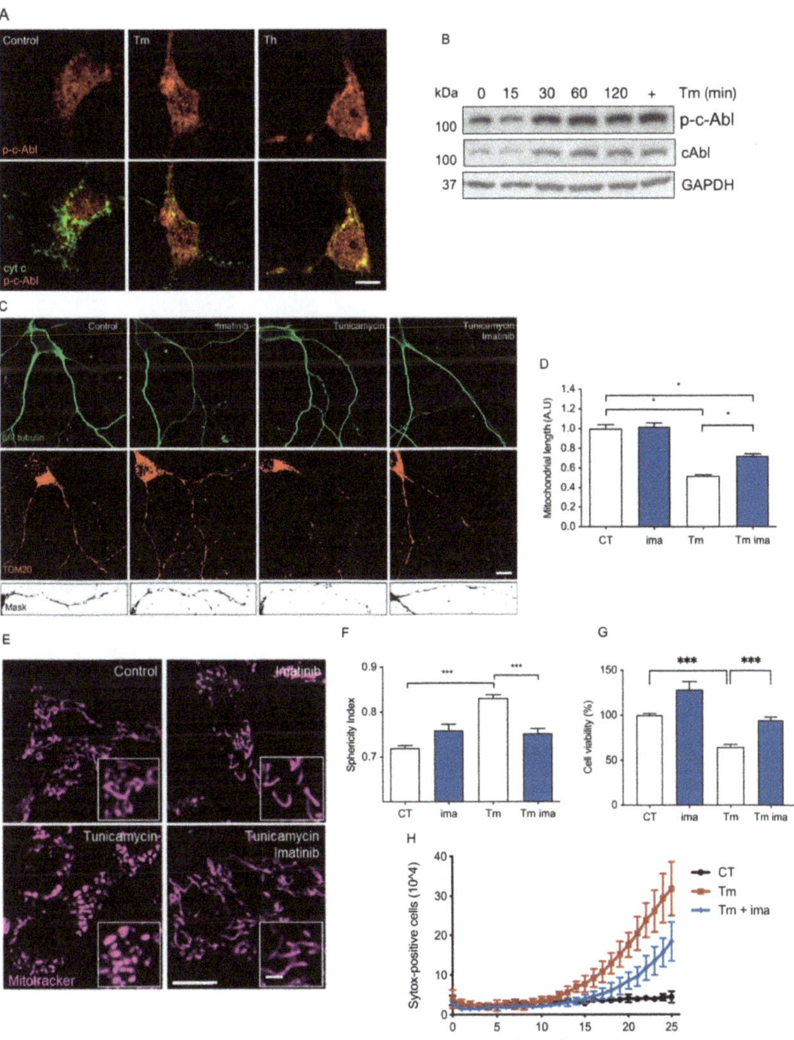

Figure 1. c-Abl is activated and localized in mitochondria in response to ER stress and collaborates in mitochondrial fragmentation. (**A**) Hippocampal primary neurons of 7 days in vitro were exposed to tunicamycin (Tm) or thapsigargin (Th) for 2 h and the immunodetection against phospho-c-Abl (red) was analyzed regarding mitochondrial localization with cytochrome c (green). Scale bar: 5 μm. (**B**) p-c-Abl and ER stress markers are increased after Tm treatment in hippocampal primary neurons. (**C**) Mitochondrial morphology of primary hippocampal neurons detecting TOM20 (red) and βIII tubulin (green) in treatments with Tm or Imatinib (Ima) for 10 h. Scale bar: 5 μm. (**D**) Quantification

of mitochondrial length from neuronal processes in experiments performed in (**C**). (**E**) Mitochondrial morphology of MEFs detecting Mitotracker (magenta) in treatments with Tm or Ima for 10 h. Scale bar: 10 µm; inset: 5 µm. (**F**) Mitochondrial sphericity from experiments performed in (**E**). (**G**) MTT assay in MEFs in treatments with Tm or Ima for 10 h. (**H**) Cytox assay in MEFs in treatments with Tm or Ima for the indicated times. One-way ANOVA with Bonferroni post-test. Results are presented as mean ± SEM. $p * < 0.05$, $p *** < 0.001$.

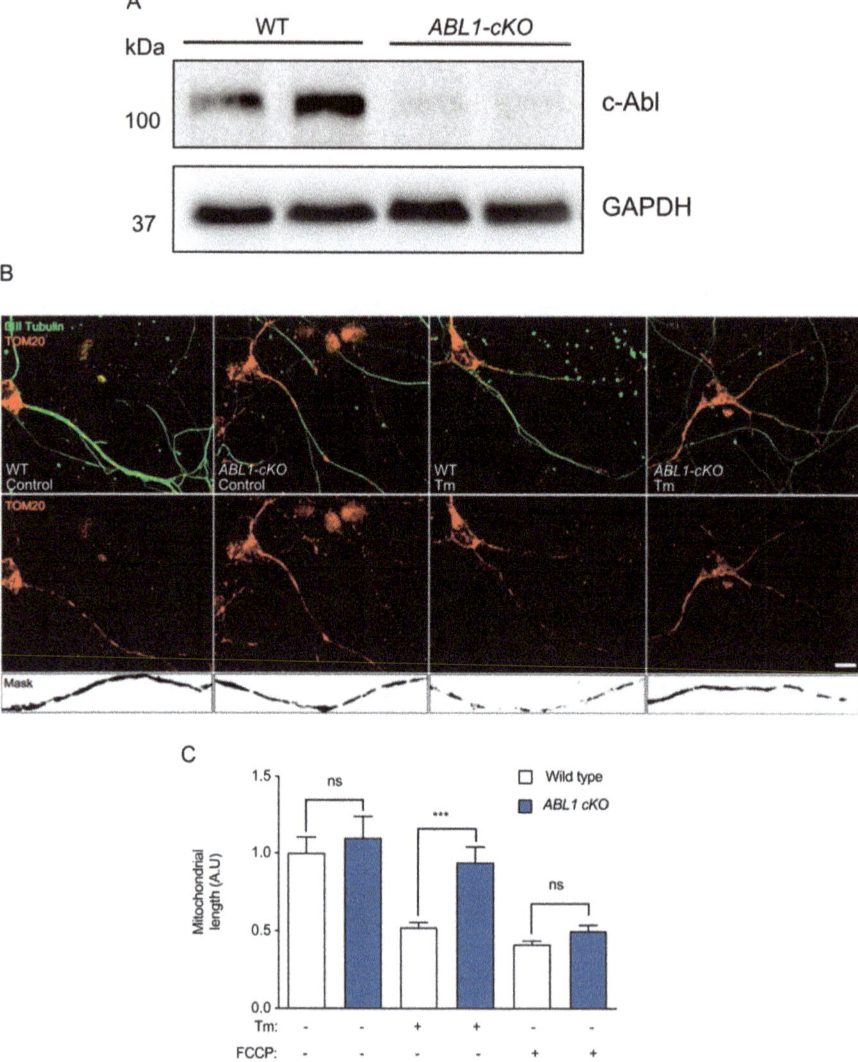

Figure 2. c-Abl regulates mitochondrial morphology in hippocampal primary neurons in response to ER stress. (**A**) c-Abl expression in primary neurons from WT and c-Abl-deficient (ABL1-cKO) mice. (**B**) Representative images of WT and ABL1-cKO primary hippocampal neurons treated with Tm for 8 h and immunostained for anti βIII tubulin (green) as the neuronal marker and TOM20 (red) as the mitochondrial marker. Scale bar: 5 µm. (**C**) Quantification of mitochondrial length from neuronal processes from experiments performed in (**B**). One-way ANOVA with Bonferroni post-test. Results are presented as mean ± SEM. ns: not significant, $p *** < 0.001$.

3.2. c-Abl Activation Induces Mitochondrial Fragmentation

To further explore the significance of c-Abl to mitochondrial dynamics, we performed gain-of-function experiments to study c-Abl activity in the absence of stress. To achieve this, we generated a c-Abl variant capable to activate its kinase function in a tamoxifen-inducible manner by fusing a C-terminal truncated c-Abl (constitutively active) to the ERT2 domain that blocks its kinase activity unless 4-hydroxy-tamoxifen (tamoxifen) is present (c-Abl WT ERT2). We also generated a kinase-dead variant (c-Abl KD ERT2) by substituting two specific amino acids (L285P/K290R) (Figure 3A). Then, to study the effect of c-Abl on mitochondria dynamics, we stably expressed both variants of c-Abl in plasmids carrying a mCherry reporter in MEF cells stably expressing the Cerulean mitochondrial marker [49]. As a control, we checked the activity of c-Abl by monitoring the phosphorylation of a well-known substrate CrkII (p-CrkII) at tyrosine 221 [50]. Hydrogen peroxide was also used as a positive control to activate endogenous c-Abl (Figure 3B). Tamoxifen-treated cells expressing the kinase-active c-Abl variant demonstrated a clear increase in the phosphorylation of CrkII, whereas the kinase-dead variant did not affect its phosphorylation (Figure 3B), demonstrating the specificity of our experimental system. We then analyzed the mitochondrial sphericity in MEF cells expressing the active or kinase-dead variants of c-Abl. We analyzed changes in tamoxifen-stimulated cells and compared mitochondrial sphericity with control cells carrying an empty vector (Figure 3C,D). Control cells exhibited tubular mitochondria (empirically minimal fragmentation), while FCCP treatment was used as a positive control to induce the maximal empirical mitochondrial fragmentation. Almost 80% of cells with stimulated c-Abl activity showed dramatic mitochondrial fragmentation, while the kinase-dead variant exhibited no significant fragmentation (around 25%), relative to control and FCCP (Figure 3C,D). These findings suggest that c-Abl activity is necessary for mitochondrial fragmentation.

3.3. c-Abl Phosphorylates the Mitochondria Fusion Protein MFN2 on Y269

To explore possible molecular mechanisms explaining the consequences of c-Abl activation on mitochondrial morphology, we investigated the potential interaction between c-Abl and components of the machinery controlling mitochondrial dynamics.

Using an in silico analysis based on the MitoCarta 3.0 dataset of 1136 mitochondrial proteins [51], we identified a group of proteins predicted to undergo phosphorylation by c-Abl (Supplementary Figure S1A and Table 1). Specifically, we identified a possible interaction between c-Abl and the outer mitochondrial membrane GTPase MFN2, which is essential for mitochondrial fusion [52–54]. We then tested whether c-Abl is capable to form a protein complex with MFN2. Endogenous c-Abl was immunoprecipitated from tunicamycin-treated MEF cells, and its association with MFN2 was assessed by Western blot analysis. Remarkably, an interaction between c-Abl and MFN2 was observed preferentially under ER stress (Figure 4A). Similar results were obtained when endogenous MFN2 was immunoprecipitated (Figure 4A). Interestingly, this interaction was reduced by the treatment of cells with imatinib (Figure 4A). We then stably reconstituted double-knockout MFN1/2 (MFN1/2 DKO) MEF cells with FLAG-tagged WT MFN2 (FLAG-MFN2) at levels that were similar to endogenous protein (Figure S2A), and we induced ER stress in the presence or absence of imatinib. We detected increased c-Abl levels in FLAG immunoprecipitates after treating cells with tunicamycin, an interaction that was reduced after the administration of imatinib (Figure 4B). These results suggest that the activation of c-Abl is required to form a complex with MFN2 and in response to ER stress.

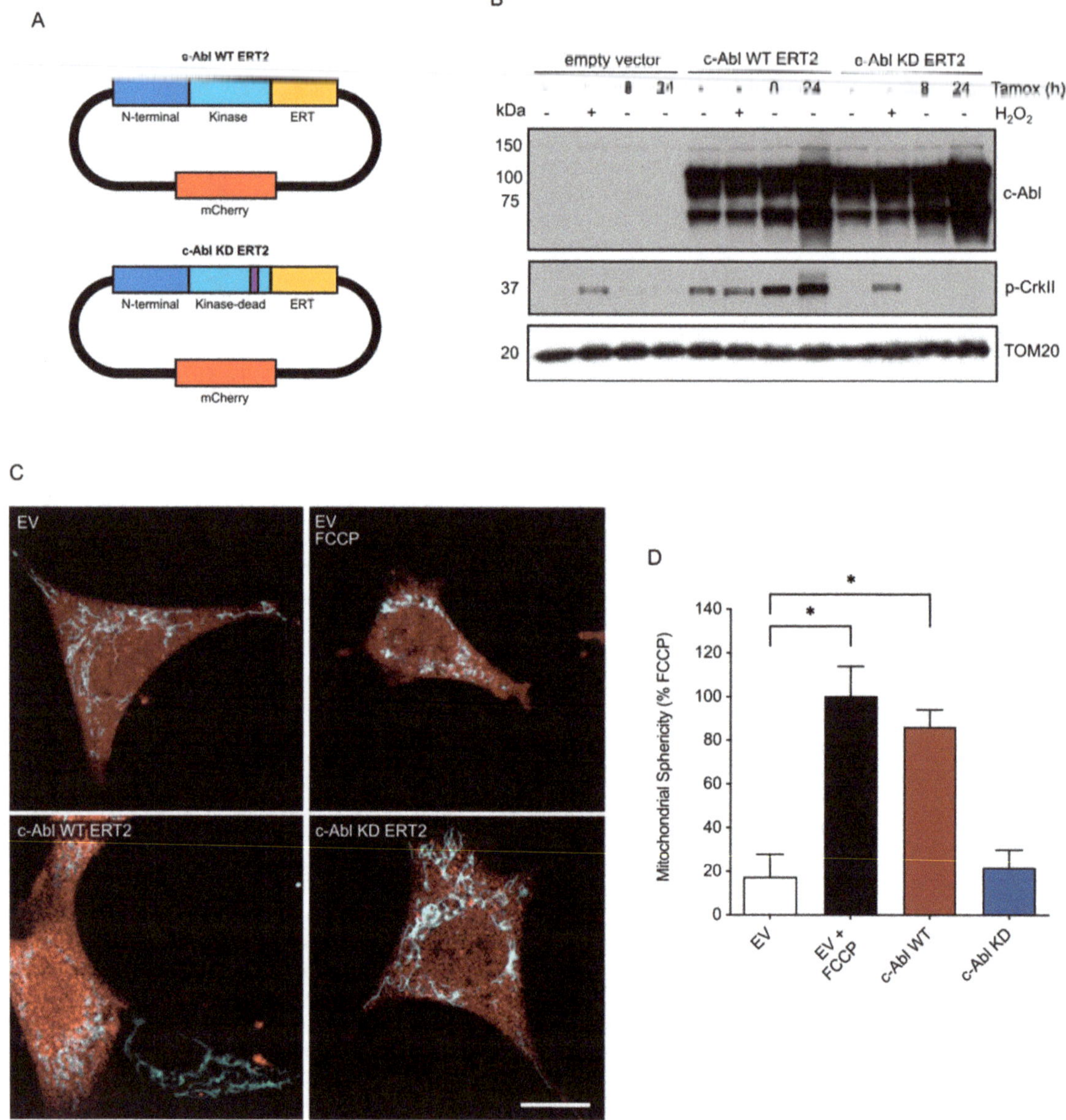

Figure 3. c-Abl activity induces mitochondrial fragmentation. (**A**) Generation of a tamoxifen-inducible system for constitutively active c-Abl (c-Abl WT ERT2) or kinase-dead (c-Abl KD ERT2) variants. (**B**) Activity of c-Abl variants in MEFs after ER stress was checked by immunoblot of c-Abl and a well-known substrate of c-Abl, p-Crk II. (**C**) Confocal microscopy of MEF cells overexpressing mitochondrial Cerulean-OMM (cyan) and the mCherry (red) empty vector (EV), c-Abl WT ERT2, or c-Abl KD ERT2 system, revealing the mitochondrial morphology under stimulation with tamoxifen. Scale bar: 20 μm. (**D**) Quantification of mitochondrial sphericity expressed as the percentage of sphericity relative to FCCP treatment in MEFs expressing the EV. One-way ANOVA with Bonferroni post-test. Results are presented as mean ± SEM. p * < 0.05.

Table 1. Sites in mitochondrial proteins predicted to undergo phosphorylation by c-Abl.

Protein ID *	Protein Name	Residue	Sequence −7/+7	Kinase-Dependent Predicted Value **	Hydrophobicity-Dependent Predicted Value ***	Score
O95140	MFN2	Y269	ASASEPEYMEEVRRQ	11.267	−1.527	17.205
Q8IWA4	MFN1	Y248	ASASEPEYMEDVRRQ	10.706	−1.527	16.348
Q9BXK5	B2L13	Y213	LESEEEEYPGITAED	10.014	−1.333	13.349
O75323	NIPS2	Y187	PRSGPNIYELRSYQL	11.430	−1.120	12.802
Q5THJ4	VP13D	Y2873	TNLEHQIYARAEVKT	12.250	−1.040	12.740
Q8IWA4	MFN1	Y40	SHFVEATYKNPELDR	10.319	−1.213	12.517
O00429	DNM1L	Y266	TDSIRDEYAFLQKKY	9.849	−1.220	12.016
Q5THJ4	VP13D	Y768	TQFSDDEYKTPLATP	9.884	−1.180	11.663
Q969Q5	RAB24	Y70	DTAGSERYEAMSRIY	10.385	−1.093	11.351
O00429	DNM1L	Y101	LHTKNKLYTDFDEIR	9.714	−1.153	11.200
Q5THJ4	VP13D	Y1189	GMANREKYGRKIATA	11.108	−0.987	10.964
Q9BUR5	MIC26	Y43	KVDELSLYSVPEGQS	13.270	−0.667	8.851
Q96HS1	PGAM5	Y224	ARQEEDSYEIFICHA	12.029	−0.680	8.180
P57105	SYJ2B	Y43	VSNDSGIYVSRIKEN	10.196	−0.767	7.820
Q6UXV4	MIC27	Y44	KPEQLPIYTAPPLQS	10.986	−0.700	7.690
Q9BPW8	NIPS1	Y262	GWDENVYYTVPLVRH	14.185	−0.533	7.561
Q9BPW8	NIPS1	Y87	KPEYLDAYNSLTEAV	11.845	−0.600	7.107
Q16611	BAK	Y108	QPTAENAYEYFTKIA	9.754	−0.720	7.023
O43236	SEPT4	Y318	EHFGIKIYQFPDCDS	10.827	−0.593	6.420
Q14318	FKBP8	Y187	GPQGRSPYIPPHAAL	9.852	−0.627	6.177
Q8NAN2	MIGA1	Y152	KGSQVCNYANGGLFS	12.228	−0.333	4.072
O43865	SAHH2	Y28	EIEDAEKYSFMATVT	9.989	−0.347	3.466
Q07817	B2CL1	Y120	HITPGTAYQSFEQVV	9.800	−0.167	1.637
O75323	NIPS2	Y264	GWEELVYYTVPLIQE	11.294	0.020	−0.226
O43236	SEPT4	Y228	CWKPVAEYIDQQFEQ	10.305	N/A	N/A
O43236	SEPT4	Y407	RETHYENYRAQCIQS	10.99	N/A	N/A
O43865	SAHH2	Y470	ALALIELYNAPEGRY	9.678	N/A	N/A
O60313	OPA1	Y637	THVIENIYLPAAQTM	9.659	N/A	N/A
O75323	NIPS2	Y89	KPECLEAYNKICQEV	9.56	N/A	N/A
O75323	NIPS2	Y231	FSQIGQLYMVHHLWA	9.553	N/A	N/A
Q5HYI7	MTX3	Y29	ESLVVMAYAKFSGAP	11.314	N/A	N/A
Q5THJ4	VP13D	Y3589	QDNRQLYYENFIYIA	9.867	N/A	N/A
Q5THJ4	VP13D	Y3861	LTGINVHYTQLATSH	10.98	N/A	N/A
Q5THJ4	VP13D	Y4369	NYAKSLYYEQQLMLR	10.067	N/A	N/A
Q6UXV4	MIC27	Y18	TMPAGLIYASVSVHA	10.63	N/A	N/A

* Protein ID in UniProt database. ** Value obtained after FASTA sequence analysis in GPS 5.0 software. *** Hydrophobicity value of the residue reported by PhosphoNET online tool. N/A: Not applicable.

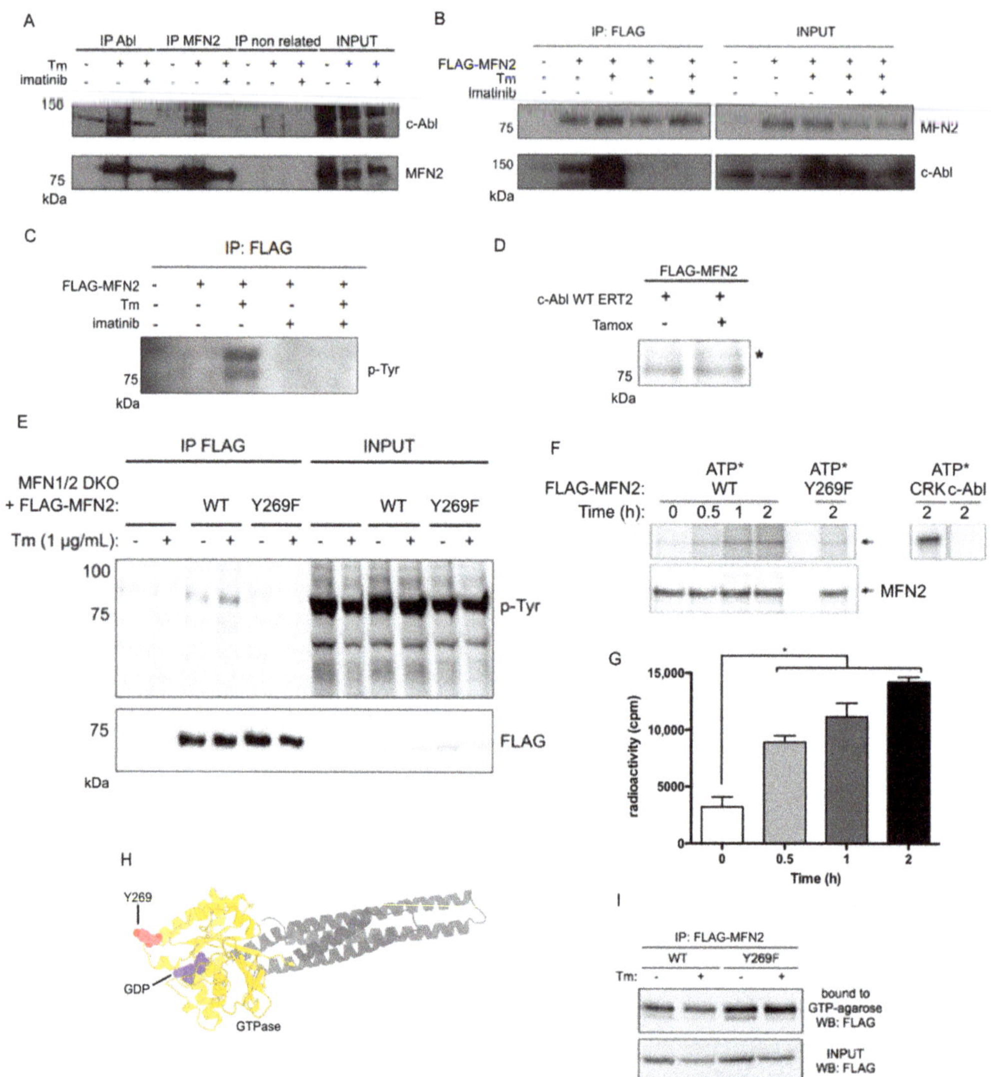

Figure 4. Activated c-Abl phosphorylates MFN2 in Y269. (**A**) Co-immunoprecipitation assay against endogenous levels for MFN2 and c-Abl in Tm or Ima treatments in MEF cells. (**B**) Co-immunoprecipitation assay for MEF cells over-expressing FLAG-MFN2 using agarose-anti-FLAG beads and detecting c-Abl in Tm or Ima treatments in MEF cells. (**C**) Phospho-tyrosine immunodetection against immunoprecipitated FLAG-MFN2 after Tm or Ima treatments in MEF cells. (**D**) In vitro phosphorylation with P32 orthophosphoric acid in living MEF cells and immunoprecipitation of FLAG-Mfn2 after tamoxifen (tamox)-stimulated c-Abl WT ERT2. (**E**) Phospho-tyrosine detection for MEF cells over-expressing either FLAG-MFN2 WT or FLAG-MFN2 Y269F variants in response to Tm. (**F**) In vitro phosphorylation of immunoprecipitated MFN2-FLAG, P32 gamma ATP (2.5 µCi), and recombinant c-Abl. (**G**) Scintillation counting of radiolabeled MFN2-FLAG after indicated times. (**H**) Modelling of Y269 residue in the MFN2 GTPase domain. (**I**) FLAG-MFN2 was immunoprecipitated and then incubated with GTP-agarose beads for 60 min at 30 °C. The bound proteins were analyzed via immunoblotting using FLAG antibody. One-way ANOVA with Bonferroni post-test. Results are presented as mean ± SEM. $p * < 0.05$.

Although MFN2 has been reported to be phosphorylated at Ser27 in response to cellular stress [55], the phosphorylation in tyrosine residues of MFN2 remains unclear. To address this, we first examined whether MFN2 was phosphorylated at tyrosine in cells under ER stress. FLAG-MFN2 reconstituted in MFN1/2 KO cells expressing FLAG-MFN2 was treated with tunicamycin, and FLAG immunoprecipitated to assess phospho-tyrosine levels. Cells undergoing ER stress showed increased detection in tyrosine phosphorylation of MFN2, detected by the phospho-tyrosine antibody 4G10, and this was reduced in cells treated with imatinib (Figure 4C). The tyrosine phosphorylation of MFN2 was further evaluated in living cells by the addition of ^{32}P-labeled orthophosphoric acid into the media of FLAG-MFN2-overexpressing MEFs using our tamoxifen-inducible system. Again, the expression of c-Abl kinase activity increased the radioactive signal of immunoprecipitated FLAG-MFN2 (Figure 4D), suggesting that MFN2 can be phosphorylated by active c-Abl in living cells.

Our in silico analysis provided the highest score for phosphorylation to the amino acid residue Y269 in the human peptide sequence 261-ASASEPEYMEEVRRQ-277 of MFN2 (Table 1), which is highly conserved in a wide range of species (Supplementary Figure S1B). Thus, we performed site-directed mutagenesis to generate a Y269F mutant. Both WT and Y269F FLAG-MFN2 versions were expressed at similar levels (Figure S2A,B). We then treated these cells with tunicamycin and measured the levels of phosphorylated tyrosine in MFN2. As expected, phospho-tyrosine levels were detected in FLAG-immunoprecipitants from MFN2-WT-expressing cells but not in cells expressing the Y269F point mutant (Figure 4E). Then, to examine whether c-Abl directly phosphorylates MFN2, we performed an in vitro assay using purified recombinant proteins. Purified recombinant c-Abl kinase was incubated with FLAG-MFN2-WT, FLAG-MFN2-Y269F, or GST-CRKII (used as a positive target for c-Abl phosphorylation) in the presence of [γ-^{32}P]ATP. We observed a significant time-dependent increase in FLAG-MFN2 phosphorylation measured via both autoradiography (Figure 4F) and scintillation counting (Figure 4G). The phosphorylation of FLAG-MFN2 WT was detected as early as 5 min of incubation and increased up to 120 min. In contrast, the phosphorylation of mutant FLAG-MFN2-Y269F remained at baseline levels until the endpoint. MFN2 fusion activity relies on the function of its GTPase domain [56]. In addition, it has been reported that the specific phosphorylation of MFN2 at Ser27 under cellular stress regulates mitochondrial fragmentation [55]. Using the Protein Imager tool [57], we modeled the Y269-containing GTPase domain of MFN2 (Figure 4H). This model predicted a close proximity of Y269 to the GTP-binding site. Based on this observation, we then addressed whether the Y269 residue of MFN2 affects its nucleotide-binding affinity upon ER stress using a GTP-agarose bead immunoprecipitation assay. While GTP binding by MFN2-WT was reduced upon ER stress, the Y269F mutant displayed a higher affinity for GTP, an activity that was unaffected by ER stress (Figure 4I). This suggests that MFN2-Y269F may remain in an active conformation for longer time than the WT form. We then measured the MFN2 hydrolytic activity and did not detect a significant difference in the GTP hydrolysis rate between MFN2 WT and the mutant form (Figure S3). Importantly, it is known that a change in the GTP hydrolysis rate is not necessary for mitochondria fusion and MFN2 activity [58]. Thus, the phosphorylation in Y269 appears to decrease the affinity of MFN2 for GTP, and this may reduce its ability to promote mitochondrial fusion. Altogether, these results indicate that MFN2 is directly phosphorylated by c-Abl kinase to control mitochondrial fragmentation.

3.4. c-Abl Mitochondria Localization Requires MFN2 Y269

We then investigated whether the interaction with MFN2 is necessary for c-Abl to localize in the mitochondria. To test this, we used FLAG-MFN2-WT and FLAG-MFN2-Y269F constructs on MEF cells, and then we determined the mitochondrial localization of c-Abl after tunicamycin treatment. Using stochastic optical reconstruction microscopy (STORM), we observed that the individual molecules of c-Abl were distributed diffusely with respect to mitochondria under basal conditions in cells expressing MFN2-WT or

MFN2-Y269F (Figure 5A,B). Interestingly, ER stress induced a significant increase of c-Abl molecules and in large clusters at the mitochondria, a pattern that was absent from FLAG-MFN2-Y269F-expressing cells (Figure 5A,B). We then evaluated the proximity between c-Abl and MFN2 molecules after the induction of ER stress using STORM. In FLAG-MFN2-WT cells, the proximity within 50 nm was significantly increased after tunicamycin treatment. Importantly, this effect was ablated in FLAG-MFN2-Y269F cells (Figure 5C,D). Thus, c-Abl mitochondrial localization in response to ER stress requires MFN2 Y269.

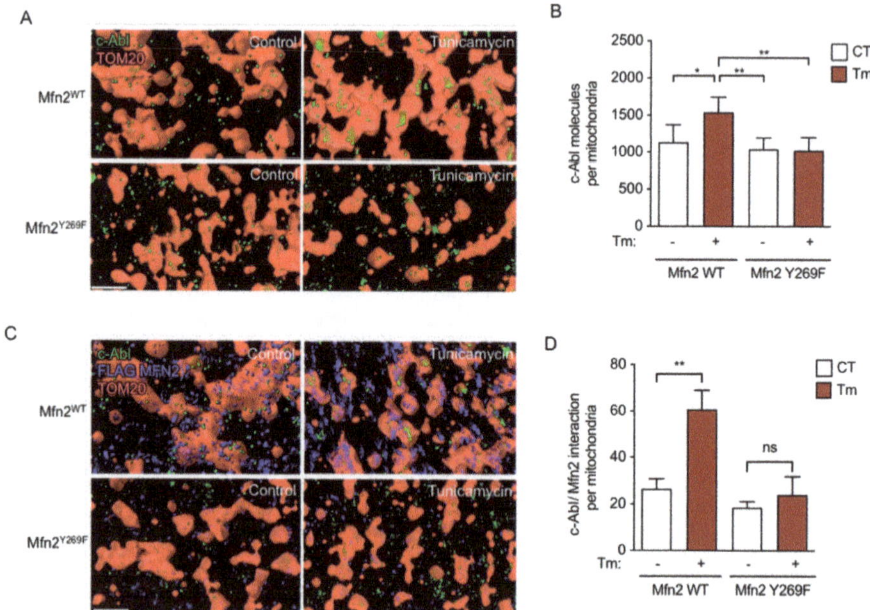

Figure 5. Activated c-Abl colocalizes with mitochondria dependent on MFN2 interaction. (**A**) Super-resolution microscopy revealing the c-Abl (green) localization in TOM20 (red)-stained FLAG-MFN2 WT or FLAG-MFN2 Y269F MEFs. Scale bar: 1 μm. (**B**) Quantification of c-Abl molecules as observed in (**A**). (**C**) Super-resolution microscopy of c-Abl (green) colocalization with FLAG-MFN2 (blue) and TOM20 (red) in FLAG-MFN2 WT or FLAG-MFN2 Y269F MEFs. Scale bar: 1 μm. (**D**) Quantification of c-Abl colocalizing with FLAG-MFN2. One-way ANOVA with Bonferroni post-test. Results are presented as mean $\pm$ SEM. ns: not significant, p * < 0.05, p ** < 0.01.

Then, we studied whether the phosphorylation of MFN2 on Y269 by c-Abl is important to mitochondrial morphology. To analyze this, we used MEF cells that expressed FLAG-MFN2-WT or FLAG-MFN2-Y269F, and then stained them with TOM20 to study mitochondria morphology (Figure 6A). In control treatment, both cell lines exhibited normal tubular mitochondria morphology, whereas ER stress triggered significant fragmentation in 70% of FLAG-MFN2-WT-expressing cells but only in 45% of FLAG-MFN2-Y269F cells (Figure 6A,B), suggesting that the association between c-Abl and MFN2 is necessary to trigger mitochondria morphology impairments in cells under ER stress. Mitochondrial dynamics have the potential to influence mitochondrial function. In our observations, we found that ER stress did not affect the oxygen consumption rate (OCR) response (Figure S4A,B). However, it did lead to a reduction in the extracellular acidification rate (ECAR) in MFN2 WT MEFs (Figure S4C), indicating that the cells were still capable of oxygen consumption, but with altered glycolytic activity. In the case of the MFN Y269F mutant MEFs, treatment with tunicamycin did not produce changes in either OCR (Figure S4A,B) or ECAR (Figure S4C), suggesting that these cells could rely on both oxidative phosphorylation and glycolysis even when subjected to ER stress. Furthermore, our previous findings revealed that imatinib treatment protects against tunicamycin-induced cell death (Figure 1G). Similar results were observed using more specific apoptotic markers including cleaved caspase-3 both by Western blot (Figure 6C) and immunofluorescence (Figure 6D,E) and TUNEL assay (Figure 6F,G). Altogether, these results showed that the phosphorylation of MFN2 on Y269 by c-Abl is important to induce mitochondrial fragmentation and apoptosis in cells under ER stress.

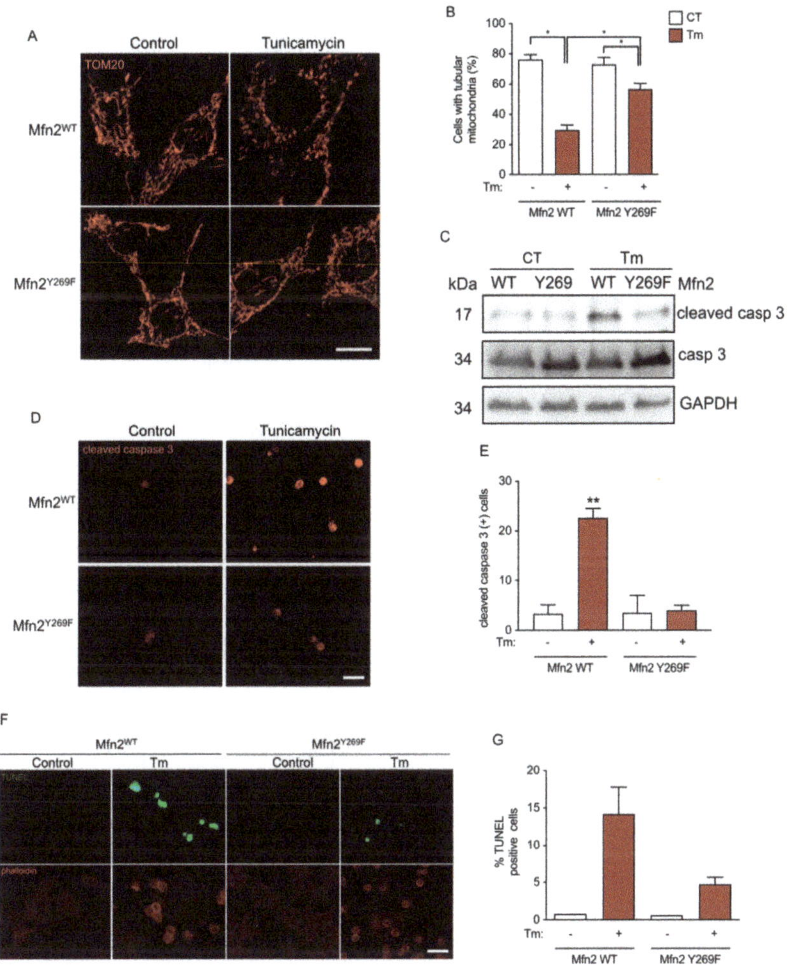

Figure 6. Mfn2 collaborates in the mitochondrial fragmentation and the apoptotic response in a Y269 residue-dependent manner. (**A**) Mfn2WT- and Mfn2Y269F-expressing cells were stained for TOM20 (red) and treated with vehicle (control) or exposed to Tm for 10 h. Scale bar: 20 μm. (**B**) The percentage of cells exhibiting tubular mitochondria in (**A**) was quantified. (**C**) Western blot against cleaved caspase 3 from Mfn2WT- and Mfn2Y269F-expressing cells exposed to Tm for 18 h. (**D**) Mfn2WT- and Mfn2Y269F-expressing cells were stained for cleaved caspase 3 (red) and treated with vehicle (control) or exposed to Tm for 18 h. Scale bar: 40 μm. (**E**) The number of cells per field exhibiting tubular mitochondria in (**D**) was quantified. (**F**) Mfn2WT- and Mfn2Y269F-expressing cells were developed for TUNEL staining (green) and phalloidin (red), and treated with vehicle (control) or exposed to Tm for 18 h. Scale bar: 40 μm. (**G**) The percentage of cells positive for TUNEL in (**F**) was quantified. One-way ANOVA with Bonferroni post-test. Results are presented as mean ± SEM. $p * < 0.05$, $p ** < 0.01$.

3.5. c-Abl Tyrosine Kinase Inhibition Reduces Abnormal Mitochondria and Increases the Survival in an ALS Animal Model

ER stress has been extensively described as a pathological mechanism in neurodegenerative diseases associated with abnormal protein aggregation, including ALS [59]. Interestingly, c-Abl has been shown to contribute to the neurodegenerative cascade observed in several neurodegenerative diseases where ER and oxidative stress has been involved on the pathological causes [21–29,31,32,34–41]. To assess the significance of our findings to disease conditions affecting the nervous system, we explored the involvement of c-Abl in mitochondrial fragmentation in cellular and animal models of ALS. First, we expressed ALS-associated mutant SOD1-G85R or WT SOD1 fused to EGFP in primary hippocampal neurons at 7 days post differentiation. As expected, the expression of mutant SOD1 resulted in protein aggregation within hippocampal neurons, whereas the transfection of the WT SOD1 protein showed no such aggregation (Figure 7A). Also, neurons overexpressing mutant SOD1 exhibited higher levels of phosphorylated c-Abl at Y412 (Figure 7A), which was inhibited with imatinib (Figure 7A). We then analyzed mitochondrial fragmentation induced by mutant SOD1. Remarkably, imatinib significantly prevented mitochondrial fragmentation in neurons overexpressing mutant SOD1 compared with the WT form (Figure 7B,C).

Moreover, we evaluated the effect of pharmacological inhibition of c-Abl with nilotinib on the SOD1 G93A ALS mice. First, we quantified the percentage of morphologically disrupted mitochondria, characterized by an altered external mitochondrial membrane with fragmented or swollen cristae/matrix, in sciatic nerves from WT or SOD1 G93A mice fed with either control or nilotinib diet (Figure 7D). The results showed an increase of disrupted mitochondria in SOD1 G93A compared to control littermates. In addition, nilotinib treatment decreased the number of disrupted mitochondria in SOD1 G93A mice, while nilotinib did not influence the mitochondria of WT mice (Figure 7E). These results suggest that c-Abl inhibition with nilotinib may have a beneficial effect on altered mitochondrial morphology in SOD1 G93A mice.

To determine the protective effects of nilotinib in SOD1 G93A mice, we quantified the number of NeuN-positive neurons in the ventral horn of the lumbar spinal cord (Figure 7F). As previously described [60], we found a lower amount of NeuN cells in SOD1 G93A mice than in control littermates (Figure 7G). Interestingly, treatment with nilotinib prevented the NeuN-positive cell loss in SOD1 G93A mice, while control mice did not exhibit significant changes (Figure 7G).

When we analyzed the progression of the disease in control, nilotinib, or the recently described c-Abl inhibitor neurotinib- [35,45] fed SOD1 G93A mice, the results showed that neurotinib administration led to a significant delay in disease onset compared to the control group ($p = 0.0081$), whereas nilotinib did not exhibit a significant effect ($p = 0.2253$). The median onset of symptoms was 152 days in the neurotinib-treated mice, compared to 139 days in the control group (Figure S5A). The median survival time was also slightly longer in the neurotinib group (165.5 days) compared to the control group (163 days), although this difference was not significant ($p = 0.1237$) (Figure S5B). Overall, the results suggest that pharmacological c-Abl inhibition has a beneficial effect on the phenotype of the SOD1 G93A ALS mouse model, in delaying disease onset and improving neuronal survival.

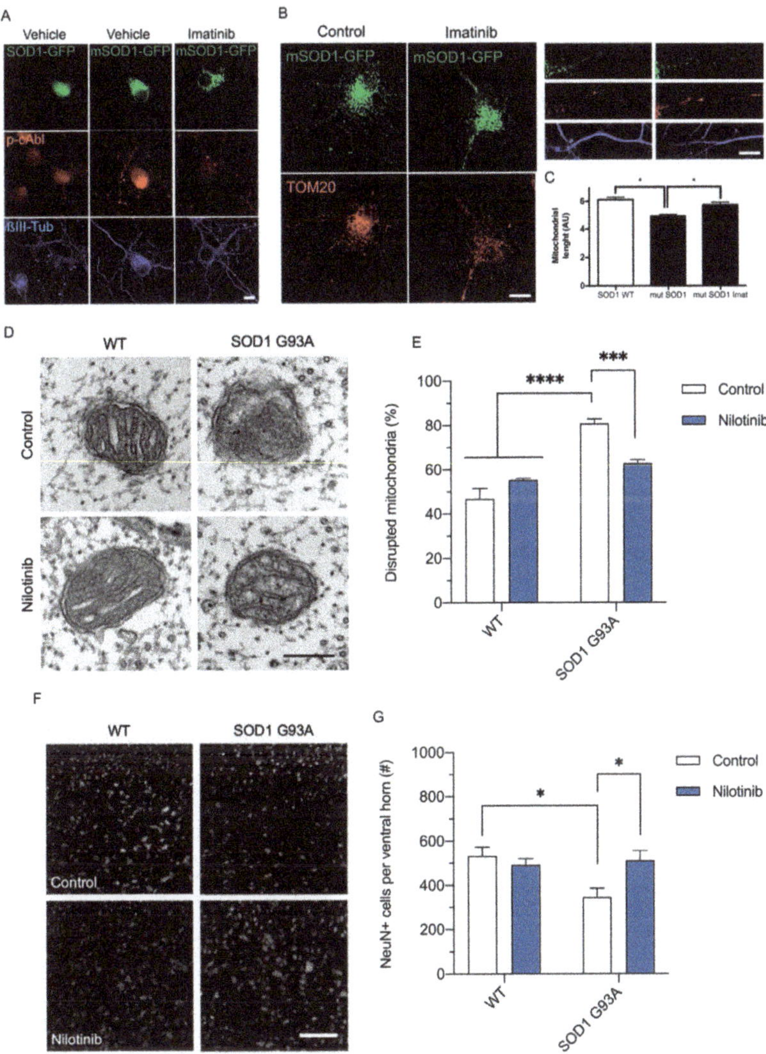

Figure 7. c-Abl regulates mitochondrial status in ALS models. (**A**) Confocal microscopy of primary hippocampal neurons overexpressing SOD1 WT fused to GFP (SOD1-GFP) or SOD1 G85R fused to GFP (mSOD1-GFP) in green, and stained for p-c-Abl (red) and βIII tubulin (blue), treated with vehicle or imatinib. Scale bar: 5 μm. (**B**) Confocal microscopy of primary hippocampal neurons overexpressing SOD1-GFP or mSOD1-GFP in green, and stained for TOM20 (red) both in soma (left side of the panel) and processes (right side of the panel) stained for and βIII tubulin (blue). Scale bar: 5 μm. (**C**) Mitochondrial length in neuronal processes as in (**B**). (**D**) Representative TEM images of mitochondria in the sciatic nerve from WT control-fed (n = 4) and nilotinib-fed (n = 4) mice and SOD1 G93A control-fed (n = 5) and nilotinib-fed (n = 5) mice. Scale bar = 500 nm. (**E**) Percentage of swollen mitochondria characterized by a disrupted external mitochondrial membrane with fragmented or swollen cristae/matrix in the sciatic nerve from (**D**). (**F**) Representative fluorescence images of NeuN in the ventral horn of the lumbar spinal cord in WT control-fed (n = 6) and nilotinib-fed (n = 5) mice, and SOD1 G93A control-fed (n = 5) and nilotinib-fed (n = 5) mice. Scale bar: 150 μm. (**G**) Graph shows the number of NeuN positive cells in 300,000 μm^2 from (**F**). Two-way ANOVA with Tukey's multiple comparisons test. Results are presented as mean ± SEM. p * < 0.05, p *** < 0.001, p **** < 0.0001.

4. Discussion

Mitochondrial dynamics influence the morphology and impact on various processes of mitochondrial homeostasis. In addition to the close relationship with the cellular bioenergetic capacity and buffering intracellular Ca^{2+}, among other functions, mitochondria interact with the endoplasmic reticulum and react to cellular stress to favor apoptosis after severe injuries. Mitochondrial dynamics allow cells to adapt themselves to physiological and environmental fluctuations. Considering the high functionality of mitochondria, it is not unexpected that the machinery for fusion and fission follows rigorous regulations to keep cellular stability [56,61,62]. However, the mechanisms controlling the mitochondrial dynamic machinery are just partially understood. In this work, we unveiled a new molecular mechanism for c-Abl kinase in mitochondria as a modulator of the mitochondrial dynamics through MFN2 in response to cellular stress. We described that activated c-Abl promotes the mitochondrial fragmentation mediated through the phosphorylation of MFN2 at Y269 in response to ER stress.

The function for activated c-Abl under cellular stress has been mainly associated with the transcription of proapoptotic genes through the factors p73 and MST1 [20–22]. Additionally, it has been described that c-Abl is activated in response to ER stress, oxidative stress, and genotoxic stress [17,22,63,64], and translocates to the mitochondria where it promotes the outer mitochondrial membrane permeabilization and apoptosis [3]. However, the mechanism through which c-Abl could exert this function is unknown. Here, we focused on a potential mechanism for c-Abl in mitochondria. Given that ER stress leads to c-Abl activation, and it also affects mitochondrial dynamics [53], we investigated a potential role for c-Abl kinase in mitochondrial dynamics in response to ER stress.

In our first approach, we found that the mitochondrial fragmentation was reduced with imatinib, a specific c-Abl inhibitor, in response to ER stress, which agrees with previous studies in animal models concerning neurodegenerative disorders, where imatinib treatment had proven benefits, ameliorating symptoms [65]. Surprisingly, the expression of constitutively active c-Abl kinase by the tamoxifen-inducible system led to a dramatic mitochondrial fragmentation, while the expression of the kinase-dead variant did not induce changes on the morphology, demonstrating that activated c-Abl kinase is sufficient to induce mitochondrial fragmentation. Furthermore, c-Abl-deficient neurons exposed to tunicamycin revealed a significantly reduced mitochondrial fragmentation. Hence, using different strategies, our data indicates a novel function for c-Abl kinase collaborating in the mitochondrial fragmentation induced by cellular stress.

Through STORM microscopy, we provided detailed evidence for the c-Abl localization at mitochondria. In basal conditions, we found that c-Abl molecules were apparently randomly distributed, and interestingly, c-Abl molecules were dramatically enriched in mitochondria with a clustered-like pattern in response to ER stress. Through biochemical studies, it has been described that mitochondrial fractions are enriched in activated c-Abl under ER stress, and furthermore, c-Abl collaborates with the outer mitochondrial membrane permeabilization and apoptosis [3,66], supporting our results. In addition, it was reported that activated c-Abl exposes its N-terminal myristoyl group, which permits its localization with membranes [66] and could explain our observations of enriched c-Abl at mitochondria.

We explored potential targets of c-Abl that could be related to their association with mitochondrial dynamics, and found that c-Abl interacts with and phosphorylates the GTPase MFN2. Post-translational modifications on MFN2 have been previously reported in the literature. The phosphorylation of MFN2 at Ser27 by the JNK kinase was identified in response to cellular stress [55]; also, PINK1 phosphorylates MFN2 at Thr111 and Ser442, favoring its ubiquitination by the E3 ubiquitin ligase Parkin, in response to cellular stress [67]. Interestingly, those post-translational modifications reported for MFN2 are described as leading to mitochondrial fragmentation in response to cellular stress. Recently, it has been described that c-Src tyrosine kinase could regulate ER–mitochondrion interactions through the phosphorylation of the C-terminal tail of MFN2 [68]. Hence, our results, revealing the

MFN2 phosphorylation by c-Abl kinase in response to ER stress, are consistent with this tendency, as described also with the phosphorylation of MFN1 by ERK in response to DNA damage [69,70].

A structural model prepared with the Protein Imager tool suggests that the Y269 amino acid residue is localized close to the GTP binding site and exposed to the cytoplasm, in the GTPase domain. Interestingly, the Y269 of MFN2 interacts with the N161 of the counterpart MFN2 monomer in the interface of the dimer. This interaction is crucial for mitochondrial elongation, as living cells seem quite sensitive to mutations in the interface [71]. These findings suggest that the Y269 residue plays a pivotal role in mediating protein–protein interactions within the MFN2 dimer interface, ultimately influencing mitochondrial fusion dynamics.

We found that the non-phosphorylable MFN2 Y269F exhibits an increased GTP affinity, which could suggest an augmented exchange from GDP to GTP. Additionally, we did not observe changes in the hydrolytic capacity of GTP between MFN2 WT and MFN2 Y269F, which could be in agreement with our data, since it was reported that changes in the hydrolytic capacity for MFN2 do not alter the mitochondrial fusion rates [58].

Through the STORM super-resolution microscopy, we detected that c-Abl colocalizes with MFN2 preferentially under stress, and interestingly, it was prevented with the expression of the mutant MFN2 Y269F. It seems that the Y269 residue of MFN2 is important to facilitate the c-Abl localization and enrichment in the mitochondrial membrane. It has been reported that MFN2 is enriched at mitochondria-associated membranes (MAMs) [72–74]. In this regard, whether c-Abl is localized at MAMs in response to ER stress remains to be determined.

We explored the biological significance of the mutant MFN2 Y269F to the mitochondrial morphology. Interestingly, the cells expressing MFN2 Y269F exhibited resistance to promote mitochondrial fragmentation in response to ER stress, suggesting that the phosphorylation of MFN2 at Y269 is required in the reduction in its pro-fusion function in response to ER stress. Our results suggest that the increased association with c-Abl could lead to the phosphorylation of MFN2 in response to ER stress. It has been reported that the phosphorylation of MFN2 at Ser27 is required to promote mitochondrial fragmentation in cells under genotoxic stress [55]. The use of other MFN2 mutants such as phospho-mimetic MFN2 Y269E could be included in future experiments to explore more deeply the contribution on mitochondrial morphology; however, our approach confirms that the Y269 participates in the c-Abl–MNF2 interaction. Finally, we evaluated potential changes in the apoptotic response in MFN2 Y269F-expressing cells. Our findings show reduced levels for cleaved caspase 3 and TUNEL activation in MFN2 Y269F-expressing cells after an extended exposure to ER stress. During ER stress, c-Abl is activated, but its role in mitochondria remains unknown [3]. Our model reveals that c-Abl promotes a reduced function pro-fusion for MFN2, and it also favors the apoptotic response. Interestingly, post-translational modifications in MFN1 and MFN2 in response to cellular stress facilitate the apoptotic response [55,67,70], supporting the role for c-Abl promoting mitochondrial fragmentation mediated by its association with MFN2 in response to ER stress.

The activation of c-Abl in ALS motoneurons has been associated with oxidative stress [39]. Bosutinib, a Src/c-Abl inhibitor, reduced the misfolded mutant SOD1 protein levels, improved the expression of mitochondrial genes, and modestly extended G93A ALS mouse model survival [40]. Our results with nilotinib showed a coherent preservation of the mitochondrial morphology, and although we did not observe a significant increase in life span, the novel allosteric inhibitor neurotinib delays the onset of symptoms in the G93A mouse model, resembling the effects described for the c-Abl inhibitor dasatinib that also delays motor neuron degeneration in the G93A mouse model [41].

Overall, our study provides a view into the complex relationship between c-Abl kinase and mitochondrial dynamics mediated by MFN2 phosphorylation at Y269 during ER stress, mainly in the context of ALS. The correlation between in vitro and in vivo results reinforces the significance of c-Abl activity in mediating mitochondrial responses to stress conditions and the use of c-Abl inhibitors for targeting mitochondrial dysfunction in ALS.

5. Conclusions

In conclusion, our research highlights c-Abl kinase in regulating mitochondrial dynamic under cell stress. Specifically, we identified a new molecular mechanism by which c-Abl phosphorylates the mitochondrial fusion protein MFN2 at Y269 in response to ER stress, leading to mitochondrial fragmentation and apoptosis.

Our results point to a therapeutic potential of c-Abl modulation in neurodegenerative diseases like ALS, where ER and oxidative stress contribute to cell death. Moreover, the pharmacological inhibition of c-Abl in an ALS mouse model improves mitochondrial health and delays the onset of symptoms, opening avenues for further research and potential treatments in neurodegenerative conditions linked to cellular stress and mitochondrial dysfunction.

Supplementary Materials: The following supporting information can be downloaded at: https://www.mdpi.com/article/10.3390/antiox12112007/s1, Figure S1: In silico analysis to determine potential c-Abl mitochondrial targets; Figure S2: The stable overexpression of FLAG-Mfn2 restitutes the mitochondrial morphology in Mfn1/2- deficient MEF cells; Figure S3: The GTP hydrolytic rate is not altered in mutant Mfn2; Figure S4: Mitochondrial activity in MFN WT and Y269F mutant MEF; Figure S5: The pharmacological inhibition of c-Abl improves ALS phenotype in SOD1 G93A mice.

Author Contributions: A.M., C.M.L. and A.R.A. contributed to the conception and design of the study; A.M., C.V. and C.M.L. performed the cell biology and microscopy and the experiments. A.M. and C.M.L. performed the studies in SOD mice treated with c-Abl inhibitors diets; N.L. helped with the hippocampal neuron's cultures and experiments; F.L., P.F., C.G. and M.M.K. helped with the c-Abl-ERT2 constructs experiments and STORM microscopy. A.E.D. synthesized nilotinib and neurotinib chow; A.M., A.R.A., C.H., B.v.Z. and C.M.L. wrote the first version of the manuscript and discussed the results; A.M., C.M.L. and G.I.C. analyze data and generate the figures; J.J.M., N.C.I., S.Z. and D.R.G. wrote sections of the manuscript. All authors have read and agreed to the published version of the manuscript.

Funding: This work was supported by the Agencia Nacional de Investigación y Desarrollo (ANID): Fondecyt grant [1161065 and 12011668 (A.R.A.) and 1230337 (S.Z.)], Fondef grant [ID21I10347 and D10E1077 (A.R.A. and S.Z.) and ID1ID22I10120 (C.H.)], and Program ICM-ANID-ICN2021 045 (A.R.A). Part of this work was supported by Intramural funding of the National Center for Advancing Translational Sciences (J.J.M.), the R35CA23160 grant (D.R.G.) from NCI, ANID-Proyecto Basal AFB170005 (A.R.A. & N.I.) and ANID-EXPLORADOR 13220203 (B.v.Z.), A.M. acknowledges support from ANID-CONICYT, Beca Doctorado Nacional (24121672), ISN-CAEN, and FONDECYT (3190638). C.H. acknowledges support from ANID-FONDAP program 15150012, U.S. Air Force Office of Scientific Research FA9550-21-1-0096, Department of Defense grant W81XWH2110960, and Swiss Consolidation Grant -The Leading House for the Latin American Region.

Institutional Review Board Statement: The animal study protocol was approved by the Bioethics and Care of Laboratory Animals Committee of the Pontificia Universidad Católica de Chile (Protocol #150721002 and 190610036) for studies involving animals.

Informed Consent Statement: Not applicable.

Data Availability Statement: The data presented in this study are available on request from the corresponding author. The data are not publicly available due to the proprietary nature of the drug neurotinib and its associated patent WO2019/173761 A1, which restricts sharing the data without prior authorization from the corresponding author and a valid, justified request.

Acknowledgments: We thank Verónica Eisner for scientific support. We thank Romina Gozalvo, Ludwig Amigo, and Jorge Ibañez for technical support. We thank the Advanced Microscopy Facility UMA UC, and the CIBEM Facility for animal housing and care. Cartoons for graphical abstracts were created with the open-source software Inkscape.

Conflicts of Interest: The authors declare that this research was conducted in the absence of any commercial or financial relationships that could be construed as a potential conflict of interest. The funders had no role in the design of the study; in the collection, analyses, or interpretation of data; in the writing of the manuscript; or in the decision to publish the results. Neurotinib is under the patent WO2019/173761 A1. D.R.G. consults and/or has equity in Ventus Pharmaceuticals and Sonata Therapeutics, and receives support from Horizon Pharmaceuticals.

References

1. Hetz, C.; Chevet, E.; Oakes, S.A. Proteostasis Control by the Unfolded Protein Response. *Nat. Cell Biol.* **2015**, *17*, 829–838. [CrossRef]
2. Carreras-Sureda, A.; Pihán, P.; Hetz, C. The Unfolded Protein Response: At the Intersection between Endoplasmic Reticulum Function and Mitochondrial Bioenergetics. *Front. Oncol.* **2017**, *7*, 55. [CrossRef]
3. Ito, Y.; Pandey, P.; Mishra, N.; Kumar, S.; Narula, N.; Kharbanda, S.; Saxena, S.; Kufe, D. Targeting of the C-Abl Tyrosine Kinase to Mitochondria in Endoplasmic Reticulum Stress-Induced Apoptosis. *Mol. Cell Biol.* **2001**, *21*, 6233–6242. [CrossRef] [PubMed]
4. Welch, P.; Wang, J. A C-Terminal Protein-Binding Domain in the Retinoblastoma Protein Regulates Nuclear c-Abl Tyrosine Kinase in the Cell Cycle. *Cell* **1993**, *75*, 779–790. [CrossRef]
5. Sawyers, C.L.; McLaughlin, J.; Goga, A.; Havlik, M.; Witte, O. The Nuclear Tyrosine Kinase C-Abl Negatively Regulates Cell Growth. *Cell* **1994**, *77*, 121–131. [CrossRef]
6. Yuan, Z.-M.; Huang, Y.; Whang, Y.; Sawyers, C.; Weichselbaum, R.; Kharbanda, S.; Kufe, D. Role for C-Abl Tyrosine Kinase in Growth Arrest Response to DNA Damage. *Nature* **1996**, *382*, 272–274. [CrossRef]
7. Koleske, A.J.; Gifford, A.M.; Scott, M.L.; Nee, M.; Bronson, R.T.; Miczek, K.A.; Baltimore, D. Essential Roles for the Abl and Arg Tyrosine Kinases in Neurulation. *Neuron* **1998**, *21*, 1259–1272. [CrossRef]
8. Zukerberg, L.R.; Patrick, G.N.; Nikolic, M.; Humbert, S.; Wu, C.-L.; Lanier, L.M.; Gertler, F.B.; Vidal, M.; Van Etten, R.A.; Tsai, L.-H. Cables Links Cdk5 and C-Abl and Facilitates Cdk5 Tyrosine Phosphorylation, Kinase Upregulation, and Neurite Outgrowth. *Neuron* **2000**, *26*, 633–646. [CrossRef] [PubMed]
9. Zhu, J.; Wang, J.Y.J. Death by Abl: A Matter of Location. *Curr. Top. Dev. Biol.* **2004**, *59*, 165–192. [CrossRef] [PubMed]
10. Wang, J.Y.J. The Capable ABL: What Is Its Biological Function? *Mol. Cell Biol.* **2014**, *34*, 1188–1197. [CrossRef] [PubMed]
11. Bradley, W.D.; Koleske, A.J. Regulation of Cell Migration and Morphogenesis by Abl-Family Kinases: Emerging Mechanisms and Physiological Contexts. *J. Cell Sci.* **2009**, *122*, 3441–3454. [CrossRef] [PubMed]
12. Khatri, A.; Wang, J.; Pendergast, A.M. Multifunctional Abl Kinases in Health and Disease. *J. Cell Sci.* **2016**, *129*, 9–16. [CrossRef]
13. Motaln, H.; Rogelj, B. The Role of C-Abl Tyrosine Kinase in Brain and Its Pathologies. *Cells* **2023**, *12*, 2041. [CrossRef] [PubMed]
14. Gonfloni, S.; Maiani, E.; Di Bartolomeo, C.; Diederich, M.; Cesareni, G. Oxidative Stress, DNA Damage, and c-Abl Signaling: At the Crossroad in Neurodegenerative Diseases? *Int. J. Cell Biol.* **2012**, *2012*, 683097. [CrossRef] [PubMed]
15. Kumar, S.; Mishra, N.; Raina, D.; Saxena, S.; Kufe, D. Abrogation of the Cell Death Response to Oxidative Stress by the C-Abl Tyrosine Kinase Inhibitor STI571. *Mol. Pharmacol.* **2003**, *63*, 276–282. [CrossRef] [PubMed]
16. Constance, J.E.; Woessner, D.W.; Matissek, K.J.; Mossalam, M.; Lim, C.S. Enhanced and Selective Killing of Chronic Myelogenous Leukemia Cells with an Engineered BCR-ABL Binding Protein and Imatinib. *Mol. Pharm.* **2012**, *9*, 3318–3329. [CrossRef]
17. Zhou, L.; Zhang, Q.; Zhang, P.; Sun, L.; Peng, C.; Yuan, Z.; Cheng, J. C-Abl-Mediated Drp1 Phosphorylation Promotes Oxidative Stress-Induced Mitochondrial Fragmentation and Neuronal Cell Death. *Cell Death Dis.* **2017**, *8*, e3117. [CrossRef]
18. Pan, B.; Yang, L.; Wang, J.; Wang, Y.; Wang, J.; Zhou, X.; Yin, X.; Zhang, Z.; Zhao, D. C-Abl Tyrosine Kinase Mediates Neurotoxic Prion Peptide-Induced Neuronal Apoptosis via Regulating Mitochondrial Homeostasis. *Mol. Neurobiol.* **2014**, *49*, 1102–1116. [CrossRef]
19. Tsai, K.K.C.; Yuan, Z.-M. C-Abl Stabilizes P73 by a Phosphorylation-Augmented Interaction. *Cancer Res.* **2003**, *63*, 3418–3424.
20. Xiao, L.; Chen, D.; Hu, P.; Wu, J.; Liu, W.; Zhao, Y.; Cao, M.; Fang, Y.; Bi, W.; Zheng, Z.; et al. The C-Abl-MST1 Signaling Pathway Mediates Oxidative Stress-Induced Neuronal Cell Death. *J. Neurosci.* **2011**, *31*, 9611–9619. [CrossRef]
21. Cancino, G.I.; Toledo, E.M.; Leal, N.R.; Hernandez, D.E.; Yévenes, L.F.; Inestrosa, N.C.; Alvarez, A.R. STI571 Prevents Apoptosis, Tau Phosphorylation and Behavioural Impairments Induced by Alzheimer's β-Amyloid Deposits. *Brain* **2008**, *131*, 2425–2442. [CrossRef] [PubMed]
22. Alvarez, A.R.; Sandoval, P.C.; Leal, N.R.; Castro, P.U.; Kosik, K.S. Activation of the Neuronal C-Abl Tyrosine Kinase by Amyloid-β-Peptide and Reactive Oxygen Species. *Neurobiol. Dis.* **2004**, *17*, 326–336. [CrossRef] [PubMed]
23. Ko, H.S.; Lee, Y.; Shin, J.-H.; Karuppagounder, S.S.; Gadad, B.S.; Koleske, A.J.; Pletnikova, O.; Troncoso, J.C.; Dawson, V.L.; Dawson, T.M. Phosphorylation by the C-Abl Protein Tyrosine Kinase Inhibits Parkin's Ubiquitination and Protective Function. *Proc. Natl. Acad. Sci. USA* **2010**, *107*, 16691–16696. [CrossRef]
24. Imam, S.Z.; Zhou, Q.; Yamamoto, A.; Valente, A.J.; Ali, S.F.; Bains, M.; Roberts, J.L.; Kahle, P.J.; Clark, R.A.; Li, S. Novel Regulation of Parkin Function through C-Abl-Mediated Tyrosine Phosphorylation: Implications for Parkinson's Disease. *J. Neurosci.* **2011**, *31*, 157–163. [CrossRef]
25. Hebron, M.L.; Lonskaya, I.; Moussa, C.E.-H. Nilotinib Reverses Loss of Dopamine Neurons and Improves Motor Behavior via Autophagic Degradation of -Synuclein in Parkinson's Disease Models. *Hum. Mol. Genet.* **2013**, *22*, 3315–3328. [CrossRef] [PubMed]

26. Karuppagounder, S.S.; Brahmachari, S.; Lee, Y.; Dawson, V.L.; Dawson, T.M.; Ko, H.S. The C-Abl Inhibitor, Nilotinib, Protects Dopaminergic Neurons in a Preclinical Animal Model of Parkinson's Disease. *Sci. Rep.* **2014**, *4*, 4874. [CrossRef]

27. Mahul-Mellier, A.-L.; Fauvet, B.; Gysbers, A.; Dikiy, I.; Oueslati, A.; Georgeon, S.; Lamontanara, A.J.; Bisquertt, A.; Eliezer, D.; Masliah, E.; et al. C-Abl Phosphorylates α-Synuclein and Regulates Its Degradation: Implication for α-Synuclein Clearance and Contribution to the Pathogenesis of Parkinson's Disease. *Hum. Mol. Genet.* **2014**, *23*, 2858–2879. [CrossRef]

28. Brahmachari, S.; Ge, P.; Lee, S.H.; Kim, D.; Karuppagounder, S.S.; Kumar, M.; Mao, X.; Shin, J.H.; Lee, Y.; Pletnikova, O.; et al. Activation of Tyrosine Kinase C-Abl Contributes to α-Synuclein–Induced Neurodegeneration. *J. Clin. Investig.* **2016**, *126*, 2970–2988. [CrossRef]

29. Lee, S.; Kim, S.; Park, Y.J.; Yun, S.P.; Kwon, S.-H.; Kim, D.; Kim, D.Y.; Shin, J.S.; Cho, D.J.; Lee, G.Y.; et al. The C-Abl Inhibitor, Radotinib HCl, Is Neuroprotective in a Preclinical Parkinson's Disease Mouse Model. *Hum. Mol. Genet.* **2018**, *27*, 2344–2356. [CrossRef]

30. Marín, T.; Valls, C.; Jerez, C.; Huerta, T.; Elgueta, D.; Vidal, R.L.; Alvarez, A.R.; Cancino, G.I. The C-Abl/P73 Pathway Induces Neurodegeneration in a Parkinson's Disease Model. *IBRO Neurosci. Rep.* **2022**, *13*, 378–387. [CrossRef]

31. Abushouk, A.I.; Negida, A.; Elshenawy, R.A.; Zein, H.; Hammad, A.M.; Menshawy, A.; Mohamed, W.M.Y. C-Abl Inhibition; A Novel Therapeutic Target for Parkinson's Disease. *CNS Neurol. Disord. Drug Targets* **2018**, *17*, 14–21. [CrossRef]

32. Jing, Z.; Caltagarone, J.; Bowser, R. Altered Subcellular Distribution of C-Abl in Alzheimer's Disease. *J. Alzheimer's Dis.* **2009**, *17*, 409–422. [CrossRef]

33. Cancino, G.I.; Perez de Arce, K.; Castro, P.U.; Toledo, E.M.; von Bernhardi, R.; Alvarez, A.R. C-Abl Tyrosine Kinase Modulates Tau Pathology and Cdk5 Phosphorylation in AD Transgenic Mice. *Neurobiol. Aging* **2011**, *32*, 1249–1261. [CrossRef] [PubMed]

34. Vargas, L.M.; Leal, N.; Estrada, L.D.; González, A.; Serrano, F.; Araya, K.; Gysling, K.; Inestrosa, N.C.; Pasquale, E.B.; Alvarez, A.R. EphA4 Activation of C-Abl Mediates Synaptic Loss and LTP Blockade Caused by Amyloid-β Oligomers. *PLoS ONE* **2014**, *9*, e92309. [CrossRef]

35. León, R.; Gutiérrez, D.A.; Pinto, C.; Morales, C.; de la Fuente, C.; Riquelme, C.; Cortés, B.I.; González-Martin, A.; Chamorro, D.; Espinosa, N.; et al. C-Abl Tyrosine Kinase down-Regulation as Target for Memory Improvement in Alzheimer's Disease. *Front. Aging Neurosci.* **2023**, *15*, 1180987. [CrossRef]

36. La Barbera, L.; Vedele, F.; Nobili, A.; Krashia, P.; Spoleti, E.; Latagliata, E.C.; Cutuli, D.; Cauzzi, E.; Marino, R.; Viscomi, M.T.; et al. Nilotinib Restores Memory Function by Preventing Dopaminergic Neuron Degeneration in a Mouse Model of Alzheimer's Disease. *Prog. Neurobiol.* **2021**, *202*, 102031. [CrossRef]

37. Alvarez, A.R.; Klein, A.; Castro, J.; Cancino, G.I.; Amigo, J.; Mosqueira, M.; Vargas, L.M.; Yevenes, L.F.; Bronfman, F.C.; Zanlungo, S. Imatinib Therapy Blocks Cerebellar Apoptosis and Improves Neurological Symptoms in a Mouse Model of Niemann-Pick Type C Disease. *FASEB J.* **2008**, *22*, 3617–3627. [CrossRef] [PubMed]

38. Contreras, P.S.; Tapia, P.J.; González-Hódar, L.; Peluso, I.; Soldati, C.; Napolitano, G.; Matarese, M.; Las Heras, M.; Valls, C.; Martinez, A.; et al. C-Abl Inhibition Activates TFEB and Promotes Cellular Clearance in a Lysosomal Disorder. *iScience* **2020**, *23*, 101691. [CrossRef] [PubMed]

39. Rojas, F.; Gonzalez, D.; Cortes, N.; Ampuero, E.; Hernández, D.E.; Fritz, E.; Abarzua, S.; Martinez, A.; Elorza, A.A.; Alvarez, A.; et al. Reactive Oxygen Species Trigger Motoneuron Death in Non-Cell-Autonomous Models of ALS through Activation of c-Abl Signaling. *Front. Cell Neurosci.* **2015**, *9*, 203. [CrossRef]

40. Imamura, K.; Izumi, Y.; Watanabe, A.; Tsukita, K.; Woltjen, K.; Yamamoto, T.; Hotta, A.; Kondo, T.; Kitaoka, S.; Ohta, A.; et al. The Src/c-Abl Pathway Is a Potential Therapeutic Target in Amyotrophic Lateral Sclerosis. *Sci. Transl. Med.* **2017**, *9*, eaaf3962. [CrossRef]

41. Katsumata, R.; Ishigaki, S.; Katsuno, M.; Kawai, K.; Sone, J.; Huang, Z.; Adachi, H.; Tanaka, F.; Urano, F.; Sobue, G. C-Abl Inhibition Delays Motor Neuron Degeneration in the G93A Mouse, an Animal Model of Amyotrophic Lateral Sclerosis. *PLoS ONE* **2012**, *7*, e46185. [CrossRef]

42. Kaech, S.; Banker, G. Culturing Hippocampal Neurons. *Nat. Protoc.* **2006**, *1*, 2406–2415. [CrossRef]

43. Tanis, K.Q.; Veach, D.; Duewel, H.S.; Bornmann, W.G.; Koleske, A.J. Two Distinct Phosphorylation Pathways Have Additive Effects on Abl Family Kinase Activation. *Mol. Cell Biol.* **2003**, *23*, 3884–3896. [CrossRef]

44. Tait, S.W.G.; Parsons, M.J.; Llambi, F.; Bouchier-Hayes, L.; Connell, S.; Muñoz-Pinedo, C.; Green, D.R. Resistance to Caspase-Independent Cell Death Requires Persistence of Intact Mitochondria. *Dev. Cell* **2010**, *18*, 802–813. [CrossRef]

45. Marín, T.; Dulcey, A.E.; Campos, F.; de la Fuente, C.; Acuña, M.; Castro, J.; Pinto, C.; Yañez, M.J.; Cortez, C.; McGrath, D.W.; et al. C-Abl Activation Linked to Autophagy-Lysosomal Dysfunction Contributes to Neurological Impairment in Niemann-Pick Type A Disease. *Front. Cell Dev. Biol.* **2022**, *10*, 844297. [CrossRef]

46. Gill, C.; Phelan, J.P.; Hatzipetros, T.; Kidd, J.D.; Tassinari, V.R.; Levine, B.; Wang, M.Z.; Moreno, A.; Thompson, K.; Maier, M.; et al. SOD1-Positive Aggregate Accumulation in the CNS Predicts Slower Disease Progression and Increased Longevity in a Mutant SOD1 Mouse Model of ALS. *Sci. Rep.* **2019**, *9*, 6724. [CrossRef] [PubMed]

47. Rybka, V.; Suzuki, Y.J.; Gavrish, A.S.; Dibrova, V.A.; Gychka, S.G.; Shults, N.V. Transmission Electron Microscopy Study of Mitochondria in Aging Brain Synapses. *Antioxidants* **2019**, *8*, 171. [CrossRef] [PubMed]

48. Greuber, E.K.; Smith-Pearson, P.; Wang, J.; Pendergast, A.M. Role of ABL Family Kinases in Cancer: From Leukaemia to Solid Tumours. *Nat. Rev. Cancer* **2013**, *13*, 559–571. [CrossRef] [PubMed]

49. Llambi, F.; Moldoveanu, T.; Tait, S.W.G.; Bouchier-Hayes, L.; Temirov, J.; McCormick, L.L.; Dillon, C.P.; Green, D.R. A Unified Model of Mammalian BCL-2 Protein Family Interactions at the Mitochondria. *Mol. Cell* **2011**, *44*, 517–531. [CrossRef] [PubMed]
50. Amoui, M.; Miller, W.T. The Substrate Specificity of the Catalytic Domain of Abl Plays an Important Role in Directing Phosphorylation of the Adaptor Protein Crk. *Cell Signal* **2000**, *12*, 637–643. [CrossRef] [PubMed]
51. Rath, S.; Sharma, R.; Gupta, R.; Ast, T.; Chan, C.; Durham, T.J.; Goodman, R.P.; Grabarek, Z.; Haas, M.E.; Hung, W.H.W.; et al. MitoCarta3.0: An Updated Mitochondrial Proteome Now with Sub-Organelle Localization and Pathway Annotations. *Nucleic Acids Res.* **2021**, *49*, D1541–D1547. [CrossRef]
52. Chen, H.; Detmer, S.A.; Ewald, A.J.; Griffin, E.E.; Fraser, S.E.; Chan, D.C. Mitofusins Mfn1 and Mfn2 Coordinately Regulate Mitochondrial Fusion and Are Essential for Embryonic Development. *J. Cell Biol.* **2003**, *160*, 189–200. [CrossRef]
53. Chan, D.C. Mitochondria: Dynamic Organelles in Disease, Aging, and Development. *Cell* **2006**, *125*, 1241–1252. [CrossRef] [PubMed]
54. Filadi, R.; Pendin, D.; Pizzo, P. Mitofusin 2: From Functions to Disease. *Cell Death Dis.* **2018**, *9*, 330. [CrossRef] [PubMed]
55. Leboucher, G.P.; Tsai, Y.C.; Yang, M.; Shaw, K.C.; Zhou, M.; Veenstra, T.D.; Glickman, M.H.; Weissman, A.M. Stress-Induced Phosphorylation and Proteasomal Degradation of Mitofusin 2 Facilitates Mitochondrial Fragmentation and Apoptosis. *Mol. Cell* **2012**, *47*, 547–557. [CrossRef]
56. Westermann, B. Mitochondrial Fusion and Fission in Cell Life and Death. *Nat. Rev. Mol. Cell Biol.* **2010**, *11*, 872–884. [CrossRef]
57. Tomasello, G.; Armenia, I.; Molla, G. The Protein Imager: A Full-Featured Online Molecular Viewer Interface with Server-Side HQ-Rendering Capabilities. *Bioinformatics* **2020**, *36*, 2909–2911. [CrossRef] [PubMed]
58. Neuspiel, M.; Zunino, R.; Gangaraju, S.; Rippstein, P.; McBride, H. Activated Mitofusin 2 Signals Mitochondrial Fusion, Interferes with Bax Activation, and Reduces Susceptibility to Radical Induced Depolarization. *J. Biol. Chem.* **2005**, *280*, 25060–25070. [CrossRef]
59. Medinas, D.B.; González, J.V.; Falcon, P.; Hetz, C. Fine-Tuning ER Stress Signal Transducers to Treat Amyotrophic Lateral Sclerosis. *Front. Mol. Neurosci.* **2017**, *10*, 216. [CrossRef]
60. Kirby, A.J.; Palmer, T.; Mead, R.J.; Ichiyama, R.M.; Chakrabarty, S. Caudal–Rostral Progression of Alpha Motoneuron Degeneration in the SOD1G93A Mouse Model of Amyotrophic Lateral Sclerosis. *Antioxidants* **2022**, *11*, 983. [CrossRef]
61. Knott, A.B.; Perkins, G.; Schwarzenbacher, R.; Bossy-Wetzel, E. Mitochondrial Fragmentation in Neurodegeneration. *Nat. Rev. Neurosci.* **2008**, *9*, 505–518. [CrossRef]
62. Mourier, A.; Motori, E.; Brandt, T.; Lagouge, M.; Atanassov, I.; Galinier, A.; Rappl, G.; Brodesser, S.; Hultenby, K.; Dieterich, C.; et al. Mitofusin 2 Is Required to Maintain Mitochondrial Coenzyme Q Levels. *J. Cell Biol.* **2015**, *208*, 429–442. [CrossRef]
63. Dufey, E.; Bravo-San Pedro, J.M.; Eggers, C.; González-Quiroz, M.; Urra, H.; Sagredo, A.I.; Sepulveda, D.; Pihán, P.; Carreras-Sureda, A.; Hazari, Y.; et al. Genotoxic Stress Triggers the Activation of IRE1α-Dependent RNA Decay to Modulate the DNA Damage Response. *Nat. Commun.* **2020**, *11*, 2401. [CrossRef]
64. Morita, S.; Villalta, S.A.; Feldman, H.C.; Register, A.C.; Rosenthal, W.; Hoffmann-Petersen, I.T.; Mehdizadeh, M.; Ghosh, R.; Wang, L.; Colon-Negron, K.; et al. Targeting ABL-IRE1α Signaling Spares ER-Stressed Pancreatic β Cells to Reverse Autoimmune Diabetes. *Cell Metab.* **2017**, *25*, 883–897. [CrossRef] [PubMed]
65. Colicelli, J. ABL Tyrosine Kinases: Evolution of Function, Regulation, and Specificity. *Sci. Signal* **2010**, *3*, re6. [CrossRef]
66. Nagar, B.; Hantschel, O.; Young, M.A.; Scheffzek, K.; Veach, D.; Bornmann, W.; Clarkson, B.; Superti-Furga, G.; Kuriyan, J. Structural Basis for the Autoinhibition of C-Abl Tyrosine Kinase. *Cell* **2003**, *112*, 859–871. [CrossRef] [PubMed]
67. Chen, Y.; Dorn, G.W. PINK1-Phosphorylated Mitofusin 2 Is a Parkin Receptor for Culling Damaged Mitochondria. *Science* **2013**, *340*, 471–475. [CrossRef]
68. Zhang, P.; Ford, K.; Sung, J.H.; Suzuki, Y.; Landherr, M.; Moeller, J.; Chaput, I.; Polina, I.; Kelly, M.; Nieto, B.; et al. C-Src-Dependent Phosphorylation of Mfn2 Regulates Endoplasmic Reticulum-Mitochondria Tethering. *bioRxiv* **2022**. [CrossRef]
69. Burté, F.; Carelli, V.; Chinnery, P.F.; Yu-Wai-Man, P. Disturbed Mitochondrial Dynamics and Neurodegenerative Disorders. *Nat. Rev. Neurol.* **2015**, *11*, 11–24. [CrossRef] [PubMed]
70. Pyakurel, A.; Savoia, C.; Hess, D.; Scorrano, L. Extracellular Regulated Kinase Phosphorylates Mitofusin 1 to Control Mitochondrial Morphology and Apoptosis. *Mol. Cell* **2015**, *58*, 244–254. [CrossRef]
71. Li, Y.-J.; Cao, Y.-L.; Feng, J.-X.; Qi, Y.; Meng, S.; Yang, J.-F.; Zhong, Y.-T.; Kang, S.; Chen, X.; Lan, L.; et al. Structural Insights of Human Mitofusin-2 into Mitochondrial Fusion and CMT2A Onset. *Nat. Commun.* **2019**, *10*, 4914. [CrossRef] [PubMed]
72. de Brito, O.M.; Scorrano, L. Mitofusin 2 Tethers Endoplasmic Reticulum to Mitochondria. *Nature* **2008**, *456*, 605–610. [CrossRef] [PubMed]
73. de Brito, O.M.; Scorrano, L. Mitofusin-2 Regulates Mitochondrial and Endoplasmic Reticulum Morphology and Tethering: The Role of Ras. *Mitochondrion* **2009**, *9*, 222–226. [CrossRef] [PubMed]
74. Filadi, R.; Greotti, E.; Turacchio, G.; Luini, A.; Pozzan, T.; Pizzo, P. Mitofusin 2 Ablation Increases Endoplasmic Reticulum–Mitochondria Coupling. *Proc. Natl. Acad. Sci. USA* **2015**, *112*, E2174–E2181. [CrossRef]

antioxidants

Article

Modulation of Kynurenic Acid Production by N-acetylcysteine Prevents Cognitive Impairment in Adulthood Induced by Lead Exposure during Lactation in Mice

Paulina Ovalle Rodríguez [1,2,†], Daniela Ramírez Ortega [1,†], Tonali Blanco Ayala [1,†], Gabriel Roldán Roldán [3], Gonzalo Pérez de la Cruz [4], Dinora Fabiola González Esquivel [1], Saúl Gómez-Manzo [5], Laura Sánchez Chapul [6], Aleli Salazar [7], Benjamín Pineda [7] and Verónica Pérez de la Cruz [1,*]

[1] Neurochemistry and Behavior Laboratory, National Institute of Neurology and Neurosurgery "Manuel Velasco Suárez", Mexico City 14269, Mexico; paulina.ovalle.rodriguez@gmail.com (P.O.R.); drmz_ortega@hotmail.com (D.R.O.); tblanco@innn.edu.mx (T.B.A.); dinora.gonzalez@innn.edu.mx (D.F.G.E.)

[2] Posgrado en Ciencias Bioquímicas, Universidad Nacional Autónoma de México, Ciudad Universitaria, Unidad de Posgrado, Mexico City 04510, Mexico

[3] Laboratorio de Neurobiología de la Conducta, Departamento de Fisiología, Facultad de Medicina, Universidad Nacional Autónoma de México, Mexico City 04510, Mexico; gabaergico@gmail.com

[4] Department of Mathematics, Faculty of Sciences, Universidad Nacional Autónoma de México UNAM, Mexico City 04510, Mexico; gonzalo.perez@ciencias.unam.mx

[5] Laboratorio de Bioquímica Genética, Instituto Nacional de Pediatría, Secretaría de Salud, Mexico City 04530, Mexico; saulmanzo@ciencias.unam.mx

[6] Neuromuscular Diseases Laboratory, Clinical Neurosciences Division, National Institute of Rehabilitation "Luis Guillermo Ibarra Ibarra", Mexico City 14389, Mexico; lausanchez@inr.gob.mx

[7] Neuroimmunology Department, National Institute of Neurology and Neurosurgery "Manuel Velasco Suárez", Mexico City 14269, Mexico; aleli.salazar@innn.edu.mx (A.S.); benjamin.pineda@innn.edu.mx (B.P.)

[*] Correspondence: veped@yahoo.com.mx; Tel.: +52-5556063822 (ext. 2006)

[†] These authors contributed equally to this work.

Citation: Ovalle Rodríguez, P.; Ramírez Ortega, D.; Blanco Ayala, T.; Roldán Roldán, G.; Pérez de la Cruz, G.; González Esquivel, D.F.; Gómez-Manzo, S.; Sánchez Chapul, L.; Salazar, A.; Pineda, B.; et al. Modulation of Kynurenic Acid Production by N-acetylcysteine Prevents Cognitive Impairment in Adulthood Induced by Lead Exposure during Lactation in Mice. *Antioxidants* **2023**, *12*, 2035. https://doi.org/10.3390/antiox12122035

Academic Editor: Waldo Cerpa

Received: 4 November 2023
Revised: 16 November 2023
Accepted: 22 November 2023
Published: 23 November 2023

Abstract: Lead (Pb^{2+}) exposure during early life induces cognitive impairment, which was recently associated with an increase in brain kynurenic acid (KYNA), an antagonist of NMDA and alpha-7 nicotinic receptors. It has been described that N-acetylcysteine (NAC) favors an antioxidant environment and inhibits kynurenine aminotransferase II activity (KAT II, the main enzyme of KYNA production), leading to brain KYNA levels decrease and cognitive improvement. This study aimed to investigate whether the NAC modulation of the brain KYNA levels in mice ameliorated Pb^{2+}-induced cognitive impairment. The dams were divided into four groups: Control, Pb^{2+}, NAC, and Pb^{2+}+NAC, which were given drinking water or 500 ppm lead acetate in the drinking water ad libitum, from 0 to 23 postnatal days (PNDs). The NAC and Pb^{2+}+NAC groups were simultaneously fed NAC (350 mg/day) in their chow from 0 to 23 PNDs. At PND 60, the effect of the treatment with Pb^{2+} and in combination with NAC on learning and memory performance was evaluated. Immediately after behavioral evaluation, brain tissues were collected to assess the redox environment; KYNA and glutamate levels; and KAT II activity. The NAC treatment prevented the long-term memory deficit exhibited in the Pb^{2+} group. As expected, Pb^{2+} group showed redox environment alterations, fluctuations in glutamate levels, and an increase in KYNA levels, which were partially avoided by NAC co-administration. These results confirmed that the excessive KYNA levels induced by Pb^{2+} were involved in the onset of cognitive impairment and could be successfully prevented by NAC treatment. NAC could be a tool for testing in scenarios in which KYNA levels are associated with the induction of cognitive impairment.

Keywords: heavy metals; kynurenic acid; cognition; N-acetylcysteine

1. Introduction

Lead (Pb^{2+}) is a widely utilized metal in the industry due to its malleability, conductivity, and ductility and usefulness as raw material in countless processes [1–3]. Due to the non-biodegradable nature of Pb^{2+}, it can accumulate in soil, rivers, lakes, and air, which are the main sources of Pb^{2+} exposure for humans. The World Health Organization established that there is no safe concentration of Pb^{2+} in the blood, as even a low amount of Pb^{2+} has been related to cognitive impairment, particularly in children [4–6]. This heavy metal can enter the organism by ingestion, inhalation, and absorption, and once inside, it is distributed to the organs and stored in the teeth and bones [7–12]. The central nervous system (CNS) is a target of Pb^{2+} toxicity, leading to long-term consequences [5,13,14]. Several clinical and experimental studies have observed that Pb^{2+} effects depend on the period, duration, and route of metal exposure, being more severe and lasting longer when it occurs during the neurodevelopmental period and/or early life, during which several processes occur, such as neurogenesis, neural migration, differentiation, synaptic pruning, neuronal plasticity, and neuronal connections establishment [15–18].

Several detrimental effects induced by Pb^{2+} exposure, such as calcium-signaling alteration, energetic homeostasis disruption, oxidative stress, inflammation, and neurotransmitter fluctuations, can trigger long-lasting cognitive impairment [19–23]. We previously proposed that kynurenic acid (KYNA) could play a relevant role in Pb^{2+} neurotoxicity [24]. KYNA is an endogenous metabolite derived from the tryptophan catabolism through the kynurenine pathway (KP), mainly synthesized in astrocytes by the kynurenine aminotransferase II (KAT II) and considered a neuromodulator due its inhibitory effects on NMDA and α-7-nicotinic receptors [25–27]. In this context, we previously showed that an increase in brain KYNA levels correlated with long-term memory impairment in adult mice exposed to Pb^{2+} during the lactation period (0–23 PNDs) [24]. Accordingly, several studies have demonstrated that the elevation of brain KYNA levels during gestation or early postnatal life led to memory deficits in adulthood [28–31]. These studies showed that the stimulation of brain KYNA neosynthesis (administrating its precursor L-kynurenine or inhibiting the long branch of KP) resulted in a reduction in extracellular glutamate levels [27,32,33], while its inhibition induced the opposite effect [26]. Additionally, other important neurotransmitters, such as dopamine, acetylcholine, and gamma-aminobutyric acid, increased when brain KYNA was reduced [25,33–38]. Taken together, this experimental evidence demonstrated that the manipulation of endogenous KYNA levels had critical effects on different neurotransmitter systems. However, to our knowledge, the influence of brain KYNA modulation within the context of heavy metal exposure during early life had not been previously described.

In this study, N-acetylcysteine (NAC), a mucolytic and a redox modulator, was used as a tool to regulate KYNA production since it had been previously shown that NAC inhibited KAT II activity in human and rat brain tissues [39–41]. It is worth mentioning that NAC had been tested for various neurodegenerative and psychiatric diseases, showing beneficial effects on cognitive performance [42–45]. Furthermore, several preclinical studies showed NAC effectiveness in ameliorating the brain damage induced by diverse neurotoxic agents; however, its protective effects have always been related to the induction of glutathione (GSH) synthesis promotion as well as to its scavenging and anti-inflammatory properties. In view of previous findings that had shown that the subchronic administration of NAC prevented memory impairment induced by increased brain KYNA levels [46], we aimed to explore whether the inhibitory modulation of brain KYNA levels by NAC administration during early-life Pb^{2+} exposure could prevent cognitive impairments in adulthood.

2. Materials and Methods

2.1. Materials

L-kynurenine (kyn), kynurenic acid (KYNA), NAC, pyruvate, pyridoxal-5-phosphate (P5P), glutathione reduced form (GSH), oxidized glutathione (GSSG), O-phtaldehyde (OPA), N-Ethylmaleimide (NEM), diethylenetriamine pentaacetate (DTPA), glucose-6-phosphate (G-6P), and glucose-6-phosphate dehydrogenase (G-6PDH) were obtained from Sigma-Aldrich Company (St. Louis, MO, USA). All other chemicals were of the highest commercially available purity. Solutions were prepared using deionized water obtained from a Milli-Q (Millipore, Burlington, MA, USA) purifier system.

2.2. Animals

Female C57 mice were housed 1:1 with a male mouse in individual acrylic cages. Successful mating was confirmed by the presence of mouse sperm on vaginal swabs. Once pregnancy was confirmed, the male was removed from the cage. Dams were housed individually until the day of birth and remained with their pups until weaning at postnatal day (PND) 23. At birth, litters were randomly assigned to one of four groups: (1) Control, (2) Pb^{2+}, (3) NAC, and (4) Pb^{2+} + NAC. At weaning, pups were separated by sex and housed by treatment. At PND 60, and once all the treatments were completed, pups were evaluated behaviorally and biochemically.

2.3. Lead Exposure and N-acetyl-L-cysteine Treatment

Dams and their offspring were randomly divided into 4 groups: (1) Control group, received normal drinking water; (2) Pb^{2+}, received 500 ppm lead acetate in drinking water; (3) NAC, received 350 mg/day of NAC in the food; and (4) Pb^{2+} + NAC, received 500 ppm lead acetate in drinking water and 350 mg/day of NAC in the food. The treatment period was from 0 to 23 PNDs. At weaning, the treatments were withdrawn and replaced with normal drinking water and a normal chow diet.

2.4. Behavioral Test

2.4.1. Buried Food Location Test (BFLT)

BFLT is an adaptation of the model described by Lehmkuhl et al. for olfactory dysfunction evaluation [47]. This test consists of two sessions: training and memory retrieval testing, both performed in an acrylic box (42 × 30 cm), covered with a 2 cm deep layer of sawdust. Mice were habituated twice to the box for 10 min, 24 h, and 48 h, before training. The box had spatial clues (black geometric figures of 8 × 8 cm, placed at a height of 13 cm in the middle of each side of the box). Mice were also habituated to eat sugary pellets ("fruit loops") in their home cages for seven days before training in order to eliminate food neophobia. Animals were fasted 24 h before the training session, with ad libitum water access. The training session, or learning phase, consisted of 6 trials (2 min inter-trial interval) and a "0" trial, during which the "fruit loops" were present. In trial "0", a fruit loop was placed in plain sight on the sawdust in a fixed position, where the mouse could find it and gnaw on it for a few seconds. If mice were unable to find the fruit loop within 180 s in the 0 trial, they were gently guided to it; in each trial, the mice were allowed to gnaw the pellet for at least 5 s. In the next 6 trials, the pellet was buried 1 cm under the sawdust in the same fixed quadrant of the box (the location of the pellet was the same in all trials). At the end of each trial, the animals were returned to their home cage, the testing area was cleaned with a 10% ethanol solution, and the sawdust was removed to eliminate odoriferous marks left by the mouse in the previous trial. After 24 h, long-term memory was evaluated in a retention test where the pellet was no longer present; mice were allowed to freely explore the testing box for 180 s. The time spent reaching the precise target location of the buried food (same location as during the training session) was recorded, and the exploration time spent in the target location quadrant was measured. All sessions were video recorded, allowing offline analysis with tracking software (ImageJ 1.54d version,

Bethesda, MD, USA). These results were expressed as the time (in seconds) and distance (in centimeters) to reach the target.

2.4.2. Novel Object Recognition Test (NOR)

The NOR test is based on a behavioral phenomenon called novelty preference in rodents [48]. NOR test was performed in an acrylic box (42 × 42 × 36 cm) covered with white paper and a 2 cm deep layer of sawdust where mice were habituated for 10 min, 24 h, and 48 h, prior to test. The test consisted of two phases: training and probing. In the training phase, two identical objects (A and A′) were placed at the center of the box, equally spaced. The mice were released into the box, starting from the center of the wall with their backs to the objects, and allowed to explore them for 5 min. The first part of the probe phase (short-term memory evaluation) was carried out two hours after the training session, one of the familiar objects was changed for a novel object "B", and mice were reintroduced into the box and allowed to explore for another 5 min. The second probe phase to evaluate long-term memory was performed 24 h later in the same way, but object "B" was changed to a novel object "C". The sessions were recorded, and the exploration time (sniffing and manipulating the objects) was registered. The results were expressed as a recognition index (time exploring a novel object/time exploring both objects × 100).

2.5. Tissue Collection and Treatment

Immediately after behavioral testing, mice were euthanized by decapitation. Brains were rapidly removed and placed on ice. Brain tissue (20 mg) was treated immediately for GSH-level quantification, and the remaining tissue was rapidly frozen for posterior analysis. To quantify kynurenic acid levels, tissues were homogenized in water (1:10, w/v), deproteinized with 30 μL of perchloric acid (PCA), and finally, centrifuged at 14,600× g for 10 min.

2.6. GSH and GSSG Levels

Brain tissue was homogenized (1:10, p/v) in Buffer A (154 mM KCl, 5 mM DTPA, and 0.1 M potassium phosphate buffer (PPB) pH 6.8), and then, Buffer B (20 mM ascorbic acid, 10 mM DTPA, 40 mM HCl and 10% trichloroacetic acid) was added (1:1 buffer A). Homogenates were centrifuged at 14,000× g for 10 min, and supernatants were filtered (0.22 μm). For GSH determination, OPA was added to the supernatant where it formed an isoindole with GSH that could be fluorometrically detected. To quantify GSSG levels, extra steps were performed: (1) GSH neutralization by mixing supernatants with NEM (7.5 mM), and (2) GSSG levels were reduced to GSH with 100 mM of sodium hydrosulfite; finally, GSSG was detected as GSH for its quantification as an isoindole when adding OPA. The final product was quantified fluorometrically at 370 nm excitation and 420 nm emission in a Synergy HTX microplate reader spectrophotometer (BioTek Instruments, Winooski, VT, USA). GSH and GSSH levels were expressed as nmoles/g of tissue.

2.7. Lipoperoxidation

Brain tissue was used to evaluate lipid peroxidation through thiobarbituric-acid-reactive species (TBA-RS). Briefly, tissues were homogenized in Buffer Krebs (NaCl (19 mM), KCl (5 mM), $CaCl_2$ (2 mM), $MgSO_4$ (1.2 mM), glucose (5 mM), NaH_2PO_4 (13 mM), and Na_2HPO_4 (3 mM); pH 7.4). A total of 250 μL of TBA solution (100 mL: 2.54 mL HCl, 0.375 g of TBA, and 15 g of trichloroacetic acid (TCA)) were mixed with 125 μL of homogenate, in a final volume of 500 μL. Then, these homogenates were boiled for 15 min. Subsequently, they were placed on ice for 5 min and centrifugated 12,000× g for 5 min at 4 °C. Finally, MDA-TBA chromogen was determined at 532 nm using a microplate reader (Synergy™ HTX, Biotek Instruments, Winooski, VT, USA).

2.8. Kynurenic Acid Determination

KYNA levels were quantified fluorometrically (344 nm excitation and 398 nm emission) in a Series 200a detector (Perkin Elmer, Waltham, MA, USA). A total of 50 µL of supernatants were isocratically eluted at a flow rate of 1 mL/min onto a 3 µm C18 reverse-phase column (ZORBAX Eclipse XDB 5 µm, 4.6 mm × 150 mm; Agilent, Santa Clara, CA, USA), using a mobile phase consisting of 50 mM sodium acetate, 250 mM zinc acetate, and 3% acetonitrile; the pH 6.2. KYNA had a retention time of ~7 min.

2.9. Kynurenine Aminotransferase Activity

KAT II activity was evaluated in brain tissue homogenized 1:10 in homogenization buffer (Tris-base acetate buffer (0.5 M, pH 8), pyridoxal phosphate (P5P, 50 mM), and 2-mercaptoethanol (775 µL)). A total of 100 microliters of homogenates were mixed with 100 µL of the reaction cocktail, consisting of L-Kyn (100 µM), pyruvate (1 mM), and P5P (80 µM) in Tris-acetate buffer (150 mM, pH 7.4), and then incubated for 2 h at 37 °C in a shaking bath. A total of 20 microliters of trichloroacetic acid (50%) plus 1 mL of HCl (0.1 M) were added to samples to stop the reaction. Then, samples were centrifuged at 14,000× g for 10 min. KYNA, the enzymatic product of KAT II, was detected by fluorescence, as mentioned previously.

2.10. Glutamate Quantification

Briefly, brain tissue was homogenized in 40 volumes of 85% methanol with a Teflon homogenizer. The homogenates were centrifuged at 4000× g for 10 min at 4 °C. The supernatant was used to perform derivatization with OPA (50%/50%). The samples were eluted on a C18, 3 µm column (ZORBAX Eclipse XDB 5 µm, 4.6 × 150 mm; Agilent, Santa Clara, CA, USA) with a mobile phase consisting of 0.29% glacial acetic and 1.5% de tetrahydroflurane at pH 5.9. Glutamate levels were quantified using a reverse-phase HPLC method with a fluorescence detector (Perkin Elmer series 200a, Waltham, MA, USA) at an excitation wavelength of 390 nm and emission wavelength of 460 nm.

2.11. Protein Quantification

The Lowry method was used to quantify protein levels. Briefly, 10 µL homogenate (1:20 v/v with water) was used, C solution (1 mL, A + B solutions) was added (A (sodium tartrate (0.2%), Na_2CO_3 (2%) and NaOH (0.4%)) and B ($Cu(SO_4)_3$ (0.5%)), and then samples were incubated at room temperature for 10 min. Then, 100 µL of Folin's solution (50% v/v with water) was added. Samples were mixed and incubated at room temperature for 30 min. Synergy™ HTX microplate reader (Biotek Instruments, Winooski, VT, USA) was used to determine the absorbance at 550 nm.

2.12. Statistical Analysis

All data were expressed as the mean ± SEM. Comparisons between groups were performed using the Kruskal–Wallis test with Dunn's test for multiple pairwise comparisons. Pairwise comparisons of novel object-recognition-test data were analyzed using the Wilcoxon signed-rank test. Spearman's correlation was used to assess the association between variables. Statistical significance was set at $p < 0.05$. All statistics were calculated with Graph Prism 9.1.0. (GraphPad, San Diego, CA, USA).

3. Results

We had previously observed that mice exposure to Pb^{2+} during the lactation period induced cognitive impairment in adulthood, along with a rise in brain KYNA levels [24]. Taking into consideration that NAC is a modulator of KYNA production due to its KAT II inhibitory activity, we decided to test it as a tool to prevent the cognitive impairment induced by this heavy metal. Using the same experimental Pb^{2+} exposure protocol during the lactation period, the strategy consisted of co-administering NAC with Pb^{2+} and then evaluating the cognitive performance, glutamate, and redox environment, as well as the KYNA levels and KAT II activity.

3.1. NAC Attenuates Memory Dysfunction Induced by Pb^{2+} Exposure during Early Life

The first step was to use the BFLT to evaluate the cognitive performance of the adult mice (60 PNDs) that were exposed to Pb^{2+} alone or in combination with NAC during the lactation period. The acquisition session (Figure 1A,B) consisted of training mice to learn where the buried food was located. We found no significant differences between the experimental groups during the training session. However, in the Pb^{2+} group, it was observed that the animals showed a less pronounced tendency to learn the target location, since their final time during training was not reduced as markedly as in the other groups, which possibly suggested an impairment in the learning process. Interestingly, the NAC+Pb^{2+} group behaved more similarly to the controls than to the Pb^{2+} group. Furthermore, when long-term memory was evaluated (24 h after training), mice of the Pb^{2+} group took almost the same time (45 ± 8 s) and travel distance to reach the target (156 ± 37 cm) as before training (first trials of the acquisition phase), while the control group had similar times to those observed during the last trials of the acquisition phase (9.3 ± 2.4 s and 59.9 ± 2.5 cm, respectively), confirming that the early postnatal exposure to this heavy metal induced cognitive impairment. The NAC group showed no changes, as compared to the control group, when assessing long-term memory. For those administered with Pb^{2+} and NAC simultaneously, they showed improved long-term memory performance by reducing the time and distance required to reach the target (12 ± 2.2 s and 65.7 ± 8.5 cm, respectively), as compared to the Pb^{2+} group (Figure 1C and 1D, respectively).

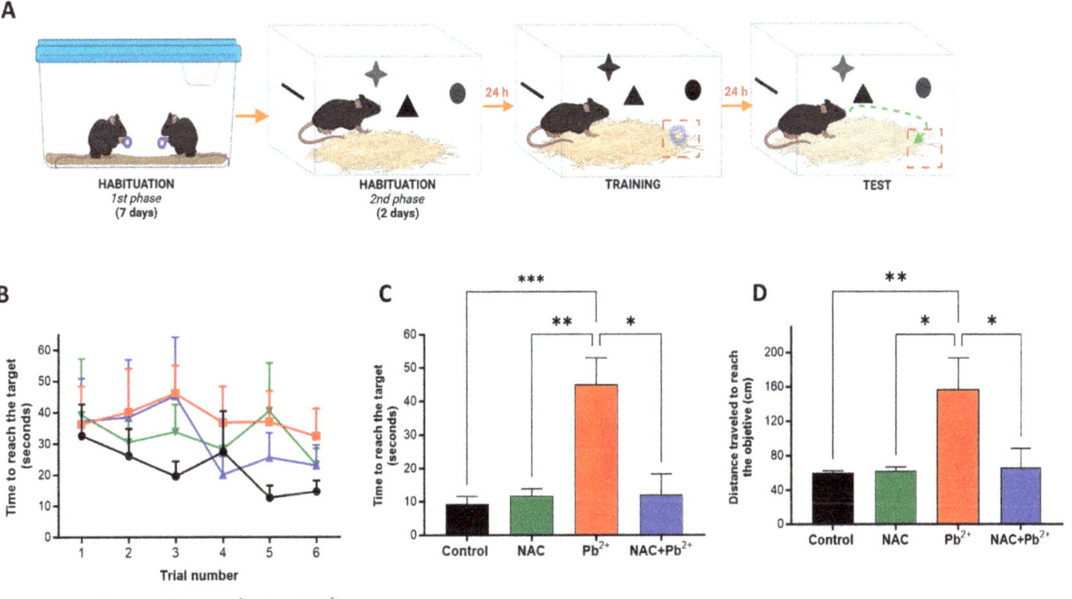

Figure 1. Effect of NAC administration on memory impairment induced by Pb^{2+} exposure. Learning (**B**) and memory (time to reach the target) (**C**) and distance to reach the target (**D**) were evaluated through the buried food location test (**A**) in all experimental groups at 60 PNDs. Data are the mean $\pm$ SEM (n = 8–10); * $p < 0.01$, ** $p < 0.001$, and *** $p < 0.0001$, based on the Kruskal–Wallis test with Dunn's test for pairwise comparisons.

The novel object recognition (NOR) test was another strategy used to evidence the NAC effect on the cognitive impairment induced by Pb^{2+} exposure. This test was based on the innate preference of the mice for novelty [49]. We first confirmed that mice from all the groups spent an equal amount of time exploring two identical objects, A and A' (Figure 2, sample phase). Next, when the short-term memory was evaluated, all groups except for the Pb^{2+} group showed that the mice were able to distinguish the novel object from the familiar one, while the Pb^{2+} group was unable to discriminate between both objects though they had the same exposure in the sample phase (Figure 2, short-term memory (STM)). When the long-term memory (LTM) was evaluated, as expected, the control group explored the novel object for longer periods of time than the familiarized object as evidenced by a significant increase in the discrimination index, while the Pb^{2+} group could not discriminate between objects (Figure 2, LTM). The long-term memory impairment induced by Pb^{2+} exposure was abolished by the simultaneous administration of NAC, indicating that NAC had prevented deficits in long-term memory.

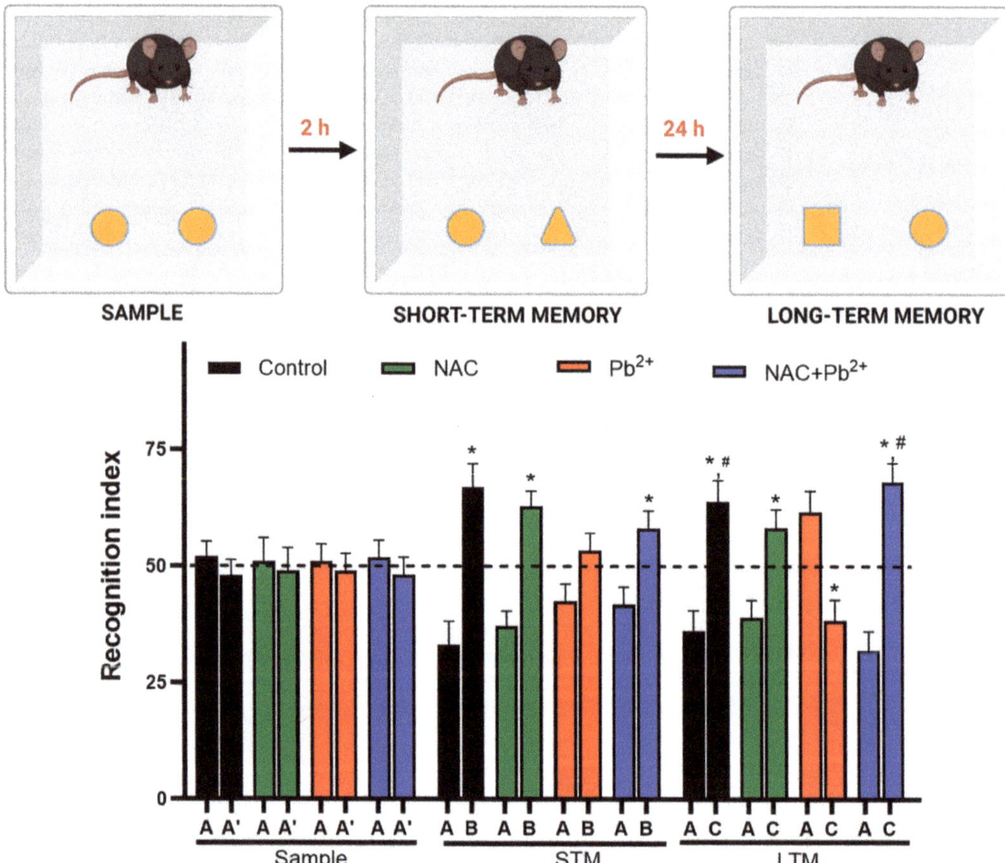

Figure 2. Effect of simultaneous administration of NAC and Pb^{2+} during lactation on the evaluation of short-term memory (STM) and long-term memory (LTM) through novel object recognition test. The recognition index (time exploring novel object/time exploring both objects × 100) was calculated for 8–10 animals per group. The data represent mean ± SEM; * $p < 0.05$ between the novel and familiar objects for each group and phase based on the Wilcoxon signed-rank test; and [#] $p < 0.001$ vs. Pb^{2+} based on the Kruskal–Wallis test with Dunn's test for pairwise comparisons.

3.2. Fluctuations in Brain KYNA Levels Are Prevented by NAC Administration

As previously described, the cognitive impairment induced by the Pb^{2+} administration correlated with an elevation of the brain KYNA levels; here, we aimed to examine whether NAC could prevent Pb^{2+}-induced cognitive impairment by reducing the brain KYNA levels inhibiting KAT II activity. Therefore, we first determined the effect of NAC on the KYNA levels and the KAT II activity in those adult mice (60 PNDs) that were exposed to Pb^{2+} during lactation. As shown in Figure 3A, no significant changes in the KAT II activity were observed in the NAC and Pb^{2+} groups, as compared to the controls. Conversely, in the NAC+Pb^{2+} group, a significant reduction (around 40%) in the KAT II activity was found in the brain tissue. As expected, in the Pb^{2+} group, the brain KYNA levels increased (about 2-fold), as compared to the control group, whereas a non-significant reduction was found in the brain KYNA levels in the NAC group. Furthermore, the increase in the brain KYNA levels induced by Pb^{2+} exposure was prevented by NAC treatment, and this effect could have been partially due to the KAT II inhibition shown in the same group.

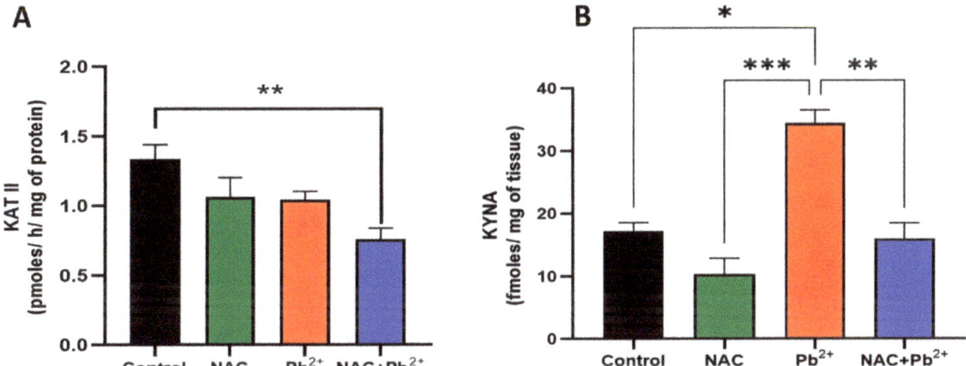

Figure 3. Effect of simultaneous administration of NAC and Pb^{2+} on KAT II activity and brain KYNA levels in mice at 60 PNDs. NAC and Pb^{2+} were administrated during the lactation period; after that, the animals were administrated tap water and a standard mice diet until 60 PNDs. The brain cortex was used to evaluate KAT II activity (**A**) and KYNA levels (**B**). Data represents mean $\pm$ SEM of 7–10 animals per group. * $p < 0.01$, ** $p < 0.001$, and *** $p < 0.0001$ based on the Kruskal-Wallis test with Dunn's test for pairwise comparisons.

Based on our data that NAC could prevent the cognitive impairment induced by Pb^{2+}, we speculated if the pro-cognitive effect exerted by NAC on Pb^{2+}-toxicity was directly correlated to the brain KYNA-level modulation. To answer this question, we analyzed the learning and memory performance parameters (time and distance to reach the target) obtained from the BFLT results of all the groups with their respective brain KYNA levels (Figure 4). A positive association between the brain KYNA levels and the long-term memory performance was found, thus suggesting that NAC improved cognitive performance, at least in part, by modulating the brain KYNA levels.

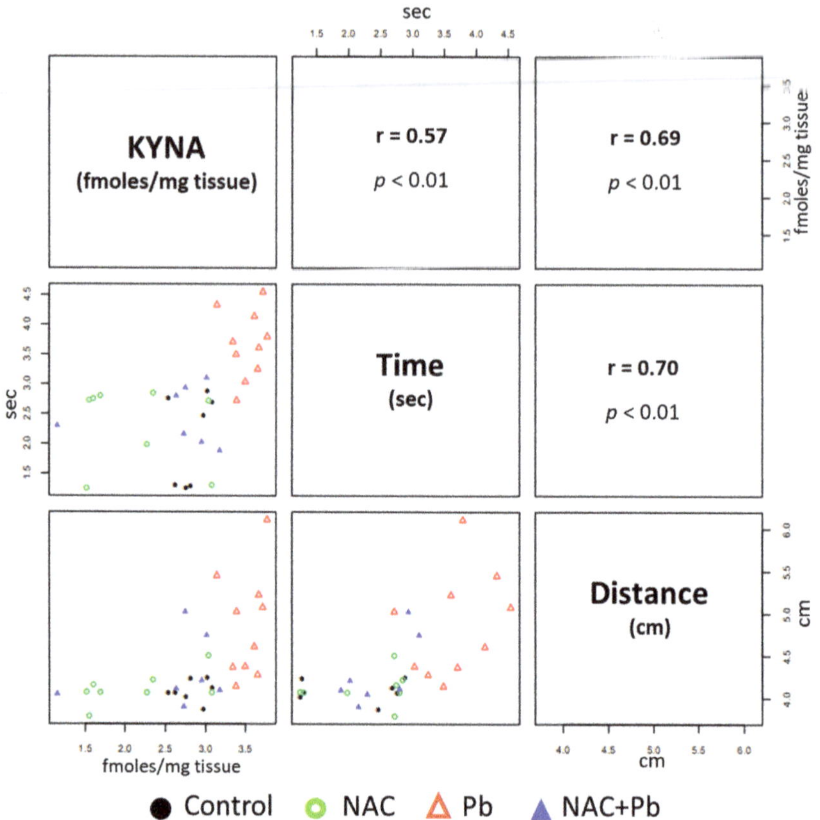

Figure 4. Correlation between long-term memory and brain KYNA levels. The lower triangular matrix contains the scatterplot for each pair of variables for all groups (Control: black circle; NAC: green circle; Pb^{2+}: red triangle; and NAC+Pb^{2+}: blue triangle). The upper triangular matrix contains the Spearman's rank correlation coefficient (r) and its associated *p*-value (*p*) (n = 7–10 per group).

3.3. Effect of NAC Co-Administration on Brain Glutamate Level Fluctuations Induced by Pb^{2+} Exposure

We had confirmed the involvement of KYNA as a part of the mechanisms by which Pb^{2+} exposure induces cognitive impairment. We also knew that this metabolite was a neuromodulator of glutamatergic transmission since previous in vivo studies had indicated that in the hippocampus and prefrontal cortex, where KYNA had decreased glutamate levels, presumably via the inhibition of $\alpha7$ nicotinic receptors (nAChRs) [32,33]. As shown in Figure 5, the Pb^{2+} group showed a trend of decreasing glutamate levels while NAC itself did not have any effect on the glutamate levels. Interestingly, the simultaneous administration of NAC with Pb^{2+} exposure had increased the glutamate levels, as compared to the Pb^{2+}, suggesting that NAC had prevented a reduction in the marginal glutamate levels induced by Pb^{2+}.

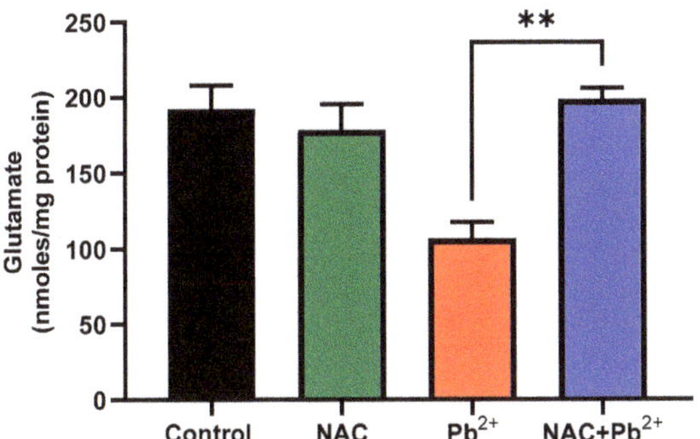

Figure 5. Effect of NAC on glutamate levels when exposed to Pb^{2+} during lactation. Brain levels of glutamate were evaluated in all the experimental groups at 60 PNDs. Data represent mean ± SEM of 4–5 animals per group. ** $p < 0.001$ based on the Kruskal-Wallis test with Dunn's test for pairwise comparisons.

3.4. Redox Environment Alteration Induced by Pb^{2+} Exposure Is Prevented by NAC Administration

In addition, since the KYNA formation could be carried out via non-canonical pathways that involved a cellular redox state, we expected that the pro-oxidant effects of Pb^{2+} would increase the amount of ROS during Pb^{2+} administration and, confirming our previous data, induce non-enzymatic KYNA production [50,51]. Comparatively, here, our hypothesis was that NAC treatment, given its antioxidant profile during Pb^{2+} administration, would reduce the concentration of these ROS and, thus, the non-enzymatic production of KYNA. To address this objective and considering the short half-life of these ROS, we assessed a GSH/GSSH ratio and lipid peroxidation levels across all experimental groups (Figure 6A and 6B, respectively). The GSH/GSSG ratio was marginally reduced in the Pb^{2+} group (around 38%) while the simultaneous administration of Pb^{2+} with NAC increased the GSH/GSSG ratio, as compared to the Pb^{2+} group. Upon evaluating the lipid peroxidation, a prominent oxidative stress marker, the Pb^{2+} group exhibited an approximately twofold increase, as compared to the control group. However, this oxidative upsurge was effectively neutralized when NAC was administered in conjunction with Pb^{2+}, underscoring NAC's potential in mitigating the Pb^{2+}-induced oxidative stress and its consequent impact on the cellular redox environment and the non-enzymatic KYNA production.

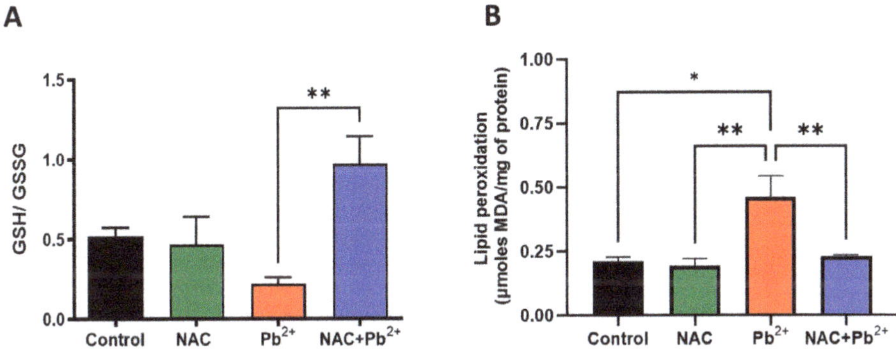

Figure 6. Effect of NAC in the pro-oxidant environment induced by Pb^{2+}. Brain levels of GSH/GSSG ratio (**A**) and lipid peroxidation (**B**) were evaluated in all the experimental groups at 60 PNDs. Data represent mean ± SEM of 4–5 animals per group. * $p < 0.01$ and ** $p < 0.001$ based on Kruskal-Wallis test with Dunn's test for pairwise comparison.

4. Discussion

The goal of the present study was to investigate whether the NAC modulation of the brain KYNA levels could mitigate the cognitive impairment induced by Pb^{2+} exposure during early postnatal life. This approach on the modulation of brain KYNA levels was based on our previous findings that had shown a correlation between a pronounced increase in the brain KYNA levels and the Pb^{2+}-induced long-term memory impairment during the lactation period, in a mouse model [24].

The important role of KYNA on neurotransmission has been extensively demonstrated in different rodent models; hence, we now know that KYNA influences GABAergic, cholinergic, glutamatergic, and dopaminergic neurotransmission [25,36–38,52]. However, among this range of neuromodulatory functions, the effect of KYNA on glutamatergic neurotransmission is the most thoroughly understood. Experimental evidence has shown that fluctuations in the brain KYNA levels substantially reduced the extracellular glutamate levels in the different brain areas, including those related to cognitive processes, such as the hippocampus and the prefrontal cortex [27,32,33]. In fact, KYNA has been used as a tool in several experimental models to reduce or block glutamate-mediated neurotransmission. Specifically, experimental manipulations to increase the brain levels of KYNA have shown that nanomolar or low micromolar elevations of this metabolite induced a wide range of cognitive impairments, including disrupting hippocampus-mediated contextual learning and memory; working memory; and contextual fear memory [32,53]. In addition to pharmacological manipulations, some exogenous stimuli have been known to induce an increase in the brain KYNA levels. Similarly, we previously demonstrated that Pb^{2+} exposure during lactation had increased the brain levels of KYNA when evaluated at 23 PNDs and 60 PNDs [24]. Our data suggested that the elevation in the KYNA levels could be a mechanism by which Pb^{2+} exposure during lactation induced cognitive impairment in adult mice (60 PNDs). Thus, if the elevated brain KYNA levels were a key mechanism in the induction of cognitive deficits, reducing the KYNA levels in this same model could prevent the observed alterations in learning and memory during adulthood.

To reduce the KYNA levels, we decided to use NAC as a pharmacological tool, as this compound had been previously shown to inhibit the main enzyme for KYNA synthesis, KAT II [39]. Therefore, to address the effect of reducing the KYNA levels by NAC on Pb^{2+}-induced cognitive deficits, we first confirmed that Pb^{2+} induced long-term memory alterations in adult mice when exposed during lactation. As expected, when NAC was co-administered with Pb^{2+}, it successfully mitigated the Pb^{2+}-induced cognitive impairment in both cognitive tests performed. The next question was to investigate whether this cognitive improvement could be related to a reduction in the brain KYNA levels modulated by NAC.

Under our experimental conditions and as we had previously shown, Pb^{2+} exposure during lactation induced a substantial increase in KYNA levels when the brain tissue was examined after the cognitive testing at 60 PNDs. This increase was successfully prevented by the NAC administration, as shown in the brain tissue of those animals co-administered with NAC during Pb^{2+} exposure. In this group, KAT II activity was significantly reduced in the brain tissue; this could partly explain the reduction in the KYNA levels. Moreover, when we analyzed Pb^{2+} and NAC groups separately, no significant effect on the KATII activity was shown [24]. In the case of the NAC+Pb^{2+} group, the potential explanation for our findings was that NAC had previously been demonstrated to reduce glial cell activation, mainly reactive astrocytes, which were the major cells expressing KAT II [54]. Accordingly, in cultured astrocytes, it has been shown that GFAP, a protein marker for reactive astrocytes, was not as overexpress after Pb^{2+} treatment as it was in response to Pb^{2+}-induced neuronal damage, indicating that astrogliosis was more likely a secondary reaction to this event [55]. Therefore, if neuronal damage was prevented, astrogliosis would also be reduced. Additionally, NAC has been observed to promote neuronal differentiation, thereby potentially improving the neuron/glia ratio, which could have subsequent effects on the KAT II activity due to its cellular distribution [56]. In the NAC-only group, it was also important to consider that the observed effect on the KAT II activity was measured

30 days after the cessation of the NAC treatment; we could not rule out a direct inhibitory effect on KAT II in this experimental group during the NAC administration (lactation). Future experiments addressing the activity of this enzyme at different times of the treatment should be performed to determine the mechanism of the KYNA reduction by NAC more precisely under these basal conditions.

Therefore, these results not only confirmed that the NAC administration during the lactation period could efficiently prevent an increase in the brain KYNA levels induced by Pb^{2+} exposure, but that this effect was also sustained into adulthood, where it translated into the ameliorating cognitive impairments observed in the Pb^{2+} group. This was consistent with the experimental evidence showing that KAT II inhibition, pharmacologically or by genetic manipulation (i.e., KATII knockout mice), had pro-cognitive effects [32,57]. Moreover, consistent with our findings, it was reported that NAC supplementation had prevented cognitive impairment associated with transient increases in the brain KYNA levels [46].

The detrimental cognitive effects of increased brain KYNA during gestational or early postnatal stages have been believed to stem from its ability to modulate glutamatergic neurotransmission [29]. Glutamate, the predominant excitatory neurotransmitter in the CNS, mediated 70–90% of synaptic transmission and played a vital role in learning and memory, as well as synaptic plasticity [58,59]. As mentioned previously, there has been plenty of evidence suggesting that elevated brain KYNA reduced glutamate levels (around 30–50%) [27,32,33], as observed in this study where postnatal Pb^{2+} exposure resulted in elevated brain KYNA levels while the glutamate in the brain tissue decreased. The NAC administration had normalized both the brain KYNA and glutamate fluctuations induced by Pb^{2+}. Our results also confirmed that the pro-cognitive effects of the KYNA reduction induced by NAC preserved the glutamate levels in the brain tissue, thus allowing the prevention of the alterations in neurotransmission and, consequently, in the cognitive performance impairments related to the excessive Pb^{2+}-induced KYNA production. On the other hand, it is noteworthy to mention that most of the pro-cognitive experimental manipulations of KYNA had been described under physiological conditions. Here, we described that the reduction in the KYNA levels, even within the context of Pb^{2+}-induced neurotoxicity, could be beneficial for cognitive processes. Furthermore, if we considered that the detrimental effects of Pb^{2+} exposure on the CNS have been well documented, particularly when exposure had occurred in early life, which is a critical period for establishing proper communication between brain cells [60–62], the pro-cognitive effect induced by the KYNA manipulation in the lactation period during Pb^{2+} exposure was more relevant. This kind of modulation could also potentially be translatable to humans since the prenatal concentrations of KYNA in the human brain are much higher than during adulthood, suggesting a key role as a direct modulator of glutamatergic neurotransmission in the developing brain [63].

Separately, we should mention that NAC is also a modulator of the redox environment [64,65]. This antioxidant profile was relevant within the context of non-canonical production of KYNA, where its synthesis was promoted by the direct interaction of its precursor, L-kynurenine, with ROS and free radicals [50]. Given that Pb^{2+} exposure has been known to promote oxidative stress by increasing ROS and disrupting the antioxidant balance, it is conceivable that KYNA could be produced via this non-canonical pathway. To verify whether these pro-oxidants conditions were promoting this pathway, we evaluated markers of oxidative stress, including lipid peroxidation and a GSH/GSSG ratio. As anticipated, NAC prevented the Pb^{2+}-induced lipid peroxidation and significantly increased the GSH/GSSG ratio, as compared to the Pb^{2+} group. These findings suggested that NAC helped to mitigate the ROS levels, consequently reducing the likelihood of KYNA being produced through the non-canonical pathway.

In this context, it had been observed that the pre-treatment with NAC before exposure to a pro-oxidant agent enabled brain tissue to better withstand oxidative damage, as compared to when no supplementation had been provided [46]. This indicated that subchronic NAC supplementation may prime or adapt the brain environment to better cope with subsequent adverse effects. Consistent with this, the simultaneous administration of NAC and Pb^{2+} resulted in reduced brain malondialdehyde levels (around 54% vs. Pb^{2+}), the upregulation of SOD gene expression, the downregulation of cell-death-related genes, and the improvement of the GSH/GSSG ratio [66,67]. Furthermore, in vitro studies had demonstrated that NAC had the capacity to bind Pb^{2+} [68,69]. This interaction may have contributed to the elimination of Pb^{2+} through feces and urine [70], providing another mechanism through which NAC exerted its protective effects against Pb^{2+}-induced oxidative stress and the subsequent KYNA production.

It is crucial to note that both cognitive performance and biochemical test were evaluated as the long-term effects of early-life KYNA modulation. As we mentioned previously, early-life Pb^{2+} exposure disrupts several essential brain processes [60–62]. Previous studied had shown that mice exposed to Pb^{2+} during early life exhibited increased brain KYNA levels and, later, demonstrated cognitive impairments in adulthood [24]. This suggested that even mild increases in the brain KYNA levels, initiated by Pb^{2+} exposure in the earliest postnatal days, may interfere with the establishment of neural networks. In our study, the NAC supplementation prevented the Pb^{2+} -induced increases in the brain KYNA, correlating with improved cognitive performance following Pb^{2+} exposure. These findings indicated that KYNA production was a mechanism through which Pb^{2+} induced neurotoxicity and highlighted KYNA modulation as a significant pathway through which NAC exerted its pro-cognitive effects. While numerous studies have underscored the protective effects of NAC in cognitive dysfunction across various experimental models and human pathologies, often attributing these benefits to its antioxidant or anti-inflammatory properties [71–74], our study confirmed KYNA modulation as a crucial mechanism [46], even in a pro-oxidant context, underlying NAC's pro-cognitive effects.

As detailed in this paper, numerous studies have sought to unravel the effects of NAC on Pb^{2+} neurotoxicity [67,75–77]. These studies collectively highlighted NAC's ability to improve cellular redox status, mitigate inflammatory responses, improve mitochondrial bioenergetics, support neurotransmission, and protect against Pb^{2+}-induced neuronal cell death. Consequently, a pertinent question arose: could the KYNA modulation observed in this study be a secondary effect of NAC's influence on these various factors? In our research, we have established a direct correlation between KYNA levels and cognitive outcomes, underscoring KYNA's significant role in the neurotoxicity triggered by Pb^{2+} exposure. Given that NAC has demonstrated efficacy in counteracting the factors that influence KYNA formation, it is plausible that NAC could hinder certain pathways leading to the synthesis of this tryptophan metabolite. However, it is important to acknowledge a limitation of our study: NAC is not a specific inhibitor of KAT II. This non-specificity makes it challenging to definitively ascertain how NAC precisely modulates only brain KYNA levels. Despite this limitation, our findings clearly indicated that NAC reduced KYNA levels, and this reduction was associated with cognitive improvements in the presence of Pb^{2+}. Another notable limitation was that all the parameters evaluated in this study were determined long-term, at 60 PNDs, following the cessation of both the Pb^{2+} exposure and the NAC treatment, at 23 PND. Future studies should aim to investigate KYNA modulation's impact, both during and immediately after Pb^{2+} exposure.

5. Conclusions

The comprehensive results of our study underscored the pivotal role of brain KYNA in mediating cognitive impairments induced by Pb^{2+} exposure during early life. We have established a direct correlation between elevated KYNA levels and compromised cognitive function, highlighting the neurotoxic potential of this tryptophan metabolite when dysregulated by heavy metal exposure. Also, our findings pave the way for further exploration of NAC as a multifaceted therapeutic agent, particularly in scenarios where dysregulated KYNA levels are implicated in cognitive impairments. The potential for NAC to modulate KYNA production, alongside its well-documented antioxidant and anti-inflammatory properties, positions it as a valuable candidate for intervention in a spectrum of cognitive disorders.

Author Contributions: Conceptualization, V.P.d.l.C. and B.P.; methodology, P.O.R., D.R.O., T.B.A., A.S., S.G.-M., L.S.C. and D.F.G.E.; validation, D.F.G.E., L.S.C., G.R.R. and S.G.-M.; formal analysis, G.P.d.l.C.; investigation, T.B.A., G.P.d.l.C., V.P.d.l.C. and P.O.R.; resources, V.P.d.l.C.; writing—original draft preparation, P.O.R., D.R.O., T.B.A. and V.P.d.l.C.; writing—review and editing, P.O.R., D.R.O., T.B.A., G.R.R., D.F.G.E., G.P.d.l.C., L.S.C., S.G.-M., A.S., B.P. and V.P.d.l.C.; visualization, V.P.d.l.C., P.O.R., A.S. and T.B.A.; supervision, D.F.G.E., G.R.R. and B.P. All authors have read and agreed to the published version of the manuscript.

Funding: This research was funded by CONACYT Grant 286885 (to V.P.C.).

Institutional Review Board Statement: Experimental protocol (Number 111/19) was approved by the Institutional Research Committee of the National Institute of Neurology and Neurosurgery and by the Institutional Committee for the care and use of laboratory animals. All procedures with animals were carried out according to regulations specified by the Bioethical Committee, and the Mexican Regulation for the production, care, and use of laboratory animals (NOM-159 062-ZOO-1999).

Informed Consent Statement: Not applicable.

Data Availability Statement: Data used to support the findings of this study are available from the corresponding author upon request.

Acknowledgments: Paulina Ovalle Rodríguez is a scholarship holder of CONACyT-México (1035837) in the Programa de Maestría en Ciencias Bioquímicas at the Universidad Nacional Autónoma de México.

Conflicts of Interest: The authors declare no conflict of interest.

References

1. Wani, A.L.; Ara, A.; Usmani, J.A. Lead toxicity: A review. *Interdiscip. Toxicol.* **2015**, *8*, 55–64. [CrossRef] [PubMed]
2. Lin, G.Z.; Wu, F.; Yan, C.H.; Li, K.; Liu, X.Y. Childhood lead poisoning associated with traditional Chinese medicine: A case report and the subsequent lead source inquiry. *Clin. Chim. Acta* **2012**, *413*, 1156–1159. [CrossRef] [PubMed]
3. Seki, Y.; Ohta, H. Lead. *Nihon Rinsho* **1999**, *57*, 290–293. [PubMed]
4. Rawat, P.S.; Singh, S.; Mahdi, A.A.; Mehrotra, S. Environmental lead exposure and its correlation with intelligence quotient level in children. *J. Trace Elem. Med. Biol.* **2022**, *72*, 126981. [CrossRef] [PubMed]
5. Rocha, A.; Trujillo, K.A. Neurotoxicity of low-level lead exposure: History, mechanisms of action, and behavioral effects in humans and preclinical models. *Neurotoxicology* **2019**, *73*, 58–80. [CrossRef] [PubMed]
6. Maidoumi, S.; Ouaziz, C.R.; Ouisselsat, M.; El Maouaki, A.; Loukid, M.; Lekouch, N.; Pineau, A.; Ahami, A.; Sedki, A. Iron deficiency and cognitive impairment in children with low blood lead levels. *Toxicol. Rep.* **2022**, *9*, 1681–1690. [CrossRef] [PubMed]
7. Mager, E.M.; Brix, K.V.; Grosell, M. Influence of bicarbonate and humic acid on effects of chronic waterborne lead exposure to the fathead minnow (*Pimephales promelas*). *Aquat. Toxicol.* **2010**, *96*, 135–144. [CrossRef] [PubMed]
8. Ahangar, H.; Karimdoost, A.; Salimi, A.; Akhgari, M.; Phillips, S.; Zamani, N.; Hassanpour, N.; Kolahi, A.A.; Krieger, G.R.; Hassanian-Moghaddam, H. Environmental assessment of pediatric Lead exposure in Tehran; a prospective cross-sectional study. *BMC Public Health* **2021**, *21*, 1437. [CrossRef]
9. Charkiewicz, A.E.; Backstrand, J.R. Lead Toxicity and Pollution in Poland. *Int. J. Environ. Res. Public. Health* **2020**, *17*, 4385. [CrossRef]
10. Harding, L.E.; Harris, M.L.; Elliott, J.E. Heavy and trace metals in wild mink (*Mustela vison*) and river otter (*Lontra canadensis*) captured on rivers receiving metals discharges. *Bull. Environ. Contam. Toxicol.* **1998**, *61*, 600–607. [CrossRef]
11. Papanikolaou, N.C.; Hatzidaki, E.G.; Belivanis, S.; Tzanakakis, G.N.; Tsatsakis, A.M. Lead toxicity update. A brief review. *Med. Sci. Monit.* **2005**, *11*, RA329–RA336. [PubMed]
12. Zhang, T.; Li, Q.; Yang, X.; Zheng, D.; Deng, H.; Zeng, Z.; Yu, J.; Wang, Q.; Shi, Y.; Wang, S.; et al. Pb contaminated soil from a lead-acid battery plant immobilized by municipal sludge and raw clay. *Environ. Technol.* **2023**, 1–13. [CrossRef] [PubMed]
13. Sanders, T.; Liu, Y.; Buchner, V.; Tchounwou, P.B. Neurotoxic effects and biomarkers of lead exposure: A review. *Rev. Environ. Health* **2009**, *24*, 15–45. [CrossRef] [PubMed]
14. Moreira, F.R.; Moreira, J.C. Effects of lead exposure on the human body and health implications. *Rev. Panam. Salud Publica* **2004**, *15*, 119–129. [CrossRef]
15. Delgado, C.F.; Ullery, M.A.; Jordan, M.; Duclos, C.; Rajagopalan, S.; Scott, K. Lead Exposure and Developmental Disabilities in Preschool-Aged Children. *J. Public. Health Manag. Pract.* **2018**, *24*, e10–e17. [CrossRef] [PubMed]

16. Swaringen, B.F.; Gawlik, E.; Kamenov, G.D.; McTigue, N.E.; Cornwell, D.A.; Bonzongo, J.J. Children's exposure to environmental lead: A review of potential sources, blood levels, and methods used to reduce exposure. *Environ. Res.* **2022**, *204*, 112025. [CrossRef] [PubMed]

17. Vorvolakos, T.; Arseniou, S.; Samakouri, M. There is no safe threshold for lead exposure: Alpha literature review. *Psychiatriki* **2016**, *27*, 204–214. [CrossRef]

18. Ramirez Ortega, D.; Gonzalez Esquivel, D.F.; Blanco Ayala, T.; Pineda, B.; Gomez Manzo, S.; Marcial Quino, J.; Carrillo Mora, P.; Perez de la Cruz, V. Cognitive Impairment Induced by Lead Exposure during Lifespan: Mechanisms of Lead Neurotoxicity. *Toxics* **2021**, *9*, 23. [CrossRef]

19. Rafalowska, U.; Struzynska, L.; Dabrowska-Bouta, B.; Lenkiewicz, A. Is lead toxicosis a reflection of altered energy metabolism in brain synaptosomes? *Acta Neurobiol. Exp.* **1996**, *56*, 611–617.

20. Antonio, M.T.; Lopez, N.; Leret, M.L. Pb and Cd poisoning during development alters cerebellar and striatal function in rats. *Toxicology* **2002**, *176*, 59–66. [CrossRef]

21. Li, Y.; Cai, W.; Ai, Z.; Xue, C.; Cao, R.; Dong, N. Protective effects of sinomenine hydrochloride on lead-induced oxidative stress, inflammation, and apoptosis in mouse liver. *Environ. Sci. Pollut. Res. Int.* **2023**, *30*, 7510–7521. [CrossRef] [PubMed]

22. Nam, S.M.; Choi, S.H.; Cho, H.J.; Seo, J.S.; Choi, M.; Nahm, S.S.; Chang, B.J.; Nah, S.Y. Ginseng Gintonin Attenuates Lead-Induced Rat Cerebellar Impairments during Gestation and Lactation. *Biomolecules* **2020**, *10*, 385. [CrossRef] [PubMed]

23. Upadhyay, K.; Viramgami, A.; Bagepally, B.S.; Balachandar, R. Association between blood lead levels and markers of calcium homeostasis: A systematic review and meta-analysis. *Sci. Rep.* **2022**, *12*, 1850. [CrossRef] [PubMed]

24. Ramirez Ortega, D.; Ovalle Rodriguez, P.; Pineda, B.; Gonzalez Esquivel, D.F.; Ramos Chavez, L.A.; Vazquez Cervantes, G.I.; Roldan Roldan, G.; Perez de la Cruz, G.; Diaz Ruiz, A.; Mendez Armenta, M.; et al. Kynurenine Pathway as a New Target of Cognitive Impairment Induced by Lead Toxicity During the Lactation. *Sci. Rep.* **2020**, *10*, 3184. [CrossRef] [PubMed]

25. Beggiato, S.; Antonelli, T.; Tomasini, M.C.; Tanganelli, S.; Fuxe, K.; Schwarcz, R.; Ferraro, L. Kynurenic acid, by targeting alpha7 nicotinic acetylcholine receptors, modulates extracellular GABA levels in the rat striatum in vivo. *Eur. J. Neurosci.* **2013**, *37*, 1470–1477. [CrossRef] [PubMed]

26. Konradsson-Geuken, A.; Wu, H.Q.; Gash, C.R.; Alexander, K.S.; Campbell, A.; Sozeri, Y.; Pellicciari, R.; Schwarcz, R.; Bruno, J.P. Cortical kynurenic acid bi-directionally modulates prefrontal glutamate levels as assessed by microdialysis and rapid electrochemistry. *Neuroscience* **2010**, *169*, 1848–1859. [CrossRef] [PubMed]

27. Carpenedo, R.; Pittaluga, A.; Cozzi, A.; Attucci, S.; Galli, A.; Raiteri, M.; Moroni, F. Presynaptic kynurenate-sensitive receptors inhibit glutamate release. *Eur. J. Neurosci.* **2001**, *13*, 2141–2147. [CrossRef]

28. Alexander, K.S.; Wu, H.Q.; Schwarcz, R.; Bruno, J.P. Acute elevations of brain kynurenic acid impair cognitive flexibility: Normalization by the alpha7 positive modulator galantamine. *Psychopharmacology* **2012**, *220*, 627–637. [CrossRef]

29. Pocivavsek, A.; Wu, H.Q.; Elmer, G.I.; Bruno, J.P.; Schwarcz, R. Pre- and postnatal exposure to kynurenine causes cognitive deficits in adulthood. *Eur. J. Neurosci.* **2012**, *35*, 1605–1612. [CrossRef]

30. Chess, A.C.; Simoni, M.K.; Alling, T.E.; Bucci, D.J. Elevations of endogenous kynurenic acid produce spatial working memory deficits. *Schizophr. Bull.* **2007**, *33*, 797–804. [CrossRef]

31. Akagbosu, C.O.; Evans, G.C.; Gulick, D.; Suckow, R.F.; Bucci, D.J. Exposure to kynurenic acid during adolescence produces memory deficits in adulthood. *Schizophr. Bull.* **2012**, *38*, 769–778. [CrossRef] [PubMed]

32. Pocivavsek, A.; Wu, H.Q.; Potter, M.C.; Elmer, G.I.; Pellicciari, R.; Schwarcz, R. Fluctuations in endogenous kynurenic acid control hippocampal glutamate and memory. *Neuropsychopharmacology* **2011**, *36*, 2357–2367. [CrossRef] [PubMed]

33. Wu, H.Q.; Pereira, E.F.; Bruno, J.P.; Pellicciari, R.; Albuquerque, E.X.; Schwarcz, R. The astrocyte-derived alpha7 nicotinic receptor antagonist kynurenic acid controls extracellular glutamate levels in the prefrontal cortex. *J. Mol. Neurosci.* **2010**, *40*, 204–210. [CrossRef] [PubMed]

34. Amori, L.; Wu, H.Q.; Marinozzi, M.; Pellicciari, R.; Guidetti, P.; Schwarcz, R. Specific inhibition of kynurenate synthesis enhances extracellular dopamine levels in the rodent striatum. *Neuroscience* **2009**, *159*, 196–203. [CrossRef] [PubMed]

35. Linderholm, K.R.; Andersson, A.; Olsson, S.; Olsson, E.; Snodgrass, R.; Engberg, G.; Erhardt, S. Activation of rat ventral tegmental area dopamine neurons by endogenous kynurenic acid: A pharmacological analysis. *Neuropharmacology* **2007**, *53*, 918–924. [CrossRef]

36. Beggiato, S.; Tanganelli, S.; Fuxe, K.; Antonelli, T.; Schwarcz, R.; Ferraro, L. Endogenous kynurenic acid regulates extracellular GABA levels in the rat prefrontal cortex. *Neuropharmacology* **2014**, *82*, 11–18. [CrossRef] [PubMed]

37. Zmarowski, A.; Wu, H.Q.; Brooks, J.M.; Potter, M.C.; Pellicciari, R.; Schwarcz, R.; Bruno, J.P. Astrocyte-derived kynurenic acid modulates basal and evoked cortical acetylcholine release. *Eur. J. Neurosci.* **2009**, *29*, 529–538. [CrossRef]

38. Wright, C.J.; Rentschler, K.M.; Wagner, N.T.J.; Lewis, A.M.; Beggiato, S.; Pocivavsek, A. Time of Day-Dependent Alterations in Hippocampal Kynurenic Acid, Glutamate, and GABA in Adult Rats Exposed to Elevated Kynurenic Acid During Neurodevelopment. *Front. Psychiatry* **2021**, *12*, 734984. [CrossRef]

39. Blanco-Ayala, T.; Sathyasaikumar, K.V.; Uys, J.D.; Perez-de-la-Cruz, V.; Pidugu, L.S.; Schwarcz, R. N-Acetylcysteine Inhibits Kynurenine Aminotransferase II. *Neuroscience* **2020**, *444*, 160–169. [CrossRef]

40. Kasperczyk, S.; Dobrakowski, M.; Kasperczyk, A.; Romuk, E.; Rykaczewska-Czerwinska, M.; Pawlas, N.; Birkner, E. Effect of N-acetylcysteine administration on homocysteine level, oxidative damage to proteins, and levels of iron (Fe) and Fe-related proteins in lead-exposed workers. *Toxicol. Ind. Health* **2016**, *32*, 1607–1618. [CrossRef]

41. Mokra, D.; Mokry, J.; Barosova, R.; Hanusrichterova, J. Advances in the Use of N-Acetylcysteine in Chronic Respiratory Diseases. *Antioxidants* **2023**, *12*, 1713. [CrossRef] [PubMed]

42. Bradlow, R.C.J.; Berk, M.; Kalivas, P.W.; Back, S.E.; Kanaan, R.A. The Potential of N-Acetyl-L-Cysteine (NAC) in the Treatment of Psychiatric Disorders. *CNS Drugs* **2022**, *36*, 451–482. [CrossRef] [PubMed]

43. Tardiolo, G.; Bramanti, P.; Mazzon, E. Overview on the Effects of N-Acetylcysteine in Neurodegenerative Diseases. *Molecules* **2018**, *23*, 3305. [CrossRef] [PubMed]

44. Atlas, D. Emerging therapeutic opportunities of novel thiol-amides, NAC-amide (AD4/NACA) and thioredoxin mimetics (TXM-Peptides) for neurodegenerative-related disorders. *Free Radic. Biol. Med.* **2021**, *176*, 120–141. [CrossRef] [PubMed]

45. Arakawa, M.; Ito, Y. N-acetylcysteine and neurodegenerative diseases: Basic and clinical pharmacology. *Cerebellum* **2007**, *6*, 308–314. [CrossRef]

46. Blanco Ayala, T.B.; Ramirez Ortega, D.R.; Ovalle Rodriguez, P.O.; Pineda, B.; Perez de la Cruz, G.P.; Gonzalez Esquivel, D.G.; Schwarcz, R.; Sathyasaikumar, K.V.; Jimenez Anguiano, A.J.; Perez de la Cruz, V.P. Subchronic N-acetylcysteine Treatment Decreases Brain Kynurenic Acid Levels and Improves Cognitive Performance in Mice. *Antioxidants* **2021**, *10*, 147. [CrossRef] [PubMed]

47. Lehmkuhl, A.M.; Dirr, E.R.; Fleming, S.M. Olfactory assays for mouse models of neurodegenerative disease. *J. Vis. Exp.* **2014**, *90*, e51804. [CrossRef]

48. Ennaceur, A.; Delacour, J. A new one-trial test for neurobiological studies of memory in rats. 1: Behavioral data. *Behav. Brain Res.* **1988**, *31*, 47–59. [CrossRef]

49. Lueptow, L.M. Novel Object Recognition Test for the Investigation of Learning and Memory in Mice. *J. Vis. Exp.* **2017**, *126*, e55718. [CrossRef]

50. Ramos-Chavez, L.A.; Lugo Huitron, R.; Gonzalez Esquivel, D.; Pineda, B.; Rios, C.; Silva-Adaya, D.; Sanchez-Chapul, L.; Roldan-Roldan, G.; Perez de la Cruz, V. Relevance of Alternative Routes of Kynurenic Acid Production in the Brain. *Oxid. Med. Cell Longev.* **2018**, *2018*, 5272741. [CrossRef]

51. Blanco Ayala, T.; Lugo Huitron, R.; Carmona Aparicio, L.; Ramirez Ortega, D.; Gonzalez Esquivel, D.; Pedraza Chaverri, J.; Perez de la Cruz, G.; Rios, C.; Schwarcz, R.; Perez de la Cruz, V. Alternative kynurenic acid synthesis routes studied in the rat cerebellum. *Front. Cell Neurosci.* **2015**, *9*, 178. [CrossRef] [PubMed]

52. Rassoulpour, A.; Wu, H.Q.; Ferre, S.; Schwarcz, R. Nanomolar concentrations of kynurenic acid reduce extracellular dopamine levels in the striatum. *J. Neurochem.* **2005**, *93*, 762–765. [CrossRef] [PubMed]

53. Chess, A.C.; Landers, A.M.; Bucci, D.J. L-kynurenine treatment alters contextual fear conditioning and context discrimination but not cue-specific fear conditioning. *Behav. Brain Res.* **2009**, *201*, 325–331. [CrossRef] [PubMed]

54. Rodrigues, F.S.; Franca, A.P.; Broetto, N.; Furian, A.F.; Oliveira, M.S.; Santos, A.R.S.; Royes, L.F.F.; Fighera, M.R. Sustained glial reactivity induced by glutaric acid may be the trigger to learning delay in early and late phases of development: Involvement of p75(NTR) receptor and protection by N-acetylcysteine. *Brain Res.* **2020**, *1749*, 147145. [CrossRef] [PubMed]

55. Tiffany-Castiglioni, E. Cell culture models for lead toxicity in neuronal and glial cells. *Neurotoxicology* **1993**, *14*, 513–536. [PubMed]

56. Qian, H.R.; Yang, Y. Neuron differentiation and neuritogenesis stimulated by N-acetylcysteine (NAC). *Acta Pharmacol. Sin.* **2009**, *30*, 907–912. [CrossRef] [PubMed]

57. Potter, M.C.; Elmer, G.I.; Bergeron, R.; Albuquerque, E.X.; Guidetti, P.; Wu, H.Q.; Schwarcz, R. Reduction of endogenous kynurenic acid formation enhances extracellular glutamate, hippocampal plasticity, and cognitive behavior. *Neuropsychopharmacology* **2010**, *35*, 1734–1742. [CrossRef]

58. Mozafari, R.; Karimi-Haghighi, S.; Fattahi, M.; Kalivas, P.; Haghparast, A. A review on the role of metabotropic glutamate receptors in neuroplasticity following psychostimulant use disorder. *Prog. Neuropsychopharmacol. Biol. Psychiatry* **2023**, *124*, 110735. [CrossRef]

59. Abraham, W.C.; Jones, O.D.; Glanzman, D.L. Is plasticity of synapses the mechanism of long-term memory storage? *NPJ Sci. Learn.* **2019**, *4*, 9. [CrossRef]

60. Regan, C.M. Lead-impaired neurodevelopment. Mechanisms and threshold values in the rodent. *Neurotoxicol. Teratol.* **1989**, *11*, 533–537. [CrossRef]

61. Hon, K.L.; Fung, C.K.; Leung, A.K. Childhood lead poisoning: An overview. *Hong Kong Med. J.* **2017**, *23*, 616–621. [CrossRef] [PubMed]

62. Baranowska-Bosiacka, I.; Gutowska, I.; Rybicka, M.; Nowacki, P.; Chlubek, D. Neurotoxicity of lead. Hypothetical molecular mechanisms of synaptic function disorders. *Neurol. Neurochir. Pol.* **2012**, *46*, 569–578. [CrossRef] [PubMed]

63. Beal, M.F.; Swartz, K.J.; Isacson, O. Developmental changes in brain kynurenic acid concentrations. *Brain Res. Dev. Brain Res.* **1992**, *68*, 136–139. [CrossRef] [PubMed]

64. Raghu, G.; Berk, M.; Campochiaro, P.A.; Jaeschke, H.; Marenzi, G.; Richeldi, L.; Wen, F.Q.; Nicoletti, F.; Calverley, P.M.A. The Multifaceted Therapeutic Role of N-Acetylcysteine (NAC) in Disorders Characterized by Oxidative Stress. *Curr. Neuropharmacol.* **2021**, *19*, 1202–1224. [CrossRef] [PubMed]

65. Samuni, Y.; Goldstein, S.; Dean, O.M.; Berk, M. The chemistry and biological activities of N-acetylcysteine. *Biochim. Biophys. Acta* **2013**, *1830*, 4117–4129. [CrossRef] [PubMed]

66. Jaafarzadeh, M.; Mahjoob Khaligh, R.; Mohsenifar, Z.; Shabani, A.; Rezvani Gilkalaei, M.; Rajabi Keleshteri, S.; Beigi Harchegani, A. Protecting Effects of N-acetyl Cysteine Supplementation Against Lead and Cadmium-Induced Brain Toxicity in Rat Models. *Biol. Trace Elem. Res.* **2022**, *200*, 1395–4403. [CrossRef] [PubMed]

67. Pedroso, T.F.; Oliveira, C.S.; Fonseca, M.M.; Oliveira, V.A.; Pereira, M.E. Effects of Zinc and N-Acetylcysteine in Damage Caused by Lead Exposure in Young Rats. *Biol. Trace Elem. Res.* **2017**, *180*, 275–284. [CrossRef] [PubMed]

68. Chen, W.; Ercal, N.; Huynh, T.; Volkov, A.; Chusuei, C.C. Characterizing N-acetylcysteine (NAC) and N-acetylcysteine amide (NACA) binding for lead poisoning treatment. *J. Colloid. Interface Sci.* **2012**, *371*, 144–149. [CrossRef]

69. Sisombath, N.S.; Jalilehvand, F. Similarities between N-Acetylcysteine and Glutathione in Binding to Lead(II) Ions. *Chem. Res. Toxicol.* **2015**, *28*, 2313–2324. [CrossRef]

70. Ottenwalder, H.; Simon, P. Differential effect of N-acetylcysteine on excretion of the metals Hg, Cd, Pb and Au. *Arch. Toxicol.* **1987**, *60*, 401–402. [CrossRef]

71. Kitamura, Y.; Ushio, S.; Sumiyoshi, Y.; Wada, Y.; Miyazaki, I.; Asanuma, M.; Sendo, T. N-Acetylcysteine Attenuates the Anxiety-Like Behavior and Spatial Cognition Impairment Induced by Doxorubicin and Cyclophosphamide Combination Treatment in Rats. *Pharmacology* **2021**, *106*, 286–293. [CrossRef] [PubMed]

72. Garcia-Serrano, A.M.; Vieira, J.P.P.; Fleischhart, V.; Duarte, J.M.N. Taurine and N-acetylcysteine treatments prevent memory impairment and metabolite profile alterations in the hippocampus of high-fat diet-fed female mice. *Nutr. Neurosci.* **2023**, *26*, 1090–1102. [CrossRef] [PubMed]

73. Smaga, I.; Pomierny, B.; Krzyzanowska, W.; Pomierny-Chamiolo, L.; Miszkiel, J.; Niedzielska, E.; Ogorka, A.; Filip, M. N-acetylcysteine possesses antidepressant-like activity through reduction of oxidative stress: Behavioral and biochemical analyses in rats. *Prog. Neuropsychopharmacol. Biol. Psychiatry* **2012**, *39*, 280–287. [CrossRef] [PubMed]

74. Ikonne, U.S.; Vann, P.H.; Wong, J.M.; Forster, M.J.; Sumien, N. Supplementation with N-Acetyl Cysteine Affects Motor and Cognitive Function in Young but Not Old Mice. *J. Nutr.* **2019**, *149*, 463–470. [CrossRef]

75. Aykin-Burns, N.; Franklin, E.A.; Ercal, N. Effects of N-acetylcysteine on lead-exposed PC-12 cells. *Arch. Environ. Contam. Toxicol.* **2005**, *49*, 119–123. [CrossRef] [PubMed]

76. Penugonda, S.; Ercal, N. Comparative evaluation of N-acetylcysteine (NAC) and N-acetylcysteine amide (NACA) on glutamate and lead-induced toxicity in CD-1 mice. *Toxicol. Lett.* **2011**, *201*, 1–7. [CrossRef] [PubMed]

77. Nehru, B.; Kanwar, S.S. N-acetylcysteine exposure on lead-induced lipid peroxidative damage and oxidative defense system in brain regions of rats. *Biol. Trace Elem. Res.* **2004**, *101*, 257–264. [CrossRef]

Review

Mitochondria-Targeted Antioxidant Therapeutics for Traumatic Brain Injury

Hiren R. Modi, Sudeep Musyaju, Meaghan Ratcliffe, Deborah A. Shear, Anke H. Scultetus and Jignesh D. Pandya *

Brain Trauma Neuroprotection (BTN) Branch, Center for Military Psychiatry and Neuroscience, Walter Reed Army Institute of Research (WRAIR), 503 Robert Grant Avenue, Silver Spring, MD 20910, USA; hiren.r.modi.ctr@health.mil (H.R.M.); sudeep.musyaju.ctr@health.mil (S.M.); meaghan.l.ratcliffe.ctr@health.mil (M.R.); deborah.a.shear.civ@health.mil (D.A.S.); anke.h.scultetus2.civ@health.mil (A.H.S.)
* Correspondence: jignesh.d.pandya.civ@health.mil; Tel.: +1-301-319-3067

Abstract: Traumatic brain injury (TBI) is a major global health problem that affects both civilian and military populations worldwide. Post-injury acute, sub-acute, and chronic progression of secondary injury processes may contribute further to other neurodegenerative diseases. However, there are no approved therapeutic options available that can attenuate TBI-related progressive pathophysiology. Recent advances in preclinical research have identified that mitochondria-centric redox imbalance, bioenergetics failure and calcium dysregulation play a crucial role in secondary injury progression after TBI. Mitochondrial antioxidants play an important role in regulating redox homeostasis. Based on the proven efficacy of preclinical and clinical compounds and targeting numerous pathways to trigger innate antioxidant defense, we may be able to alleviate TBI pathology progression by primarily focusing on preserving post-injury mitochondrial and cerebral function. In this review, we will discuss novel mitochondria-targeted antioxidant compounds, which offer a high capability of successful clinical translation for TBI management in the near future.

Keywords: traumatic brain injury; mitochondria; free radicals; oxidative stress; cell death; antioxidants; therapeutics; neuroprotection

Citation: Modi, H.R.; Musyaju, S.; Ratcliffe, M.; Shear, D.A.; Scultetus, A.H.; Pandya, J.D. Mitochondria-Targeted Antioxidant Therapeutics for Traumatic Brain Injury. *Antioxidants* **2024**, *13*, 303. https://doi.org/10.3390/antiox13030303

Academic Editor: Alessandra Napolitano

Received: 15 January 2024
Revised: 9 February 2024
Accepted: 21 February 2024
Published: 29 February 2024

1. Introduction

Traumatic brain injury (TBI) is caused by a mechanical blow, penetration, bump, or jolt to the head subsequently leading to tissue and cellular damage, and ultimately resulting in alteration of physiological and behavioral functions. TBI represents a major contributor to morbidity and mortality amongst civilian and military populations across the world. There were over 69,000 TBI-related deaths in the United States alone in 2021, accounting for about 190 deaths per day [1]. TBI also has a big global impact, with annual TBI incidence estimated to be 27 to 69 million [2,3]. These injuries have both short-term and long-term effects on individuals, their families, and society and their financial cost is enormous. Many survivors live with significant disabilities, resulting in major socioeconomic burden. The economic impact of TBI in the United States is estimated to be about USD 76.5 billion for survivors [4,5]. Clinically, TBI is categorized as mild, moderate, or severe injury based on the Glasgow Coma Scale (GCS) scores range between 3 to 15, with a lower score indicating more severe brain damage and a poorer prognosis. The GCS describes the level of consciousness of an individual after acute brain trauma [6,7]. Nevertheless, across all TBI severities, the consequences of TBI may lead to long-term disability, including cognitive and motor function limitations/impairments, and decreased psychosocial health.

TBI-induced neuronal tissue damage manifests in primary and secondary injuries. The primary injury stems from the initial mechanical impacts to the brain [8]. The primary injury is considered an irreversible injury resulting from brain tissue compression, displacement, stretching, shearing, tearing, crushing of the brain parenchyma, brain hemorrhage, and blood–brain barrier (BBB) disruption [8]. Following the primary mechanical insult, the downstream sequelae of molecular events activate complex secondary injury pathophysiological cascades such as excitotoxicity, intracranial hypertension, edema, elevated calcium, metabolic dysregulation, mitochondrial dysfunction (energy crisis, antioxidant depletion, and free-radical generation), inflammation, and ischemic injury, which occur at the acute (i.e., minutes to hours) and sub-acute (i.e., hours to weeks) phases of progressive TBI [9–12]. Consequently, brain functions are first disrupted at the injury site and subsequently disrupted at distal interconnected regions. Despite the advancements in TBI research, the precise mechanisms leading to the progression of TBI pathophysiology are yet to be fully elucidated.

The chronic progression of post-TBI secondary injury responses (i.e., weeks to years) further affects TBI patients' neuronal ability to maintain their long-term physiological and behavioral functions. TBI progression is linked to the etiology of many neurodegenerative diseases such as Alzheimer's disease (AD), Amyotrophic lateral sclerosis (ALS), Huntington's disease (HD), Multiple sclerosis (MS), and Parkinson's disease (PD) [13]. However, the specific epidemiological factors and pathophysiological mechanisms that underlie this association between TBI and specific neurodegenerative pathologies remain unclear. Notably, reports indicate a 63–96% increased risk of all-cause dementia following TBI [14]. Additionally, the risk of PD may go up by at least 1.8 times following moderate to severe TBI [14,15]. Meta analysis indicates an increased risk of ALS following TBI [16]. Additionally, patients with TBI have a higher risk of developing MS [17]. Furthermore, a World War II study suggested that early-adulthood TBI increases the likelihood of developing AD later in life by 2.3 to 4.5 times, respectively, for moderate and severe injuries [14,15]. These reports suggest that TBI is a major risk factor for the onset of neurodegenerative disorders in later life (Figure 1).

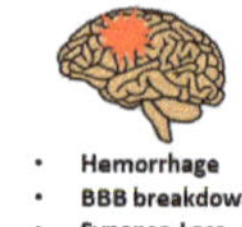

Figure 1. Progressive pathology of TBI. The primary mechanical impact may lead to brain damage in the form of brain hemorrhage, blood–brain barrier (BBB) breakdown and synapse loss, thereby subsequent secondary injury processes beginning immediately after post-injury, and sometimes may sustained over lifetime. During these acute (minutes to hours) and sub-acute (hours to weeks) processes, mitochondria-centric mechanisms play a key role in the further progression of TBI pathology. During the chronic (weeks to years) period of secondary brain injury, neurological behavior deficits in terms of cognitive and motor functions are evident, and may further contribute to neurological diseases such as AD, PD, HD, ALS and MS.

Much of our understanding of the pathobiology of TBI has arisen from animal models that simulate features of human TBI. Multiple preclinical TBI models, including models of penetrating traumatic brain injury (PTBI), controlled cortical impact (CCI) injury, blast-induced traumatic brain injury (BTBI) and closed head injury (CHI) have ascertained that mitochondrial dysfunction is a common and immediate indicator of cellular damage [18–21] that may even play a critical role in secondary excitotoxic post-injury events. Several detailed previous reports have highlighted preclinical models and cellular mechanisms of TBI [22–30]. Mitochondria-centered cellular mechanisms involve calcium homeostasis, energy homeostasis, and redox homeostasis. Their imbalance subsequently may prompt downstream cellular processes such as cell death pathways and neuronal death and alter behavior outcomes in TBI.

Mitochondria are key organelles in all eukaryotic cells and play a central role in cellular energy homeostasis through the metabolism of carbohydrates, fats, and/or proteins. Brain cells manage higher cellular energy (i.e., adenosine triphosphate, ATP) demands by oxidizing their metabolic substrates through respiration and oxidative phosphorylation. Mitochondrial dysfunction following TBI has been shown to be devastating for neuronal cell survival [31–34]. Several time-course studies of mitochondrial bioenergetics in preclinical models of TBI have suggested that mitochondrial energy failure is the key pathological event that is initiated immediately (e.g., within 30 min) and remains evident for up to 2 weeks after injury [11,12,33–36].

Interestingly, under normal physiological conditions, intracellular calcium levels are modulated by mitochondria to maintain cellular homeostasis at a certain threshold; however, rapid increase in cellular calcium following TBI may lead to excitotoxicity. Under physiological conditions during ATP production, mitochondria also maintain calcium homeostasis and regulate mitochondrial permeability transition (MPT) pore formation. Several reports have identified impaired mitochondrial calcium-buffering capacity and early mitochondrial MPT opening following acute and sub-acute phases of TBI [37]. Therefore, mitochondrial MPT is considered as the "biological on/off switch" that determines the fate of cells in response to noxious excitotoxic stimuli [38–41].

Additionally, mitochondrial dysfunction in response to secondary injury elevates the oxidative stress response. Post-TBI oxidative damage leads to structural functional alteration in cellular and subcellular components. This, coupled with the impairment of mitochondrial bioenergetics, initiates a vicious cycle of free-radical formation and apoptosis. In this review, we evaluate the detailed mechanisms of mitochondrial redox homeostasis and discuss potential antioxidant strategies to mitigate oxidative damage following TBI. The aim of this review is to provide a comprehensive overview of antioxidant therapy for TBI to the scientific research community, categorized into different classes, and systematically discussed in the following sections.

2. Mitochondrial Redox Mechanisms in TBI

Mitochondria are vital organelles present in all eukaryotic cells, that consume approximately 98% of body's total oxygen supply. Efficiently utilizing this oxygen, mitochondria produce energy through oxidative phosphorylation processes linked by respiration via the electron transport chain (ETC) complex proteins. In normal physiological conditions, oxygen slippage estimated from 3 to 5% can occur during the utilization of oxygen in the mitochondrial ETC complexes I and III [42–44]. This oxygen slippage results in the generation of the superoxide radical ($O_2^{\bullet-}$), a highly reactive unstable singlet oxygen, which, in turn, can lead to the production of other reactive oxygen species (ROS). Mitochondria also contain antioxidants to manage elevated levels of free radicals as part of normal repair mechanisms. Highly reactive superoxide ($O_2^{\bullet-}$) is rapidly converted into less-reactive ROS, hydrogen peroxide (H_2O_2) by superoxide dismutase (SOD), which is then further decomposed/neutralized into water by catalase (CAT) or peroxiredoxin (Prx) or thioredoxin (Trx) complex enzyme systems. However, $O_2^{\bullet-}$ also generates hydroxyl ($^{\bullet}OH$) radicals through the Fenton reaction, which can be further converted to peroxynitrite ($ONOO^-$), and subsequently other reactive nitrogen species (RNS) such as nitrogen dioxide and peroxynitrous acid (ONOOH). Normally, mitochondria maintain redox homeostasis with antioxidant activities to scavenge ROS–RNS [45,46]. However, under pathophysiological conditions, elevated levels of ROS–RNS have been observed as early as 30 min after TBI [47,48].

Moreover, these harmful ROS–RNS molecules can further oxidize and damage cellular proteins, lipids, nucleic acids, and extracellular matrix components. The oxidized protein adducts, 3-nitrotyrosine (3-NT), protein carbonylation (PC), and the lipid peroxidation adduct 4-hydroxynonenal (4-HNE) are the hallmarks of peroxynitrite-mediated oxidative stress. Additionally, ROS–RNS may further induce nuclear and mitochondrial DNA damage and affect gene expression responses. ROS–RNS overproduction may inducedamage to ETC subunits, which may further exaggerate the vicious cycle of mitochondrial dysfunction.

Following TBI, higher redox footprints with respect to elevated free radicals (ROS–RNS), together with altered lipid, protein, and DNA adducts have been observed during the acute and sub-acute phases of secondary injury. If not mitigated, these elevated redox mediators may further contribute to other chronic neurological disease pathologies. The protective antioxidant defense system has potential to mitigate the TBI-induced oxidative stress response is further discussed in detail (Figure 2).

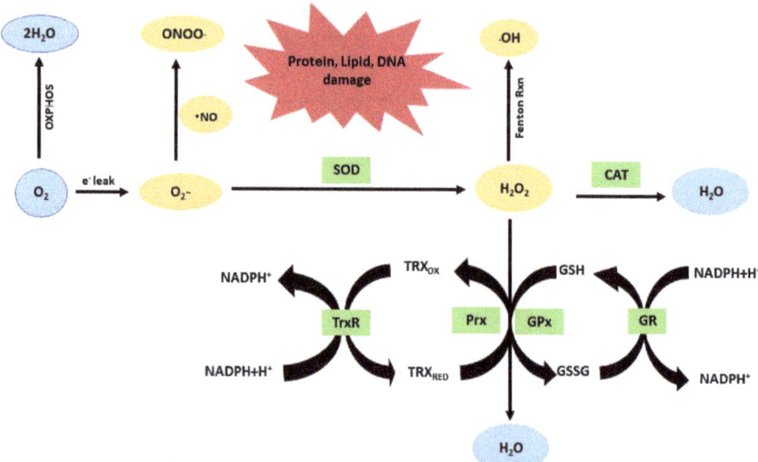

Figure 2. Generation and scavenging of reactive oxygen species (ROS) using the antioxidant defense system. Electrons released from the mitochondrial ETC and produced by NADPH oxidases are the major source of endogenous reactive oxygen species. The oxidative stress generated via this mechanism can be countered via the antioxidant defense system. Oxygen (O_2) is reduced to super-oxide ($O_2\bullet^-$), which can be reduced to hydrogen peroxide (H_2O_2) by superoxide dismutase (SOD). Nitric oxide radicals ($\bullet NO$) form the potent oxidant peroxynitrite ($ONOO^-$) following reaction with $O_2\bullet^-$. The H_2O_2 can undergo the Fenton reaction and transformed into hydroxyl radicals ($\bullet OH$) or reduced to water (H_2O) by catalase (CAT), or the glutathione (GSH)/glutathione peroxidase (GPx) or peroxiredoxin (Prx) systems. The oxidized form of thioredoxin (Trx) is reduced back by the reaction with Trx reductase (TrxR), while that of Grx is reduced back by GSH and terminal NADPH oxidation. An oxidized GSH (GSSG) is reduced back to two GSH molecules through the enzymatic reaction of GSH reductase (GR). Both Trx and Grx reduce protein disulfides. These enzymatic antioxidant defense systems counter free-radical-induced stress, and maintain cellular redox homeostasis. The reactive oxygen species may further lead to protein, lipid, and DNA damage.

3. Mitochondrial Antioxidants in TBI

TBI is a highly heterogeneous condition, with patients exhibiting diverse patterns of injury, severity, and outcomes. Oxidative stress plays a crucial role in the development of acute brain injury and acts as a key mediator in the secondary injury cascade of TBI pathology. Oxidative stress leading to oxidative damage represents a state where oxygen levels, combined with oxygen-derived free radicals overwhelm the scavenging antioxidant system.

Following TBI, excitotoxicity occurs, where an excess of Ca^{2+} further promotes ROS–RNS production. The increased ROS–RNS levels, coupled with depleted antioxidant levels after TBI, lead to elevated oxidative stress, wherein protective mechanisms, such as antioxidants, fail to control these radicals, resulting in oxidative stress and subsequent neuronal death. Post TBI, various complications such as brain edema, mitochondrial dysfunction,

BBB breakdown, sensory–motor dysfunction, and secondary neuronal injury have been proposed to be linked to oxidative stress [49–51]. Our recent findings indicate decreased mitochondrial antioxidants and increased oxidative stress markers during the acute phase of TBI [37], a trend supported by numerous researchers highlighting the role of oxidative stress following TBI [33,52–57]. This underscores oxidative stress redox mechanisms as a valid therapeutic target for TBI. Furthermore, antioxidant intervention emerges as a logical therapeutic approach for achieving neuroprotection after TBI. Both elevated free-radical-mediated oxidative stress and depleted endogenous antioxidant responses have been observed during acute and sub-acute phases preclinically; and only limited reports have noted duringchronic phase of TBI [37,52,58].

Antioxidants are substances which scavenge or neutralize free radicals in cells, thereby prevent oxidative damage. Antioxidants may be able to reduce the risk of the onset of chronic diseases. Antioxidant therapy emerge as a novel approach to preventing and treating neurodegenerative conditions where oxidative stress acts as a major contributing factor to the pathogenesis and/or progression of the diseases. There are two main approaches by which antioxidant levels can be replenished in brain cells, and these may serve as options for therapeutic interventions to limit free-radical generation and oxidative stress responses. This, in turn, improves the balance of redox homeostasis after brain trauma. In an injured brain, antioxidants may be able to modulate redox mechanisms through (a) scavenging or detoxifying excessive ROS–RNS using natural or synthetic antioxidants and restrict free-radical overproduction, and (b) modulating cell signaling pathways that favor endogenous antioxidant synthesis and balanced redox homeostasis.

This review highlights each category of antioxidants that may serve as future therapeutic options to restrict/stimulate the mechanisms listed above and favor balanced redox homeostasis following the secondary injury phases of TBI. Unfortunately, there are no FDA-approved treatment options that are currently available to restrain multifaceted TBI pathophysiology, leaving a critical gap unfilled. Therefore, more preclinical research efforts are warranted to identify novel therapeutic targets. Additionally, repetitive injuries aggravate secondary injuries and lead to early neurological deficit. Collaborative efforts between preclinical and clinical communities under regulatory guidance of the FDA are ongoing to conduct better-designed clinical studies, and gain rapid approvals of therapeutic products for TBI. This review is intended to provide an overview on comprehensive information about antioxidant therapy for TBI to the scientific research community, classified into different categories (Figure 3), and discussed below.

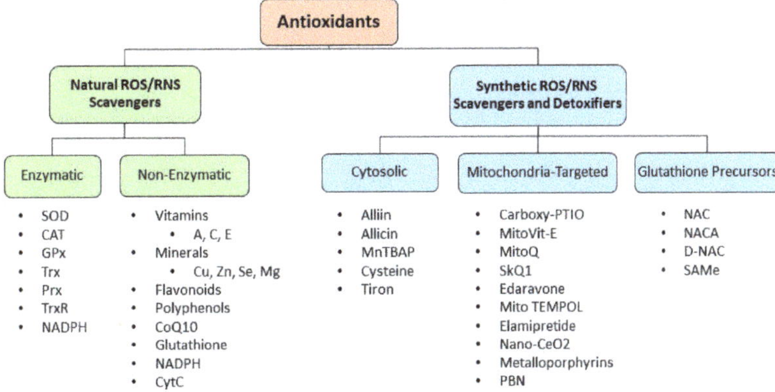

Figure 3. Classification of different antioxidants discussed in the current review. As shown in this illustration, antioxidants are further classified into natural and synthetic ROS–RNS scavengers and detoxifiers. Among the natural antioxidants, they are further sub-divided into enzymatic and non-enzymatic forms, whereas in the group of synthetic antioxidants, they are further sub-divided into three categories, namely non-targeted cytosolic, mitochondria-targeted, and glutathione precursors,

based on their sub-cellular target, pharmacological properties, and abundance. Examples of each category of antioxidants are listed and discussed in detail.

4. ROS–RNS Scavengers

The ROS–RNS scavengers/detoxifiers are further categorized as natural (e.g., endogenous enzymatic or non-enzymatic) and synthetic (e.g., drug molecules and dietary supplements), as described below.

4.1. Natural ROS–RNS Scavengers

Among natural enzymatic antioxidants, the mitochondria-specific superoxide dismutase (SOD) isoform (e.g., manganese SOD or Mn-SOD) plays a critical role in scavenging mitochondrial production of $O_2^{\bullet-}$ at the mitochondrial ETC complex I and III sites [59,60]. Another isoform of SOD (i.e., cytosolic copper–zinc SOD or Cu-Zn-SOD) scavenges $O_2^{\bullet-}$ in the cytosolic compartment. The mitochondrial ETC is a primary site of $O_2^{\bullet-}$ generation; therefore, scavenging $O_2^{\bullet-}$ at the mitochondrial ETC level offers the greatest benefit. Catalase (CAT) is an additional important ROS scavenger that neutralizes H_2O_2 and converts it into water [61]. The redoxin family, including peroxiredoxin (Prx) and thioredoxin (Trx) enzyme systems, together with other non-enzymatic reducing cofactors, nicotinamide adenine dinucleotide phosphate (NADPH) and/or glutathione, also plays an important role in scavenging ROS [37,61]. They help to convert H_2O_2 into water (Table 1, and Figure 2).

Chronic exposure to oxidative stress may further lead to activation of the first line of antioxidant defense by increasing SOD-, CAT-, and GPx-mediated protective feedback mechanisms that may be able to help mitigate oxidative stress responses to some extent. Interestingly, we observed a depletion of SOD protein expression during the acute period following PTBI [37]. Other studies have also detected diminished SOD activity during the acute phase of TBI that remained low for at least several weeks post TBI [60,62]. Depletion of SOD after TBI could make injured tissue more susceptible to increased $O_2^{\bullet-}$ formation, amplifying post-injury oxidative damage over time. In contrast, studies have reported an increase in CAT during the acute and sub-acute phases following TBI; however, the precise mechanism by which brain injury leads to increased CAT protein expression is currently unknown [37]. Earlier studies have reported decreased SOD, CAT, and GPx antioxidant enzyme activity in AD patients [63,64]. Interestingly, in an AD mouse model, the overexpression of the SOD protein showed great promise in relation to improvement in neurological outcomes [65]. Therefore, in therapeutic applications, SOD and SOD mimetics have great potential to serve as a drug to ameliorate TBI-related oxidative damage.

The other class of natural antioxidants are the non-enzymatic antioxidants, which are mainly acquired from dietary sources; these are also called natural ROS scavengers. The most common dietarily derived antioxidants are Vitamin A (retinol), Vitamin C (ascorbate), Vitamin E (α-tocopherol), carotenoids (carotene, zeaxanthin, lutein, lycopene, cryptoxanthin, retinoids), polyphenols, and flavonoids, among others.

Vitamin A is obtained from dietary sources such as green and yellow vegetables, dairy products, fruits, and meats. Vitamin A can act as a chain-breaking antioxidant by combining with reactive radicals before these radicals can propagate peroxidation in the lipid phase of the cell and generate H_2O_2 [66]. Likewise, Vitamin C is acquired from dietary sources such as fruits and vegetables, and is available as a dietary supplement. Vitamin C has been used as an antioxidant to treat mitochondrial diseases; additionally, it can act as an electron transfer mediator to bypass complex III in combination with Vitamin K at the ETC [67,68]. The oxidized form of Vitamin C is transported into the mitochondria via glucose transporter 1, which helps to maintain a healthy mitochondrial membrane potential and inhibits mitochondrial membrane depolarization [69]. Additionally, Vitamin C facilitates electron movement, favoring energy production [70]. Vitamin E is mainly found in vegetable oil and its derivatives, nuts and seeds. Vitamin E interrupts the chain reaction of oxidant generation and oxidative damage by capturing free radicals.

Table 1. Natural ROS scavengers.

ROS Scavengers	Properties and Mechanisms of Action
Enzymatic ROS scavengers	
Superoxide dismutase (SOD)	Enzyme. Converts superoxide radicals into oxygen and H_2O_2.
Catalase (CAT)	Enzyme in the peroxisomes. Neutralizes H_2O_2 in water.
Glutathione peroxidase (GPx) Thioredoxin system: Thioredoxin (Trx), Peroxiredoxin (Prx), Thioredoxin reductase (TrxR)	Thiol-dependent enzymatic antioxidants. Neutralize H_2O_2 and are recycled by nicotinamide adenine dinucleotide phosphate (NADPH) as a cofactor.
Non-enzymatic ROS scavengers	
Vitamin A (retinol) or carotenoids	Fat-soluble antioxidant. Donates electrons to neutralize free radicals.
Vitamin E (tocopherols and tocotrienols)	Fat-soluble antioxidant. Scavenges lipid peroxyl radicals.
Vitamin C (ascorbic acid)	Water-soluble antioxidant. Donates electrons to neutralize free radicals. Scavenge superoxide.
Carotenoids	Found in various fruits and vegetables. Of the ~600 types of carotenoids, some can synthesize Vitamin A. Neutralizers of ROS.
Polyphenols	Ubiquitously present in fruits and vegetables. Free-radical scavenger.
Flavonoids	Phytochemicals present in plants, fruits, and vegetables. Scavengers of ROS.
Pycnogenol (PYC)	Combination of bioflavonoids with robust capacity to scavenge free radicals.
Alliin	Found in both natural and synthetic compounds. A bioactive compound derived from garlic. Superoxide scavenger.
Allicin	Synthesized from alliin. Inhibits superoxide, nitric oxide (NO) and hydroxyl radicals.
Minerals (copper, zinc and selenium, magnesium)	Precursors to antioxidants that help regulate free radicals.
Coenzyme Q10 (CoQ10), coenzyme Q (CoQ)	Lipid antioxidant. Essential component of the ETC. Protects cells from oxidative damage.
Glutathione	Tripeptide. Detoxifies ROS. Maintains redox balance.
NADPH	NADPH, as a cofactor independently and with redoxins, plays a crucial role in ROS detoxification.
Cytochrome C	Endogenous heme protein located in mitochondria. Oxidized cytochrome C is able to scavenge superoxide radicals.

Carotenoids are another class of antioxidants, and their main dietary sources are red vegetables and fruits (carrots, tomatoes, apricots, plums) and green leafy vegetables (spinach, kale). Indeed, carotenoids are important precursors of Vitamin A [71]. Carotenoids are very efficient quenchers of singlet oxygen and potent scavengers of other ROS–RNS. Similarly, polyphenol antioxidants (flavanols, anthocyanins, isoflavones, phenolic acid), mainly found in fruits (apples, berries, grapes), vegetables (celery, kale, onions), legumes (beans, soybeans), nuts, wine, tea, coffee, and cocoa, can be obtained from nutritional sources. Polyphenol acts as an antioxidant via a direct ROS-scavenging mechanism and the modulation of antioxidant enzymes. Flavonoids are phenolic structures containing natural substances mainly found in fruits, vegetables, grains, bark, roots, stems, flowers, tea, and wine. Flavonoids exert antioxidant, anti-inflammatory, and anti-cholinesterase activities. Flavonoids act as potent inhibitors for several enzymes, such as xanthine oxidase (XO), cyclo-oxygenase (COX), lipoxygenase, and phosphoinositide 3-kinase [72,73]. Pycnogenol (PYC) is a combination of bioflavonoids that is extracted from the bark of the French maritime pine tree (Pinus maritima), and has a robust capacity to scavenge free radicals.

The neuroprotective effects of PYC have been explored in a rodent model of TBI [74]. Additionally, alliin, a garlic-derivative compound, reacts with $O_2^{\bullet-}$ and scavenges by utilizing the xanthine/xanthine oxidase system [75]. Allicin, a derivative of alliin, also inhibits $O_2^{\bullet-}$, nitric oxide ($NO^{\bullet}$), and hydroxyl ($^{\bullet}OH$) radical production [75–77].

Some antioxidants are produced by cells that chelate and/or bind to redox metals, thus protecting the cells against oxidative stress indirectly. Micronutrients such as metal and trace elements (zinc, iron, selenium, and copper) possess antioxidant properties. Supplementation with either selenium or zinc has been found to restore the alterations of mitochondrial parameters, including ETC enzymes and antioxidant enzymes, in several diseases [78].

The membrane-bound coenzyme Q10 (CoQ10) is an important antioxidant that is part of the mitochondrial ETC. It shuttles electrons from complexes I/II to complex III. CoQ10 prevents the generation of free radicals and modifications of proteins, lipids, and DNA. Thus, CoQ10 markedly regulates the cellular redox balance.

Among other non-enzymatic ROS scavengers, one of the key cellular antioxidants is glutathione (e.g., γ-L-glutamyl-L-cysteinylglycine). Glutathione is synthesized from the amino acids L-cysteine, L-glutamic acid, and glycine. Glutathione is an important antioxidant which reacts with ROS using thiol-SH groups of cysteine. Glutathione is a ubiquitously distributed tripeptide antioxidant abundantly present in all cells in millimolar concentrations (~5 mM) [79]. The reduced form of glutathione (i.e., GSH) is involved in various cell functions, including the detoxification of oxidized amino acids/proteins, the biosynthesis of proteins and DNA precursors, amino acid transport, and the maintenance of redox balance. During this process, the endogenously generated oxidized glutathione (GSSG) can be recycled back to GSH by the endogenous Grx system. The GSH/GSSG ratio remains an important indicator of redox homeostasis and imbalance in cell oxidative metabolism.

Another antioxidant, NADPH (e.g., nicotinamide dinucleotide phosphate), works closely with glutathione and other redoxin enzymes to protect against ROS–RNS-induced cell damage. In redoxin systems, NADPH serves as a cofactor, used for transferring and preserving redox potential for multiple antioxidants such as glutathione, Prx, and Trx. This NADPH-induced conversion reactivates the functions of antioxidant molecules. We found that NADPH levels significantly decreased following TBI [37]. This reinforces the importance of exogenous NADPH treatment following TBI to increase the effectiveness of antioxidant proteins as the scavengers of oxidants. Additionally, in cells, endogenous cytochrome C (Cyt C) may act as an $O_2^{\bullet-}$ scavenger since it is reduced by $O_2^{\bullet-}$ and oxidized by H_2O_2 [80]. Cyt C seems to be an ideal antioxidant since Cyt C can regenerate and avoid being damaged during antioxidant reactions [81].

Additionally, several dietary or nutritional supplements serve as conventional (non-targeted) antioxidants in cells. However, all of these natural antioxidants have limited effectiveness in scavenging mitochondrial ROS–RNS and oxidative stress due to their limited ability to cross the mitochondrial biomembranes [82,83]. In the next section, we compile a list of mitochondria-targeted synthetic antioxidants, which may serve as better options to combat ROS–RNS and oxidative stress and may offer neuroprotection.

4.2. Synthetic ROS–RNS Scavengers

Novel synthetic ROS–RNS scavengers targeted towards preventing or minimizing oxidative damage have contributed new insights into potential neuroprotective therapies (Table 2). Superoxide ($O_2^{\bullet-}$) scavengers are important antioxidants due to their ability to mitigate oxidative stress during the acute post-injury phase. One such synthetic compound is Mn (III) tetrakis (4-benzoic acid) porphyrin (MnTBAP), an ROS scavenger. MnTBAP is both an SOD mimetic and peroxynitrite ($ONOO^-$) scavenger [84,85]. Other compounds that can donate electrons to $O_2^{\bullet-}$ are ascorbic acid, cysteine (via the sulfhydryl group), tiron, and carboxy-PTIO (a nitric oxide scavenger), which can also react with superoxide radicals [86–94]. Tiron is a Vitamin E-analog antioxidant that can enhance NF-κB-dependent gene transcription with an anti-apoptotic effect [95]. Carboxy-PTIO is an imidazole-derived free-radical scavenging compound that inactivates $NO^\bullet$ and NO_2, subsequently reacting with water to form nitrite and nitrate. Phenelzine (PZ) is an FDA-approved drug for the management of treatment-resistant depression, panic disorder, and social anxiety disorder that functions as an MAO inhibitor [96]. PZ has aldehyde-scavenging properties. PZ administration was also shown to significantly improve mitochondrial respiration following TBI [96].

Table 2. Synthetic ROS scavengers and detoxifiers.

ROS Scavengers and Detoxifiers	Properties and Mechanisms of Action
Non-targeted compounds	
MnTBAP	$O_2^{\bullet-}$ scavenger. Possesses SOD- and catalase-like activity. Also scavenges $ONOO^-$.
Cysteine	Amino acid. $O_2^{\bullet-}$ scavenger.
Tiron	Reduced and oxidized Tiron species. Reacts with $O_2^{\bullet-}$ radical.
Carboxy-PTIO	Specific NO scavenger. Reacts with $O_2^{\bullet-}$ radical.
Phenelzine	FDA-approved drug. MAO inhibitor. Aldehyde-scavenging properties partially protect against oxidative damage.
Mitochondria-targeted compounds	
MitoVit-E	Vitamin E attached to TPP. Reduces mitochondrial oxidative damage.
MitoQ	CoQ10 derivative linked with TPP. Scavenges mitochondrial ROS.
Plastoquinone (SkQ1)	Targeted antioxidant. Scavenges mitochondrial ROS.
Edaravone	Used clinically as a neuroprotective compound. Reduces oxidative damage and lipid peroxidation.
Mito TEMPOL	Cell permeable, stable nitroxide. SOD mimetic.
Elamipretide (SS-31)	Cationic tetrapeptide freely permeable to the mitochondria. Reduces the production of toxic ROS.
Cerium oxide nanoparticles (Nano-CeO2)	Cerium atoms linked by oxygen atoms. Scavengers of ROS.
Metalloporphyrins	Manganese and iron complexes. Synthetic catalytic antioxidants that mimic the body's own antioxidant enzymes.
Phenyl-tert-butylnitrone (PBN)	Nitroxide radical. ROS-scavenging properties.
Glutathione precursors	
NAC	A cysteine prodrug. Replenishes intracellular glutathione level.
NACA	N-acetyl cysteine (NAC) analog. Glutathione precursor.
D-NAC	Dendrimer-tagged NAC. Serves as a prodrug to synthesize glutathione.
S-adenosyl methionine (SAMe)	SAMe is processed stepwise into cysteine synthesis, and ultimately synthesize glutathione.

Moreover, to overcome the limited effectiveness of natural ROS scavengers, several synthetic mitochondrial ROS scavengers have been designed to cross the BBB and accumulate in neuronal mitochondria. These compounds are formulated to target mitochondria at the injured region to neutralize ROS and promote the mitigation of oxidative damage, together with improving bioenergetic function. The development of antioxidants capable of restoring mitochondrial function following brain injury is highly significant since redox homeostasis dysregulation is a critical factor in the cell death pathway during the acute, sub-acute, and chronic phases of TBI. Also, specific mitochondrial targeting leads to more precise and effective mitigation of redox homeostasis. We have compiled a list of such covalently modified compounds in this table.

The synthetic mitochondrial-targeted Vitamin E compound (MitoVit-E) is created by covalently attaching natural Vitamin E (α-tocopherol) to a triphenylphosphonium (TPP^+) cation. MitoVit-E facilitates the accumulation of TPP^+ in the mitochondrial matrix against the negatively charged mitochondrial membrane potential ($\Delta\Psi m$). This unique feature makes MitoVit-E an effective mitochondria-targeted ROS scavenger. By utilizing the concentration gradients of $\Delta\Psi$m, MitoVit-E decreases ROS production and apoptosis in aortic endothelial cells via peroxide-induced oxidative stress and apoptosis [97,98]. One disadvantage of Vitamin E is that it is not a catalytic antioxidant and therefore its scavenging activity is not regenerated. Another well-studied mitochondria-targeted antioxidant is mitoquinone (MitoQ), a ubiquinone derivative conjugated to the TPP^+ cation that serves as a potent reactive oxygen species (ROS) scavenger. MitoQ has structural similarity with endogenous components of the mitochondrial ETC ubiquinone; therefore, it may help in assisting efficient electron transfer through the ETC [99,100]. The active form of MitoQ, i.e., ubiquinolis able to scavenge ROS and is being modified into its inactive form, ubiquinone. This inactive form is continuously recycled back into its active form by the mitochondrial complex II. This reduction–oxidation cycle enables MitoQ to maintain an efficient chain-breaking antioxidant capacity. MitoQ treatment has been shown to inhibit mitochondrial oxidative damage in rodent models of cardiac ischemia and reperfusion injury [101]. The antioxidant properties of MitoQ were further demonstrated in several preclinical models of TBI, where it increased the activity of antioxidant enzymes and reduced oxidative damage [102,103]. Plastoquinonyl-decyl-triphenylphosphonium bromide (SkQ1) is another class of mitochondria-targeted antioxidant [104]. In the case of SkQ1, the phosphorus cation bound to three phenyl rings (TPP^+) is conjugated to plastoquinol via a decyl linker. The binding of this cation to the phenyls ensures the ability of SkQ1 to penetrate membranes [105]. A positive electrical charge leads to a thousand-fold accumulation of SkQ1 in the mitochondrial membrane's inner layer [105]. SkQ1 is able to reduce cardiac ischemic injury, and is well known for lipid peroxidation inhibition [106–109].

Edaravone is a free-radical scavenger that can quench hydroxyl radicals and hydroxyl radical-dependent lipid peroxidation. It is an FDA-approved compound for the treatment of acute ischemic strokes and amyotrophic lateral sclerosis (ALS) [110]. Additionally, edaravone has shown promising beneficial effects in a wide range of diseases, such as PD, AD, atherosclerosis, chronic heart failure, and diabetes mellitus [111–114]. Other potential synthetic antioxidants designed to reduce oxidative damage effectively include Mito TEMPOL, elamipretide (SS-31), cerium oxide nanoparticles (Nano-CeO_2), metalloporphyrins, and phenyl-tert-butylnitrone (PBN) [115–121]. However, their roles in TBI have not yet been investigated. Research on the evaluation of ROS–RNS scavengers in TBI is ongoing, and the field continues to explore novel approaches and compounds to mitigate oxidative stress and improve behavioral outcomes following TBI [52,58,122,123]. Indeed, several preclinical studies have shown the therapeutic efficacy of mitochondria-targeted antioxidants by improving cognitive and functional recovery post TBI [122,123]. Thus, this strategy may offer new hope for treating TBI patients.

Amongst synthetic ROS scavengers and detoxifiers, novel precursors of glutathione play a significant role. Glutathione, a ubiquitous reducing sulfhydryl tripeptide, plays a major role in ROS–RNS detoxification. Many studies have reported a depletion of glutathione and its precursors, namely cysteine, methionine, and glycine, in brain tissue and cerebrospinal fluid (CSF) following TBI [37,49,50,124]. Therefore, several strategies have explored boosting glutathione levels following TBI to protect neurons against oxidative damage. One approach is to administer glutathione directly. Glutathione injections have been used in the past to boost glutathione levels in blood and skin; however, there was no systemic study available to prove its efficacy [125]. Direct enhancement of glutathione comes with its own challenges like short half-life, absorption, BBB permeability, and limited brain bioavailability [125].

The de novo synthesis of glutathione is primarily controlled by the cellular concentration of cysteine. In keeping with this, NAC and its analogs, such as the cysteine supplement, are effective at raising levels of glutathione in various neurological diseases and injuries, preclinically and clinically [126–129]. Therefore, various glutathione prodrugs or antioxidant supplements to boost innate glutathione levels have been investigated.

N-acetyl cysteine (NAC) is perhaps the most widely studied glutathione precursor to act as an antioxidant. NAC has been approved by the FDA for treating hepatotoxic doses of acetaminophen (Tylenol). Additionally, NAC has been widely used because of its mucolytic effects, taking part in the therapeutic protocols of cystic fibrosis. Over the past decade, studies have documented the positive outcome of NAC treatment for many CNS diseases, including TBI [130]. Additionally, NAC's ability to replenish glutathione, maintain cellular homeostasis, and support mitochondrial function has been successfully demonstrated in TBI [131–135].

Clinical treatment with NAC has been shown to upregulate glutathione-centered pathways in the CSF of severe TBI pediatric patients (ClinicalTrials.gov NCT01322009) [136–138]. NAC treatment was evaluated in U.S. service members who had been exposed to a blast-induced mild TBI [139]: the outcome of this study demonstrated NAC as safe and effective pharmaceutical agent for acute countermeasure. NAC treatment has beneficial effects on the injury severity, and resolution of post-traumatic sequelae of blast-induced mild TBI (ClinicalTrials.gov NCT00822263) [139]. Furthermore, NAC's neuroprotective effects are mediated by both antioxidant and anti-inflammatory mechanisms [140–143]. These multimodal neuroprotective properties of NAC may confer significant benefits on the complex and heterogenous nature of TBI pathology.

The BBB permeability of NAC is limited by its physiochemical properties, such as its acidic nature and negative charge [144,145]. Notably, numerous studies evaluating the neuroprotective properties of NAC have yielded inconsistent results, which may be due to its low bioavailability [144,145]. A potential strategy for overcoming the low bioavailability of NAC is to use an NAC analog where the carboxyl group of NAC is neutralized, thus making it more hydrophobic and increasing its BBB permeability [146]. In this regard, the preparation of NAC analogs, such as N-acetylcysteine amide (NACA), is very attractive and may have advantages over NAC in treating CNS pathologies due to the improved stability and bioavailability [147]. For instance, there are studies reporting the neuroprotective efficacy of NACA in neurological diseases including PD, AD, and HIV-associated neurological disorders [148–150]. In the same line of effort, we have demonstrated that NACA effectively reduces oxidative damage, maintains the glutathione level, and improves mitochondrial bioenergetics following TBI [128,129,151]. Other studies have reported similar outcomes in spinal cord injury (SCI) patients [129]. Thus, NACA may offer neuronal protection by reducing oxidative stress and supporting cellular pathways to limit mitochondrial dysfunctions following TBI.

To enhance NAC's bioavailability and address neurological conditions, researchers are investigating an alternative intranasal route for its direct delivery to the CNS via neuronal pathways, thus minimizing the BBB permeability issues [152]. However, the optimal dosing regimen for this intranasal route of NAC administration still needs to be further investigated at the preclinical level for TBI.

Recently, researchers have used nanoparticle delivery systems, such as dendrimers, to ensure targeted and effective drug delivery to the CNS. Hydroxyl-terminated polyamidoamine (PAMAM) dendrimer, a dendrimer linked with NAC (D-NAC), has shown to be a promising route of drug delivery to injury sites within the brain. In particular, D-NAC has been investigated as a drug delivery system to target cells involved in neuroinflammation [153]. In the presence of a brain injury, D-NAC traverses the BBB and localizes specifically in activated microglia and astrocytes, and the extent of its uptake correlates with the extent of the injury [154,155]. D-NAC also has been shown to be effective in improving myelination and motor functions in cerebral palsy [156,157]. The protective role of D-NAC has been established in ischemic brain injury, asphyxia brain injury [158–160], and other CNS pathologies like choroidal and retinal neovascularization [161]. Collectively, novel dendrimer-based delivery methods, such as D-NAC, appear to be promising avenues for targeting therapeutic agents in CNS diseases.

Similarly, another compound that aids in restoring glutathione synthesis by recycling its precursor cysteine is S-adenosyl methionine (SAMe). SAMe has been studied for its potential neuroprotective efficacy in several CNS diseases [162,163]. Besides providing amino acids during methyltransferase reactions for glutathione synthesis, SAMe serves as a key metabolite in many biochemical reactions, and is available as a dietary supplement. Depletion of methionine and its crucial metabolites has been reported in TBI; therefore, restoring methionine metabolites with SAMe supplementation may improve its outcome [124].

A thorough understanding of methods to replenish glutathione and the application of innovative technology to advance targeted therapy in research is critically important when considering therapies to combat TBI secondary pathogenesis. Similarly, the development of neuroprotective formulations to enhance signaling pathways to upsurge innate antioxidants as a potential tool for the therapeutic treatment of neurological diseases represents an important goal for current neuroscience research.

5. Signaling Pathway Modulators for Cellular Antioxidant Synthesis

The initiation of redox homeostasis originates from extracellular or intracellular signals via nuclear receptors and mitochondria-mediated pathways. There are intra- and extracellular signaling pathways that activate the protective mechanisms that particularly trigger the endogenous synthesis of antioxidants. Inducers such as ROS, oxidative stress, mitophagy, apoptosis, excitotoxicity, ischemic insults, calcium, neurotransmitters, exercise, or therapeutic treatment (agonists/antagonists) may trigger the onset of signal transduction via modulating several transcription factors in the nucleus, thereby activating gene expression of downstream protein expression. More specific to the current review topic, there are several inducers listed below that may be able to modulate notable antioxidant signaling pathways, such as the Nrf2, AKT, SIRT1, PGC1α, and mTOR signaling pathways (Table 3). The Nrf2 pathway centers around the broad-reaching transcription factor Nrf2, which modulates the transcription of a myriad of endogenous antioxidants. Protein kinase B, a serine/threonine kinase (AKT), is the main mediator of the downstream effector protein phosphoinositide 3-kinase (PI3K). AKT serves as the central component in numerous signaling pathways regulating cell metabolism, growth, proliferation, and survival. Thus, activating AKT can help preserve typical mitochondrial function across several disease conditions [164]. Additionally, AKT regulates Nrf2 to affect the transcription of pro- and antioxidant enzymes and maintain the cellular redox state [165]. Likewise, SIRT1 is a deacetylase that controls the expression of a multitude of antioxidants and oxidative stress modulators like PGC-1α, which plays a major role in the antioxidant defense system. The rapamycin (mTOR) signaling pathway integrates both intracellular and extracellular signals, and serves as a regulator of cellular metabolism, growth, proliferation, and survival. These pathways, in turn, modulate gene expression and the protein biosynthesis of downstream targets, such as antioxidants and mitochondrial biogenesis proteins. Thus, these pathways may be able to modulate ROS–RNS levels, keep the redox balance in check, and maintain cellular integrity. An overview of cell signaling pathways favoring cellular antioxidants synthesis and neuroprotection is illustrated in detail (Figure 4) and discussed below.

Table 3. Signaling pathway modulators.

Pathway Modulators	Properties and Mechanisms of Action
Nrf2 activators	
Omaveloxolone (RTA-408)	Synthetic compound. FDA-approved for the treatment of FA. Prevents Nrf2 degradation.
Dimethyl fumarate (DMF)	Synthetic compound. Activates the Nrf2 pathway and AKT pathway.
Curcumin	Derived from turmeric. Activates the Nrf2 pathway.
Sulforaphane	Naturally found in cruciferous vegetables. Activates Nrf2 by inhibiting Keap1.
Epigallocatechin gallate (EGCG)	Abundant in green tea. Activates the Nrf2 pathway and has antioxidant and anti-inflammatory properties.
Quercetin	Present in various fruits, vegetables and grains. Activates Nrf2 and SIRT1.
Oltipraz	Synthetic compound. Activates Nrf2 by modifying Keap1.
Bardoxolone methyl	Synthetic compound. Activates the Nrf2 pathway.

Table 3. *Cont.*

Pathway Modulators	Properties and Mechanisms of Action
SIRT1, PGC-1α, and mTOR modulators	
Resveratrol	Natural polyphenol compound. Most-relevant SIRT1 and mTOR modulator, AKT activator, Nrf2 activator and PGC-1α activator.
Naringenin	Natural citrus flavonoid. Modulates SIRT1.
SRT2104	Synthetic compound. SIRT1 activator.
1,4-dihydropyridine derivative	Synthetic compound. SIRT1 activator.
Naphthofuran derivative	Synthetic compound. SIRT1 activator.
Bisarylaniline derivative	New synthetic analog. SIRT1 activator.
Berberine	Small molecule isolated from various plants, mainly used in Chinese traditional medicine. PGC-1α activator.
Metformin	Anti-diabetic drug. Activator of AMPK, which further regulates PGC-1α.
Rapamycin/Sirolimus	Bacterial origin natural product. mTOR inhibitor and increases antioxidant defense.
Everolimus	Newly developed mTOR inhibitor. Rapamycin analog.
Temsirolimus	Newly developed mTOR inhibitor. Rapamycin analog.

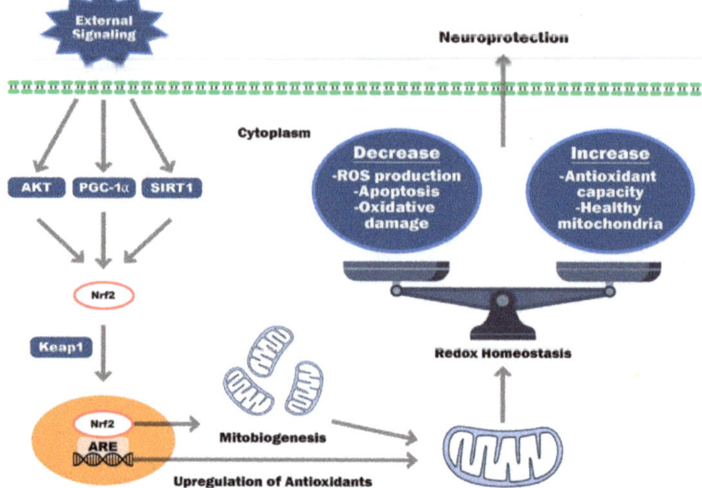

Figure 4. Illustration of signaling pathway modulators involved in endogenous cellular antioxidant synthesis. Extracellular stimulants such as mitophagy, ROS, and oxidative damage may lead to the activation of cell signaling pathways such as AKT, PGC-1α, mTOR, and SIRT. These signaling pathways are involved in upregulating endogenous antioxidant homeostasis by activating nuclear antioxidant response element (ARE) signaling via the activation of common transcription factors Nrf-2 and Keap1. Activation of the nuclear ARE gene upregulates mitochondrial biogenesis. Additionally, ARE gene expression activation may also leads to activation of several mitochondrial antioxidant transcription factors, thereby protein biosynthesis and protect cells against external stimuli. Increased antioxidant levels further balance redox homeostasis by decreasing cellular ROS, oxidative stress, and apoptotic cell death response together by improving the cellular antioxidant capacity and overall health of mitochondria. By activating these pathways, therapeutic compounds may be further able to offer neuroprotection following TBI and CNS diseases.

5.1. Nrf2 Activators

One of the main cellular signaling and oxidative stress defense pathways is the nuclear factor erythroid 2 (Nrf2)-dependent transcriptional mechanism. Nrf2 is responsible for regulating an extensive panel of antioxidant enzymes involved in the detoxification of oxidative stress. Several strategies have been proposed to activate this pathway to counter ROS production and promote neuroprotection. Nrf2 is a transcription factor responsible for regulating the expression of various downstream genes that modulate the oxidative stress response through regulating the antioxidant response element (ARE). Thus, Nrf2-targeted genes affect many vital antioxidants through ARE gene regulation, such as SOD, CAT, and GPX, among others, which help in combatting ROS [166]. Additionally, Nrf2 downregulation supports the decreased efficiency of mitochondrial oxidative phosphorylation. It is widely thought that the Nrf2 pathway plays an important role in TBI pathogenesis, and even in other neurological diseases like AD and PD [167–169]. In a mouse model of TBI, Nrf2 was found to be downregulated in cortical tissue, leading to increased oxidative stress, inflammation, and apoptosis [168]. Therefore, targeting and activating Nrf2 signaling is a potential novel target following the oxidative stress-centered pathology of TBI. Owing to its excellent therapeutic potential in CNS diseases in both preclinical and clinical settings, recently, several Nrf2 activators have been approved by the FDA.

Omaveloxolone (RTA-408) and dimethyl fumarate (DMF) are both FDA-approved Nrf2 activators used to treat various neurological conditions like FA and MS. Other Nrf2 activators like curcumin, sulforaphane, epigallocatechin gallate (EGCG), quercetin, oltipraz, and bardoxolone methyl also hold promise as therapeutic agents due to their antioxidant properties [170,171].

RTA-408 is one of the FDA-approved Nrf2 activators for treating FA, a progressive neurodegenerative condition. RTA-408 affects the Nrf2 pathway by preventing Nrf2 ubiquitination and degradation, leading to Nrf2 translocation to the nucleus and increased antioxidant expression. RTA-408 therapy has been found to enhance mitochondrial function and improve neurological symptoms, cognitive impairment, and neuroinflammation in multiple preclinical and clinical models of CNS conditions like epilepsy [172]. RTA-408's efficacy in minimizing CNS pathology and its mitochondria-protective properties makes it a potential candidate for treating TBI and other neurological diseases [173–175].

Similarly, dimethyl fumarate (DMF) is another Nrf2 activator approved by the FDA for treating MS [176,177]. DMF influences the Nrf2 pathway by modifying Keap1, thus promoting Nrf2 nuclear translocation. DMF also activates AKT pathways and thus promotes neuroprotection [178]. DMF upregulates several antioxidants including glutathione, bolstering the downstream antioxidant capacity in CNS conditions like MS, cerebral edema, TBI, and intracerebral hemorrhage [179–183]. DMF treatment in clinical settings has shown long-term efficacy in reducing relapse rates and minimizing lesion formation in relapsing forms of MS [184]. Furthermore, DMF treatment was found to increase neuronal mitochondrial biogenesis via Nrf2 regulation along with improved mitochondrial function and neurological symptoms in a preclinical model of MS [185]. Additionally, DMF has shown to improve cognitive functions in animal models of AD and PD [186]. Together, this evidence emphasizes that DMF has broader therapeutic applicationfor MS, and other neurodegenerative diseases.

Several Nrf2 activators listed here exhibit potential health benefits. Curcumin, found in turmeric, a commonly used spice in Indian cuisine and in traditional medicine, activates the Nrf2 pathway, leading to increased antioxidant and detoxifying enzyme production [187]. Curcumin has shown to be effective against cancer, cardiovascular diseases, and various metabolic and neurological conditions [188]. Sulforaphane, another compound, is mainly found in cruciferous vegetables such as broccoli, cabbage, and brussels sprouts. It activates Nrf2 by inhibiting the protein Keap1 [189]. It similarly enhances endogenous antioxidants and detoxifying enzymes. Sulforaphane has shown therapeutic potential against neurodegenerative diseases [190]. Epigallocatechin gallate (EGCG) is a catechin present in green tea. It activates Nrf2, and it is known for its antioxidant and health-promoting properties [191]. Research suggests that ECCG may help protect neurons from oxidative damage and improve cognitive function [192]. Quercetin, found in fruits and vegetables, is an Nrf2 pathway activator that reduces inflammation and improves antioxidant defense mechanisms. The therapeutic effects of quercetin have been investigated in cancer, cardiovascular diseases, and neurodegenerative conditions [193,194]. Oltipraz, a synthetic compound, activates Nrf2 and reduces oxidative stress in cancer, and additionally has shown neuroprotective benefits [195]. Bardoxolone methyl, another synthetic compound, activates the Nrf2 pathway and stimulates antioxidant enzyme production [196,197]. It has been found to be effective against various disease conditions. Although Nrf2 compounds have shown promising protective effects in various health conditions, further research is warranted to confirm and fully understand their safety and efficacy.

5.2. SIRT, PGC-1α, AKT, and mTOR Modulators

There are other critical regulatory mechanisms in redox homeostasis, such as the Silent Information Regulator (SIRT) genes, also known as Sirtuins, which stimulate antioxidant expression of several enzymes. One of the members of the SIRT family, SIRT1, is a nicotinamide adenine dinucleotide (NAD^+)-dependent deacetylase that plays a wide range of roles in transcriptional regulation, inflammation, cell survival, and repair mechanisms. It guards against oxidative stress by activating the gene transcription of peroxisome proliferator-activated receptor gamma coactivator-1α (PGC-1α) via the removal of the acetyl group [198]. PGC-1α is a transcriptional coactivator that is able to upregulate mitochondrial biogenesis, and plays a central role in regulating the oxidative stress defense [199]. SIRT1 is described as a complex target for multiple strategies addressed for the prevention/treatment of several chronic age-related diseases and CNS diseases. Natural and synthetic SIRT1 modulators have been examined. This review examines compounds of a natural origin that have recently been found to upregulate SIRT1 activity, such as polyphenolic products in fruits, vegetables, and plants, including resveratrol, quercetin, and curcumin.

Resveratrol is a natural polyphenol found in various plant sources, such as grapes, berries, and peanuts, which acts as an antioxidant by activating SIRT1. SIRT1 is involved in various cellular processes, including mitochondrial biogenesis [200–202]. Resveratrol also activates the Nrf2 signaling pathway to ameliorate oxidative stress and improve mitochondrial function [203]. Moreover, resveratrol activates the PI3K/AKT pathway. On the other hand, resveratrol modulates the recently identified mammalian target of the rapamycin (mTOR) and Janus kinase/signal transducer and activator of transcription (Jak/STAT) pathways to enhance antioxidant defense and positively modulate mitochondrial function. Resveratrol has been suggested to influence mitochondrial dynamics by modulating the balance between mitochondrial fusion and fission, thus regulating mitophagy. Proper regulation of fusion/fission processes is crucial for maintaining mitochondrial health and function. Therefore, resveratrol helps preserve mitochondrial integrity.

Moreover, resveratrol has been shown to be beneficial in neurological diseases like AD, PD, HD, and ALS [201,202,204]. The evidence supports resveratrol's role in attenuating TBI-associated behavioral abnormalities, brain edema, and pathophysiology [205–210]. In a TBI preclinical model, resveratrol improved mitochondrial biogenesis and function by activating the PGC-1α signaling pathway [210]. PGC-1α is central modulator of cell metabolism, where it regulates mitochondrial biogenesis and oxidative metabolism, and controls the expression of antioxidants. It is important to note that while numerous preclinical studies and some clinical trials have explored the potential benefits of resveratrol, the findings are often mixed, and the optimal dose and duration of resveratrol supplementation for specific health conditions remain areas of ongoing research. Other compounds like berberine and metformin which activate PGC-1α may be more useful in neurodegenerative diseases conditions [211–214].

Despite the potential positive health benefits of resveratrol, it exhibits low CNS bioavailability. This unfavorable pharmacokinetic profile of natural SIRT1 modulators has prompted the development of novel compounds that can positively modulate SIRT1 activity and display better neuroprotective efficacy profiles. Numerous synthetic SIRT1 modulators have been formulated, such as SRT2104, 1,4-dihydropyridine derivative, naphthofuran derivative, and bisarylaniline derivative. However, studies to confirm the pharmacokinetic profiles of these compounds are ongoing. These compounds may have implications in CNS therapeutic development. Additionally, mTOR modulators like everolimus and temsirolimus might regulate ROS through mTOR-mediated antioxidant defense [215]. The mTOR signaling pathway is at the core of many metabolic activities; its activation improves oxidative stress adaptation by activating Nrf2-associated antioxidant signaling [216].

6. Challenges and Future Approach

Despite significant strides in characterizing TBI pathophysiology and identifying therapeutic interventions, the landscape is marred by numerous clinical trial failures, even after promising preclinical success [217–220]. Several conceptual and methodological issues have undoubtedly contributed to the hitches in translating the preclinical results of antioxidant therapy to a clinical setting. Major challenges lie in the inherent heterogeneity of traumatic brain injuries, the complex and multifaced pathology of brain injuries, the limited information on their molecular pathology, the clinical predictiveness/relevance of animal models, the adequacy of pharmacological methodology, the ill-defined category of TBI, and the outcome measures used. Herein, we reviewed some of these critical problems and potential solutions.

The complex and multifaceted nature of TBI pathophysiology complicates treatment, rendering it challenging to address comprehensively with a single targeted drug. A drug targeting multiple components of the secondary TBI cascade may have superior potency compared to a drug that has a single target. For instance, the Nrf2 pathway activator discussed above broadly modulates intracellular and mitochondria-mediated oxidative and inflammatory responses and may support multiple innate defense mechanisms against TBI pathology. Therefore, any drug with pleiotropic mechanisms of action may be advantageous for TBI research [221].

It has been suggested that the complex pathophysiology of TBI may even possibly be addressed through a combination of therapeutic interventions [222]. The need for integrated multitargeted treatments for TBI has been recognized [223]. At the mitochondrial level, we have identified significant impairment of multitargeted homeostasis, including bioenergetics, calcium, apoptosis, and redox mechanisms, post TBI [11,37]. Providing acetate supplements such as glyceryl triacetate (GTA) and acetyl-L-carnitine (ALC) to boost energy production could contribute to neuronal repair and recovery in the energy deprivation-related pathophysiology of TBI [224,225]. Combining acetate therapy with antioxidants may have additive or synergetic mitochondrial mechanism-targeted neuroprotective efficacy compared to monotherapy in attenuating TBI pathology or promoting recovery. Thus, an effective approach to interrupt post-injury oxidative brain damage might involve the combined treatment of antioxidants with mechanistically complementary energy substrates that simultaneously provide a boost in their antioxidant capacity. Ideally, numerous combination therapies should undergo preclinical testing, with the best combinations chosen for further clinical exploration. An efficient and validated screening platform for candidate therapeutics, sensitive and clinically relevant biomarkers and outcome measures, and standardization and data sharing across centers would greatly facilitate the development of successful combination therapies for TBI [221].

There remains a strong need for rigorous studies to understand the temporal profile of oxidative injury mechanisms following preclinical heterogeneous models of TBI, which may identify novel targets for evaluating neuroprotective therapeutics. As the pathophysiology of secondary injury evolves over time, antioxidant interventions must be able to adapt to evolution in the molecular causes of injury; each compound is likely to have a unique therapeutic time window based on the molecular timeline of secondary injury, during which it is most effective and outside of which it may lack significant benefit [222]. Thus, it is crucial to determine the most efficacious therapeutic window for initiating each antioxidant based on its physiochemical properties and molecular targets in TBI.

Many pharmacological methodological issues have limited the clinical application of antioxidant therapies. Failure to demonstrate sufficient CNS penetration, inadequate dose optimization, or failure to show effectiveness with the treatment delays common in human studies represent some key issues. Nanotechnology, including dendrimers and structural modifications like TPP, discussed earlier, offers excellent potential to increase the efficiency and efficacy of antioxidant therapeutics as their customizable size, stealthy chemistry, and multifunctionality allow them to enhance drug penetration through the BBB. One strategy to improve the delivery of antioxidants to the brain involves the use of the nose-to-brain route, with administration of the antioxidant in specific nasal formulations and its passage to the CNS mainly through the olfactory nerve route [226,227].

Outcomes between individuals following TBI greatly vary, making antioxidant treatment or other treatments for TBI so challenging [228]. The "one-size-fits-all" approach to TBI medicine that has been followed for many years is questionable. Due to this, many researchers have begun to investigate the possibility of using precision medicine techniques to address TBI treatment [228]. TheFDA-approved novel biomarkers for TBI screening, such as GFAP and UCH-L1, which are released from the brain into the bloodstream within 12 h of injury [229]. Notably, personalized stratification based on recently discovered biomarkers can account for individual variability, forming a practical tool that can be used to assist clinical decision making for early TBI diagnosis, and evaluation of therapeutics intervention. This approach holds the potential to overcome the challenges posed by TBI heterogeneity, offering a more tailored and effective strategy for treating TBI patients. An increased understanding of additional biomarkers across the TBI spectrum is needed to improve antioxidant precision medicine in TBI. We stress the importance of further research into this area to improve the clinical efficacy of antioxidant therapy for TBI in the future.

7. Holistic Approach to Improve TBI Outcomes

A holistic approach to provide support that looks at the whole person, not just their CNS health, should be taken into consideration for TBI management. TBI alone or in combination with polytraumatic injuries (i.e., TBI + polytrauma) heavily impacts the body, damaging the brain tissue and shifting homeostasis in many bodily systems such as the immune system, GI system, lungs, heart, and gut microbiota [11,12,230]. This systemic insult can result in changes throughout the body that can increase morbidity and even mortality following TBI [231]. Herein, we reviewed a bidirectional relationship between the gut microbiome and the brain, which also plays a role in TBI-associated pathology. Damage to the brain alters the composition of the microbiome; the altered microbiome affects TBI severity, neuroplasticity, and metabolic pathways through various bacterial metabolites [232]. Significant changes in the gut microbiome within two hours following a TBI was demonstrated in rats, and dysbiosis persisted throughout the study period of 7 days [233]. Furthermore, gut dysbiosis was associated with neuronal loss 3 months after TBI [234]. Notably, emerging research indicates a potential link between the gut microbiome and neurological health [232,233]. The interaction of the CNS and gut signaling pathways includes chemical, neural, metabolic, immune, and endocrine routes, and imbalances in these pathways have been associated with neurological disorders like PD, MS, and AD [235]. Therefore, microbiota manipulation has been proposed as a treatment target for such diseases [236]. While this field of research evolves, maintaining a healthy gut through diet and lifestyle may positively impact outcomes following TBI.

The gut–brain axis suggests that a bidirectional communication between the gut and the brain may influence neurological conditions. A balanced microbiome may contribute to antioxidant production, potentially influencing our body's ability to combat oxidative damage [237]. Recently, it has been shown that the intake of antioxidant compounds might modulate the composition of beneficial microbial species in the gut, and these commensal bacteria often exhibit antioxidant properties [238]. Thus, the antioxidant supplements and balanced microbiome complement each other due to their mutualistic associations. Probiotic-derived metabolites such as butyrate, propionate, and acetate may serve as alternative energy sources for an injured brain and may improve mitochondrial function following TBI [230,239]. Further supporting the benefit of antioxidants, polyphenol antioxidants such as quercetin, resveratrol, and flavonoid intervention have shown to selectively inhibit pathogenic bacteria in the gut [240]. Additionally, short-chain fatty acids (SCFAs), the main metabolites produced in the colon by bacterial fermentation may contribute to host energy production and ROS modulation [241]. Furthermore, the gut microbiota has been shown to regulate key transcriptional co-activators, transcription factors and enzymes involved in mitochondrial biogenesis, such as the PGC-1α, SIRT1, and AMPK genes [241]. Thus, metabolites produced by commensal gut microbiota, including the beneficial SCFAs, might influence key mitochondrial functions related to TBI pathobiology such as energy production, mitochondrial biogenesis, and redox balance, making them a potential therapeutic target.

Due to the high energy demands exist during the repair of an injured brain; and growing our understanding of brain-gut microbiota crosstalks for the host's overall health, we have briefly highlighted the existence of interactions between the brain, gut microbiota and mitochondrial redox homeostasis. However, the underlying mechanisms through which antioxidants might influence the gut–brain axis to exert neuroprotection in TBI is yet to be fully elucidated. This knowledge gap is of paramount clinical significance.

8. Conclusions

Emerging evidence indicates that mitochondrial homeostasis is central to the secondary injury cascade in TBI pathology, which lacks approved therapy. Loss of this homeostasis, including redox imbalance, excitotoxicity, calcium overload, bioenergetics failure, and apoptosis, are the main participants in mitochondria-centered damage following TBI, contributing to neuronal death and long-term neurobehavioral sequelae. Thus, mitochondria-targeted antioxidant strategies in TBI have been increasingly studied, as their maintenance could potentially preserve neuronal homeostasis and crucial brain functions. Properly selecting mitochondria-targeted antioxidants, greater understanding of the underlying injury mechanisms, better-tailored treatments, and the application of novel pharmacological methodology offer new insights into the successful management of TBI, and its translation from bench to bed. Therefore, the antioxidants reviewed here could be a viable therapeutic option to minimize secondary damage and improve the quality of life after TBI. However, further research using antioxidants as a treatment for TBI is necessary in order to move towards adding them into routine care for TBI.

Author Contributions: All authors contributed to the discussion and writing of the manuscript. All authors have read and agreed to the published version of the manuscript.

Funding: The research conducted here was supported by the US Army Combat Casualty Care Research Program (CCCRP) H_001_2018_WRAIR (FY18-23) and ongoing research support CO240012_WRAIR (FY24-26).

Informed Consent Statement: This material has been reviewed by the Walter Reed Army Institute of Research. There are no objections to its presentation and/or publication. The opinions or assertions contained herein are the private views of the authors, and are not to be construed as official, or as reflecting true views of the Department of the Army or the Department of Defense.

Acknowledgments: No experimental studies were conducted to generate any preclinical or clinical data for this manuscript. We thank visual information specialist Christopher S. Nititham, Department of Strategic Communications at WRAIR for providing outstanding graphical design support for this work.

Conflicts of Interest: All authors have declared no conflicts of interest.

References

1. CDC. National Center for Health Statistics. 2022. Available online: https://wonder.cdc.gov/mcd.html (accessed on 8 September 2023).
2. James, S.L.; Theadom, A.; Ellenbogen, R.G.; Bannick, M.S.; Montjoy-Venning, W.; Lucchesi, L.R.; Abbasi, N.; Abdulkader, R.; Abraha, H.N.; Adsuar, J.C.; et al. Global, regional, and national burden of traumatic brain injury and spinal cord injury, 1990-2016: A systematic analysis for the Global Burden of Disease Study 2016. *Lancet Neurol.* **2019**, *18*, 56–87. [CrossRef]
3. Dewan, M.C.; Rattani, A.; Gupta, S.; Baticulon, R.E.; Hung, Y.-C.; Punchak, M.; Agrawal, A.; Adeleye, A.O.; Shrime, M.G.; Rubiano, A.M.; et al. Estimating the global incidence of traumatic brain injury. *J. Neurosurg.* **2018**, *130*, 1080–1097. [CrossRef] [PubMed]
4. Finkelstein, E.A.; Corso, P.S.; Miller, T.R. *The Incidence and Economic Burden of Injuries in the United States*; Oxford University Press: New York, NY, USA, 2006. [CrossRef]
5. Coronado, V.G.; McGuire, L.C.; Faul, M.; Sugerman, D.E.; Pearson, W.S. Traumatic brain injury epidemiology and public health issues. *Brain Inj. Med. Princ. Pract.* **2012**, *84*, 84–100.
6. Nagalakshmi; Sagarkar, S.; Sakharkar, A.J. Epigenetic mechanisms of traumatic brain injuries. *Prog. Mol. Biol. Transl. Sci.* **2018**, *157*, 263–298.
7. Mena, J.H.; Sanchez, A.I.; Rubiano, A.M.; Peitzman, A.B.; Sperry, J.L.; Gutierrez, M.I.M.; Puyana, J.C. Effect of the Modified Glasgow Coma Scale Score Criteria for Mild Traumatic Brain Injury on Mortality Prediction: Comparing Classic and Modified Glasgow Coma Scale Score Model Scores of 13. *J. Trauma* **2011**, *71*, 1185–1192; Discussion 1193. [CrossRef] [PubMed]
8. Alves, W.M.; Marshall, L.F. Traumatic brain injury. In *Handbook of Neuroemergency Clinical Trials*; Elsevier: Amsterdam, The Netherlands, 2006; pp. 61–79.
9. Başkaya, M.K.; Doğan, A.; Rao, A.M.; Dempsey, R.J. Neuroprotective effects of citicoline on brain edema and blood—Brain barrier breakdown after traumatic brain injury. *J. Neurosurg.* **2000**, *92*, 448–452. [CrossRef]
10. Chodobski, A.; Zink, B.J.; Szmydynger-Chodobska, J. Blood–Brain Barrier Pathophysiology in Traumatic Brain Injury. *Transl. Stroke Res.* **2011**, *2*, 492–516. [CrossRef]

11. Pandya, J.D.; Leung, L.Y.; Hwang, H.M.; Yang, X.; Deng-Bryant, Y.; Shear, D.A. Time-Course Evaluation of Brain Regional Mitochondrial Bioenergetics in a Pre-Clinical Model of Severe Penetrating Traumatic Brain Injury. *J. Neurotrauma* **2021**, *38*, 2323–2334. [CrossRef]
12. Pandya, J.D.; Leung, L.Y.; Flerlage, W.J.; Gilsdorf, J.S.; Bryant, Y.D.; Shear, D. Comprehensive profile of acute mitochondrial dysfunction in a preclinical model of severe penetrating TBI. *Front. Neurol.* **2019**, *10*, 605. [CrossRef]
13. Li, J.; O, W.; Li, W.; Jiang, Z.-G.; Ghanbari, H.A. Oxidative Stress and Neurodegenerative Disorders. *Int. J. Mol. Sci.* **2013**, *14*, 24438–24475. [CrossRef]
14. Brett, B.L.; Gardner, R.C.; Godbout, J.; Dams-O'connor, K.; Keene, C.D. Traumatic Brain Injury and Risk of Neurodegenerative Disorder. *Biol. Psychiatry* **2022**, *91*, 498–507. [CrossRef] [PubMed]
15. Plassman, B.; Havlik, R.; Steffens, D.; Helms, M.; Newman, T.; Drosdick, D.; Phillips, C.; Gau, B.; Welsh–Bohmer, K.; Burke, J.; et al. Documented head injury in early adulthood and risk of Alzheimer's disease and other dementias. *Neurology* **2000**, *55*, 1158–1166. [CrossRef] [PubMed]
16. Liu, G.; Ou, S.; Cui, H.; Li, X.; Yin, Z.; Gu, D.; Wang, Z. Head Injury and Amyotrophic Lateral Sclerosis: A Meta-Analysis. *Neuroepidemiology* **2021**, *55*, 11–19. [CrossRef] [PubMed]
17. Kang, J.-H.; Lin, H.-C. Increased Risk of Multiple Sclerosis after Traumatic Brain Injury: A Nationwide Population-Based Study. *J. Neurotrauma* **2012**, *29*, 90–95. [CrossRef] [PubMed]
18. Hubbard, W.B.; Joseph, B.; Spry, M.; Vekaria, H.J.; Saatman, K.E.; Sullivan, P.G. Acute Mitochondrial Impairment Underlies Prolonged Cellular Dysfunction after Repeated Mild Traumatic Brain Injuries. *J. Neurotrauma* **2019**, *36*, 1252–1263. [CrossRef]
19. Kilbaugh, T.J.; Karlsson, M.; Byro, M.; Bebee, A.; Ralston, J.; Sullivan, S.; Duhaime, A.-C.; Hansson, M.J.; Elmér, E.; Margulies, S.S. Mitochondrial bioenergetic alterations after focal traumatic brain injury in the immature brain. *Exp. Neurol.* **2015**, *271*, 136–144. [CrossRef] [PubMed]
20. Pandya, J.D.; Pauly, J.R.; Nukala, V.N.; Sebastian, A.H.; Day, K.M.; Korde, A.S.; Maragos, W.F.; Hall, E.D.; Sullivan, P.G. Post-Injury Administration of Mitochondrial Uncouplers Increases Tissue Sparing and Improves Behavioral Outcome following Traumatic Brain Injury in Rodents. *J. Neurotrauma* **2007**, *24*, 798–811. [CrossRef]
21. Sullivan, P.G.; Rabchevsky, A.G.; Keller, J.N.; Lovell, M.; Sodhi, A.; Hart, R.P.; Scheff, S.W. Intrinsic differences in brain and spinal cord mitochondria: Implication for therapeutic interventions. *J. Comp. Neurol.* **2004**, *474*, 524–534. [CrossRef]
22. Andriessen, T.M.J.C.; Jacobs, B.; Vos, P.E. Clinical characteristics and pathophysiological mechanisms of focal and diffuse traumatic brain injury. *J. Cell. Mol. Med.* **2010**, *14*, 2381–2392. [CrossRef]
23. Bullock, M.R.; Povlishock, J.T. Guidelines for the management of severe traumatic brain injury. Editor's Commentary. *J. Neurotrauma* **2007**, *24* (Suppl. S1), 2 p preceding S1. [CrossRef]
24. Maas, A.I.R.; Stocchetti, N.; Bullock, R. Moderate and severe traumatic brain injury in adults. *Lancet Neurol.* **2008**, *7*, 728–741. [CrossRef] [PubMed]
25. Narayan, R.K.; Michel, M.E.; Ansell, B.; Baethmann, A.; Biegon, A.; Bracken, M.B.; Bullock, M.R.; Choi, S.C.; Clifton, G.L.; Contant, C.F.; et al. Clinical Trials in Head Injury. *J. Neurotrauma* **2002**, *19*, 503–557. [CrossRef] [PubMed]
26. Povlishock, J.T.; Katz, D.I. Update of Neuropathology and Neurological Recovery After Traumatic Brain Injury. *J. Head Trauma Rehabilit.* **2005**, *20*, 76–94. [CrossRef] [PubMed]
27. Prins, M.; Greco, T.; Alexander, D.; Giza, C.C. The pathophysiology of traumatic brain injury at a glance. *Dis. Model. Mech.* **2013**, *6*, 1307–1315. [CrossRef] [PubMed]
28. Vespa, P.; Bergsneider, M.; Hattori, N.; Wu, H.-M.; Huang, S.-C.; Martin, N.A.; Glenn, T.C.; McArthur, D.L.; Hovda, D.A. Metabolic Crisis without Brain Ischemia is Common after Traumatic Brain Injury: A Combined Microdialysis and Positron Emission Tomography Study. *J. Cereb. Blood Flow Metab.* **2005**, *25*, 763–774. [CrossRef] [PubMed]
29. Yoshino, A.; Hovda, D.A.; Kawamata, T.; Katayama, Y.; Becker, D.P. Dynamic changes in local cerebral glucose utilization following cerebral conclusion in rats: Evidence of a hyper- and subsequent hypometabolic state. *Brain Res.* **1991**, *561*, 106–119. [CrossRef] [PubMed]
30. Xiong, Y.; Mahmood, A.; Chopp, M. Animal models of traumatic brain injury. *Nat. Rev. Neurosci.* **2013**, *14*, 128–142. [CrossRef] [PubMed]
31. Lifshitz, J.; Sullivan, P.G.; Hovda, D.A.; Wieloch, T.; McIntosh, T.K. Mitochondrial damage and dysfunction in traumatic brain injury. *Mitochondrion* **2004**, *4*, 705–713. [CrossRef]
32. Sullivan, P.G.; Krishnamurthy, S.; Patel, S.P.; Pandya, J.D.; Rabchevsky, A.G. Temporal Characterization of Mitochondrial Bioenergetics after Spinal Cord Injury. *J. Neurotrauma* **2007**, *24*, 991–999. [CrossRef]
33. Singh, I.N.; Sullivan, P.G.; Deng, Y.; Mbye, L.H.; Hall, E.D. Time Course of Post-Traumatic Mitochondrial Oxidative Damage and Dysfunction in a Mouse Model of Focal Traumatic Brain Injury: Implications for Neuroprotective Therapy. *J. Cereb. Blood Flow Metab.* **2006**, *26*, 1407–1418. [CrossRef]
34. Xiong, Y.; Gu, Q.; Peterson, P.; Muizelaar, J.; Lee, C. Mitochondrial Dysfunction and Calcium Perturbation Induced by Traumatic Brain Injury. *J. Neurotrauma* **1997**, *14*, 23–34. [CrossRef] [PubMed]
35. Gilmer, L.K.; Roberts, K.N.; Sullivan, P.G.; Miller, K.; Scheff, S. Early mitochondrial dysfunction after cortical contusion injury. *J. Neurotrauma* **2009**, *219*, 1. [CrossRef] [PubMed]
36. Pandya, J.D.; Pauly, J.R.; Sullivan, P.G. The optimal dosage and window of opportunity to maintain mitochondrial homeostasis following traumatic brain injury using the uncoupler FCCP. *Exp. Neurol.* **2009**, *218*, 381–389. [CrossRef]

37. Pandya, J.D.; Musyaju, S.; Modi, H.R.; Cao, Y.; Flerlage, W.J.; Huynh, L.; Kociuba, B.; Visavadiya, N.P.; Kobeissy, F.; Wang, K.; et al. Comprehensive evaluation of mitochondrial redox profile, calcium dynamics, membrane integrity and apoptosis markers in a preclinical model of severe penetrating traumatic brain injury. *Free. Radic. Biol. Med.* **2023**, *198*, 44–58. [CrossRef] [PubMed]
38. Crompton, M. The mitochondrial permeability transition pore and its role in cell death. *Biochem. J.* **1999**, *341 Pt 2*, 233–249. [CrossRef]
39. Halestrap, A.P. Mitochondrial calcium in health and disease. *Biochim. Biophys. Acta (BBA)—Bioenerg.* **2009**, *1787*, 1289–1290. [CrossRef] [PubMed]
40. Halestrap, A.P. What is the mitochondrial permeability transition pore? *J. Mol. Cell. Cardiol.* **2009**, *46*, 821–831. [CrossRef]
41. Sullivan, P.G.; Rabchevsky, A.G.; Waldmeier, P.C.; Springer, J.E. Mitochondrial permeability transition in CNS trauma: Cause or effect of neuronal cell death? *J. Neurosci. Res.* **2005**, *79*, 231–239. [CrossRef]
42. Chance, B.; Sies, H.; Walker, C.L.; Pomatto, L.C.D.; Tripathi, D.N.; Davies, K.J.A.; Ninsontia, C.; Phiboonchaiyanan, P.P.; Kiratipaiboon, C.; Chanvorachote, P.; et al. Hydroperoxide metabolism in mammalian organs. *Physiol. Rev.* **1979**, *59*, 527–605. [CrossRef]
43. Halliwell, B.; Gutteridge, J.M.C. Oxygen toxicity, oxygen radicals, transition metals and disease. *Biochem. J.* **1984**, *219*, 1–14. [CrossRef]
44. Halliwell, B. Reactive oxygen species and the central nervous system. *J. Neurochem.* **1992**, *59*, 1609–1623. [CrossRef] [PubMed]
45. Berg, J.M.; Tymoczko, J.L.; Stryer, L. *Biochemistry*, 5th ed.; W. H. Freeman and Company: New York, NY, USA, 2002.
46. Chance, B.; Williams, G.R. The respiratory chain and oxidative phosphorylation. *Adv. Enzymol. Relat. Subj. Biochem.* **1956**, *17*, 65–134. [PubMed]
47. Deng, Y.; Thompson, B.M.; Gao, X.; Hall, E.D. Temporal relationship of peroxynitrite-induced oxidative damage, calpain-mediated cytoskeletal degradation and neurodegeneration after traumatic brain injury. *Exp. Neurol.* **2007**, *205*, 154–165. [CrossRef] [PubMed]
48. Singh, I.N.; Sullivan, P.G.; Hall, E.D. Peroxynitrite-mediated oxidative damage to brain mitochondria: Protective effects of peroxynitrite scavengers. *J. Neurosci. Res.* **2007**, *85*, 2216–2223. [CrossRef] [PubMed]
49. Bayir, H.; E Kagan, V.; Tyurina, Y.Y.; Tyurin, V.; A Ruppel, R.; Adelson, P.D.; Graham, S.H.; Janesko, K.; Clark, R.S.B.; Kochanek, P.M. Assessment of Antioxidant Reserves and Oxidative Stress in Cerebrospinal Fluid after Severe Traumatic Brain Injury in Infants and Children. *Pediatr. Res.* **2002**, *51*, 571–578. [CrossRef] [PubMed]
50. Ansari, M.A.; Roberts, K.N.; Scheff, S.W. Oxidative stress and modification of synaptic proteins in hippocampus after traumatic brain injury. *Free. Radic. Biol. Med.* **2008**, *45*, 443–452. [CrossRef] [PubMed]
51. Hall, E.D.; Detloff, M.R.; Johnson, K.; Kupina, N.C. Peroxynitrite-Mediated Protein Nitration and Lipid Peroxidation in a Mouse Model of Traumatic Brain Injury. *J. Neurotrauma* **2004**, *21*, 9–20. [CrossRef]
52. Hall, E.D.; Vaishnav, R.A.; Mustafa, A.G. Antioxidant Therapies for Traumatic Brain Injury. *Neurotherapeutics* **2010**, *7*, 51–61. [CrossRef]
53. Hall, E.D.; Wang, J.A.; Miller, D.M. Relationship of nitric oxide synthase induction to peroxynitrite-mediated oxidative damage during the first week after experimental traumatic brain injury. *Exp. Neurol.* **2012**, *238*, 176–182. [CrossRef]
54. Hill, R.L.; Kulbe, J.R.; Singh, I.N.; Wang, J.A.; Hall, E.D. Synaptic Mitochondria are More Susceptible to Traumatic Brain Injury-induced Oxidative Damage and Respiratory Dysfunction than Non-synaptic Mitochondria. *Neuroscience* **2018**, *386*, 265–283. [CrossRef]
55. Abdul-Muneer, P.M.; Chandra, N.; Haorah, J. Interactions of Oxidative Stress and Neurovascular Inflammation in the Pathogenesis of Traumatic Brain Injury. *Mol. Neurobiol.* **2015**, *51*, 966–979. [CrossRef] [PubMed]
56. Petronilho, F.; Feier, G.; de Souza, B.; Guglielmi, C.; Constantino, L.S.; Walz, R.; Quevedo, J.; Dal-Pizzol, F. Oxidative Stress in Brain According to Traumatic Brain Injury Intensity. *J. Surg. Res.* **2010**, *164*, 316–320. [CrossRef] [PubMed]
57. Cornelius, C.; Crupi, R.; Calabrese, V.; Graziano, A.; Milone, P.; Pennisi, G.; Radak, Z.; Calabrese, E.J.; Cuzzocrea, S. Traumatic Brain Injury: Oxidative Stress and Neuroprotection. *Antioxid. Redox Signal.* **2013**, *19*, 836–853. [CrossRef] [PubMed]
58. Fernández-Gajardo, R.; Matamala, J.M.; Carrasco, R.; Gutiérrez, R.; Melo, R.; Rodrigo, R. Novel Therapeutic Strategies for Traumatic Brain Injury: Acute Antioxidant Reinforcement. *CNS Drugs* **2014**, *28*, 229–248. [CrossRef] [PubMed]
59. Kitada, M.; Xu, J.; Ogura, Y.; Monno, I.; Koya, D. Manganese Superoxide Dismutase Dysfunction and the Pathogenesis of Kidney Disease. *Front. Physiol.* **2020**, *11*, 755. [CrossRef] [PubMed]
60. Suthammarak, W.; Somerlot, B.H.; Opheim, E.; Sedensky, M.; Morgan, P.G. Novel interactions between mitochondrial superoxide dismutases and the electron transport chain. *Aging Cell* **2013**, *12*, 1132–1140. [CrossRef] [PubMed]
61. Holley, A.K.; Clair, D.K.S.; Huang, C.-L.; Yokomise, H.; Miyatake, A.; Setyawati, M.I.; Tay, C.Y.; Leong, D.T.; Schumacher, B.; Gartner, A.; et al. Watching the watcher: Regulation of p53 by mitochondria. *Futur. Oncol.* **2009**, *5*, 117–130. [CrossRef] [PubMed]
62. DeKosky, S.T.; Taffe, K.M.; Abrahamson, E.E.; Dixon, C.E.; Kochanek, P.M.; Ikonomovic, M.D. Time Course Analysis of Hippocampal Nerve Growth Factor and Antioxidant Enzyme Activity following Lateral Controlled Cortical Impact Brain Injury in the Rat. *J. Neurotrauma* **2004**, *21*, 491–500. [CrossRef]
63. Pappolla, M.A.; Omar, R.A.; Kim, K.S.; Robakis, N.K. Immunohistochemical evidence of oxidative [corrected] stress in Alzheimer's disease. *Am. J. Pathol.* **1992**, *140*, 621–628.
64. Zemlan, F.P.; Thienhaus, O.J.; Bosmann, H.B. Superoxide dismutase activity in Alzheimer's disease: Possible mechanism for paired helical filament formation. *Brain Res.* **1989**, *476*, 160–162. [CrossRef]

65. Massaad, C.A.; Washington, T.M.; Pautler, R.G.; Klann, E. Overexpression of SOD-2 reduces hippocampal superoxide and prevents memory deficits in a mouse model of Alzheimer's disease. *Proc. Natl. Acad. Sci. USA* **2009**, *106*, 13576–13581. [CrossRef] [PubMed]
66. Palace, V.P.; Khaper, N.; Qin, Q.; Singal, P.K. Antioxidant potentials of vitamin A and carotenoids and their relevance to heart disease. *Free. Radic. Biol. Med.* **1999**, *26*, 746–761. [CrossRef] [PubMed]
67. Gonzalez, M.J.; Miranda-Massari, J.R.; Olalde, J. Vitamin C and mitochondrial function in health and exercise. In *Molecular Nutrition and Mitochondria*; Elsevier: Amsterdam, The Netherlands, 2023; pp. 225–242.
68. Eleff, S.; Kennaway, N.G.; Buist, N.R.; Darley-Usmar, V.M.; A Capaldi, R.; Bank, W.J.; Chance, B. 31P NMR study of improvement in oxidative phosphorylation by vitamins K3 and C in a patient with a defect in electron transport at complex III in skeletal muscle. *Proc. Natl. Acad. Sci. USA* **1984**, *81*, 3529–3533. [CrossRef] [PubMed]
69. Kc, S.; Càrcamo, J.M.; Golde, D.W. Vitamin C enters mitochondria via facilitative glucose transporter 1 (Gluti) and confers mitochondrial protection against oxidative injury. *FASEB J.* **2005**, *19*, 1657–1667. [CrossRef] [PubMed]
70. González, M.J.; Miranda, J.R.; Riordan, H.D. Vitamin C as an Ergogenic Aid. *J. Orthomol. Med.* **2005**, *20*, 100–102.
71. U.S. Department of Health & Human Services. Vitamin A and Carotenoids. Fact Sheet for Health Professionals. Available online: https://ods.od.nih.gov/factsheets/VitaminA-HealthProfessional/ (accessed on 29 April 2022).
72. Hayashi, T.; Sawa, K.; Kawasaki, M.; Arisawa, M.; Shimizu, M.; Morita, N. Inhibition of cow's milk xanthine oxidase by flavonoids. *J. Nat. Prod.* **1988**, *51*, 345–348. [CrossRef] [PubMed]
73. Panche, A.N.; Diwan, A.D.; Chandra, S.R. Flavonoids: An overview. *J. Nutr. Sci.* **2016**, *5*, e47. [CrossRef] [PubMed]
74. Ansari, M.A.; Roberts, K.N.; Scheff, S.W. Dose- and Time-Dependent Neuroprotective Effects of Pycnogenol® following Traumatic Brain Injury. *J. Neurotrauma* **2013**, *30*, 1542–1549. [CrossRef]
75. Chung, L.Y. The Antioxidant Properties of Garlic Compounds: Allyl Cysteine, Alliin, Allicin, and Allyl Disulfide. *J. Med. Food* **2006**, *9*, 205–213. [CrossRef]
76. Schwartz, I.F.; Hershkovitz, R.; Iaina, A.; Gnessin, E.; Wollman, Y.; Chernichowski, T.; Blum, M.; Levo, Y.; Schwartz, D. Garlic attenuates nitric oxide production in rat cardiac myocytes through inhibition of inducible nitric oxide synthase and the arginine transporter CAT-2 (cationic amino acid transporter-2). *Clin. Sci.* **2002**, *102*, 487–493. [CrossRef]
77. Nadeem, M.S.; Kazmi, I.; Ullah, I.; Muhammad, K.; Anwar, F. Allicin, an Antioxidant and Neuroprotective Agent, Ameliorates Cognitive Impairment. *Antioxidants* **2021**, *11*, 87. [CrossRef] [PubMed]
78. Adebayo, O.L.; Adenuga, G.A.; Sandhir, R. Selenium and zinc protect brain mitochondrial antioxidants and electron transport chain enzymes following postnatal protein malnutrition. *Life Sci.* **2016**, *152*, 145–155. [CrossRef] [PubMed]
79. Pizzorno, J. Glutathione! *Integr. Med.* **2014**, *13*, 8–12.
80. Korshunov, S.S.; Krasnikov, B.F.; O Pereverzev, M.; Skulachev, V.P. The antioxidant functions of cytochrome *c*. *FEBS Lett.* **1999**, *462*, 192–198. [CrossRef] [PubMed]
81. Pereverzev, M.O.; Vygodina, T.V.; Konstantinov, A.A.; Skulachev, V.P. Cytochrome c, an ideal antioxidant. *Biochem. Soc. Trans.* **2003**, *31 Pt 6*, 1312–1315. [CrossRef] [PubMed]
82. Sokol, R.J.; McKim, J.M.; Goff, M.; Ruyle, S.Z.; Devereaux, M.W.; Han, D.; Packer, L.; Everson, G. Vitamin E reduces oxidant injury to mitochondria and the hepatotoxicity of taurochenodeoxycholic acid in the rat. *Gastroenterology* **1998**, *114*, 164–174. [CrossRef] [PubMed]
83. Gilgun-Sherki, Y.; Melamed, E.; Offen, D. Oxidative stress induced-neurodegenerative diseases: The need for antioxidants that penetrate the blood brain barrier. *Neuropharmacology* **2001**, *40*, 959–975. [CrossRef]
84. Batinić-Haberle, I.; Cuzzocrea, S.; Rebouças, J.S.; Ferrer-Sueta, G.; Mazzon, E.; Di Paola, R.; Radi, R.; Spasojević, I.; Benov, L.; Salvemini, D. Pure MnTBAP selectively scavenges peroxynitrite over superoxide: Comparison of pure and commercial MnTBAP samples to MnTE-2-PyP in two different models of oxidative stress injuries, SOD-specific E. coli model and carrageenan-induced pleurisy. *Free. Radic. Biol. Med.* **2009**, *46*, 192. [CrossRef]
85. Zahmatkesh, M.; Kadkhodaee, M.; Moosavi, S.M.S.; Jorjani, M.; Kajbafzadeh, A.; Golestani, A.; Ghaznavi, R. Beneficial effects of MnTBAP, a broad-spectrum reactive species scavenger, in rat renal ischemia/reperfusion injury. *Clin. Exp. Nephrol.* **2005**, *9*, 212–218. [CrossRef]
86. Kim, J.-H.; Jang, H.-J.; Cho, W.-Y.; Yeon, S.-J.; Lee, C.-H. In vitro antioxidant actions of sulfur-containing amino acids. *Arab. J. Chem.* **2020**, *13*, 1678–1684. [CrossRef]
87. Nandi, A.; Chatterjee, I.B. Scavenging of superoxide radical by ascorbic acid. *J. Biosci.* **1987**, *11*, 435–441. [CrossRef]
88. E Niki, E. Action of ascorbic acid as a scavenger of active and stable oxygen radicals. *Am. J. Clin. Nutr.* **1991**, *54*, S1119–S1124. [CrossRef] [PubMed]
89. Hussain, S.; Slikker, W., Jr.; Ali, S.F. Role of Metallothionein and other Antioxidants in Scavenging Superoxide Radicals and their Possible Role in Neuroprotection. *Neurochem. Int.* **1996**, *29*, 145–152. [CrossRef] [PubMed]
90. Schneider, M.P.; Delles, C.; Schmidt, B.M.; Oehmer, S.; Schwarz, T.K.; Schmieder, R.E.; John, S. Superoxide scavenging effects of N-acetylcysteine and vitamin C in subjects with essential hypertension. *Am. J. Hypertens.* **2005**, *18*, 1111–1117. [CrossRef] [PubMed]
91. Taiwo, F.A. Mechanism of tiron as scavenger of superoxide ions and free electrons. *Spectroscopy* **2008**, *22*, 491–498. [CrossRef]
92. Wright, V.P.; Klawitter, P.F.; Iscru, D.F.; Merola, A.J.; Clanton, T.L. Superoxide scavengers augment contractile but not energetic responses to hypoxia in rat diaphragm. *J. Appl. Physiol.* **2005**, *98*, 1753–1760. [CrossRef] [PubMed]

93. Pfeiffer, S.; Leopold, E.; Hemmens, B.; Schmidt, K.; Werner, E.R.; Mayer, B. Interference of Carboxy-PTIO with Nitric Oxide- and Peroxynitrite-Mediated Reactions. *Free. Radic. Biol. Med.* **1997**, *22*, 787–794. [CrossRef]

94. Lilley, E.; Gibson, A. Antioxidant protection of NO-induced relaxations of the mouse anococcygeus against inhibition by superoxide anions, hydroquinone and carboxy-PTIO. *Br. J. Pharmacol.* **1996**, *119*, 432–438. [CrossRef]

95. Yang, J.; Su, Y.; Richmond, A. Antioxidants tiron and N-acetyl-L-cysteine differentially mediate apoptosis in melanoma cells via a reactive oxygen species-independent NF-kappaB pathway. *Free Radic. Biol. Med.* **2007**, *42*, 1369–1380. [CrossRef]

96. Hill, R.L.; Singh, I.N.; Wang, J.A.; Kulbe, J.R.; Hall, E.D. Protective effects of phenelzine administration on synaptic and non-synaptic cortical mitochondrial function and lipid peroxidation-mediated oxidative damage following TBI in young adult male rats. *Exp. Neurol.* **2020**, *330*, 113322. [CrossRef]

97. Smith, R.A.J.; Porteous, C.M.; Coulter, C.V.; Murphy, M.P. Selective targeting of an antioxidant to mitochondria. *Eur. J. Biochem.* **1999**, *263*, 709–716. [CrossRef] [PubMed]

98. Davidson, S.M. Endothelial mitochondria and heart disease. *Cardiovasc. Res.* **2010**, *88*, 58–66. [CrossRef] [PubMed]

99. Murphy, M.P. Targeting lipophilic cations to mitochondria. *Biochim. Biophys. Acta (BBA)—Bioenerg.* **2008**, *1777*, 1028–1031. [CrossRef] [PubMed]

100. Murphy, M.P.; Smith, R.A. Targeting Antioxidants to Mitochondria by Conjugation to Lipophilic Cations. *Annu. Rev. Pharmacol. Toxicol.* **2007**, *47*, 629–656. [CrossRef] [PubMed]

101. Graham, D.; Huynh, N.N.; Hamilton, C.A.; Beattie, E.; Smith, R.A.; Cochemé, H.M.; Murphy, M.P.; Dominiczak, A.F. Mitochondria-targeted antioxidant MitoQ10 improves endothelial function and attenuates cardiac hypertrophy. *Hypertension* **2009**, *54*, 322–328. [CrossRef] [PubMed]

102. Zhou, J.; Wang, H.; Shen, R.; Fang, J.; Yang, Y.; Dai, W.; Zhu, Y.; Zhou, M. Mitochondrial-targeted antioxidant MitoQ provides neuroprotection and reduces neuronal apoptosis in experimental traumatic brain injury possibly via the Nrf2-ARE pathway. *Am. J. Transl. Res.* **2018**, *10*, 1887–1899. [PubMed]

103. Tabet, M.; El-Kurdi, M.; Haidar, M.A.; Nasrallah, L.; Reslan, M.A.; Shear, D.; Pandya, J.D.; El-Yazbi, A.F.; Sabra, M.; Mondello, S.; et al. Mitoquinone supplementation alleviates oxidative stress and pathologic outcomes following repetitive mild traumatic brain injury at a chronic time point. *Exp. Neurol.* **2022**, *351*, 113987. [CrossRef] [PubMed]

104. Liberman, E.A.; Topaly, V.P.; Tsofina, L.M.; Jasaitis, A.A.; Skulachev, V.P. Mechanism of Coupling of Oxidative Phosphorylation and the Membrane Potential of Mitochondria. *Nature* **1969**, *222*, 1076–1078. [CrossRef]

105. Skulachev, V.P.; Vyssokikh, M.Y.; Chernyak, B.V.; Averina, O.A.; Andreev-Andrievskiy, A.A.; Zinovkin, R.A.; Lyamzaev, K.G.; Marey, M.V.; Egorov, M.V.; Frolova, O.J.; et al. Mitochondrion-targeted antioxidant SkQ1 prevents rapid animal death caused by highly diverse shocks. *Sci. Rep.* **2023**, *13*, 4326. [CrossRef]

106. Amemiya, S.; Kamiya, T.; Nito, C.; Inaba, T.; Kato, K.; Ueda, M.; Shimazaki, K.; Katayama, Y. Anti-apoptotic and neuroprotective effects of edaravone following transient focal ischemia in rats. *Eur. J. Pharmacol.* **2005**, *516*, 125–130. [CrossRef]

107. Toyoda, K.; Fujii, K.; Kamouchi, M.; Nakane, H.; Arihiro, S.; Okada, Y.; Ibayashi, S.; Iida, M. Free radical scavenger, edaravone, in stroke with internal carotid artery occlusion. *J. Neurol. Sci.* **2004**, *221*, 11–17. [CrossRef] [PubMed]

108. Abe, K.; Yuki, S.; Kogure, K. Strong attenuation of ischemic and postischemic brain edema in rats by a novel free radical scavenger. *Stroke* **1988**, *19*, 480–485. [CrossRef] [PubMed]

109. Yamamoto, T.; Yuki, S.; Watanabe, T.; Mitsuka, M.; Saito, K.-I.; Kogure, K. Delayed neuronal death prevented by inhibition of increased hydroxyl radical formation in a transient cerebral ischemia. *Brain Res.* **1997**, *762*, 240–242. [CrossRef]

110. Homma, T.; Kobayashi, S.; Sato, H.; Fujii, J. Edaravone, a free radical scavenger, protects against ferroptotic cell death in vitro. *Exp. Cell Res.* **2019**, *384*, 111592. [CrossRef] [PubMed]

111. Yuan, W.J.; Yasuhara, T.; Shingo, T.; Muraoka, K.; Agari, T.; Kameda, M.; Uozumi, T.; Tajiri, N.; Morimoto, T.; Jing, M.; et al. Neuroprotective effects of edaravone-administration on 6-OHDA-treated dopaminergic neurons. *BMC Neurosci.* **2008**, *9*, 75. [CrossRef]

112. Jiao, S.-S.; Yao, X.-Q.; Liu, Y.-H.; Wang, Q.-H.; Zeng, F.; Lu, J.-J.; Liu, J.; Zhu, C.; Shen, L.-L.; Liu, C.-H.; et al. Edaravone alleviates Alzheimer's disease-type pathologies and cognitive deficits. *Proc. Natl. Acad. Sci. USA* **2015**, *112*, 5225–5230. [CrossRef] [PubMed]

113. Xi, H.; Akishita, M.; Nagai, K.; Yu, W.; Hasegawa, H.; Eto, M.; Kozaki, K.; Toba, K. Potent free radical scavenger, edaravone, suppresses oxidative stress-induced endothelial damage and early atherosclerosis. *Atherosclerosis* **2007**, *191*, 281–289. [CrossRef]

114. Higashi, Y.; Jitsuiki, D.; Chayama, K.; Yoshizumi, M. Edaravone (3-Methyl-1-Phenyl-2-Pyrazolin-5-one), A Novel Free Radical Scavenger, for Treatment of Cardiovascular Diseases. *Recent Pat. Cardiovasc. Drug Discov.* **2006**, *1*, 85–93. [CrossRef]

115. Shetty, S.; Anushree, U.; Kumar, R.; Bharati, S. Mitochondria-targeted antioxidant, mito-TEMPO mitigates initiation phase of N-Nitrosodiethylamine-induced hepatocarcinogenesis. *Mitochondrion* **2021**, *58*, 123–130. [CrossRef]

116. Nhu, N.T.; Xiao, S.-Y.; Liu, Y.; Kumar, V.B.; Cui, Z.-Y.; Lee, S.-D. Neuroprotective Effects of a Small Mitochondrially-Targeted Tetrapeptide Elamipretide in Neurodegeneration. *Front. Integr. Neurosci.* **2021**, *15*, 747901. [CrossRef]

117. Rzigalinski, B.A.; Carfagna, C.S.; Ehrich, M. Cerium oxide nanoparticles in neuroprotection and considerations for efficacy and safety. *Wiley Interdiscip. Rev. Nanomed. Nanobiotechnol.* **2017**, *9*, e1444. [CrossRef] [PubMed]

118. Liang, L.-P.; Waldbaum, S.; Rowley, S.; Huang, T.-T.; Day, B.J.; Patel, M. Mitochondrial oxidative stress and epilepsy in SOD2 deficient mice: Attenuation by a lipophilic metalloporphyrin. *Neurobiol. Dis.* **2012**, *45*, 1068–1076. [CrossRef] [PubMed]

119. Peterson, S.L.; Purvis, R.S.; Griffith, J.W. Comparison of Neuroprotective Effects Induced by α-Phenyl-N-tert-butyl nitrone (PBN) and N-tert-Butyl-α-(2 sulfophenyl) nitrone (S-PBN) in Lithium-Pilocarpine Status Epilepticus. *NeuroToxicology* **2005**, *26*, 969–979. [CrossRef] [PubMed]
120. Deletraz, A.; Zéamari, K.; Hua, K.; Combes, M.; Villamena, F.A.; Tuccio, B.; Callizot, N.; Durand, G. Substituted α-phenyl and α-naphthlyl-N-tert-butyl nitrones: Synthesis, spin-trapping, and neuroprotection evaluation. *J. Org. Chem.* **2020**, *85*, 6073–6085. [CrossRef] [PubMed]
121. Chamorro, B.; Diez-Iriepa, D.; Merás-Sáiz, B.; Chioua, M.; García-Vieira, D.; Iriepa, I.; Hadjipavlou-Litina, D.; López-Muñoz, F.; Martínez-Murillo, R.; Gonzàlez-Nieto, D.; et al. Synthesis, antioxidant properties and neuroprotection of α-phenyl-tert-butylnitrone derived HomoBisNitrones in in vitro and in vivo ischemia models. *Sci. Rep.* **2020**, *10*, 14150. [CrossRef] [PubMed]
122. Di Pietro, V.; Yakoub, K.M.; Caruso, G.; Lazzarino, G.; Signoretti, S.; Barbey, A.K.; Tavazzi, B.; Lazzarino, G.; Belli, A.; Amorini, A.M. Antioxidant Therapies in Traumatic Brain Injury. *Antioxidants* **2020**, *9*, 260. [CrossRef] [PubMed]
123. Davis, C.K.; Vemuganti, R. Antioxidant therapies in traumatic brain injury. *Neurochem. Int.* **2022**, *152*, 105255. [CrossRef] [PubMed]
124. Dash, P.K.; Hergenroeder, G.W.; Jeter, C.B.; Choi, H.A.; Kobori, N.; Moore, A.N. Traumatic Brain Injury Alters Methionine Metabolism: Implications for Pathophysiology. *Front. Syst. Neurosci.* **2016**, *10*, 36. [CrossRef]
125. Sonthalia, S.; Daulatabad, D.; Sarkar, R. Glutathione as a skin whitening agent: Facts, myths, evidence and controversies. *Indian J. Dermatol. Venereol. Leprol.* **2016**, *82*, 262–272. [CrossRef]
126. Holmay, M.J.B.; Terpstra, M.; Coles, L.D.; Mishra, U.; Ahlskog, M.B.; Öz, G.; Cloyd, J.C.; Tuite, P.J. N-acetylcysteine Boosts Brain and Blood Glutathione in Gaucher and Parkinson Diseases. *Clin. Neuropharmacol.* **2013**, *36*, 103–106. [CrossRef]
127. Alkandari, A.F.; Madhyastha, S.; Rao, M.S. N-Acetylcysteine Amide against Aβ-Induced Alzheimer's-like Pathology in Rats. *Int. J. Mol. Sci.* **2023**, *24*, 12733. [CrossRef] [PubMed]
128. Pandya, J.D.; Readnower, R.D.; Patel, S.P.; Yonutas, H.M.; Pauly, J.R.; Goldstein, G.A.; Rabchevsky, A.G.; Sullivan, P.G. N-acetylcysteine amide confers neuroprotection, improves bioenergetics and behavioral outcome following TBI. *Exp. Neurol.* **2014**, *257*, 106–113. [CrossRef] [PubMed]
129. Patel, S.P.; Sullivan, P.G.; Pandya, J.D.; Goldstein, G.A.; VanRooyen, J.L.; Yonutas, H.M.; Eldahan, K.C.; Morehouse, J.; Magnuson, D.S.; Rabchevsky, A.G. N-acetylcysteine amide preserves mitochondrial bioenergetics and improves functional recovery following spinal trauma. *Exp. Neurol.* **2014**, *257*, 95–105. [CrossRef] [PubMed]
130. Berk, M.; Malhi, G.S.; Gray, L.J.; Dean, O.M. The promise of N-acetylcysteine in neuropsychiatry. *Trends Pharmacol. Sci.* **2013**, *34*, 167–177. [CrossRef]
131. Baki, S.G.A.; Schwab, B.; Haber, M.; Fenton, A.A.; Bergold, P.J. Minocycline Synergizes with N-Acetylcysteine and Improves Cognition and Memory Following Traumatic Brain Injury in Rats. *PLoS ONE* **2010**, *5*, e12490. [CrossRef]
132. Hicdonmez, T.; Kanter, M.; Tiryaki, M.; Parsak, T.; Cobanoglu, S. Neuroprotective Effects of N-acetylcysteine on Experimental Closed Head Trauma in Rats. *Neurochem. Res.* **2006**, *31*, 473–481. [CrossRef] [PubMed]
133. Thomale, U.-W.; Griebenow, M.; Kroppenstedt, S.-N.; Unterberg, A.W.; Stover, J.F. The effect of N-acetylcysteine on posttraumatic changes after controlled cortical impact in rats. *Intensiv. Care Med.* **2006**, *32*, 149–155. [CrossRef] [PubMed]
134. Xiong, Y.; Peterson, P.; Lee, C. Effect of N-Acetylcysteine on Mitochondrial Function Following Traumatic Brain Injury in Rats. *J. Neurotrauma* **1999**, *16*, 1067–1082. [CrossRef]
135. Thomale, U.W.; Griebenow, M.; Kroppenstedt, S.N.; Unterberg, A.W.; Stover, J.F. The antioxidant effect of N-acethylcysteine on experimental contusion in rats. *Acta Neurochir. Suppl.* **2005**, *95*, 429–431.
136. Hagos, F.T.; Empey, P.E.; Wang, P.; Ma, X.; Poloyac, S.M.; Bayir, H.; Kochanek, P.M.; Bell, M.J.; Clark, R.S.B. Exploratory Application of Neuropharmacometabolomics in Severe Childhood Traumatic Brain Injury*. *Crit. Care Med.* **2018**, *46*, 1471–1479. [CrossRef]
137. Clark, R.S.; Empey, P.E.; Bayır, H.; Rosario, B.L.; Poloyac, S.M.; Kochanek, P.M.; Nolin, T.D.; Au, A.K.; Horvat, C.M.; Wisniewski, S.R.; et al. Phase I randomized clinical trial of N-acetylcysteine in combination with an adjuvant probenecid for treatment of severe traumatic brain injury in children. *PLoS ONE* **2017**, *12*, e0180280. [CrossRef]
138. Clark, R.S.B.; Empey, P.E.; Kochanek, P.M.; Bell, M.J. N-Acetylcysteine and Probenecid Adjuvant Therapy for Traumatic Brain Injury. *Neurotherapeutics* **2023**, *20*, 1529–1537. [CrossRef] [PubMed]
139. Hoffer, M.E.; Balaban, C.; Slade, M.D.; Tsao, J.W.; Hoffer, B. Amelioration of Acute Sequelae of Blast Induced Mild Traumatic Brain Injury by N-Acetyl Cysteine: A Double-Blind, Placebo Controlled Study. *PLoS ONE* **2013**, *8*, e54163. [CrossRef] [PubMed]
140. Khan, M.; Sekhon, B.; Jatana, M.; Giri, S.; Gilg, A.G.; Sekhon, C.; Singh, I.; Singh, A.K. Administration of N-acetylcysteine after focal cerebral ischemia protects brain and reduces inflammation in a rat model of experimental stroke. *J. Neurosci. Res.* **2004**, *76*, 519–527. [CrossRef] [PubMed]
141. Sekhon, B.; Sekhon, C.; Khan, M.; Patel, S.J.; Singh, I.; Singh, A.K. N-Acetyl cysteine protects against injury in a rat model of focal cerebral ischemia. *Brain Res.* **2003**, *971*, 1–8. [CrossRef] [PubMed]
142. Gilgun-Sherki, Y.; Rosenbaum, Z.; Melamed, E.; Offen, D. Antioxidant Therapy in Acute Central Nervous System Injury: Current State. *Pharmacol. Rev.* **2002**, *54*, 271–284. [CrossRef] [PubMed]
143. Pahan, K.; Sheikh, F.G.; Namboodiri, A.M.; Singh, I. N-Acetyl Cysteine Inhibits Induction of No Production By Endotoxin or Cytokine Stimulated Rat Peritoneal Macrophages, C6 Glial Cells and Astrocytes. *Free. Radic. Biol. Med.* **1998**, *24*, 39–48. [CrossRef]
144. Aitio, M.L. N-acetylcysteine—Passe-partout or much ado about nothing? *Br. J. Clin. Pharmacol.* **2006**, *61*, 5–15. [CrossRef]

145. Tenório, M.C.D.S.; Graciliano, N.G.; Moura, F.A.; Oliveira, A.C.M.D.; Goulart, M.O.F. N-Acetylcysteine (NAC): Impacts on Human Health. *Antioxidants* **2021**, *10*, 967. [CrossRef]

146. Chen, W.; Ercal, N.; Huynh, T.; Volkov, A.; Chusuei, C.C. Characterizing N-acetylcysteine (NAC) and N-acetylcysteine amide (NACA) binding for lead poisoning treatment. *J. Colloid Interface Sci.* **2012**, *371*, 144–149. [CrossRef]

147. Atlas, D. Emerging therapeutic opportunities of novel thiol-amides, NAC-amide (AD4/NACA) and thioredoxin mimetics (TXM-Peptides) for neurodegenerative-related disorders. *Free Radic. Biol. Med.* **2021**, *176*, 120–141. [CrossRef] [PubMed]

148. Offen, D.; Gilgun-Sherki, Y.; Barhum, Y.; Benhar, M.; Grinberg, L.; Reich, R.; Melamed, E.; Atlas, D. A low molecular weight copper chelator crosses the blood–brain barrier and attenuates experimental autoimmune encephalomyelitis. *J. Neurochem.* **2004**, *89*, 1241–1251. [CrossRef] [PubMed]

149. Grinberg, L.; Fibach, E.; Amer, J.; Atlas, D. N-acetylcysteine amide, a novel cell-permeating thiol, restores cellular glutathione and protects human red blood cells from oxidative stress. *Free. Radic. Biol. Med.* **2005**, *38*, 136–145. [CrossRef] [PubMed]

150. Bahat-Stroomza, M.; Gilgun-Sherki, Y.; Offen, D.; Panet, H.; Saada, A.; Krool-Galron, N.; Barzilai, A.; Atlas, D.; Melamed, E. A novel thiol antioxidant that crosses the blood brain barrier protects dopaminergic neurons in experimental models of Parkinson's disease. *Eur. J. Neurosci.* **2005**, *21*, 637–646. [CrossRef] [PubMed]

151. Bhatti, J.; Nascimento, B.; Akhtar, U.; Rhind, S.G.; Tien, H.; Nathens, A.; Da Luz, L.T. Systematic Review of Human and Animal Studies Examining the Efficacy and Safety of N-Acetylcysteine (NAC) and N-Acetylcysteine Amide (NACA) in Traumatic Brain Injury: Impact on Neurofunctional Outcome and Biomarkers of Oxidative Stress and Inflammation. *Front. Neurol.* **2017**, *8*, 744. [CrossRef] [PubMed]

152. Kurano, T.; Kanazawa, T.; Iioka, S.; Kondo, H.; Kosuge, Y.; Suzuki, T. Intranasal Administration of N-acetyl-L-cysteine Combined with Cell-Penetrating Peptide-Modified Polymer Nanomicelles as a Potential Therapeutic Approach for Amyotrophic Lateral Sclerosis. *Pharmaceutics* **2022**, *14*, 2590. [CrossRef]

153. Kannan, S.; Balakrishnan, B.; Nance, E.; Johnston, M.V.; Rangaramanujam, K. Nanomedicine in cerebral palsy. *Int. J. Nanomed.* **2013**, *8*, 4183–4195. [CrossRef] [PubMed]

154. Nance, E.; Zhang, F.; Mishra, M.K.; Zhang, Z.; Kambhampati, S.P.; Kannan, R.M.; Kannan, S. Nanoscale effects in dendrimer-mediated targeting of neuroinflammation. *Biomaterials* **2016**, *101*, 96–107. [CrossRef]

155. Zhang, F.; Nance, E.; Alnasser, Y.; Kannan, R.; Kannan, S. Microglial migration and interactions with dendrimer nanoparticles are altered in the presence of neuroinflammation. *J. Neuroinflamm.* **2016**, *13*, 65. [CrossRef]

156. Lesniak, W.G.; Mishra, M.K.; Jyoti, A.; Balakrishnan, B.; Zhang, F.; Nance, E.; Romero, R.; Kannan, S.; Kannan, R.M. Biodistribution of Fluorescently Labeled PAMAM Dendrimers in Neonatal Rabbits: Effect of Neuroinflammation. *Mol. Pharm.* **2013**, *10*, 4560–4571. [CrossRef]

157. Kannan, S.; Dai, H.; Navath, R.S.; Balakrishnan, B.; Jyoti, A.; Janisse, J.; Romero, R.; Kannan, R.M. Dendrimer-Based Postnatal Therapy for Neuroinflammation and Cerebral Palsy in a Rabbit Model. *Sci. Transl. Med.* **2012**, *4*, 130ra46. [CrossRef] [PubMed]

158. Nance, E.; Porambo, M.; Zhang, F.; Mishra, M.K.; Buelow, M.; Getzenberg, R.; Johnston, M.; Kannan, R.M.; Fatemi, A.; Kannan, S. Systemic dendrimer-drug treatment of ischemia-induced neonatal white matter injury. *J. Control. Release* **2015**, *214*, 112–120. [CrossRef] [PubMed]

159. Mishra, M.K.; Beaty, C.A.; Lesniak, W.G.; Kambhampati, S.P.; Zhang, F.; Wilson, M.A.; Blue, M.E.; Troncoso, J.C.; Kannan, S.; Johnston, M.V.; et al. Dendrimer Brain Uptake and Targeted Therapy for Brain Injury in a Large Animal Model of Hypothermic Circulatory Arrest. *ACS Nano* **2014**, *8*, 2134–2147. [CrossRef] [PubMed]

160. Modi, H.R.; Wang, Q.; Olmstead, S.J.; Khoury, E.S.; Sah, N.; Guo, Y.; Gharibani, P.; Sharma, R.; Kannan, R.M.; Kannan, S.; et al. Systemic administration of dendrimer N-acetyl cysteine improves outcomes and survival following cardiac arrest. *Bioeng. Transl. Med.* **2022**, *7*, e10259. [CrossRef] [PubMed]

161. Kambhampati, S.P.; Bhutto, I.A.; Wu, T.; Ho, K.; McLeod, D.S.; Lutty, G.A.; Kannan, R.M. Systemic dendrimer nanotherapies for targeted suppression of choroidal inflammation and neovascularization in age-related macular degeneration. *J. Control. Release* **2021**, *335*, 527–540. [CrossRef] [PubMed]

162. Wilson, A. S-Adenosyl Methionine (SAMe) for Depression in Adults. *Issues Ment. Heal. Nurs.* **2019**, *40*, 725–726. [CrossRef] [PubMed]

163. Guo, T.; Chang, L.; Xiao, Y.; Liu, Q. S-Adenosyl-L-Methionine for the Treatment of Chronic Liver Disease: A Systematic Review and Meta-Analysis. *PLoS ONE* **2015**, *10*, e0122124. [CrossRef] [PubMed]

164. Xie, X.; Shu, R.; Yu, C.; Fu, Z.; Li, Z. Mammalian AKT, the Emerging Roles on Mitochondrial Function in Diseases. *Aging Dis.* **2022**, *13*, 157–174. [CrossRef]

165. Li, H.; Tang, Z.; Chu, P.; Song, Y.; Yang, Y.; Sun, B.; Niu, M.; Qaed, E.; Shopit, A.; Han, G.; et al. Neuroprotective effect of phosphocreatine on oxidative stress and mitochondrial dysfunction induced apoptosis in vitro and in vivo: Involvement of dual PI3K/Akt and Nrf2/HO-1 pathways. *Free Radic. Biol. Med.* **2018**, *120*, 228–238. [CrossRef]

166. He, F.; Ru, X.; Wen, T. NRF2, a Transcription Factor for Stress Response and Beyond. *Int. J. Mol. Sci.* **2020**, *21*, 4777. [CrossRef]

167. Buendia, I.; Michalska, P.; Navarro, E.; Gameiro, I.; Egea, J.; León, R. Nrf2–ARE pathway: An emerging target against oxidative stress and neuroinflammation in neurodegenerative diseases. *Pharmacol. Ther.* **2016**, *157*, 84–104. [CrossRef] [PubMed]

168. Bhowmick, S.; D'mello, V.; Caruso, D.; Abdul-Muneer, P.M. Traumatic brain injury-induced downregulation of Nrf2 activates inflammatory response and apoptotic cell death. *J. Mol. Med.* **2019**, *97*, 1627–1641. [CrossRef] [PubMed]

169. Abdul-Muneer, P.M. Nrf2 as a Potential Therapeutic Target for Traumatic Brain Injury. *J. Integr. Neurosci.* **2023**, *22*, 81. [CrossRef] [PubMed]
170. Wu, A.-G.; Yong, Y.-Y.; Pan, Y.-R.; Zhang, L.; Wu, J.-M.; Zhang, Y.; Tang, Y.; Wei, J.; Yu, L.; Law, B.Y.-K.; et al. Targeting Nrf2-Mediated Oxidative Stress Response in Traumatic Brain Injury: Therapeutic Perspectives of Phytochemicals. *Oxidative Med. Cell. Longev.* **2022**, *2022*, 1015791. [CrossRef] [PubMed]
171. Dong, W.; Yang, B.; Wang, L.; Li, B.; Guo, X.; Zhang, M.; Jiang, Z.; Fu, J.; Pi, J.; Guan, D.; et al. Curcumin plays neuroprotective roles against traumatic brain injury partly via Nrf2 signaling. *Toxicol. Appl. Pharmacol.* **2018**, *346*, 28–36. [CrossRef] [PubMed]
172. Lalitha, A.; Nanjundeshwari, G.; Swetha, M.; Swetha, R. A Review On Omaveloxolone. *Int. J. Pharm. Sci.* **2023**, *1*, 447–456. [CrossRef]
173. Lynch, D.R.; Farmer, J.; Hauser, L.; Blair, I.A.; Wang, Q.Q.; Mesaros, C.; Snyder, N.; Boesch, S.; Chin, M.; Delatycki, M.B.; et al. Safety, pharmacodynamics, and potential benefit of omaveloxolone in Friedreich ataxia. *Ann. Clin. Transl. Neurol.* **2019**, *6*, 15–26. [CrossRef] [PubMed]
174. Lynch, D.R.; Chin, M.P.; Delatycki, M.B.; Subramony, S.H.; Corti, M.; Hoyle, J.C.; Boesch, S.; Nachbauer, W.; Mariotti, C.; Mathews, K.D.; et al. Safety and Efficacy of Omaveloxolone in Friedreich Ataxia (MOXIe Study). *Ann. Neurol.* **2021**, *89*, 212–225. [CrossRef]
175. Abeti, R.; Baccaro, A.; Esteras, N.; Giunti, P. Novel Nrf2-Inducer Prevents Mitochondrial Defects and Oxidative Stress in Friedreich's Ataxia Models. *Front. Cell. Neurosci.* **2018**, *12*, 188. [CrossRef]
176. Wingerchuk, D.M.; Carter, J.L. Multiple Sclerosis: Current and Emerging Disease-Modifying Therapies and Treatment Strategies. *Mayo Clin. Proc.* **2014**, *89*, 225–240. [CrossRef]
177. Maldonado, P.P.; Guevara, C.; Olesen, M.A.; Orellana, J.A.; Quintanilla, R.A.; Ortiz, F.C. Neurodegeneration in Multiple Sclerosis: The Role of Nrf2-Dependent Pathways. *Antioxidants* **2022**, *11*, 1146. [CrossRef]
178. Abd El-Fatah, I.M.; Abdelrazek, H.M.; Ibrahim, S.M.; Abdallah, D.M.; El-Abhar, H.S. Dimethyl fumarate abridged tauo-/amyloidopathy in a D-Galactose/ovariectomy-induced Alzheimer's-like disease: Modulation of AMPK/SIRT-1, AKT/CREB/BDNF, AKT/GSK-3β, adiponectin/Adipo1R, and NF-κB/IL-1β/ROS trajectories. *Neurochem. Int.* **2021**, *148*, 105082. [CrossRef] [PubMed]
179. Bresciani, G.; Manai, F.; Davinelli, S.; Tucci, P.; Saso, L.; Amadio, M. Novel potential pharmacological applications of dimethyl fumarate—An overview and update. *Front. Pharmacol.* **2023**, *14*, 1264842. [CrossRef] [PubMed]
180. Kunze, R.; Urrutia, A.; Hoffmann, A.; Liu, H.; Helluy, X.; Pham, M.; Reischl, S.; Korff, T.; Marti, H.H. Dimethyl fumarate attenuates cerebral edema formation by protecting the blood–brain barrier integrity. *Exp. Neurol.* **2015**, *266*, 99–111. [CrossRef] [PubMed]
181. Casili, G.; Campolo, M.; Paterniti, I.; Lanza, M.; Filippone, A.; Cuzzocrea, S.; Esposito, E. Dimethyl Fumarate Attenuates Neuroinflammation and Neurobehavioral Deficits Induced by Experimental Traumatic Brain Injury. *J. Neurotrauma* **2018**, *35*, 1437–1451. [CrossRef] [PubMed]
182. Zhao, X.; Sun, G.; Zhang, J.; Ting, S.-M.; Gonzales, N.; Aronowski, J. Dimethyl Fumarate Protects Brain From Damage Produced by Intracerebral Hemorrhage by Mechanism Involving Nrf2. *Stroke* **2015**, *46*, 1923–1928. [CrossRef] [PubMed]
183. Krämer, T.; Grob, T.; Menzel, L.; Hirnet, T.; Griemert, E.; Radyushkin, K.; Thal, S.C.; Methner, A.; Schaefer, M.K.E. Dimethyl fumarate treatment after traumatic brain injury prevents depletion of antioxidative brain glutathione and confers neuroprotection. *J. Neurochem.* **2017**, *143*, 523–533. [CrossRef] [PubMed]
184. Gold, R.; Arnold, D.L.; Bar-Or, A.; Fox, R.J.; Kappos, L.; Mokliatchouk, O.; Jiang, X.; Lyons, J.; Kapadia, S.; Miller, C. Long-term safety and efficacy of dimethyl fumarate for up to 13 years in patients with relapsing-remitting multiple sclerosis: Final ENDORSE study results. *Mult. Scler. J.* **2022**, *28*, 801–816. [CrossRef]
185. Hayashi, G.; Jasoliya, M.; Sahdeo, S.; Saccà, F.; Pane, C.; Filla, A.; Marsili, A.; Puorro, G.; Lanzillo, R.; Brescia Morra, V.; et al. Dimethyl fumarate mediates Nrf2-dependent mitochondrial biogenesis in mice and humans. *Hum. Mol. Genet.* **2017**, *26*, 2864–2873. [CrossRef]
186. Majkutewicz, I. Dimethyl fumarate: A review of preclinical efficacy in models of neurodegenerative diseases. *Eur. J. Pharmacol.* **2022**, *926*, 175025. [CrossRef]
187. Ashrafizadeh, M.; Ahmadi, Z.; Mohammadinejad, R.; Farkhondeh, T.; Samarghandian, S. Curcumin Activates the Nrf2 Pathway and Induces Cellular Protection Against Oxidative Injury. *Curr. Mol. Med.* **2020**, *20*, 116–133. [PubMed]
188. Monroy, A.; Lithgow, G.J.; Alavez, S. Curcumin and neurodegenerative diseases. *BioFactors* **2013**, *39*, 122–132. [CrossRef] [PubMed]
189. Dinkova-Kostova, A.T.; Fahey, J.W.; Kostov, R.V.; Kensler, T.W. KEAP1 and done? Targeting the NRF2 pathway with sulforaphane. *Trends Food Sci. Technol.* **2017**, *69 Pt B*, 257–269. [CrossRef]
190. Schepici, G.; Bramanti, P.; Mazzon, E. Efficacy of Sulforaphane in Neurodegenerative Diseases. *Int. J. Mol. Sci.* **2020**, *21*, 8637. [CrossRef]
191. Talebi, M.; Talebi, M.; Farkhondeh, T.; Mishra, G.; Ilgün, S.; Samarghandian, S. New insights into the role of the Nrf2 signaling pathway in green tea catechin applications. *Phytother. Res.* **2021**, *35*, 3078–3112. [CrossRef]
192. Afzal, O.; Dalhat, M.H.; Altamimi, A.S.A.; Rasool, R.; Alzarea, S.I.; Almalki, W.H.; Murtaza, B.N.; Iftikhar, S.; Nadeem, S.; Nadeem, M.S.; et al. Green Tea Catechins Attenuate Neurodegenerative Diseases and Cognitive Deficits. *Molecules* **2022**, *27*, 7604. [CrossRef]

193. Chiang, M.-C.; Tsai, T.-Y.; Wang, C.-J. The Potential Benefits of Quercetin for Brain Health: A Review of Anti-Inflammatory and Neuroprotective Mechanisms. *Int. J. Mol. Sci.* **2023**, *24*, 6328. [CrossRef] [PubMed]
194. Elumalai, P.; Lakshmi, S. Role of Quercetin Benefits in Neurodegeneration. *Adv. Neurobiol.* **2016**, *12*, 229–245.
195. Dinkova-Kostova, A.T.; Kostov, R.V.; Kazantsev, A.G. The role of Nrf2 signaling in counteracting neurodegenerative diseases. *FEBS J.* **2018**, *285*, 3576–3590. [CrossRef]
196. Zhang, D.D. Bardoxolone Brings Nrf2-Based Therapies to Light. *Antioxidants Redox Signal.* **2013**, *19*, 517–518. [CrossRef]
197. Hisamichi, M.; Kamijo-Ikemori, A.; Sugaya, T.; Hoshino, S.; Kimura, K.; Shibagaki, Y. Role of bardoxolone methyl, a nuclear factor erythroid 2-related factor 2 activator, in aldosterone- and salt-induced renal injury. *Hypertens. Res.* **2018**, *41*, 8–17. [CrossRef] [PubMed]
198. Ren, Z.; He, H.; Zuo, Z.; Xu, Z.; Wei, Z.; Deng, J. The role of different SIRT1-mediated signaling pathways in toxic injury. *Cell. Mol. Biol. Lett.* **2019**, *24*, 36. [CrossRef] [PubMed]
199. Halling, J.F.; Pilegaard, H. PGC-1α-mediated regulation of mitochondrial function and physiological implications. *Appl. Physiol. Nutr. Metab.* **2020**, *45*, 927–936. [CrossRef] [PubMed]
200. Tellone, E.; Galtieri, A.; Russo, A.; Giardina, B.; Ficarra, S. Resveratrol: A Focus on Several Neurodegenerative Diseases. *Oxidative Med. Cell. Longev.* **2015**, *2015*, 392169. [CrossRef] [PubMed]
201. Sun, A.Y.; Wang, Q.; Simonyi, A.; Sun, G.Y. Resveratrol as a Therapeutic Agent for Neurodegenerative Diseases. *Mol. Neurobiol.* **2010**, *41*, 375–383. [CrossRef] [PubMed]
202. Bastianetto, S.; Ménard, C.; Quirion, R. Neuroprotective action of resveratrol. *Biochim. Biophys. Acta (BBA)—Mol. Basis Dis.* **2015**, *1852*, 1195–1201. [CrossRef] [PubMed]
203. Kim, E.N.; Lim, J.H.; Kim, M.Y.; Ban, T.H.; Jang, I.A.; Yoon, H.E.; Park, C.W.; Chang, Y.S.; Choi, B.S. Resveratrol, an Nrf2 activator, ameliorates aging-related progressive renal injury. *Aging* **2018**, *10*, 83–99. [CrossRef]
204. Wahab, A.; Gao, K.; Jia, C.; Zhang, F.; Tian, G.; Murtaza, G.; Chen, J. Significance of Resveratrol in Clinical Management of Chronic Diseases. *Molecules* **2017**, *22*, 1329. [CrossRef]
205. Sönmez, Ü.; Sönmez, A.; Erbil, G.; Tekmen, I.; Baykara, B. Neuroprotective effects of resveratrol against traumatic brain injury in immature rats. *Neurosci. Lett.* **2007**, *420*, 133–137. [CrossRef]
206. Singleton, R.H.; Yan, H.Q.; Fellows-Mayle, W.; Dixon, C.E. Resveratrol Attenuates Behavioral Impairments and Reduces Cortical and Hippocampal Loss in a Rat Controlled Cortical Impact Model of Traumatic Brain Injury. *J. Neurotrauma* **2010**, *27*, 1091–1099. [CrossRef]
207. Salberg, S.; Yamakawa, G.; Christensen, J.; Kolb, B.; Mychasiuk, R. Assessment of a nutritional supplement containing resveratrol, prebiotic fiber, and omega-3 fatty acids for the prevention and treatment of mild traumatic brain injury in rats. *Neuroscience* **2017**, *365*, 146–157. [CrossRef] [PubMed]
208. Cong, P.; Wang, T.; Tong, C.; Liu, Y.; Shi, L.; Mao, S.; Shi, X.; Jin, H.; Liu, Y.; Hou, M. Resveratrol ameliorates thoracic blast exposure-induced inflammation, endoplasmic reticulum stress and apoptosis in the brain through the Nrf2/Keap1 and NF-κB signaling pathway. *Injury* **2021**, *52*, 2795–2802. [CrossRef] [PubMed]
209. Ateş, O.; Çayli, S.; Altinoz, E.; Gurses, I.; Yucel, N.; Sener, M.; Kocak, A.; Yologlu, S. Neuroprotection by resveratrol against traumatic brain injury in rats. *Mol. Cell. Biochem.* **2007**, *294*, 137–144. [CrossRef] [PubMed]
210. Zhou, J.; Yang, Z.; Shen, R.; Zhong, W.; Zheng, H.; Chen, Z.; Tang, J.; Zhu, J. Resveratrol Improves Mitochondrial Biogenesis Function and Activates PGC-1α Pathway in a Preclinical Model of Early Brain Injury Following Subarachnoid Hemorrhage. *Front. Mol. Biosci.* **2021**, *8*, 620683. [CrossRef]
211. Jiang, W.; Li, S.; Li, X. Therapeutic potential of berberine against neurodegenerative diseases. *Sci. China Life Sci.* **2015**, *58*, 564–569. [CrossRef] [PubMed]
212. Javadipour, M.; Rezaei, M.; Keshtzar, E.; Khodayar, M.J. Metformin in contrast to berberine reversed arsenic-induced oxidative stress in mitochondria from rat pancreas probably via Sirt3-dependent pathway. *J. Biochem. Mol. Toxicol.* **2019**, *33*, e22368. [CrossRef] [PubMed]
213. Rotermund, C.; Machetanz, G.; Fitzgerald, J.C. The Therapeutic Potential of Metformin in Neurodegenerative Diseases. *Front. Endocrinol.* **2018**, *9*, 400. [CrossRef]
214. Wang, Y.; An, H.; Liu, T.; Qin, C.; Sesaki, H.; Guo, S.; Radovick, S.; Hussain, M.; Maheshwari, A.; Wondisford, F.E.; et al. Metformin Improves Mitochondrial Respiratory Activity through Activation of AMPK. *Cell Rep.* **2019**, *29*, 1511–1523.e5. [CrossRef]
215. Fanoudi, S.; Hosseini, M.; Alavi, M.S.; Boroushaki, M.T.; Hosseini, A.; Sadeghnia, H.R. Everolimus, a mammalian target of rapamycin inhibitor, ameliorated streptozotocin-induced learning and memory deficits via neurochemical alterations in male rats. *EXCLI J.* **2018**, *17*, 999–1017. [CrossRef]
216. Shiau, J.-P.; Chuang, Y.-T.; Cheng, Y.-B.; Tang, J.-Y.; Hou, M.-F.; Yen, C.-Y.; Chang, H.-W. Impacts of Oxidative Stress and PI3K/AKT/mTOR on Metabolism and the Future Direction of Investigating Fucoidan-Modulated Metabolism. *Antioxidants* **2022**, *11*, 911. [CrossRef]
217. Stein, D.G. Embracing failure: What the Phase III progesterone studies can teach about TBI clinical trials. *Brain Inj.* **2015**, *29*, 1259–1272. [CrossRef] [PubMed]
218. Ikonomidou, C.; Turski, L. Why did NMDA receptor antagonists fail clinical trials for stroke and traumatic brain injury? *Lancet Neurol.* **2002**, *1*, 383–386. [CrossRef] [PubMed]

219. Menon, D.K. Unique challenges in clinical trials in traumatic brain injury. *Crit. Care Med.* **2009**, *37*, S129–S135. [CrossRef] [PubMed]

220. Maas, A.I.R.; Roozenbeek, B.; Manley, G.T. Clinical Trials in Traumatic Brain Injury: Past Experience and Current Developments. *Neurotherapeutics* **2010**, *7*, 115–126. [CrossRef] [PubMed]

221. Margulies, S.; Hicks, R. Combination therapies for traumatic brain injury: Prospective considerations. *J. Neurotrauma* **2009**, *26*, 925–939. [CrossRef] [PubMed]

222. Lynch, D.G.; Narayan, R.K.; Li, C. Multi-Mechanistic Approaches to the Treatment of Traumatic Brain Injury: A Review. *J. Clin. Med.* **2023**, *12*, 2179. [CrossRef] [PubMed]

223. Somayaji, M.R.; Przekwas, A.J.; Gupta, R.K. Combination Therapy for Multi-Target Manipulation of Secondary Brain Injury Mechanisms. *Curr. Neuropharmacol.* **2018**, *16*, 484–504. [CrossRef]

224. Arun, P.; Ariyannur, P.S.; Moffett, J.R.; Xing, G.; Hamilton, K.; Grunberg, N.E.; Ives, J.A.; Namboodiri, A.M. Metabolic Acetate Therapy for the Treatment of Traumatic Brain Injury. *J. Neurotrauma* **2010**, *27*, 293–298. [CrossRef]

225. Scafidi, S.; Racz, J.; Hazelton, J.; McKenna, M.C.; Fiskum, G. Neuroprotection by Acetyl-L-Carnitine after Traumatic Injury to the Immature Rat Brain. *Dev. Neurosci.* **2010**, *32*, 480–487. [CrossRef]

226. Bonferoni, M.C.; Rassu, G.; Gavini, E.; Sorrenti, M.; Catenacci, L.; Giunchedi, P. Nose-to-Brain Delivery of Antioxidants as a Potential Tool for the Therapy of Neurological Diseases. *Pharmaceutics* **2020**, *12*, 1246. [CrossRef]

227. Pandya, J.D.; Musyaju, S.; Modi, H.R.; Okada-Rising, S.L.; Bailey, Z.S.; Scultetus, A.H.; Shear, D.A. Intranasal delivery of mitochondria targeted neuroprotective compounds for traumatic brain injury: Screening based on pharmacological and physiological properties. *J. Transl. Med.* **2024**, *22*, 167. [CrossRef] [PubMed]

228. Giarratana, A.; Reddi, S.; Thakker-Varia, S.; Alder, J. Status of precision medicine approaches to traumatic brain injury. *Neural Regen. Res.* **2022**, *17*, 2166–2171. [CrossRef] [PubMed]

229. Su, Y.S.; Schuster, J.M.; Smith, D.H.; Stein, S.C. Cost-Effectiveness of Biomarker Screening for Traumatic Brain Injury. *J. Neurotrauma* **2019**, *36*, 2083–2091. [CrossRef] [PubMed]

230. Rice, M.W.; Pandya, J.D.; Shear, D.A. Gut Microbiota as a Therapeutic Target to Ameliorate the Biochemical, Neuroanatomical, and Behavioral Effects of Traumatic Brain Injuries. *Front. Neurol.* **2019**, *10*, 875. [CrossRef] [PubMed]

231. Gaddam, S.S.; Buell, T.; Robertson, C.S. Systemic manifestations of traumatic brain injury. *Handb. Clin. Neurol.* **2015**, *127*, 205–218. [PubMed]

232. Taraskina, A.; Ignatyeva, O.; Lisovaya, D.; Ivanov, M.; Ivanova, L.; Golovicheva, V.; Baydakova, G.; Silachev, D.; Popkov, V.; Ivanets, T.; et al. Effects of Traumatic Brain Injury on the Gut Microbiota Composition and Serum Amino Acid Profile in Rats. *Cells* **2022**, *11*, 1409. [CrossRef] [PubMed]

233. Zhu, C.S.; Grandhi, R.; Patterson, T.T.; Nicholson, S.E. A Review of Traumatic Brain Injury and the Gut Microbiome: Insights into Novel Mechanisms of Secondary Brain Injury and Promising Targets for Neuroprotection. *Brain Sci.* **2018**, *8*, 113. [CrossRef]

234. Celorrio, M.; Abellanas, M.A.; Rhodes, J.; Goodwin, V.; Moritz, J.; Vadivelu, S.; Wang, L.; Rodgers, R.; Xiao, S.; Anabayan, I.; et al. Gut microbial dysbiosis after traumatic brain injury modulates the immune response and impairs neurogenesis. *Acta Neuropathol. Commun.* **2021**, *9*, 40. [CrossRef]

235. Ullah, H.; Arbab, S.; Tian, Y.; Liu, C.-Q.; Chen, Y.; Li, Q.; Khan, M.I.U.; Hassan, I.U.; Li, K. The gut microbiota–brain axis in neurological disorder. *Front. Neurosci.* **2023**, *17*, 1225875. [CrossRef]

236. Dinan, T.G.; Cryan, J.F. Microbes, Immunity, and Behavior: Psychoneuroimmunology Meets the Microbiome. *Neuropsychopharmacology* **2017**, *42*, 178–192. [CrossRef]

237. Naliyadhara, N.; Kumar, A.; Gangwar, S.K.; Devanarayanan, T.N.; Hegde, M.; Alqahtani, M.S.; Abbas, M.; Sethi, G.; Kunnumakkara, A. Interplay of dietary antioxidants and gut microbiome in human health: What has been learnt thus far? *J. Funct. Foods* **2023**, *100*, 105365. [CrossRef]

238. Riaz Rajoka, M.S.; Thirumdas, R.; Mehwish, H.M.; Umair, M.; Khurshid, M.; Hayat, H.F.; Phimolsiripol, Y.; Pallarés, N.; Martí-Quijal, F.J.; Barba, F.J. Role of Food Antioxidants in Modulating Gut Microbial Communities: Novel Understandings in Intestinal Oxidative Stress Damage and Their Impact on Host Health. *Antioxidants* **2021**, *10*, 1563. [CrossRef]

239. Portincasa, P.; Bonfrate, L.; Vacca, M.; De Angelis, M.; Farella, I.; Lanza, E.; Khalil, M.; Wang, D.Q.-H.; Sperandio, M.; Di Ciaula, A. Gut Microbiota and Short Chain Fatty Acids: Implications in Glucose Homeostasis. *Int. J. Mol. Sci.* **2022**, *23*, 1105. [CrossRef]

240. Wang, X.; Qi, Y.; Zheng, H. Dietary Polyphenol, Gut Microbiota, and Health Benefits. *Antioxidants* **2022**, *11*, 1212. [CrossRef]

241. Clark, A.; Mach, N. The Crosstalk between the Gut Microbiota and Mitochondria during Exercise. *Front. Physiol.* **2017**, *8*, 319. [CrossRef] [PubMed]

 *antioxidants*

Article

Effects of L-Type Voltage-Gated Calcium Channel (LTCC) Inhibition on Hippocampal Neuronal Death after Pilocarpine-Induced Seizure

Chang-Jun Lee [1], Song-Hee Lee [1], Beom-Seok Kang [1], Min-Kyu Park [1], Hyun-Wook Yang [1], Seo-Young Woo [1], Se-Wan Park [1], Dong-Yeon Kim [1], Hyun-Ho Jeong [1], Won-Il Yang [1,2], A-Ra Kho [3,4], Bo-Young Choi [2], Hong-Ki Song [5,6], Hui-Chul Choi [6,7], Yeo-Jin Kim [5] and Sang-Won Suh [1,6,*]

[1] Department of Physiology, Hallym University College of Medicine, Chuncheon 24252, Republic of Korea; doog0716@hallym.ac.kr (C.-J.L.); 2thqml@naver.com (S.-H.L.); bskang93@hallym.ac.kr (B.-S.K.); d22029@hallym.ac.kr (M.-K.P.); akqjqtj5@hallym.ac.kr (H.-W.Y.); m22091@hallym.ac.kr (S.-Y.W.); m2022087@hallym.ac.kr (S.-W.P.); m22525@hallym.ac.kr (D.-Y.K.); wjdgusgh1021@hallym.ac.kr (H.-H.J.); wonil4u@hallym.ac.kr (W.-I.Y.)
[2] Department of Physical Education, Hallym University, Chuncheon 24252, Republic of Korea; bychoi@hallym.ac.kr
[3] Neuroregeneration and Stem Cell Programs, Institute for Cell Engineering, Johns Hopkins University School of Medicine, Baltimore, MD 21205, USA; akho3@jhu.edu
[4] Department of Neurology, Johns Hopkins University School of Medicine, Baltimore, MD 21205, USA
[5] Department of Neurology, Kangdong Sacred Heart Hospital, Seoul 05355, Republic of Korea; hksong0@hallym.oc.kr (H.-K.S.); yjhelena@hanmail.net (Y.-J.K.)
[6] Hallym Institute of Epilepsy Research, Chuncheon 24252, Republic of Korea; dohchi@hallym.ac.kr
[7] Department of Neurology, Hallym University Chuncheon Sacred Heart Hospital, Chuncheon 24253, Republic of Korea
* Correspondence: swsuh@hallym.ac.kr

Citation: Lee, C.-J.; Lee, S.-H.; Kang, B.-S.; Park, M.-K.; Yang, H.-W.; Woo, S.-Y.; Park, S.-W.; Kim, D.-Y.; Jeong, H.-H.; Yang, W.-I.; et al. Effects of L-Type Voltage-Gated Calcium Channel (LTCC) Inhibition on Hippocampal Neuronal Death after Pilocarpine-Induced Seizure. *Antioxidants* **2024**, *13*, 389. https://doi.org/10.3390/antiox13040389

Academic Editor: Waldo Cerpa

Received: 17 January 2024
Revised: 15 March 2024
Accepted: 15 March 2024
Published: 24 March 2024

Abstract: Epilepsy, marked by abnormal and excessive brain neuronal activity, is linked to the activation of L-type voltage-gated calcium channels (LTCCs) in neuronal membranes. LTCCs facilitate the entry of calcium (Ca^{2+}) and other metal ions, such as zinc (Zn^{2+}) and magnesium (Mg^{2+}), into the cytosol. This Ca^{2+} influx at the presynaptic terminal triggers the release of Zn^{2+} and glutamate to the postsynaptic terminal. Zn^{2+} is then transported to the postsynaptic neuron via LTCCs. The resulting Zn^{2+} accumulation in neurons significantly increases the expression of nicotinamide adenine dinucleotide phosphate (NADPH) oxidase subunits, contributing to reactive oxygen species (ROS) generation and neuronal death. Amlodipine (AML), typically used for hypertension and coronary artery disease, works by inhibiting LTCCs. We explored whether AML could mitigate Zn^{2+} translocation and accumulation in neurons, potentially offering protection against seizure-induced hippocampal neuronal death. We tested this by establishing a rat epilepsy model with pilocarpine and administering AML (10 mg/kg, orally, daily for 7 days) post-epilepsy onset. We assessed cognitive function through behavioral tests and conducted histological analyses for Zn^{2+} accumulation, oxidative stress, and neuronal death. Our findings show that AML's LTCC inhibition decreased excessive Zn^{2+} accumulation, reactive oxygen species (ROS) production, and hippocampal neuronal death following seizures. These results suggest amlodipine's potential as a therapeutic agent in seizure management and mitigating seizures' detrimental effects.

Keywords: seizure; amlodipine; L-type voltage-gated calcium channel; zinc; neuronal death; oxidative stress

1. Introduction

Epilepsy is a neurological disorder characterized by abnormal electrical activity in the brain, resulting in recurrent seizures. While the exact causes and mechanisms of epilepsy remain incompletely understood, significant progress has been made in understanding some of the contributing factors [1–3]. Seizures disrupt ion concentrations in the brain, including potassium and calcium, leading to the depolarization of neighboring neurons and the increased release of neuromodulators like zinc, which contribute to abnormal brain activity. Seizures can also damage various brain cells, such as astrocytes and microglia, disrupt microtubules, compromise the blood–brain barrier, and induce reactive oxidative stress. Zinc accumulation has also been observed in certain cases of epilepsy [4–6].

L-type voltage-gated calcium channels (LTCCs) play a critical role in regulating calcium influx in smooth muscle cells and neurons. The $\alpha1$ subunit of LTCCs forms the channel pore and controls its opening. Upon membrane depolarization, the $\alpha1$ subunit allows the entry of calcium ions (Ca^{2+}) and other ions such as zinc (Zn^{2+}) [7,8]. There are four isoforms of LTCCs, namely Cav1.1 ($\alpha1S$), Cav1.2 ($\alpha1C$), Cav1.3 ($\alpha1D$), and Cav1.4 ($\alpha1F$). In the brain, Cav1.2 and Cav1.3 are the predominant isoforms, with Cav1.2 being particularly abundant and playing a significant role in brain LTCCs. Cav1.2 has been associated with hippocampal long-term potentiation (LTP), a form of synaptic plasticity linked to learning and memory. It also participates in activity-dependent gene transcription [7–13].

Zinc is an essential mineral that plays crucial roles in various physiological functions in the body, including cell division, development, and DNA synthesis. Adequate zinc levels are necessary for optimal brain functioning and memory formation. However, disruptions in zinc homeostasis can have negative effects on brain function. Zinc deficiency can impair cognitive functions, particularly short-term memory. Conversely, excessive accumulation of zinc in brain cells can be detrimental. Following brain injuries such as seizures, ischemia, or trauma, an increase in neuronal death is observed. In these conditions, zinc accumulation within brain cells is believed to contribute to neuronal damage and cell death [14–17]. Zinc released during seizures can translocate to the intracellular space of postsynaptic neurons through various ion channels, including NMDA (*N*-methyl-D-aspartate) and AMPA (α-amino-3-hydroxy-5-methyl-4-isoxazolepropionic acid) receptors, as well as L-type voltage-gated calcium channels (LTCCs). This influx of zinc into postsynaptic neurons can increase intracellular Zn^{2+} concentration [18,19]. Elevated intracellular Zn^{2+} levels can interact with NADPH oxidase, an enzyme involved in the production of reactive oxygen species (ROS), such as superoxide radicals. This interaction can promote ROS production within neurons. ROS are highly reactive molecules that can cause oxidative stress and damage cellular components, including proteins, lipids, and DNA. Increased ROS production and oxidative stress resulting from zinc-induced interactions can have deleterious effects on neurons, ultimately leading to neuronal death. This process has been implicated in various neurodegenerative disorders, including Alzheimer's disease, Parkinson's disease, and ischemic brain injury [20,21].

Amlodipine is a dihydropyridine (DHP)-type drug commonly used as a calcium channel blocker (CCB). It specifically acts on L-type voltage-gated calcium channels (LTCCs) in various tissues, including smooth muscle cells in blood vessels and cardiac myocytes [22,23]. Amlodipine is primarily indicated for hypertension and angina. By reducing peripheral vascular resistance and improving coronary blood flow, amlodipine lowers blood pressure and relieves angina symptoms. Recent research has explored the potential use of amlodipine beyond vascular diseases and into the field of brain diseases [24,25].

During seizures, there is excessive activation of neurons and other cells in the brain, leading to an overload of calcium ions (Ca^{2+}) and zinc ions (Zn^{2+}). This overload can contribute to cell death and neuronal damage. The excessive influx of Ca^{2+} and Zn^{2+} into brain cells is facilitated by calcium channels, including L-type voltage-gated calcium channels (LTCCs). Amlodipine, as a dihydropyridine (DHP)-type calcium channel blocker, specifically binds to and blocks the Cav1.2 $\alpha1$ subunit, which is the main component of LTCCs. By inhibiting the activity of Cav1.2 channels, amlodipine can effectively reduce the influx of calcium ions into brain cells. Additionally, by blocking LTCCs, amlodipine may indirectly affect the accumulation of Zn^{2+} in brain cells. As previously mentioned, excessive activation of neurons and cells during seizures can lead to the release of zinc ions, and the entry of Zn^{2+} into cells is facilitated by calcium channels, including LTCCs. By inhibiting LTCCs, amlodipine may help reduce the influx of Zn^{2+} into brain cells, potentially mitigating the detrimental effects of zinc overload [26–28].

Some reports have suggested the neuroprotective effects of amlodipine in central nervous system diseases and its potential use as an anticonvulsant in acute seizures [29–33]. Based on these findings, we hypothesize that treatment with amlodipine after pilocarpine-induced seizures in an animal model may lead to a reduction in oxidative stress, zinc accumulation, astrocyte and microglia over-activation, blood–brain barrier damage, and microtubule damage, and to an increase in cognitive function.

We hypothesize that amlodipine may exert a neuroprotective effect in seizure conditions. To test this hypothesis, we employed a controlled experimental design involving four groups: a seizure-vehicle group, a seizure-amlodipine group, and two sham controls. Our method included administering amlodipine post-seizure induction, followed by quantitative assessments using NeuN (Neuronal Nuclear) and DAPI (4′,6-diamidino-2-phenylindole) staining, to evaluate neuronal survival and integrity. Our findings suggest that amlodipine treatment leads to an increase in NeuN-positive cells compared to the seizure-vehicle group, indicating a potential protective effect against seizure-induced neuronal damage. However, these numbers did not reach the levels observed in the sham groups, suggesting a partial mitigation effect.

2. Materials and Methods

2.1. Ethics Statement and Care of Experimental Animals

The present study adhered to ethical guidelines for animal research and was approved by the Animal Research Committee at the College of Medicine at Hallym University (protocol number: Hallym 2021-39). Adult male Sprague Dawley rats (SD-Rats) were obtained from DBL Co. in Chungcheongbuk-do, Republic of Korea. The rats used in the experiment were approximately 8 weeks old and had an initial weight of 300–350 g. Strict control was maintained over environmental conditions, including a temperature of 20 ± 2 °C, humidity levels of $55\% \pm 5\%$, and a 12 h light/dark cycle.

2.2. Seizure Induction

Seizures were induced in rats through the administration of pilocarpine (25 mg/kg, i.p.) [34–37]. Prior to the injection of pilocarpine, lithium chloride (LiCl, 127 mg/kg, i.p.) was intraperitoneally administered, to enhance the action of muscarinic receptors. Additionally, scopolamine (2 mg/kg, i.p.) was injected 30 min before pilocarpine to reduce saliva production. The severity of status epilepticus (SE) was assessed using the Racine stage method, which categorizes seizures into five stages based on the following observed behaviors: 1. facial movement, 2. head nodding, 3. forelimb clonus, 4. rearing, and 5. falling. Following the occurrence of forelimb clonus and rearing, our pilocarpine-induced seizure model consistently reached a seizure severity of over phase 4, characterized by rearing, as per the Racine scoring system. This confirms the successful induction of seizures in our experimental setup [38,39]. Diazepam (10 mg/kg, i.p.), a commonly used antiepileptic medication for terminating or suppressing seizures, was administered one hour later. This experimental design allowed for the controlled induction and assessment of the severity of status epilepticus in the rats, followed by the administration of diazepam to manage and control the seizures.

2.3. Amlodipine Administration

The animals were divided into four groups: sham-vehicle, sham-amlodipine, seizure-vehicle, and seizure-amlodipine. In the amlodipine groups, amlodipine was administered orally at a dose of 10 mg/kg, dissolved in 0.9% saline. The first dose of amlodipine or saline was administered 1 h after seizure induction, and the treatment was continued daily for 7 days. For the behavior test to check cognitive function, the seizure groups received amlodipine or saline 1 h after seizure induction, and the administration was performed once a day for 12 days. For the TSQ (N-(6-methoxy-8-quinolyl)-para-toluene sulfonamide) staining procedure to check zinc accumulation, the seizure-amlodipine group received amlodipine (10 mg/kg, p.o., Daewoong, Dr.Reddy's lab, Korea) 1 h after seizure induction, while the seizure-vehicle group received an equivalent volume of 0.9% saline at the same time [40].

2.4. Brain Sample Preparation

The rats that experienced seizures were euthanized at two time points, 24 h and 7 days after seizure induction. Anesthesia was induced by administering urethane (1.5 g/kg, i.p.). To preserve the brain tissue, the animals were perfused with 0.9% saline, followed by 4% paraformaldehyde. The brains were carefully extracted and post-fixed in 4% paraformaldehyde for 1 h. Following fixation, the brains were transferred to a 30% sucrose solution and allowed to sink until they reached the desired consistency. After two days, the brains were sectioned into 30μm thick slices using a cryostat microtome (CM1850; Leica, Wetzlar, Germany).

2.5. Detection of Zinc Accumulation

After a 24 h period following seizure induction, the rats were anesthetized with urethane (1.5 g/kg, i.p.), and the brains were harvested without perfusion. To preserve the brain tissue, the brains were rapidly frozen in dry ice for approximately 1 min and then stored at −80 °C. Using a cryostat, brain samples were cut into 10 μm thick slices. These sliced sections were immediately mounted onto pre-coated slides (Fisher Scientific, Pittsburgh, PA, USA) and allowed to dry at room temperature for 1 h. Subsequently, the samples were stained with a 0.001% solution of TSQ (N-(6-methoxy-8-quinolyl)-para-toluene sulfonamide) obtained from Molecular Probes, Eugene, OR, USA [40]. The staining process lasted for 1 min. After staining, the samples were washed with a 0.9% saline solution for 1 min. The stained samples were observed using a fluorescence microscope (Olympus, Tokyo, Japan) equipped with UV light under a 360 nm wavelength and a 500 nm long-pass filter. This allowed for the visualization of the TSQ-positive cells, indicating zinc accumulation. To quantify the TSQ-positive cells, blind quantification was performed. The cells were counted without prior knowledge of the experimental conditions, ensuring an unbiased assessment of the results.

2.6. Detection of Oxidative Stress in the Hippocampal Region

For the 4-hydroxyl-2-nonenal (4HNE) immunohistochemistry assay, brain tissue samples were washed three times in 0.01 M phosphate-buffered saline (PBS) for 10 min each. A pretreatment step was performed to eliminate any blood present in the brain tissue's blood vessels. The tissue samples were sequentially treated with distilled water, 90% methanol, and 30% hydrogen peroxide for 15 min each. After each treatment, the samples were washed three times for 10 min each in 0.1 M PBS. The brain tissue samples were then incubated overnight at 4 °C with a 4HNE-specific primary antibody solution. The primary antibody used was mouse anti-4HNE serum, obtained from Alpha Diagnostic Intl. Inc., San Antonio, TX, USA, diluted at a ratio of 1:500 in PBS containing 0.3% Triton X-100. Following the overnight incubation, the brain tissue samples were washed three times for 10 min each in 0.01 M PBS. Next, the samples were incubated with a secondary antibody solution containing donkey anti-mouse IgG conjugated with Alexa-Fluor-594 at a dilution of 1:250, obtained from Invitrogen, Grand Island, NY, USA. This incubation step was carried out at room temperature for 2 h. After incubation with the secondary antibody, the brain tissue samples were washed three times for 10 min each in 0.01 M PBS. Finally, the tissue samples were mounted onto slides, cover-slipped (Fisher Scientific, Pittsburgh, PA, USA) using DPX (Sigma-Aldrich Co., St. Louis, MO, USA) mounting medium obtained from Sigma-Aldrich Co., St. Louis, MO, USA, and observed using a fluorescence microscope from Olympus, Tokyo, Japan. To quantify the fluorescence intensity of the 4HNE staining, we used the ImageJ software (version 1.47c; NIH, Bethesda, MD, USA), developed by the National Institutes of Health, Bethesda, MD, USA, to quantify the fluorescence intensity of the 4HNE staining. Specifically, we analyzed the intensity within defined regions of interest in the tissue sections. The software allowed us to calculate the mean fluorescence intensity for these regions, and we then averaged these values to represent the overall fluorescence intensity for each experimental group.

2.7. Immunofluorescence Assay

To evaluate the impact of amlodipine on neuronal cells, an immunofluorescence assay was conducted. Brain tissue samples underwent a series of procedures, outlined as follows: First, the samples were washed in 0.01 M PBS for 10 min, and this process was repeated three times. Subsequently, a pretreatment solution, composed of distilled water, 90% methanol, and 30% hydrogen peroxide, was applied to the brain tissues for 15 min. After the pretreatment, the tissues were washed in 0.01 M PBS for 10 min, and this step was repeated three times. The brain tissue samples were then incubated overnight at 4 °C with primary antibodies. The specific primary antibodies used, along with their respective dilutions, in this study were as follows: rabbit anti-MAP2 (1:200, Abcam), rabbit anti-GFAP (1:1000, Abcam, Cambridge, UK), goat anti-C3 (1:300, Invitrogen, Boston, MA, USA), goat anti-Iba1 (1:500, Abcam, Cambridge, UK), mouse anti-CD68 (1:100, Bio-rad, California, USA), mouse anti-NeuN (1:500, Millipore, Billerica, MA, USA), and rabbit anti-Cav1.2 (1:300, Alomone labs, Jerusalem BioPark, Jerusalem). All primary antibodies were diluted in PBS containing 0.3% Triton X-100 (PBS-T). Following the overnight incubation, the brain samples were washed three times for 10 min each in 0.01 M PBS. Subsequently, the brain tissue samples were subjected to a 2 h incubation with a secondary antibody solution in PBS-T. The secondary antibodies used were specifically selected to align with the host species of the primary antibodies and were conjugated with fluorescent markers. However, the precise details of the secondary antibodies were not provided. After incubation with the secondary antibodies, the brain tissue samples underwent three washes of 10 min each in 0.01 M PBS. The washed brain samples were then mounted onto slides and cover-slipped (Fisher Scientific, Pittsburgh, PA, USA) using DPX (Sigma-Aldrich Co., St. Louis, MO, USA) mounting medium and observed using a fluorescence microscope from Olympus, Tokyo, Japan. For the fluorescence intensity measurement of brain tissue samples, we utilized ImageJ software (version 1.47c; NIH, Bethesda, MD, USA). The average value of the mean fluorescence intensity was determined and reported. Overall, this immunofluorescence assay provided insights into the effects of amlodipine on various neuronal markers, including MAP2, GFAP, C3, Iba1, CD68, NeuN, and Cav1.2, in the brain tissue samples.

2.8. Immunohistochemistry Assay

To evaluate live neuron detection and blood–brain barrier (BBB) disruption in brain tissue, the following steps were conducted, according to the immunofluorescence assay protocol. First, brain tissue samples were washed and pretreated as described. Then, the samples were incubated overnight at 4 °C with a primary antibody solution, containing mouse anti-NeuN (1:500, Millipore, Billerica, MA, USA) and 0.3% Triton-X in PBS. After the overnight incubation, the brain samples were washed three times with 0.01 M PBS. Subsequently, the samples were incubated with a secondary antibody solution containing anti-mouse IgG (1:250, Vector, Burlingame, CA, USA) and 0.3% Triton-X in PBS for 2 h at room temperature. This step allowed for the detection of IgG leakage following a seizure. For the analysis of IgG leakage, brain samples underwent the same pretreatment and washing steps as mentioned above. Then, the samples were incubated with a secondary antibody solution containing anti-rat IgG (1:250, Vector Labororoid, Burlingame, CA, USA) and 0.3% Triton-X in PBS for 2 h at room temperature. After the secondary antibody incubation, an ABC (avidin–biotin complex, Vector, Burlingame, CA, USA) solution was applied to the brain samples at room temperature for 2 h, helping to amplify the signal from the primary antibody. The brain samples were then colored with a 3,3'-diaminobenzidine (DAB, Sigma-Aldrich Co., St. Louis, MO, USA) solution in 0.01 M PBS buffer for 2 min. This reaction produced a brown color, as the DAB reacted with the ABC complex. The colored brain samples were mounted onto slides (Fisher Scientific, Pittsburgh, PA, USA) using a Canada balsam mounting medium (Junsei Chemical, Chuo-ku, Tokyo, Japan), and they were cover-slipped. To quantify the number of NeuN-positive cells and assess IgG leakage, we have now included a detailed description of the image analysis technique used for quantifying staining intensity, which serves as an indicator of blood leakage. This includes the ImageJ software used, the parameters set for analysis, and the approach for selecting regions of interest. The image analysis software program ImageJ (version 1.47c; NIH, Bethesda, MD, USA) was employed. This software facilitated the measurement of staining intensity and the quantification of the desired markers. By following this protocol, the immunohistochemistry assay enabled the detection of live neurons in the area of the hippocampus (CA1, CA3, hilus, subiculum) using the NeuN marker and the assessment of BBB disruption by analyzing IgG leakage in all hippocampal regions.

2.9. Behavior Test
2.9.1. Barnes Maze Test

To evaluate the recovery of spatial cognitive ability after seizures, we conducted a study where rats were treated with amlodipine and subjected to the Barnes maze test. The rats, which had been epileptic for a week, were divided into two groups: one group received amlodipine treatment from the first day of seizure induction, while the other group had a rest period before starting the treatment. The Barnes maze test involved a circular board with multiple holes, with only one hole open and a cage placed underneath it. The rats were assessed over several trials, with each trial consisting of the rat being placed in a black cylinder, disoriented, and timed as they found the opening hole. After five days of testing, the rats were euthanized, and samples were stained using the previously mentioned method.

2.9.2. Adhesive Removal Test

We examined cognitive recovery post-seizure using amlodipine in rats, employing both the adhesive removal test and the Barnes maze test on the same days. In the adhesive removal test, rats had 1 cm tape pieces on their paws; if removal exceeded 2 min, we intervened. This test assessed fine motor skills and sensory perception. Multiple trials were conducted for each rat. The Barnes maze test evaluated spatial learning and memory. The combination of these tests allowed for a comprehensive assessment of cognitive function recovery, shedding light on the effects of amlodipine treatment.

2.10. Data Analysis

The data in this study are reported as the mean $\pm$ SEM. To compare the vehicle-treated and amlodipine-treated groups, statistical analyses were conducted using the Mann–Whitney U test or the Kruskal–Wallis with post hoc Bonferroni. Behavioral data were assessed for variance using an ANOVA. All data were analyzed using IBM, from the Statistical Package for the Social Sciences (IBM SPSS statistics version 25, Chicago, IL, USA) software. Moreover, a blind test approach was implemented during the statistical analysis to reduce bias.

3. Results

3.1. Amlodipine Administration Reduced Cav 1.2 Activation and Zinc Accumulation in Neurons Following Seizure

In this study, we quantified the expression of the Cav 1.2 calcium channel as an indirect indicator of altered calcium influx in neuronal cells [41,42]. Our objective was to examine the effects of amlodipine, an inhibitor of L-type voltage-gated calcium channels (LTCCs), on Cav1.2 channel expression and zinc accumulation in the hippocampal region of the brain. Cav1.2 channels play a crucial role in the influx of various ions, including Ca^{2+} and Zn^{2+}, and their activation during seizures leads to Zn^{2+} accumulation in neuronal cells.

To evaluate Cav1.2 channel expression, we performed immunofluorescence staining of brain samples using an assay that specifically targeted Cav1.2 channels in NeuN-positive cells, representing live neurons. We compared the expression levels between the group treated with amlodipine after status epilepticus (SE) and the SE group treated with a vehicle, as well as the vehicle-only groups. Our findings revealed that Cav1.2 expression was significantly higher in the CA1 and CA3 regions of the hippocampus in the vehicle-treated SE group, compared to the amlodipine-treated SE group and the vehicle-only groups. Interestingly, the expression levels of Cav1.2 channels in the SE amlodipine-treated group were comparable to those in the sham-treated group. Thus, our results demonstrate that amlodipine effectively reduced Cav1.2 channel expression during seizures (Figure 1A–D).

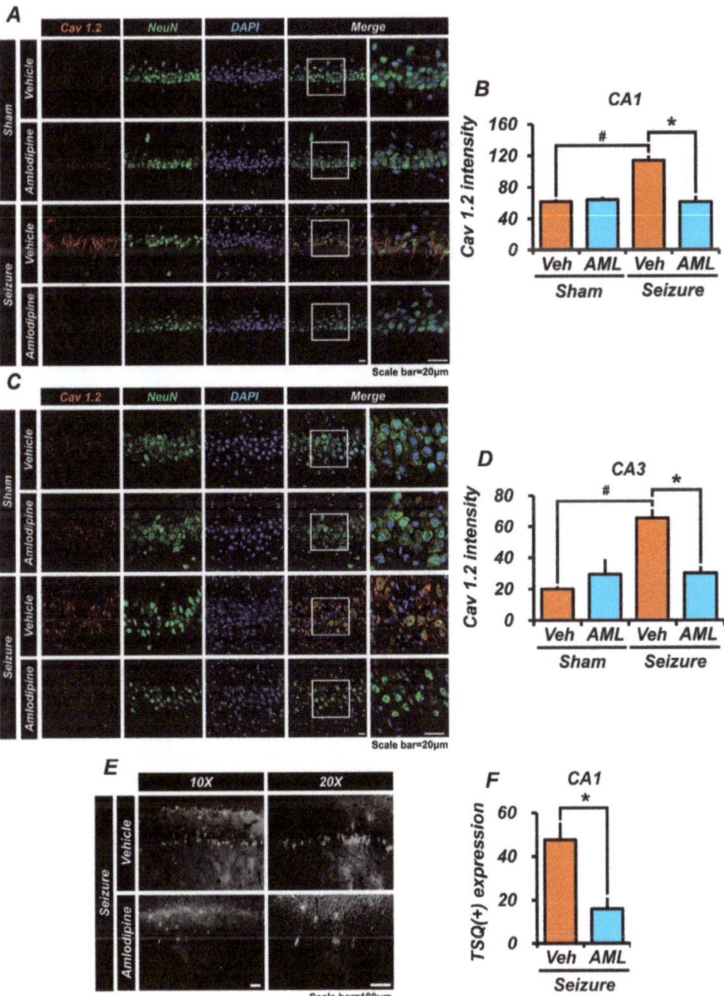

Figure 1. Amlodipine administration reduced Cav 1.2 activation and zinc accumulation in neurons following seizure. (**A,C**) Confocal micrographs of Cav1.2 (red), NeuN (green), and DAPI (blue) in the hippocampal CA1 and CA3 regions. Scale bar = 20 µm. (**B,D**) The bar graph shows the intensity of Cav1.2 in the hippocampal regions CA1 and CA3. Both graphs indicate that the intensity of Cav1.2 in the seizure-amlodipine group is similar to that of the sham groups and approximately half of that in the seizure-vehicle group, with significant differences. Additionally, there was no significant difference between the sham-vehicle group and the sham-amlodipine group. Animals were sacrificed on day 7. Cav 1.2 expression decreased in the seizure-amlodipine group, compared with seizure-vehicle

group, by about 46% in theCA1 region (seizure-vehicle group, 113.9 ± 5.9; seizure-amlodipine group, 61.0 ± 7.2; sham-vehicle group, 61.6 ± 2.7; sham-amlodipine group, 64.0 ± 3.3), and by 54% in the CA3 region (seizure-vehicle group, 65.7 ± 5.5; seizure-amlodipine group, 30.0 ± 4.2; sham-vehicle group, 20.0 ± 1.7; sham-amlodipine group, 29.3 ± 9.7). The sample sizes were n = 3 animals for each sham group and n = 5 animals for each seizure group. (Bonferroni post hoc test after Kruskal–Wallis test, CA1: chi-squared = 10.212, df = 3, p = 0.017; CA3: chi-squared = 10.471, df = 3, p = 0.015.) (**E**) Difference in the number of TSQ (+) cells between the seizure-vehicle and seizure-amlodipine groups in the CA1 region one day after the seizure. (**F**) The seizure-amlodipine group shows a significant reduction in TSQ-positive neuron cells, with a decrease of about 66%, compared to the seizure-vehicle group. Thus, amlodipine treatment after a seizure reduces Cav1.2 over-activation and zinc accumulation, compared to vehicle treatment after a seizure. Animals were sacrificed on day 1. TSQ (+) neuron cells decreased in the seizure-amlodipine group, compared with seizure-vehicle group, by about 66% in the CA1 region (seizure-vehicle group, 47.5 ± 8.0; seizure-amlodipine group, 16.0 ± 5.0). Scale bar = 100 μm (the sample sizes were n = 5 animals for the seizure-amlodipine group, n = 6 animals for the seizure-vehicle group), (Mann–Whitney U test measurement results, whole brain region z = 2.373, p = 0.018). [#] indicates significant difference between sham vehicle and seizure vehicle groups, * Significantly different from the vehicle group, p < 0.05.

Furthermore, we conducted TSQ staining to assess zinc accumulation in the hippocampal region following pilocarpine-induced seizures. Seizures result in excessive neuronal activation, leading to neuronal damage and the subsequent accumulation of zinc within neurons. By blocking LTCCs, which are ion channels in neurons, our aim was to mitigate zinc accumulation and its potential detrimental effects on neuronal survival. Neither the sham-vehicle nor the sham-amlodipine groups showed significant zinc accumulation in neurons. However, within the seizure groups, a notable zinc accumulation was observed. Crucially, the seizure-vehicle group exhibited a twofold increase in the number of neurons with positive zinc accumulation compared to the seizure-amlodipine group. These findings highlight the efficacy of amlodipine in reducing Cav1.2 channel expression and diminishing zinc accumulation in neurons, potentially contributing to neuroprotection against cell death (Figure 1E,F).

In summary, our study demonstrates that amlodipine treatment during seizures effectively downregulates Cav1.2 channel expression and limits zinc accumulation in neurons. These findings underscore the potential neuroprotective properties of amlodipine, highlighting its therapeutic value in mitigating neuronal damage associated with seizures.

3.2. Administration of Amlodipine Decreased Reactive Oxidative Stress (ROS) after Seizure

To investigate the impact of amlodipine on reactive oxidative stress (ROS) in neuronal damage, we examined the expression of ROS in the hippocampal region one week after pilocarpine-induced seizures. ROS, triggered by factors such as neuronal over-activation and ion accumulation during seizures, can cause significant harm to neuronal cells through DNA, protein, and lipid damage, ultimately leading to cell death.

To assess seizure-induced ROS expression, we performed 4HNE staining in the hippocampal region. The results demonstrated a significant expression of ROS following pilocarpine-induced seizures. However, when amlodipine was administered, it effectively reduced ROS expression in the CA1, CA3, Sub, and DG regions (Figure 2A,B).

These findings indicate that amlodipine administration can effectively decrease ROS levels following seizures, suggesting its potential role in mitigating neuronal damage caused by oxidative stress.

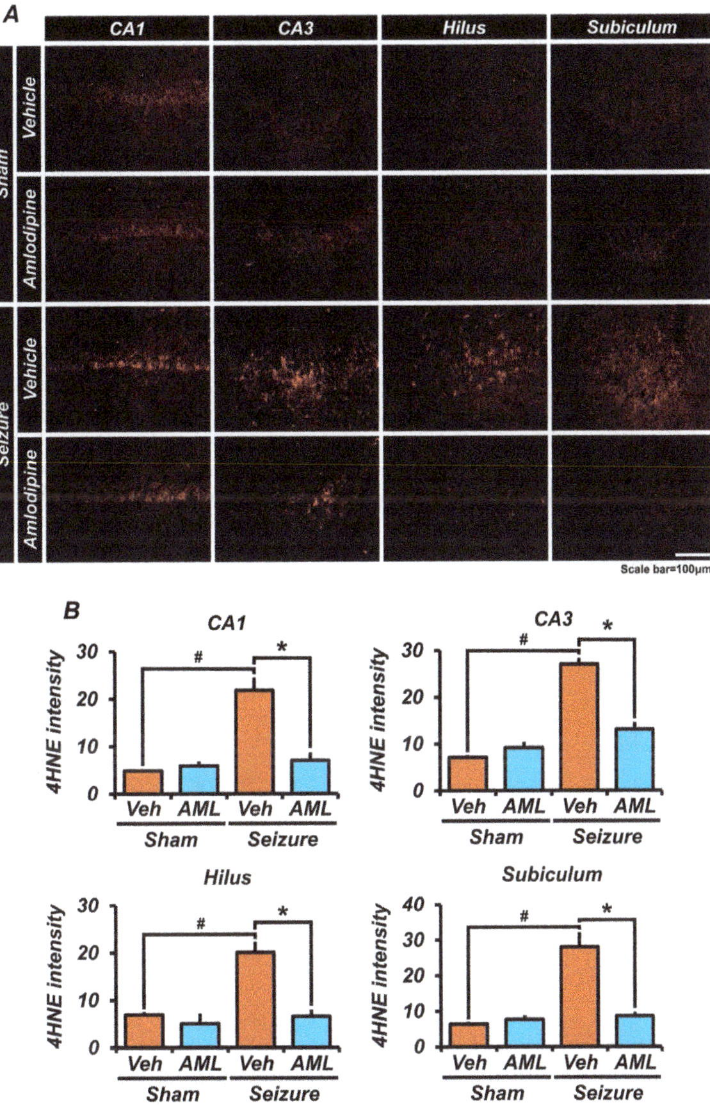

Figure 2. Administration of amlodipine decreased reactive oxidative stress (ROS) after seizure. ROS was detected using 4-hydroxy-2-nonenal (4HNE) (red) staining in the hippocampal CA1, CA3, dentate gyrus (DG), and subiculum (Sub) regions, 7 days after the seizure. (**A**) Micrographs showing 4HNE (red) in the hippocampal CA1, CA3, DG, and Sub regions. (**B**) In every region, there was no significant difference between the sham groups. There were differences in intensity between the seizure-vehicle and seizure-amlodipine groups in every region, including the CA3 region. In every region, the intensity was similar between the sham groups and the seizure-amlodipine-treated group, and there was also a significant difference between the sham-vehicle group and the seizure-vehicle group. Therefore, administering amlodipine after a seizure significantly reduced ROS, compared to administering the vehicle after a seizure. Animals were sacrificed on day 7. The expression of 4HNE decreased in the seizure-amlodipine group, compared with seizure-vehicle group, by about 68% in the CA1 region (seizure-vehicle group, 21.8 ± 3.1; seizure-amlodipine group, 7.0 ± 2.0; sham-vehicle group, 4.8 ± 0.3; sham-amlodipine group, 5.9 ± 1.0), by 52% in the CA3 region (seizure-vehicle group,

27.0 $\pm$ 2.4; seizure-amlodipine group, 13.1 $\pm$ 1.6; sham-vehicle group, 7.1 $\pm$ 0.5; sham-amlodipine group, 29.2 $\pm$ 1.3), by 69% in the hilus (seizure-vehicle group, 20.1 $\pm$ 3.4; seizure-amlodipine group, 6.6 $\pm$ 1.8; sham-vehicle group, 7.0 $\pm$ 0.6; sham amlodipine group, 5.1 $\pm$ 2.1), and by 67% in the subiculum (seizure-vehicle group, 28.0 $\pm$ 4.0; seizure-amlodipine group, 8.6 $\pm$ 1.2; sham-vehicle group, 6.3 $\pm$ 0.7; sham-amlodipine group, 7.6 $\pm$ 1.1). Scale bar = 100 μm (the sample sizes were n = 5 animals for each sham group, n = 6 animals for the seizure-vehicle group, and n = 8 animals for the seizure-amlodipine group). (Bonferroni post hoc test after Kruskal–Wallis test, CA1: chi-squared = 11.085, df = 3, p = 0.011; CA3: chi-squared = 14.367, df = 3, p = 0.002; subiculum: chi-squared = 13.485, df = 3, p = 0.004; dentate gyrus: chi-squared = 12.686, df = 3, p = 0.005). [#] indicates significant difference between sham vehicle and seizure vehicle groups, * Significantly different from the vehicle group, $p < 0.05$.

3.3. Amlodipine Administration Increased Neuron Survival after Seizure

To evaluate the potential neuroprotective effect of amlodipine in pilocarpine-induced status epilepticus (SE), we stained neuronal nuclei (NeuN) to identify and quantify neurons in various regions of the hippocampus, including the CA1, CA3, Sub, and DG regions. Our objective was to compare the number of NeuN-positive neuron cells in the amlodipine-treated group and in the vehicle-treated group, to assess neuronal survival. After one week, the analysis of stained brain samples revealed a significantly higher number of live neuron cells in the amlodipine-treated SE group, compared to the SE vehicle group. This observation suggests that administering amlodipine for one week in rats has a neuroprotective effect, mitigating neuronal death caused by pilocarpine-induced status epilepticus.

The seizure-amlodipine group had a higher number of NeuN-positive cells compared to the seizure-vehicle group. NeuN is a neuronal marker, and NeuN-positive cells are typically used to quantify neurons in various conditions. Despite this increase, the number of NeuN-positive cells in the seizure-amlodipine group was still lower than each sham group. This implies that, while amlodipine may have had a protective or restorative effect on neuron numbers post-seizure, it did not fully restore neuron numbers to the level seen in the healthy control (sham) groups (Figure 3A,B).

These findings indicate that amlodipine may play a crucial role in preserving neuronal integrity and enhancing neuronal survival during status epilepticus, highlighting its potential as a therapeutic agent for protecting against the harmful effects of seizures on neuronal cells.

3.4. Amlodipine Reduced Blood–Brain Barrier Breakdown after Seizure

To explore amlodipine's potential in safeguarding the blood–brain barrier (BBB) against pilocarpine-triggered seizures, we performed immunohistochemical examinations on brain tissues to measure serum IgG leakage, a marker of BBB impairment during seizures. Brain sections were immunostained for rat IgG, to compare leakage levels between the amlodipine- and vehicle-treated groups under seizure conditions. Our analysis revealed notably reduced IgG leakage in the amlodipine group, indicative of a significant protective effect of the drug on the BBB integrity during seizures. These results suggest that amlodipine may be instrumental in maintaining BBB health, potentially shielding the brain from deleterious agents and seizure-associated neuronal harm (Figure 4A,B).

Upon re-examination of the staining intensity data, we observed a notable variance in leakage among individual samples within the amlodipine-treated pilocarpine group. This variance is now more accurately represented in the revised graph in Figure 4, which shows the range of staining intensities observed, highlighting the significant difference in leakage when compared to the sham controls.

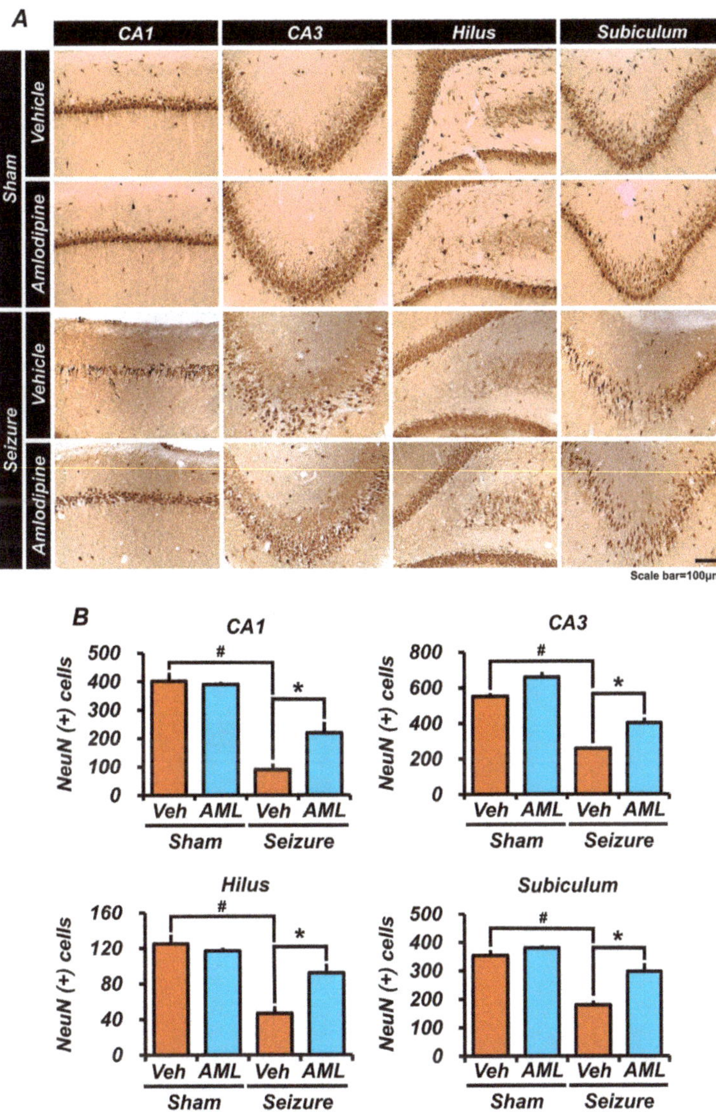

Figure 3. Amlodipine administration increased neuron survival after seizure. (**A**) Micrographs of NeuN (+) cells representing surviving cells in the hippocampal CA1, CA3, hilus, and subiculum (Sub) regions. The number of surviving neurons appears to be greater in the seizure-amlodipine group than in the seizure-vehicle group. (**B**) Cell counting in the micrographs of every region shows a significant difference between the sham-vehicle group and the seizure-vehicle group. In the CA1 and hilus regions, cell counting showed that the seizure-amlodipine group had approximately twice the number of NeuN (+) cells as the seizure-vehicle group. In the CA3 and Sub regions, the seizure-amlodipine group had more NeuN (+) cells, which was a noticeable difference from the seizure-vehicle group. Additionally, in every region, there is no significant difference in NeuN (+) cells between the sham-vehicle group and the sham-amlodipine group. Therefore, administering amlodipine after a seizure significantly increases neuron survival compared to treating with a vehicle after a seizure. Animals were sacrificed on day 7. The number of NeuN (+) cells was higher in the seizure-amlodipine group than the seizure-vehicle group by about 58% in the CA1 region (seizure-vehicle group, 89.5 ± 21.5;

seizure-amlodipine group, 215.4 ± 32.3; sham-vehicle group, 401.6 ± 22.3; sham-amlodipine group, 389.5 ± 6.9), by 34.3% in the CA3 region (seizure-vehicle group, 257.9 ± 8.7; seizure-amlodipine group, 392.8 ± 24.7; sham-vehicle group, 551.1 ± 28.3; sham-amlodipine group, 659.0 ± 19.4), by 49% in the hilus region (seizure-vehicle group, 46.7 ± 8.0; seizure-amlodipine group, 92.1 + 10.0; sham-vehicle group, 125.2 ± 11.0; sham-amlodipine group, 117.0 ± 2.8), and by 39% in the subiculum (seizure-vehicle group, 179.9 ± 14.1; seizure-amlodipine group, 297.0 ± 25.4; sham-vehicle group, 353.8 ± 14.9; sham-amlodipine group, 380.3 ± 6.6). Scale bar = 100 μm (the sample sizes were n = 5 animals for each sham group, n = 6 animals for the seizure-vehicle group, and n = 8 animals for the seizure-amlodipine group). (Bonferroni post hoc test after Kruskal–Wallis test, CA1: chi-squared = 17.649, df = 3, p = 0.001; CA3: chi-squared = 18.543, df = 3, p < 0.001; subiculum: chi-squared =15.535, df = 3, p = 0.001; dentate gyrus: chi-squared = 15.649, df = 3, p < 0.001.) [#] indicates significant difference between sham vehicle and seizure vehicle groups, * Significantly different from the vehicle group, p < 0.05.

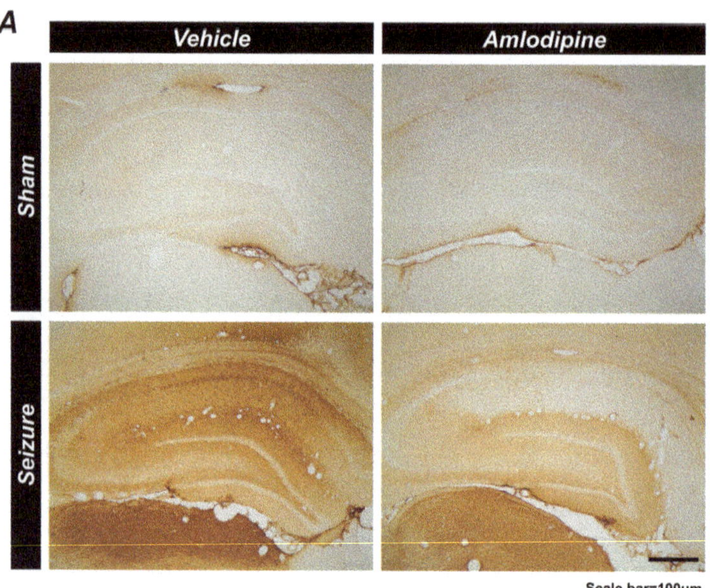

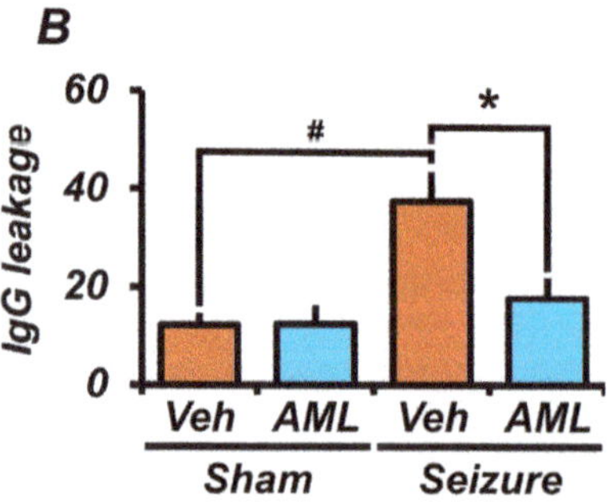

Figure 4. Amlodipine administration reduced blood–brain barrier (BBB) breakdown after seizure. (**A**,**B**) The micrographs show the amount of IgG leakage in a specific region of the whole brain. We found a significant difference between the sham-vehicle group and the seizure-vehicle group. There is also a twofold decrease in IgG leakage between the seizure-amlodipine group and the seizure-vehicle group. However, the IgG leakage in the seizure-amlodipine group is similar to that of the sham groups. Also, the sham-vehicle group shows no significant difference when compared with the sham-amlodipine

group. This indicates that administering an amlodipine treatment after a seizure significantly reduces IgG leakage. The scale bar represents a distance of 100 μm. The data are presented as mean values with the standard error of the mean (SEM) indicated. The sample sizes were n = 3 animals for each sham group, n = 5 animals for the seizure-vehicle group, and six for the seizure-amlodipine group. The difference between the seizure-amlodipine group and the seizure-vehicle group is considered statistically significant, with a p-value less than 0.05. Animals were sacrificed on day 7. IgG leakage decreased in the seizure-amlodipine group, compared with the seizure-vehicle group, by about 53% in the hippocampal area (seizure-vehicle group, 37.4 ± 6.0; seizure-amlodipine group, 17.6 ± 4.3; sham-vehicle group, 12.1 ± 2.4; sham-amlodipine group, 12.3 ± 3.8). (Bonferroni post hoc test after Kruskal–Wallis test of IgG, CA1: chi-squared = 7.763, df = 3, p = 0.051.) [#] indicates significant difference between sham vehicle and seizure vehicle groups, * Significantly different from the vehicle group, $p < 0.05$.

3.5. Amlodipine Administration Reduced Astrocyte Over-Activation after Seizure

In order to investigate the activation of astrocytes in the CA1 region of the hippocampus following pilocarpine-induced seizures, we conducted staining using glial fibrillary acidic protein (GFAP) and complement component 3 (C3). Astrocytes become activated in response to seizure-induced brain damage, leading to the expression of C3. The presence of C3 triggers synaptic death through phagocytosis by microglia or macrophages that recognize the C3 receptor, contributing to neuronal death and other seizure-related side effects.

The results of our staining analysis revealed that the seizure-vehicle group exhibited three times higher astrocyte activation than the sham groups. Similarly, the seizure-amlodipine group showed twice the level of astrocyte activation of the sham groups. Notably, there was a significant difference observed between the seizure-vehicle and seizure-amlodipine groups, in terms of astrocyte activation. The seizure-vehicle group had a lower number of DAPI-positive cells than the sham operating group. DAPI is a fluorescent stain that binds strongly to DNA, and it is commonly used in microscopy to visualize cell nuclei. This observation could imply a reduced number of cells in the seizure-vehicle group compared to the sham group (Figure 5A,B).

Furthermore, the intensity of C3 staining in the CA1 region was higher in the seizure-vehicle group compared to the other groups, while the seizure-amlodipine group exhibited similar intensity to the sham groups. This suggests that treating seizures with amlodipine significantly reduces the excessive activation of astrocytes and subsequent release of C3, which is associated with synaptic death in neurons [42].

Based on these findings, it can be concluded that amlodipine has a protective effect in mitigating astrocyte-mediated damage and neuronal death in the context of seizures. By reducing astrocyte activation and the release of C3, amlodipine may help preserve synaptic integrity and neuronal function, offering potential neuroprotective benefits during seizures.

3.6. Amlodipine Administration Reduced Microglia Over-Activation after Seizure

To evaluate the activation of microglia, we conducted staining using ionized calcium-binding adaptor molecule-1 (Iba1) and cluster of differentiation 68 (CD68), which specifically identify microglia and inflammatory microglia, respectively [43,44]. Microglia are the primary immune cells in the brain and play a crucial role in immune and inflammatory responses. However, the excessive activation of microglia after seizures can result in additional harm to neurons. In our study, we measured the intensity of Iba1 staining to assess microglia activation. The results demonstrated that the seizure-vehicle group exhibited approximately three times higher Iba1 intensity than the seizure-amlodipine group. Additionally, we measured the intensity of CD68 staining, to evaluate the presence of inflammatory microglia. The sham groups displayed similar intensities of CD68 staining, while the seizure-vehicle group showed around twice the intensity in comparison with the seizure-amlodipine group (Figure 6A,B).

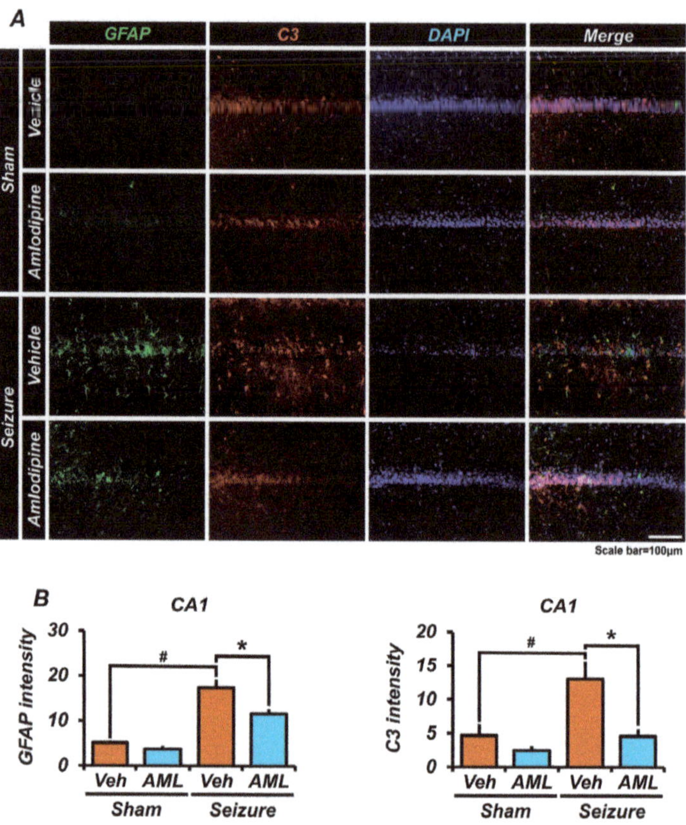

Figure 5. Amlodipine administration reduced astrocyte over-activation after seizure. (**A**) The micrographs show the staining of GFAP (green), C3 (red), and DAPI (blue) in the hippocampal CA1 region. (**B**) The bar graph illustrates the intensity of GFAP, which is significantly different in the sham-vehicle group compared with the seizure-vehicle group; there is also a significant difference between the seizure-amlodipine group and the seizure-vehicle group. The intensity of C3 in the seizure-amlodipine group is comparable to each sham group and is approximately three times lower than that of the seizure-vehicle group. There is also a significant difference between the sham-vehicle group and the seizure-vehicle group. Also, there is no significant difference in GFAP and C3 intensity among the various sham groups. This indicates that administering an amlodipine treatment after a seizure reduces the over-activation of astrocytes. The scale bar represents a distance of 100 μm. The data are presented as mean values, with the standard error of the mean (SEM) indicated. The sample sizes were five animals for each sham group, six animals for the seizure-vehicle group, and eight animals for the seizure-amlodipine group. The difference between the seizure-amlodipine group and the seizure-vehicle group is considered statistically significant, with a p-value less than 0.05. Animals were sacrificed on day 7. GFAP and C3 intensity decreased in the seizure-amlodipine group, compared with the seizure-vehicle group, by about 33% in the CA1 region for GFAP (seizure-vehicle group, 17.3 ± 1.8; seizure-amlodipine group, 11.6 ± 1.0; sham-vehicle group, 5.1 ± 0.4; sham-amlodipine group, 3.7 ± 0.8) and by 64% in the CA1 region for C3 (seizure-vehicle group, 17.3 ± 2.5; seizure-amlodipine group, 6.2 ± 1.4; sham-vehicle group, 5.5 ± 1.9; sham-amlodipine group, 12.9 ± 0.8). (Bonferroni post hoc test after Kruskal–Wallis test for GFAP, CA1: chi-squared = 12.894, df = 3, p = 0.005; Bonferroni post hoc test after Kruskal–Wallis test for C3, CA1: chi-squared = 12.246, df = 3, p = 0.007.) [#] indicates significant difference between sham vehicle and seizure vehicle groups, * Significantly different from the vehicle group, $p < 0.05$.

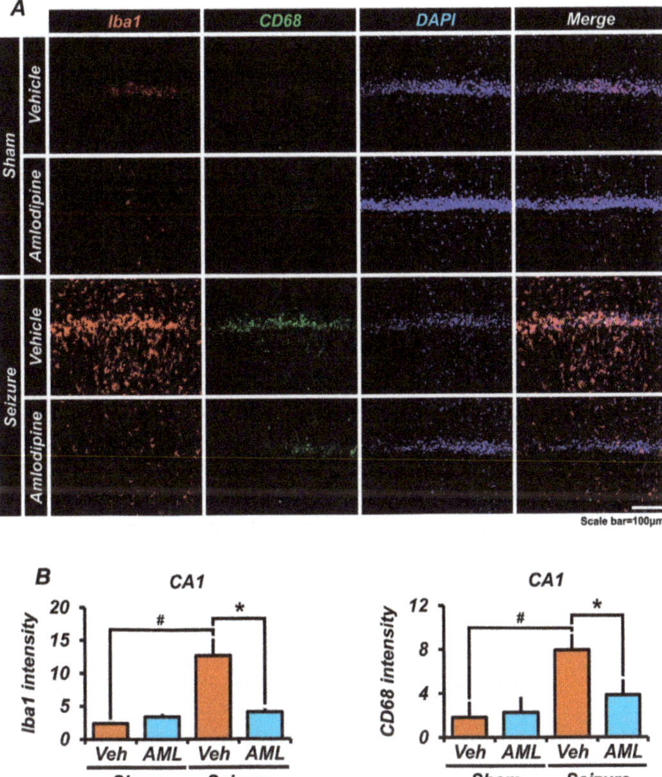

Figure 6. Amlodipine administration reduced microglia over-activation after seizure. (**A**) The micrographs depict the staining of Iba1 (red), CD68 (green), and DAPI (blue). There is a high presence of detected cells positive for both Iba1 and CD68 in the seizure-vehicle group. (**B**) The bar graph displays the intensity of Iba1 staining, which is found to be twice as high in the seizure-vehicle group compared to the other groups. The intensity of Iba1 in both the seizure-amlodipine group and each sham group is almost similar. Additionally, there is a significant difference in C3 intensity between the seizure-vehicle group and the seizure-amlodipine group, with the latter showing a slightly higher intensity compared to each sham group. Additionally, both graphs illustrate a significant difference between the sham-vehicle group and the seizure-vehicle group, while showing no significant difference among the sham groups. This indicates that administering amlodipine after a seizure reduces the over-activation of astrocytes, in comparison to the seizure-vehicle group. The scale bar represents a distance of 100 μm. The data are presented as mean values with the standard error of the mean (SEM) indicated. The sample sizes were five animals for each sham group, six animals for the seizure-vehicle group, and eight animals for the seizure-amlodipine group. The difference between the seizure-amlodipine group and the seizure-vehicle group is considered statistically significant, with a p-value less than 0.05. Animals were sacrificed on day 7. Iba1 and CD68 intensity decreased in the seizure-amlodipine group, compared with the seizure-vehicle group, by about 68% in the CA1 region for Iba1 (seizure-vehicle group, 12.6 ± 2.5; seizure-amlodipine group, 4.1 ± 0.5; sham-vehicle group, 2.4 ± 0.1; sham-amlodipine group, 3.4 ± 0.7) and by 52% in the CA1 region for CD68 (seizure-vehicle group, 7.9 ± 1.6; seizure-amlodipine group, 3.8 ± 0.7; sham-vehicle group, 1.8 ± 0.3; sham-amlodipine group, 2.3 ± 0.5). (Bonferroni post hoc test after Kruskal–Wallis test for Iba1, CA1: chi-squared = 16.884, df = 3, p = 0.001; Bonferroni post hoc test after Kruskal–Wallis test for CD68, CA1: chi-squared = 12.379, df = 3, p = 0.006.) # indicates significant difference between sham vehicle and seizure vehicle groups, * Significantly different from the vehicle group, $p < 0.05$.

These findings suggest that the treatment of seizures with amlodipine successfully inhibits microglia activation subsequent to the seizure episode. By reducing microglia activation, amlodipine may offer protective effects against secondary damage to neuronal cells following seizures.

3.7. Amlodipine Administration Reduced Microtubule Disruption after Seizure

To assess microtubule damage subsequent to seizures, we conducted staining for microtubule-associated protein 2 (MAP2) on brain samples, specifically focusing on the CA1 region of the hippocampus. Microtubules play a critical role in maintaining the cytoskeletal structure, and their disruption during seizures can lead to the increased vulnerability of brain tissue. Our results revealed that the sham groups exhibited comparable levels of MAP2 staining intensity, indicating intact microtubule structures. Conversely, the seizure groups displayed significantly lower positive intensities of MAP2, signifying that microtubule damage was a consequence of the seizures. However, in the seizure-amlodipine group, the intensity of MAP2 staining was twice that of the seizure-vehicle group. This observation suggests that amlodipine treatment was effective in preventing or mitigating microtubule damage caused by seizures (Figure 7A,B).

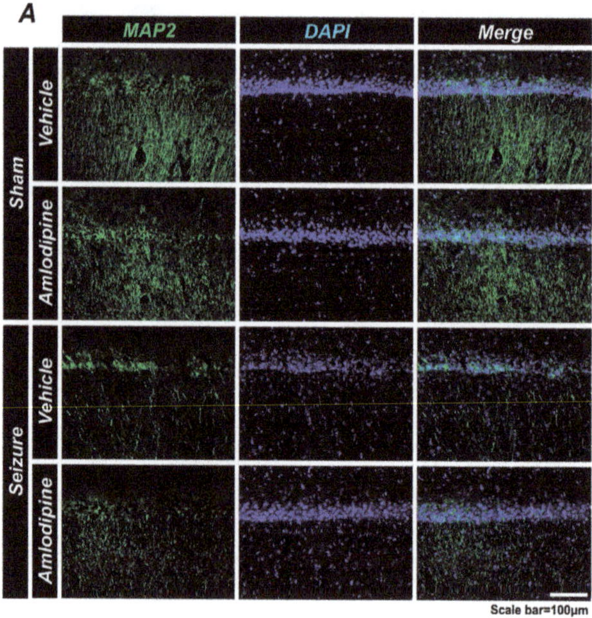

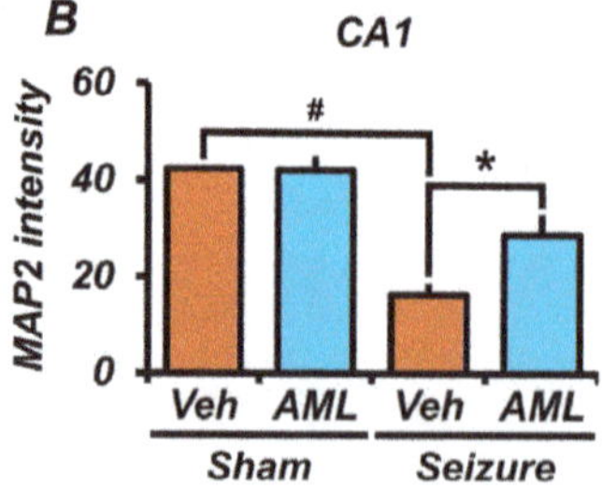

Figure 7. Amlodipine administration reduced microtubule disruption after seizure. (**A**) The micrographs illustrate the staining of MAP2 (green) and DAPI (blue). The intensity of MAP2 staining in the seizure groups is lower than in the sham groups. (**B**) The bar graph demonstrates that each sham

group exhibits a similar intensity of MAP2 staining. There is a significant difference between the sham-vehicle group and the seizure-vehicle group, and the seizure-amlodipine group has a lower intensity than the sham groups but a higher intensity than the seizure-vehicle group, with a significant difference observed. Additionally, there was no significant difference between the sham-vehicle group and the sham-amlodipine group. These results indicate that administering amlodipine after a seizure provides protection to microtubules. The scale bar represents a distance of 100 μm. The data are presented as mean values, with the standard error of the mean (SEM) indicated. The sample sizes were five animals for each sham group, six animals for the seizure-vehicle group, and eight animals for the seizure-amlodipine group. The difference between the seizure-amlodipine group and the seizure-vehicle group is considered statistically significant, with a p-value less than 0.05. Animals were sacrificed on day 7. Microtubule intensity increased in the seizure-amlodipine group, compared with the seizure-vehicle group, by about 43% in the CA1 area (seizure-vehicle group, 16.3 ± 2.4; seizure-amlodipine group, 28.9 ± 4.1; sham-vehicle group, 42.4 ± 0.7; sham-amlodipine group, 42.0 ± 2.9). (Bonferroni post hoc test after Kruskal–Wallis test, CA1: chi-squared =10.452, df = 3, p = 0.015.) [#] indicates significant difference between sham vehicle and seizure vehicle groups, * Significantly different from the vehicle group, $p < 0.05$.

3.8. Amlodipine Administration Improved Spatial Cognitive, Memory, and Cognitive Function Recovery after Seizure

To assess the recovery of spatial cognitive ability and cognitive function following seizures, we employed the Barnes maze and adhesive removal behavior tests. These tests were conducted for five days, one week after inducing seizures or during a control period. Additionally, we stained brain samples using NeuN to evaluate overall neuron survival and the neuroprotective effects of amlodipine.

Our results showed that the sham groups displayed similar patterns in both the Barnes maze and adhesive removal tests. However, in the Barnes maze, the seizure groups did not exhibit any significant differences on days 1 and 2. From day 3 onwards, the seizure-amlodipine group not only displayed better performance in reaching the opening hole but also showed significant improvements compared to the seizure-vehicle group (Figure 8A). In the adhesive removal test, the seizure-amlodipine group exhibited significantly faster performance than the seizure-vehicle group on days 4 and 5. However, there were no significant differences between the seizure-vehicle group and the seizure-amlodipine group on days 1, 2, and 3 (Figure 8B).

We also conducted NeuN staining and quantified the number of NeuN-positive cells in the CA1 and CA3 regions of the hippocampus in brain samples. We found significant differences between the seizure-amlodipine group and the seizure-vehicle group, indicating enhanced neuron survival in the amlodipine-treated group. There were no significant differences observed in the sham groups (Figure 8C,D).

Overall, our findings suggest that amlodipine treatment after seizures improves the recovery of spatial cognitive ability and cognitive function. Furthermore, amlodipine appears to have a protective effect against microtubule damage induced by seizures. By preserving the integrity of microtubules, amlodipine may contribute to the stability and functionality of brain tissue, ultimately enhancing the recovery of spatial and functional cognitive abilities after seizures.

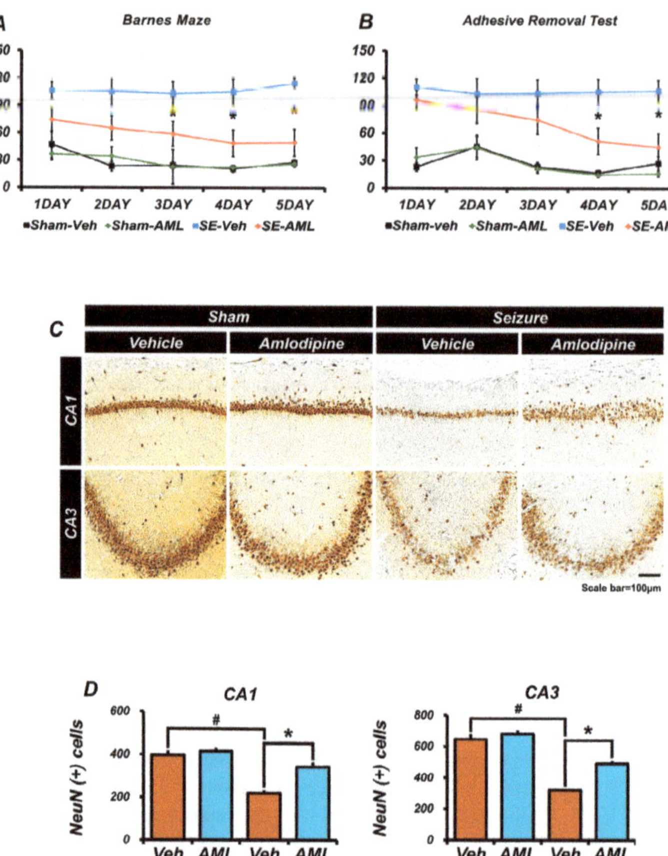

Figure 8. Amlodipine administration improved spatial cognitive, memory, and cognitive function recovery after seizure. One week following the seizures for the seizure groups, and after a recovery period for the sham groups, we conducted several tests, and the results were as follows: (**A**) Barnes maze test: This test assessed spatial cognitive ability recovery. The sham groups outperformed the seizure groups. Notably, the seizure-amlodipine group performed better than the seizure-vehicle group, with the most significant differences observed between the third and fifth day. Additionally, there was no significant difference between the sham-vehicle group and the sham-amlodipine group on any day. Statistical analyses (a repeated measures test, followed by an ANOVA) revealed significant differences, especially in group interaction effects. (Barnes maze: time x group: F = 3.268, $p < 0.021$.) (**B**) Adhesive removal test: This measured cognitive function recovery, and the seizure-amlodipine group demonstrated progressive improvement from day one to five. In contrast, the seizure-vehicle group exhibited negligible progression, with stark differences apparent on the fourth and fifth days. Additionally, there was no significant difference between the sham groups on any day. Statistical analyses (a repeated measures test, followed by an ANOVA) revealed significant differences, especially in group interaction effects (adhesive removal: time * group: F = 12.005, $p < 0.006$). (**C**) We analyzed micrographs for neuronal nuclei (NeuN)-positive cells in the hippocampal CA1 and CA3 areas. (**D**) Bar graphs showed the sham groups had similar NeuN (+) cell counts in the hippocampal regions. In contrast, there was a noticeable discrepancy in the seizure groups, with a significant difference compared to the sham-vehicle group, as the amlodipine treatment seemed to bolster recovery in spatial cognition and general cognitive functions. Also, there were no differences in the CA1 and CA3 regions between the various sham groups. Each sham group consisted of five animals,

while the seizure groups had six. Statistical significance was achieved between the seizure-amlodipine and seizure-vehicle groups, with a *p*-value below 0.05. Animals were sacrificed on day 12. The number of NeuN (+) cells was higher in the seizure-amlodipine group, compared with the seizure-vehicle group, by about 36% in the CA1 region (seizure-vehicle group, 218.2 $\pm$ 12.1; seizure-amlodipine group, 342.7 $\pm$ 21.1; sham-vehicle group, 395.8 $\pm$ 23.1; sham-amlodipine group, 414.9 $\pm$ 14.4) and by 35% in the CA3 region (seizure-vehicle group, 320.4 $\pm$ 5.3; seizure-amlodipine group, 492.5 $\pm$ 18.8; sham-vehicle group, 646.6 $\pm$ 33.5; sham-amlodipine group, 682.3 $\pm$ 24.2). (Bonferroni post hoc test after Kruskal–Wallis test: CA1: $p = 0.006$; CA3: $p = 0.004$.) This underscores the neuroprotective potential of amlodipine post-seizure. [#] indicates significant difference between sham vehicle and seizure vehicle groups, [*] Significantly different from the vehicle group, $p < 0.05$.

4. Discussion

The present study aimed to investigate the effects of amlodipine, which is known to block L-type voltage-gated calcium channels (LTCCs), on various aspects of neuronal damage and dysfunction caused by pilocarpine-induced seizures [45–47]. The results of this study revealed several positive effects of amlodipine treatment. Firstly, it was found that amlodipine treatment led to a reduction in the overexpression of Cav1.2 channels and zinc accumulation induced by seizures. These channels are involved in the entry of calcium (Ca^{2+}) and zinc (Zn^{2+}) ions into neurons during seizures, ultimately leading to neuronal damage [46,48]. The decreased expression of Cav1.2 channels in the amlodipine-treated group suggests that amlodipine inhibits the activation of these channels, potentially reducing the entry of harmful ions into neurons.

In addition, the study examined the effect of amlodipine treatment on reactive oxidative stress (ROS), which plays a significant role in neuronal damage during seizures [49–55]. It was observed that amlodipine treatment effectively reduced ROS levels in the hippocampal regions compared to the vehicle-treated group after seizures. By reducing ROS levels, amlodipine may mitigate oxidative damage to essential components in neuronal cells, such as DNA, proteins, and lipids, therefore protecting neuronal cells from harm.

Pilocarpine-induced seizures are a well-established experimental model for studying epilepsy, particularly temporal lobe epilepsy. When administered to animals, pilocarpine can induce seizures that closely resemble human temporal lobe epilepsy. These seizures often lead to neuronal death in the hippocampus, a critical brain region involved in memory and learning [5,6,34,56,57]. However, in the present study, we found that amlodipine treatment increased the survival of neurons following seizures. Staining for neuronal nuclei (NeuN) revealed a higher number of live neuron cells in the amlodipine-treated group than in the vehicle-treated group. This finding suggests that amlodipine treatment contributes to the preservation of neuronal viability during pilocarpine-induced status epilepticus.

Pilocarpine-induced seizures have been studied for their effects on blood–brain barrier (BBB) disruption, which is a critical aspect in understanding epilepsy and its complications. Research indicates that BBB breakdown can both induce epileptic seizures and result from them. This disruption is dynamic and time-dependent, and it is particularly noticeable in the acute phase of epilepsy induced by pilocarpine. Thus, we investigated whether amlodipine could prevent BBB disruption after a seizure [5,58–62]. In the present study, we found that amlodipine showed a protective effect on the blood–brain barrier (BBB) during seizures. The study examined BBB integrity using IgG leakage as an indicator, and the results showed that amlodipine treatment significantly reduced the intensity of IgG leakage, compared to the vehicle-treated group, after seizures [63–66]. This suggests that amlodipine protects against BBB disruption, preventing the passage of blood components into the brain tissue.

Research has shown that pilocarpine-induced seizures in rats led to the rapid activation of astrocytes and microglia in the brain. Astrocytes and microglia are types of glial cells in the central nervous system that play crucial roles in brain health and disease. In the context of epilepsy, such as in the pilocarpine-induced model, these cells become activated as part of the brain's response to seizures. One study reported that, following pilocarpine-induced seizures, there was a significant increase in the labeling of astrocytes and microglial cells in the hippocampal region, indicating their activation. This suggests that glial cells respond quickly to seizure activity and may be involved in the pathophysiological changes that occur in the brain during epilepsy [67–71]. In the present study, we found that amlodipine treatment reduced the over-activation of astrocytes, as evidenced by the decreased expression of glial fibrillary acidic protein (GFAP) and complement component 3 (C3) staining. Additionally, the administration of amlodipine demonstrated the ability to prevent microglia activation in response to seizures. The excessive activation of astrocytes and microglia can cause secondary damage to neurons, so by inhibiting their over-activation, amlodipine may play a protective role in preventing such detrimental effects and preserving the integrity of neuronal cells.

Several studies have shown that pilocarpine-induced epilepsy causes microtubule-associated protein 2 (MAP2) disruption in the hippocampus [72,73]. Therefore, the present study examined the impact of amlodipine on microtubule disruption following seizures. It was observed that amlodipine treatment mitigated the damage caused to microtubules during seizures. Microtubules are essential for maintaining the structural integrity and stability of brain tissue, so preserving their integrity may contribute to the overall stability and function of neurons, offering protection against neuronal damage [72,74,75].

Calcium channels, particularly LTCCs, play a crucial role in various cognitive processes including memory formation, learning, and synaptic plasticity [76–78]. The influx of calcium through these channels is a key event in the signaling pathways that underlie these cognitive functions. Aberrant calcium signaling, often observed in neurodegenerative diseases, is linked to impaired cognitive abilities. Our study's focus on amlodipine's action on LTCCs, therefore, has significant implications for understanding and potentially treating cognitive dysfunctions associated with abnormal calcium channel activity. Research has shown that pilocarpine-induced status epilepticus in rats leads to significant cognitive impairment. This impairment includes deficits in memory and spatial cognition [6,62]. Thus, the present study evaluated the effect of amlodipine treatment on cognitive ability following seizures. Behavioral tests indicated that amlodipine treatment improved the recovery of cognitive function and spatial cognition in rats. Staining for neuronal nuclei (+) cells in various regions of the hippocampus further supported the hypothesis that amlodipine treatment had a positive impact on neuronal protection and cognitive ability following seizures.

This study highlights the critical role of calcium channels in neuronal function, especially in relation to seizures and their impact on cognitive abilities. The dysregulated influx of calcium into neurons during seizures, a result of abnormal neuronal firing, leads to an increase in intracellular calcium [79–82]. This increase can cause neuronal damage or death (excitotoxicity), disrupting neural circuits essential for cognitive functions such as memory, attention, and learning [83,84]. Our findings suggest that chronic alterations in calcium channel function and expression, as a result of prolonged or repeated seizures, could contribute to the long-term cognitive deficits often observed in chronic epilepsy [85]. The potential of calcium channel blockers in mitigating these seizure-induced cognitive impairments offers a promising therapeutic avenue [86].

A considerable body of research supports the effectiveness of calcium channel blockers in reducing the severity of various seizure types. This is evidenced in studies conducted by various researchers [87,88]. The dihydropyridine class of these blockers, in particular, has been noted for its anticonvulsant properties across multiple experimental settings [89–91]. A subset of these, the LTCC blockers, such as nicardipine, have been observed to provide neuroprotection under conditions of extended depolarization [92]. Nimodipine, another compound in this class, has demonstrated efficacy in mitigating seizure effects in various animal studies, including those focusing on pilocarpine-induced seizures in rats [47,90]. While the specific mechanisms of LTCC inhibitors are not completely understood, they are known to regulate calcium flow in and out of synaptosomes, affecting neurotransmitter dynamics [93]. Nimodipine's neuroprotective effect may also play a role in moderating the excitotoxicity linked with an overproduction of free radicals during seizures induced by pilocarpine [94].

Our study provides insights into the action of amlodipine, a known L-type calcium channel (LTCC) inhibitor, across different physiological states. Particularly, we observed its varied efficacy in normal versus pathological conditions, such as those mimicking epilepsy. In pathological states characterized by elevated LTCC expression, amlodipine exhibited pronounced efficacy. This aligns with its role as an LTCC inhibitor, where heightened LTCC activity in disease states amplifies the drug's impact. Contrary to its effect in pathological states, amlodipine showed no significant effects in normal physiological conditions. This finding is pivotal, underscoring the drug's specificity and potential for targeted action in pathological states without influencing normal physiological processes. These observations emphasize the importance of context in pharmacological efficacy. Understanding how amlodipine functions differently under varying physiological conditions can guide its clinical use, ensuring effective and targeted therapy, especially in conditions with altered LTCC dynamics. In conclusion, our study highlights the context-dependent action of amlodipine, offering crucial insights for its application in therapeutic strategies that require precise modulation of calcium channels.

In summary, this study provides evidence supporting the neuroprotective effects of amlodipine in the context of pilocarpine-induced seizures. Amlodipine demonstrated the ability to reduce Cav1.2 channel overexpression and zinc accumulation, decrease reactive oxidative stress, increase neuron survival, reduce blood–brain barrier leakage, inhibit astrocyte over-activation, and protect against microtubule disruption. Furthermore, amlodipine improved the recovery of both spatial and functional cognitive ability following seizures. These findings collectively highlight the potential of amlodipine as a therapeutic intervention for mitigating neuronal damage and dysfunction associated with seizures.

While the present study provides insights into the neuroprotective effects of amlodipine in a pilocarpine-induced seizure model, it is important to consider: Firstly, the results are specific to the pilocarpine model of seizures, which may not fully replicate the complex pathophysiology of human epileptic conditions. Secondly, the findings in an animal model may not directly translate to human epilepsy treatment. Thirdly, the study may not have addressed the long-term effects and safety profile of amlodipine in the context of chronic epilepsy management. Fourthly, while the study indicates neuroprotective effects, the detailed mechanisms of how amlodipine affects neuronal functioning and epilepsy pathology might require further exploration.

5. Conclusions

The study highlights amlodipine's potential in reducing brain damage from pilocarpine-induced seizures. These findings suggest its potential as a therapeutic agent.

Author Contributions: Conceptualization: S.-W.S. and C.-J.L. Methodology: C.-J.L. Validation: C.-J.L. Formal analysis: C.-J.L. Investigation: C.-J.L., H.-W.Y., A.-R.K., S.-H.L., B.-S.K., M.-K.P., S.-Y.W., S.-W.P., D.-Y.K., H.-H.J., B.-Y.C. and W.-I.Y. Data curation: C.-J.L. Writing—original draft preparation: C.-J.L. Writing—review and editing: S.-H.L., A.-R.K. and S.-W.S. Visualization: B.-S.K., M.-K.P. and C.-J.L. Supervision: H.-C.C., H.-K.S., Y.-J.K. and S.-W.S. All authors have read and agreed to the published version of the manuscript.

Funding: This research was supported by the Hallym University Research Fund (HRF-202303-008). awarded to S.W.S.

Institutional Review Board Statement: The present study adhered to ethical guidelines for animal research, and was approved by the Animal Research Committee at the College of Hallym University of Medicine (Protocol Number: Hallym 2021-39).

Informed Consent Statement: Not applicable.

Data Availability Statement: All data generated or analyzed during this study are included in this published article.

Conflicts of Interest: The authors declare no competing interests.

References

1. Arnold, S.T.; Dodson, W.E. Epilepsy in children. *Baillieres Clin. Neurol.* **1996**, *5*, 783–802. [PubMed]
2. Goldberg, E.M.; Coulter, D.A. Mechanisms of epileptogenesis: A convergence on neural circuit dysfunction. *Nat. Rev. Neurosci.* **2013**, *14*, 337–349. [CrossRef]
3. Shen, Z.; Haragopal, H.; Li, Y.V. Zinc modulates synaptic transmission by differentially regulating synaptic glutamate homeostasis in hippocampus. *Eur. J. Neurosci.* **2020**, *52*, 3710–3722. [CrossRef]
4. Stefan, H. Pathophysiology of human epilepsy: Imaging and physiologic studies. *Curr. Opin. Neurol.* **2000**, *13*, 177–181. [CrossRef]
5. Jeong, J.H.; Lee, S.H.; Kho, A.R.; Hong, D.K.; Kang, D.H.; Kang, B.S.; Park, M.K.; Choi, B.Y.; Choi, H.C.; Lim, M.S.; et al. The Transient Receptor Potential Melastatin 7 (TRPM7) Inhibitors Suppress Seizure-Induced Neuron Death by Inhibiting Zinc Neurotoxicity. *Int. J. Mol. Sci.* **2020**, *21*, 7897. [CrossRef]
6. Lee, S.H.; Choi, B.Y.; Kho, A.R.; Hong, D.K.; Kang, B.S.; Park, M.K.; Lee, S.H.; Choi, H.C.; Song, H.K.; Suh, S.W. Combined Treatment of Dichloroacetic Acid and Pyruvate Increased Neuronal Survival after Seizure. *Nutrients* **2022**, *14*, 4804. [CrossRef]
7. Wahl-Schott, C.; Baumann, L.; Cuny, H.; Eckert, C.; Griessmeier, K.; Biel, M. Switching off calcium-dependent inactivation in L-type calcium channels by an autoinhibitory domain. *Proc. Natl. Acad. Sci. USA* **2006**, *103*, 15657–15662. [CrossRef]
8. Park, S.J.; Min, S.H.; Kang, H.W.; Lee, J.H. Differential zinc permeation and blockade of L-type Ca^{2+} channel isoforms Cav1.2 and Cav1.3. *Biochim. Biophys. Acta* **2015**, *1848*, 2092–2100. [CrossRef] [PubMed]
9. Berger, S.M.; Bartsch, D. The role of L-type voltage-gated calcium channels Cav1.2 and Cav1.3 in normal and pathological brain function. *Cell Tissue Res.* **2014**, *357*, 463–476. [CrossRef] [PubMed]
10. Striessnig, J.; Koschak, A.; Sinnegger-Brauns, M.J.; Hetzenauer, A.; Nguyen, N.K.; Busquet, P.; Pelster, G.; Singewald, N. Role of voltage-gated L-type Ca^{2+} channel isoforms for brain function. *Biochem. Soc. Trans.* **2006**, *34*, 903–909. [CrossRef] [PubMed]
11. Sinnegger-Brauns, M.J.; Huber, I.G.; Koschak, A.; Wild, C.; Obermair, G.J.; Einzinger, U.; Hoda, J.C.; Sartori, S.B.; Striessnig, J. Expression and 1,4-dihydropyridine-binding properties of brain L-type calcium channel isoforms. *Mol. Pharmacol.* **2009**, *75*, 407–414. [CrossRef]
12. Clark, N.C.; Nagano, N.; Kuenzi, F.M.; Jarolimek, W.; Huber, I.; Walter, D.; Wietzorrek, G.; Boyce, S.; Kullmann, D.M.; Striessnig, J.; et al. Neurological phenotype and synaptic function in mice lacking the CaV1.3 alpha subunit of neuronal L-type voltage-dependent Ca^{2+} channels. *Neuroscience* **2003**, *120*, 435–442. [CrossRef]
13. Moosmang, S.; Haider, N.; Klugbauer, N.; Adelsberger, H.; Langwieser, N.; Muller, J.; Stiess, M.; Marais, E.; Schulla, V.; Lacinova, L.; et al. Role of hippocampal Cav1.2 Ca^{2+} channels in NMDA receptor-independent synaptic plasticity and spatial memory. *J. Neurosci.* **2005**, *25*, 9883–9892. [CrossRef] [PubMed]
14. Frederickson, C.J.; Hernandez, M.D.; McGinty, J.F. Translocation of zinc may contribute to seizure-induced death of neurons. *Brain Res.* **1989**, *480*, 317–321. [CrossRef] [PubMed]
15. Szewczyk, B. Zinc homeostasis and neurodegenerative disorders. *Front. Aging Neurosci.* **2013**, *5*, 33. [CrossRef]
16. Choi, S.; Hong, D.K.; Choi, B.Y.; Suh, S.W. Zinc in the Brain: Friend or Foe? *Int. J. Mol. Sci.* **2020**, *21*, 8941. [CrossRef] [PubMed]
17. Sandstead, H.H. Subclinical zinc deficiency impairs human brain function. *J. Trace Elem. Med. Biol.* **2012**, *26*, 70–73. [CrossRef]
18. Morris, D.R.; Levenson, C.W. Ion channels and zinc: Mechanisms of neurotoxicity and neurodegeneration. *J. Toxicol.* **2012**, *2012*, 785647. [CrossRef]
19. Sensi, S.L.; Yin, H.Z.; Carriedo, S.G.; Rao, S.S.; Weiss, J.H. Preferential Zn^{2+} influx through Ca^{2+}-permeable AMPA/kainate channels triggers prolonged mitochondrial superoxide production. *Proc. Natl. Acad. Sci. USA* **1999**, *96*, 2414–2419. [CrossRef]
20. Bishop, G.M.; Dringen, R.; Robinson, S.R. Zinc stimulates the production of toxic reactive oxygen species (ROS) and inhibits glutathione reductase in astrocytes. *Free Radic. Biol. Med.* **2007**, *42*, 1222–1230. [CrossRef]
21. Stork, C.J.; Li, Y.V. Elevated Cytoplasmic Free Zinc and Increased Reactive Oxygen Species Generation in the Context of Brain Injury. *Acta Neurochir. Suppl.* **2016**, *121*, 347–353. [CrossRef] [PubMed]
22. Mason, R.P.; Marche, P.; Hintze, T.H. Novel vascular biology of third-generation L-type calcium channel antagonists: Ancillary actions of amlodipine. *Arterioscler. Thromb. Vasc. Biol.* **2003**, *23*, 2155–2163. [CrossRef] [PubMed]
23. Bulsara, K.G.; Cassagnol, M. *Amlodipine*; StatPearls: Treasure Island, FL, USA, 2024.
24. Nelson, M.T.; Cheng, H.; Rubart, M.; Santana, L.F.; Bonev, A.D.; Knot, H.J.; Lederer, W.J. Relaxation of arterial smooth muscle by calcium sparks. *Science* **1995**, *270*, 633–637. [CrossRef]
25. Ortner, N.J.; Striessnig, J. L-type calcium channels as drug targets in CNS disorders. *Channels* **2016**, *10*, 7–13. [CrossRef] [PubMed]
26. Narayanan, D.; Xi, Q.; Pfeffer, L.M.; Jaggar, J.H. Mitochondria control functional CaV1.2 expression in smooth muscle cells of cerebral arteries. *Circ. Res.* **2010**, *107*, 631–641. [CrossRef]
27. Catterall, W.A. Structure and regulation of voltage-gated Ca^{2+} channels. *Annu. Rev. Cell Dev. Biol.* **2000**, *16*, 521–555. [CrossRef]
28. Hockerman, G.H.; Peterson, B.Z.; Johnson, B.D.; Catterall, W.A. Molecular determinants of drug binding and action on L-type calcium channels. *Annu. Rev. Pharmacol. Toxicol.* **1997**, *37*, 361–396. [CrossRef]
29. Hirooka, Y.; Kimura, Y.; Nozoe, M.; Sagara, Y.; Ito, K.; Sunagawa, K. Amlodipine-induced reduction of oxidative stress in the brain is associated with sympatho-inhibitory effects in stroke-prone spontaneously hypertensive rats. *Hypertens. Res.* **2006**, *29*, 49–56. [CrossRef]

30. Lee, Y.J.; Park, H.H.; Koh, S.H.; Choi, N.Y.; Lee, K.Y. Amlodipine besylate and amlodipine camsylate prevent cortical neuronal cell death induced by oxidative stress. *J. Neurochem.* **2011**, *119*, 1262–1270. [CrossRef]

31. Qureshi, I.H.; Riaz, A.; Khan, R.A.; Siddiqui, A.A. Synergistic anticonvulsant effects of pregabalin and amlodipine on acute seizure model of epilepsy in mice. *Metab. Brain Dis.* **2017**, *32*, 1051–1060. [CrossRef]

32. Sathyanarayana Rao, K.N.; Subbalakshmi, N.K. An experimental study of the anticonvulsant effect of amlodipine in mice. *Singap. Med. J.* **2010**, *51*, 424–428.

33. Kaminski, R.; Jasinski, M.; Jagiello-Wojtowicz, E.; Kleinrok, Z.; Czuczwar, S.J. Effect of amlodipine upon the protective activity of antiepileptic drugs against maximal electroshock-induced seizures in mice. *Pharmacol. Res.* **1999**, *40*, 319–325. [CrossRef] [PubMed]

34. Turski, W.A.; Cavalheiro, E.A.; Schwarz, M.; Czuczwar, S.J.; Kleinrok, Z.; Turski, L. Limbic seizures produced by pilocarpine in rats: Behavioural, electroencephalographic and neuropathological study. *Behav. Brain Res.* **1983**, *9*, 315–335. [CrossRef] [PubMed]

35. Cavalheiro, E.A.; Santos, N.F.; Priel, M.R. The pilocarpine model of epilepsy in mice. *Epilepsia* **1996**, *37*, 1015–1019. [CrossRef] [PubMed]

36. Curia, G.; Longo, D.; Biagini, G.; Jones, R.S.; Avoli, M. The pilocarpine model of temporal lobe epilepsy. *J. Neurosci. Methods* **2008**, *172*, 143–157. [CrossRef]

37. Luna-Munguia, H.; Marquez-Bravo, L.; Concha, L. Longitudinal changes in gray and white matter microstructure during epileptogenesis in pilocarpine-induced epileptic rats. *Seizure* **2021**, *90*, 130–140. [CrossRef]

38. Luttjohann, A.; Fabene, P.F.; van Luijtelaar, G. A revised Racine's scale for PTZ-induced seizures in rats. *Physiol. Behav.* **2009**, *98*, 579–586. [CrossRef]

39. Ihara, Y.; Tomonoh, Y.; Deshimaru, M.; Zhang, B.; Uchida, T.; Ishii, A.; Hirose, S. Retigabine, a Kv7.2/Kv7.3-Channel Opener, Attenuates Drug-Induced Seizures in Knock-In Mice Harboring Kcnq2 Mutations. *PLoS ONE* **2016**, *11*, e0150095. [CrossRef]

40. Frederickson, C.J.; Kasarskis, E.J.; Ringo, D.; Frederickson, R.E. A quinoline fluorescence method for visualizing and assaying the histochemically reactive zinc (bouton zinc) in the brain. *J. Neurosci. Methods* **1987**, *20*, 91–103. [CrossRef]

41. Xu, J.H.; Long, L.; Tang, Y.C.; Hu, H.T.; Tang, F.R. Ca(v)1.2, Ca(v)1.3, and Ca(v)2.1 in the mouse hippocampus during and after pilocarpine-induced status epilepticus. *Hippocampus* **2007**, *17*, 235–251. [CrossRef]

42. Anekonda, T.S.; Quinn, J.F.; Harris, C.; Frahler, K.; Wadsworth, T.L.; Woltjer, R.L. L-type voltage-gated calcium channel blockade with isradipine as a therapeutic strategy for Alzheimer's disease. *Neurobiol. Dis.* **2011**, *41*, 62–70. [CrossRef] [PubMed]

43. Ito, D.; Imai, Y.; Ohsawa, K.; Nakajima, K.; Fukuuchi, Y.; Kohsaka, S. Microglia-specific localisation of a novel calcium binding protein, Iba1. *Mol. Brain Res.* **1998**, *57*, 1–9. [CrossRef] [PubMed]

44. Holness, C.L.; Simmons, D.L. Molecular cloning of CD68, a human macrophage marker related to lysosomal glycoproteins. *Blood* **1993**, *81*, 1607–1613. [CrossRef] [PubMed]

45. Berg, M.; Bruhn, T.; Frandsen, A.; Schousboe, A.; Diemer, N.H. Kainic acid-induced seizures and brain damage in the rat: Role of calcium homeostasis. *J. Neurosci. Res.* **1995**, *40*, 641–646. [CrossRef]

46. Nunez, J.L.; McCarthy, M.M. Androgens predispose males to GABAA-mediated excitotoxicity in the developing hippocampus. *Exp. Neurol.* **2008**, *210*, 699–708. [CrossRef]

47. Nascimento, V.S.; D'Alva, M.S.; Oliveira, A.A.; Freitas, R.M.; Vasconcelos, S.M.; Sousa, F.C.; Fonteles, M.M. Antioxidant effect of nimodipine in young rats after pilocarpine-induced seizures. *Pharmacol. Biochem. Behav.* **2005**, *82*, 11–16. [CrossRef]

48. Nolte, C.; Gore, A.; Sekler, I.; Kresse, W.; Hershfinkel, M.; Hoffmann, A.; Kettenmann, H.; Moran, A. ZnT-1 expression in astroglial cells protects against zinc toxicity and slows the accumulation of intracellular zinc. *Glia* **2004**, *48*, 145–155. [CrossRef]

49. Shao, Y.Y.; Li, B.; Huang, Y.M.; Luo, Q.; Xie, Y.M.; Chen, Y.H. Thymoquinone Attenuates Brain Injury via an Anti-oxidative Pathway in a Status Epilepticus Rat Model. *Transl. Neurosci.* **2017**, *8*, 9–14. [CrossRef]

50. Quincozes-Santos, A.; Bobermin, L.D.; Tramontina, A.C.; Wartchow, K.M.; Tagliari, B.; Souza, D.O.; Wyse, A.T.; Goncalves, C.A. Oxidative stress mediated by NMDA, AMPA/KA channels in acute hippocampal slices: Neuroprotective effect of resveratrol. *Toxicol. Vitr.* **2014**, *28*, 544–551. [CrossRef] [PubMed]

51. Luo, Q.; Xian, P.; Wang, T.; Wu, S.; Sun, T.; Wang, W.; Wang, B.; Yang, H.; Yang, Y.; Wang, H.; et al. Antioxidant activity of mesenchymal stem cell-derived extracellular vesicles restores hippocampal neurons following seizure damage. *Theranostics* **2021**, *11*, 5986–6005. [CrossRef] [PubMed]

52. Singh, P.K.; Saadi, A.; Sheeni, Y.; Shekh-Ahmad, T. Specific inhibition of NADPH oxidase 2 modifies chronic epilepsy. *Redox Biol.* **2022**, *58*, 102549. [CrossRef] [PubMed]

53. Pestana, R.R.; Kinjo, E.R.; Hernandes, M.S.; Britto, L.R. Reactive oxygen species generated by NADPH oxidase are involved in neurodegeneration in the pilocarpine model of temporal lobe epilepsy. *Neurosci. Lett.* **2010**, *484*, 187–191. [CrossRef] [PubMed]

54. Kovac, S.; Domijan, A.M.; Walker, M.C.; Abramov, A.Y. Seizure activity results in calcium- and mitochondria-independent ROS production via NADPH and xanthine oxidase activation. *Cell Death Dis.* **2014**, *5*, e1442. [CrossRef] [PubMed]

55. Abramov, A.Y.; Scorziello, A.; Duchen, M.R. Three distinct mechanisms generate oxygen free radicals in neurons and contribute to cell death during anoxia and reoxygenation. *J. Neurosci.* **2007**, *27*, 1129–1138. [CrossRef] [PubMed]

56. Du, F.; Eid, T.; Lothman, E.W.; Kohler, C.; Schwarcz, R. Preferential neuronal loss in layer III of the medial entorhinal cortex in rat models of temporal lobe epilepsy. *J. Neurosci.* **1995**, *15*, 6301–6313. [CrossRef] [PubMed]

57. Sankar, R.; Shin, D.H.; Liu, H.; Mazarati, A.; Pereira de Vasconcelos, A.; Wasterlain, C.G. Patterns of status epilepticus-induced neuronal injury during development and long-term consequences. *J. Neurosci.* **1998**, *18*, 8382–8393. [CrossRef] [PubMed]

58. Zucker, D.K.; Wooten, G.F.; Lothman, E.W. Blood-brain barrier changes with kainic acid-induced limbic seizures. *Exp. Neurol.* **1983**, *79*, 422–433. [CrossRef]
59. Krizanac-Bengez, L.; Mayberg, M.R.; Janigro, D. The cerebral vasculature as a therapeutic target for neurological disorders and the role of shear stress in vascular homeostasis and pathophysiology. *Neurol. Res.* **2004**, *26*, 846–853. [CrossRef]
60. Oby, E.; Janigro, D. The blood-brain barrier and epilepsy. *Epilepsia* **2006**, *47*, 1761–1774. [CrossRef]
61. Vezzani, A.; Granata, T. Brain inflammation in epilepsy: Experimental and clinical evidence. *Epilepsia* **2005**, *46*, 1724–1743. [CrossRef]
62. Lee, S.H.; Choi, B.Y.; Kim, J.H.; Kho, A.R.; Sohn, M.; Song, H.K.; Choi, H.C.; Suh, S.W. Late treatment with choline alfoscerate (l-alpha glycerylphosphorylcholine, alpha-GPC) increases hippocampal neurogenesis and provides protection against seizure-induced neuronal death and cognitive impairment. *Brain Res.* **2017**, *1654 Pt A*, 66–76. [CrossRef]
63. Dey, A.; Kang, X.; Qiu, J.; Du, Y.; Jiang, J. Anti-Inflammatory Small Molecules To Treat Seizures and Epilepsy: From Bench to Bedside. *Trends Pharmacol. Sci.* **2016**, *37*, 463–484. [CrossRef]
64. Marchi, N.; Granata, T.; Janigro, D. Inflammatory pathways of seizure disorders. *Trends Neurosci.* **2014**, *37*, 55–65. [CrossRef] [PubMed]
65. Dong, X. Current Strategies for Brain Drug Delivery. *Theranostics* **2018**, *8*, 1481–1493. [CrossRef]
66. Bowyer, J.F.; Robinson, B.; Ali, S.; Schmued, L.C. Neurotoxic-related changes in tyrosine hydroxylase, microglia, myelin, and the blood-brain barrier in the caudate-putamen from acute methamphetamine exposure. *Synapse* **2008**, *62*, 193–204. [CrossRef] [PubMed]
67. Castiglioni, A.J.; Peterson, S.L.; Sanabria, E.L.; Tiffany-Castiglioni, E. Structural changes in astrocytes induced by seizures in a mode of temporal lobe epilepsy. *J. Neurosci. Res.* **1990**, *26*, 334–341. [CrossRef]
68. Jung, K.H.; Chu, K.; Lee, S.T.; Kim, J.H.; Kang, K.M.; Song, E.C.; Kim, S.J.; Park, H.K.; Kim, M.; Lee, S.K.; et al. Region-specific plasticity in the epileptic rat brain: A hippocampal and extrahippocampal analysis. *Epilepsia* **2009**, *50*, 537–549. [CrossRef] [PubMed]
69. Norden, D.M.; Trojanowski, P.J.; Villanueva, E.; Navarro, E.; Godbout, J.P. Sequential activation of microglia and astrocyte cytokine expression precedes increased Iba-1 or GFAP immunoreactivity following systemic immune challenge. *Glia* **2016**, *64*, 300–316. [CrossRef]
70. Benson, M.J.; Manzanero, S.; Borges, K. Complex alterations in microglial M1/M2 markers during the development of epilepsy in two mouse models. *Epilepsia* **2015**, *56*, 895–905. [CrossRef]
71. Boison, D.; Steinhauser, C. Epilepsy and astrocyte energy metabolism. *Glia* **2018**, *66*, 1235–1243. [CrossRef]
72. Schartz, N.D.; Herr, S.A.; Madsen, L.; Butts, S.J.; Torres, C.; Mendez, L.B.; Brewster, A.L. Spatiotemporal profile of Map2 and microglial changes in the hippocampal CA1 region following pilocarpine-induced status epilepticus. *Sci. Rep.* **2016**, *6*, 24988. [CrossRef] [PubMed]
73. Jalava, N.S.; Lopez-Picon, F.R.; Kukko-Lukjanov, T.K.; Holopainen, I.E. Changes in microtubule-associated protein-2 (MAP2) expression during development and after status epilepticus in the immature rat hippocampus. *Int. J. Dev. Neurosci.* **2007**, *25*, 121–131. [CrossRef] [PubMed]
74. Sanchez, C.; Arellano, J.I.; Rodriguez-Sanchez, P.; Avila, J.; DeFelipe, J.; Diez-Guerra, F.J. Microtubule-associated protein 2 phosphorylation is decreased in the human epileptic temporal lobe cortex. *Neuroscience* **2001**, *107*, 25–33. [CrossRef]
75. Ballough, G.P.; Martin, L.J.; Cann, F.J.; Graham, J.S.; Smith, C.D.; Kling, C.E.; Forster, J.S.; Phann, S.; Filbert, M.G. Microtubule-associated protein 2 (MAP-2): A sensitive marker of seizure-related brain damage. *J. Neurosci. Methods* **1995**, *61*, 23–32. [CrossRef] [PubMed]
76. Ingram, D.K.; Joseph, J.A.; Spangler, E.L.; Roberts, D.; Hengemihle, J.; Fanelli, R.J. Chronic nimodipine treatment in aged rats: Analysis of motor and cognitive effects and muscarinic-induced striatal dopamine release. *Neurobiol. Aging* **1994**, *15*, 55–61. [CrossRef]
77. Batuecas, A.; Pereira, R.; Centeno, C.; Pulido, J.A.; Hernandez, M.; Bollati, A.; Bogonez, E.; Satrustegui, J. Effects of chronic nimodipine on working memory of old rats in relation to defects in synaptosomal calcium homeostasis. *Eur. J. Pharmacol.* **1998**, *350*, 141–150. [CrossRef] [PubMed]
78. Quevedo, J.; Vianna, M.; Daroit, D.; Born, A.G.; Kuyven, C.R.; Roesler, R.; Quillfeldt, J.A. L-type voltage-dependent calcium channel blocker nifedipine enhances memory retention when infused into the hippocampus. *Neurobiol. Learn. Mem.* **1998**, *69*, 320–325. [CrossRef]
79. Zhang, J.; Ding, M.P.; Liu, Z.; Xiao, B.; Li, G.L.; Zhou, F.Y. Dynamics of calcium in the hippocampal neuronal culture model of epilepsy. *Zhongguo Ying Yong Sheng Li Xue Za Zhi* **2007**, *23*, 200–203.
80. Yang, F.; Xu, G.L.; Yang, Y.Q.; Shen, D.K.; Feng, P.Z.; Wang, P.; Liu, X.G. Effect of electroacupuncture on epileptic EEG and intracellular Ca^{2+} content in the hippocampus in epilepsy rats. *Zhen Ci Yan Jiu* **2009**, *34*, 163–166.
81. Raza, M.; Blair, R.E.; Sombati, S.; Carter, D.S.; Deshpande, L.S.; DeLorenzo, R.J. Evidence that injury-induced changes in hippocampal neuronal calcium dynamics during epileptogenesis cause acquired epilepsy. *Proc. Natl. Acad. Sci. USA* **2004**, *101*, 17522–17527. [CrossRef]
82. Su, T.; Cong, W.D.; Long, Y.S.; Luo, A.H.; Sun, W.W.; Deng, W.Y.; Liao, W.P. Altered expression of voltage-gated potassium channel 4.2 and voltage-gated potassium channel 4-interacting protein, and changes in intracellular calcium levels following lithium-pilocarpine-induced status epilepticus. *Neuroscience* **2008**, *157*, 566–576. [CrossRef]

83. Wang, C.; Xie, N.; Wang, Y.; Li, Y.; Ge, X.; Wang, M. Role of the Mitochondrial Calcium Uniporter in Rat Hippocampal Neuronal Death After Pilocarpine-Induced Status Epilepticus. *Neurochem. Res.* **2015**, *40*, 1739–1746. [CrossRef]
84. Friedman, L.K.; Segal, M.; Veliskova, J. GluR2 knockdown reveals a dissociation between [Ca^{2+}]i surge and neurotoxicity. *Neurochem. Int.* **2003**, *43*, 179–189. [CrossRef]
85. Ge, Y.X.; Lin, Y.Y.; Bi, Q.Q.; Chen, Y.J. Brivaracetam Prevents the Over-expression of Synaptic Vesicle Protein 2A and Rescues the Deficits of Hippocampal Long-term Potentiation In Vivo in Chronic Temporal Lobe Epilepsy Rats. *Curr. Neurovascular Res.* **2020**, *17*, 354–360. [CrossRef] [PubMed]
86. Tan, X.; Zeng, Y.; Tu, Z.; Li, P.; Chen, H.; Cheng, L.; Tu, S.; Jiang, L. TRPV1 Contributes to the Neuroprotective Effect of Dexmedetomidine in Pilocarpine-Induced Status Epilepticus Juvenile Rats. *Biomed. Res. Int.* **2020**, *2020*, 7623635. [CrossRef] [PubMed]
87. van Luijtelaar, G.; Wiaderna, D.; Elants, C.; Scheenen, W. Opposite effects of T- and L-type Ca^{2+} channels blockers in generalized absence epilepsy. *Eur. J. Pharmacol.* **2000**, *406*, 381–389. [CrossRef]
88. Zupan, G.; Erakovic, V.; Simonic, A.; Kriz, J.; Varljen, J. The influence of nimodipine, nicardipine and amlodipine on the brain free fatty acid level in rats with penicillin-induced seizures. *Prog. Neuropsychopharmacol. Biol. Psychiatry* **1999**, *23*, 951–961. [CrossRef]
89. Meyer, F.B.; Anderson, R.E.; Sundt, T.M., Jr.; Yaksh, T.L.; Sharbrough, F.W. Suppression of pentylenetetrazole seizures by oral administration of a dihydropyridine Ca^{2+} antagonist. *Epilepsia* **1987**, *28*, 409–414. [CrossRef] [PubMed]
90. Chakrabarti, A.; Saini, H.K.; Garg, S.K. Dose-finding study with nimodipine: A selective central nervous system calcium channel blocker on aminophylline induced seizure models in rats. *Brain Res. Bull.* **1998**, *45*, 495–499. [CrossRef] [PubMed]
91. Kaminski, R.M.; Mazurek, M.; Turski, W.A.; Kleinrok, Z.; Czuczwar, S.J. Amlodipine enhances the activity of antiepileptic drugs against pentylenetetrazole-induced seizures. *Pharmacol. Biochem. Behav.* **2001**, *68*, 661–668. [CrossRef]
92. Ikegaya, Y.; Nishiyama, N.; Matsuki, N. L-type Ca^{2+} channel blocker inhibits mossy fiber sprouting and cognitive deficits following pilocarpine seizures in immature mice. *Neuroscience* **2000**, *98*, 647–659. [CrossRef] [PubMed]
93. Woodward, J.J.; Cook, M.E.; Leslie, S.W. Characterization of dihydropyridine-sensitive calcium channels in rat brain synaptosomes. *Proc. Natl. Acad. Sci. USA* **1988**, *85*, 7389–7393. [CrossRef] [PubMed]
94. Erakovic, V.; Zupan, G.; Varljen, J.; Laginja, J.; Simonic, A. Lithium plus pilocarpine induced status epilepticus--biochemical changes. *Neurosci. Res.* **2000**, *36*, 157–166. [CrossRef] [PubMed]

MDPI AG
Grosspeteranlage 5
4052 Basel
Switzerland
Tel.: +41 61 683 77 34

Antioxidants Editorial Office
E-mail: antioxidants@mdpi.com
www.mdpi.com/journal/antioxidants